MW00655149

Praise

"This is a comprehensive, organized, and compelling presentation of vaccine safety data that has accumulated after mass, indiscriminate administration of the Pfizer mRNA COVID-19 vaccines. Sadly, a large group of vaccine recipients have become injured, disabled, and many have died after the ill-advised injections. The data with histopathological evaluation at necropsy and autopsy with expert analysis is presented so you can evaluate it for yourself. Never before has there been a class of products with this wide range and extended duration of injury to the recipient. Join me and the authors in calling for all mRNA COVID-19 vaccines to be removed from human use."

—Peter A. McCullough, MD, MPH, internist and cardiologist, and coauthor of *The Courage to Face COVID-19*

"*The Pfizer Papers* by Dr. Naomi Wolf combined with project director Amy Kelly and her DailyClout team is a compelling book shining light in the midst of global darkness. *The Pfizer Papers* expose the most corrupt pharmaceutical company in world history contributing to the cartel that is responsible for killing and injuring hundreds of millions of global citizens, even targeting the most vulnerable—pregnant women, preborns, and newborns. Dr. Wolf's brilliant writing style along with her passion, conviction, data, and investigative journalism is a mandatory read for every global citizen."

—James A Thorp, MD; board certified obstetrician and gynecologist and maternal-fetal medicine subspecialist

"Naomi Wolf, in her continued quest to protect women and preserve personal liberties, turned her attention to the excesses of the COVID-19 response. This encyclopedic compendium of reports from 3,250 doctors and scientists, compiled under the leadership of project director Amy Kelly, is the ultimate resource—where readers can review the range of consequences of a vaccine rushed to market, poorly researched, and then mandated to an unsuspecting populace."

—Drew Pinsky, nationally known as "Dr. Drew," is an internist and addiction medicine specialist, television host, author, and public speaker

"*The Pfizer Papers* are a stunning introduction into a new reality—the complete corruption of the CDC, FDA, and drug companies. When a court forced the FDA to reveal Pfizer's records about its genetic "vaccine," official reports to the company documented more than 1,000 deaths and thousands more seriously harmful effects, all within the first ten weeks. This incredibly innovative and strenuous effort shows the entire "research" program to be a calamity of negligence, often for the purpose covering up "vaccine" mayhem and murder. The authors, including editor Naomi Wolf and editor and project director Amy Kelly, should get the Nobel Prize for medicine and the praises of a grateful humanity. As a physician with considerable experience reviewing and testifying in legal cases against drug companies, including Pfizer, I was nonetheless stunned and educated by the revelations of this book."

—Peter R. Breggin MD, author, *COVID-19 and the Global Predators: We Are the Prey*

"This new piece of work by Dr. Naomi Wolf and her team is staggering and needed, a critical project as it helps unpack the devastation. As part of this work, a team of doctors and scientists reviewed the 450,000 Pfizer documents released by court rulings. It is vital and a must-have document for any doctor, scientist, hospital administrator, politician, or citizen who wants to understand what happened to us via the COVID-19 mRNA injections and over the past few years. A monumental achievement documenting a great crime in history."

—Dr. Paul Elias Alexander, PhD, author of *Presidential Takedown*

"Unlike the zombies of the politico-media class for whom it is a mere catchphrase, Naomi Wolf and her extraordinary team actually did "follow the science," all 450,000 pages of it. It has led them to a dark and disturbing place, but the evidence presented here is overwhelming: what Pfizer did in alliance with Western governments was not accident or incompetence but a crime, for which those responsible should be prosecuted and jailed. The authors of this book have paid a huge price for their integrity: Naomi herself will never be on the BBC or in the *New York Times* again. But their gift to the rest of us is priceless and will grow more invaluable as the years go by. So read this book and buy a couple of copies for friends. One day they will thank you and wonder why, when it mattered, ninety-nine per cent of all the commentators and celebrities and influencers chose to look away. Thank God for the Wolf/Kelly team."

—Mark Steyn, host of *The Mark Steyn Show* and author of *After America*

"No history of the COVID years—that is, no history of the 2020s nearly anywhere worldwide—could be complete, or even make much sense, without consideration of *The Pfizer Papers.* A masterpiece of research in its own way just as daunting, and heroic, as the most thoroughgoing studies of the Holocaust, or Stalin's terror-famine in Ukraine (yet vastly more important even than such projects), this staggering translation of some 450,000 pages of internal Pfizer documents, carried out by well over 3,000 researchers, reveals in jaw-dropping detail the horrid truth about the trials of Pfizer's bioweapon—which is to say that it reveals the lethal falseness of the propaganda that has beclouded all the world (and, one might say, helped to kill the world) since 2020: that Pfizer proved its "vaccine" to be "safe and effective" (it was neither), and that only it and other such "vaccines" could "save us" from "the virus" (they could not, and certainly did not, but killed and crippled just as many people as that "virus" was reported to have done). And yet, despite its thoroughness and clarity—or, rather, because of them—this all-important study has been blacked out by the media worldwide."

—Mark Crispin Miller, professor of media, culture, and communication, NYU

The Pfizer Papers

Pfizer's Crimes Against Humanity

By The WarRoom/Daily Clout Pfizer Documents Analysts

Edited by
Naomi Wolf
With
Amy Kelly
Foreword by
Stephen K. Bannon

Skyhorse Publishing

War Room Books may be purchased in bulk at special discounts for sales promotion, corporate gifts, fund-raising, or educational purposes. Special editions can also be created to specifications. For details, contact the Special Sales Department, Skyhorse Publishing, 307 West 36th Street, 11th Floor, New York, NY 10018 or info@skyhorsepublishing.com.

War Room Books® is a registered trademark of WarRoom, LLC.

Skyhorse Publishing® is a registered trademark of Skyhorse Publishing, Inc.®, a Delaware corporation.

Visit our website at www.skyhorsepublishing.com.
Please follow our publisher Tony Lyons on Instagram @tonylyonsisuncertain

10 9 8 7 6 5 4 3 2 1

Library of Congress Cataloging-in-Publication Data is available on file.

Print ISBN: 978-1-64821-037-2
eBook ISBN: 978-1-64821-038-9

Cover design by Brian Peterson

Printed in the United States of America

Contents

Foreword by Stephen K. Bannon

The entire War Room posse sends a special shout-out to Naomi Wolf and Amy Kelly for their leadership and their diligence in investigating and creating this book. The Pfizer documents are a stunning revelation of corporate greed and dishonesty, with utter disregard for the law, and Americans' actual health.

This exhaustive work by Naomi Wolf and Amy Kelly is far more than a unique review of the emergence of the coronavirus and the subsequent epidemic that began in 2019. It is a stinging indictment of the pharmaceutical industry, both globally and in the United States. Pfizer was relentless in its pursuit of corporate profit, unhampered by ethical, moral, or patriotic considerations.

Wolf and Kelly were tireless in their pursuit of the truth behind the vaccines. With dogged research following lawsuits by Aaron Siri to force the FDA to release internal documents, they learned that Pfizer knew its vaccines risked causing heart damage in young men. Pfizer did everything it could to keep these documents secret, including stamping "FDA Confidential" on them. The book also explains, critically, the story of how Big Pharma has historically been freed from much legal liability regardless of what damage its drugs, particularly vaccines, may cause to American citizens.

Wolf and Kelly explain the PREP Act, the Public Readiness and Emergency Preparedness Act, passed by Congress and signed into law by President George W. Bush in 2005. It protects pharmaceutical companies from liability for the use of so-called "medical countermeasures" in fighting a number of events, including pandemics. In other words, it allows pharma to escape responsibility for harm.

Of course, vaccine makers lobbied hard in favor of the legislation, which has since been codified. Interestingly, then Senator Ted Kennedy, a Democrat, opposed the legislation.

Since 2019, the PREP Act has been invoked to defend against injuries and illness caused by the experimental COVID vaccines.

Under the PREP Act, an injured plaintiff is, notably, not entitled to a jury trial under any circumstances. And even if a plaintiff can manage a vaccine injury claim that might grant them access to a jury trial, they are entitled to only bring their claim in only ONE court—the United States District Court for the District of Columbia.

This deeply and thoroughly researched book is an indictment of the Trifecta—Big Pharma, mainstream media, and the DC swamp. It demonstrates how Pfizer's clinical trials for the vaccine were deeply and fatally flawed. By November 2020, the company *knew* that its vaccine was neither safe nor effective.

The Pfizer Papers proves that the U.S. Food and Drug Administration knew about the shortfalls of Pfizer's clinical trials . . . and looked the other way.

Wolf and Kelly also raise the specter of completely dishonest communication strategies from Big Pharma. What is a vaccine anyway? Pfizer decided to call this shot a "vaccine" against COVID. Was it? The definition of a vaccine used to be to prevent sickness from a disease. But did these "vaccines" prevent COVID infection? They did not.

Americans were promised they wouldn't get "as sick" or "die" if they got the "vaccine." That is not anyone's understanding of a "vaccine."

Did Dr. Edward Jenner, who created the first vaccine against smallpox in 1796 claim people would only get "a little smallpox"?

No. It was a vaccination. The so-called COVID "vaccine" was a pharmaceutical industry fraud, and this book proves it.

Did Dr. Jonas Salk and Dr. Albert Sabin, men who created vaccines against polio in the 1950s and 1960s, claim people would only get a little polio but wouldn't feel as bad?

No. Those were actual vaccines.

Introduction by Naomi Wolf

This book in your hands is the result of an extraordinary set of confluences. It also presents, in a format available in bookstores for the first time, material that has already changed history.

You are about to embark as a reader on a journey through an extraordinary story—one whose elements almost defy belief.

The Pfizer Papers is the result of a group of strangers—ordinary people with extraordinary skills, located in different places around the world, with different backgrounds and interests—who all came together, for no money or professional recompense at all; out of the goodness of their hearts, and motivated by love for true medicine and true science—to undertake a rigorous, painfully detailed, and complex research project, which spanned the years 2022 to the present, and which continues to this day.

The material they read through and analyzed involved 450,000 pages of documents, all written in extremely dense, technical language.

This far-flung, relentlessly pursued research project—under the leadership of DailyClout's COO, the remarkably gifted project director Amy Kelly—brought one of the largest and most corrupt institutions in the world, Pfizer, to its knees. This project, pursued by 3,250 strangers who worked virtually and became friends and colleagues, drove a global pharmaceutical behemoth to lose billions of dollars in revenue. It balked the plans of the most powerful politicians on earth. It bypassed the censorship of the most powerful tech companies on earth.

This is the ultimate David and Goliath story.

The story began when lawyer Aaron Siri successfully sued the Food and Drug Administration, to compel them to release "The Pfizer Documents." These are Pfizer's internal documents—as noted above, 450,000 pages in number—that detail the clinical trials Pfizer conducted in relation to its COVID mRNA injection. These trials were undertaken to secure the ultimate prize for a pharmaceutical company, the "EUA," or Emergency Use Authorization from the FDA. The FDA awarded EUA for ages 16+ to Pfizer in December 2020. The "pandemic," of course (a crisis in public health that a book of mine, *The Bodies of Others,* confirmed, involved hyped and manipulated "infections" data and skewed mortality documentation) became the pretext for the "urgency" that led the FDA to bestow EUA on Pfizer's (and Moderna's) novel drug. The EUA is the hall pass, essentially, allowing Pfizer to race right to market with a not-fully-tested product.

The Pfizer Papers also contains documentation of what happened in "post-marketing," meaning in the three months, December 2020 to February 2021, as the vaccine was rolled out upon the public. All leading spokespeople, and bought-off media, called the injection "safe and effective," reading from what was a centralized script.

Many people who took this injection, as it was launched in 2020–2021–2022 and to the present, did not realize that normal testing for safety of a new vaccine—testing that typically takes ten to twelve years—had simply been bypassed via the mechanisms of a "state of emergency" and the FDA's "Emergency Use Authorization." They did not understand that the real "testing" was in fact Pfizer and the FDA observing whatever was happening to them and their loved ones, after these citizens rolled up their sleeves and submitted to the shot. As we can never forget, many millions of these people who submitted to the injection were "mandated" to take it, facing the threat of job loss, suspension of their education, or loss of their military positions if they refused; in some US states and overseas countries, people also faced the suspension of their rights to take transportation, cross borders, go to school or college, receive certain medical procedures, or enter buildings such as churches and synagogues, restaurants and gyms—if they refused.

The FDA asked the judge in the Aaron Siri lawsuit to withhold the release of the Pfizer documents for seventy-five years. Why would a government agency wish to conceal certain material until the present generation, those affected by what is in these documents, is dead and gone? There can be no good answer to that question.

Fortunately for history, and fortunately for millions of people whose lives were saved by this decision, the judge refused the FDA's request, and compelled the release of the documents; a tranche of 55,000 pages per month.

When I heard about this, though, I was concerned as a journalist. I knew that no reporter had the bandwidth to go through material of this volume. I also understood that virtually no reporter had the training or skill sets required to understand the multidimensional, technically highly specialized language of the reports. In order to understand the reports, one would need a background in immunology; statistics; biostatistics; pathology; oncology; sports medicine; obstetrics; neurology; cardiology; pharmacology; cellular biology; chemistry; and many other specialties. In addition to doctors and scientists, in order to understand what was really happening in the Pfizer documents, you would also need people deeply knowledgeable about government and pharmaceutical industry regulatory processes; you would need people who understood the FDA approval process; you would need medical fraud specialists; and eventually, in order to understand what crimes were committed in the Papers, you would need lawyers.

I was worried that without people with all of those skill sets reading through the documents, their volume and complexity would lead them to vanish down "the memory hole."

Enter Steve Bannon, the former Naval Officer, former Goldman Sachs investment banker, former advisor to President Trump, and current host of the most popular political podcast in America and one of the most listened-to worldwide, *WarRoom*.

He and I come from opposite ends of the political spectrum. I had been a lifelong Democrat, an advisor to President Bill Clinton's reelection campaign, and to Al Gore's presidential campaign. He, of course, is a staunch Republican-turned-MAGA. I had been deplatformed in June 2021, before the Pfizer documents came out, for the crime of warning that women were reporting menstrual dysregulation upon having received the mRNA injections. As a career-long writer on women's sexual and reproductive health issues, I knew that this was a serious danger signal and that this side effect would affect fertility. (Any eighth grader should be able to foresee that as well.) Upon my having posted this warning, I was banned from Twitter, Facebook, YouTube, and other platforms. I was attacked globally, all at once, as an "anti-vaxxer" and "conspiracy theorist"; and my life as a well-known, bestselling feminist author, within the legacy media, ended. No one in that world would talk to me anymore, publish my work, or return my calls. I was un-personed.

(It turned out, upon two successful lawsuits in 2023 by Missouri and Louisiana attorneys general, that it was actually the White House, the CDC, and senior leaders of other government agencies, including the Department of Homeland Security, that unlawfully pressured Twitter and Facebook to remove that cautionary tweet of mine, to shut me down, and to "BOLO" or Be On the Lookout for similar posts. This suppression is now the subject of a pending Supreme Court decision on whether or not it violated the First Amendment.)

In this dark time in my life, to my surprise, I received a text from Steve Bannon's producer, who invited me onto *WarRoom*. I brought forward my concerns about women's reproductive health in the wake of mRNA injection, and to my surprise he was respectful, thoughtful about the implications, and took the issue very seriously. I returned again and again, to bring that and other concerns that were emerging in relation to the mRNA injections to his audience. I was relieved to have a platform on which I could share these urgent warnings. At the same time, I was sad that the Left, which was supposed to champion feminism, seemed not to care at all about serious risks to women and unborn babies. I recognized the irony that a person whom I had been taught to believe was the Devil Incarnate, actually cared more about women and babies than did all of my right-on former colleagues, including the feminist health establishment, who had always spoken so loudly about women's wellbeing and women's rights.

Given my appearances on *WarRoom* leading up to 2022, it was natural that the subject of the Pfizer documents came up on that show when the documents were released. I shared my concern that they would be lost to history due to their volume and technical language. Bannon said something like, "Well, you will crowdsource a project to read through them."

I was taken aback, as I had zero skills related to, or knowledge about how possibly to do such a thing. I answered something like, "Of course."

So, my news and opinion platform DailyClout was deluged with offers from around the world, from *WarRoom* listeners with the skill sets needed, to decipher the Pfizer documents. I was terrified. It was chaos. I had excellent people on my team. But none of us knew how to manage or even organize the deluge of emails; we did not know how to evaluate the thousands of CVs; and even once we had "onboarded" these thousands of people, in different time zones, to "the project," our inboxes became even more terrifying, as it was literally impossible to organize 3,250 experts into an organization chart that could systematically work through these documents. Emails were getting tangled or went unanswered. People asked questions we could not answer. We had no idea what structure could allow such a huge number of disparate experts to work through the vast trove of material.

A few weeks in, as I was in despair, Bannon had me on again. He asked about the progress of the project, and I replied, more upbeat than I felt, that many people had joined us, and they were starting to read. "Of course, you will begin delivering reports," he prompted. "Of course," I answered, horrified at being in so far over my head.

I have never had a corporate job, so it had not even occurred to me that a series of reports was the format that the analyses of the documents should take.

Then something happened that I can only describe as providential. We put out a call to the volunteers for a project manager, and Amy Kelly reached out. Ms. Kelly is a Six Sigma-certified project manager, with extensive experience in telecommunications and tech project management. She is also a simply inexplicably effective leader. The day that she put her hand to the chaos in the inboxes, the waters were stilled. Peace and productivity prevailed. Ms. Kelly somehow effortlessly organized the volunteers into six working groups, with a supra-committee at the head of each, and the proper work began.

I can only explain the scope and smoothness and effectiveness of the work that followed, as occurring in a state of grace.

In the two years since Ms. Kelly and the volunteers have been working together, they have gone through 2,369 documents and data files totaling hundreds of thousands of pages and have issued almost one hundred reports. I taught the volunteers to write these in a language that everyone could understand—which I thought was very important to maximize their impact. And Amy Kelly meticulously revised almost all, and edited all, of them.

The first forty-six reports appeared in a self-published format that we put out. It was very important to us that they appear in a published form that was physical, and not just digital, as we wanted something that people could hand to their doctors, their loved ones, their congressional representatives.

These forty-six reports broke huge stories. We learned that Pfizer knew within three months after rollout in December 2020, that the vaccines did not work to stop COVID. Pfizer's language was "vaccine failure" and "failure of efficacy." One of the most common "adverse events" in the Pfizer documents is "COVID."

Pfizer knew that the vaccine materials—lipid nanoparticles, an industrial fat, coated in polyethylene glycol, a petroleum byproduct; mRNA; and spike protein—did not remain in the deltoid muscle, as claimed by all spokespeople. Rather, it dispersed throughout the body in forty-eight hours "like a shotgun blast," as one of the authors, Dr. Robert Chandler, put it; it crossed every membrane in the human body—including the blood-brain barrier—and accumulated in the liver, adrenals, spleen, brain, and, if one is a woman, in the ovaries. Dr. Chandler saw no mechanism whereby those materials leave the body, so every injection appears to pack more such materials into organs.

Pfizer hired 2,400 fulltime staffers to help process "the large increase of adverse event reports" being submitted to the company's Worldwide Safety database.

Pfizer knew by April 2021 that the injections damaged the hearts of young people.

Pfizer knew by February 28, 2021—just ninety days after the public rollout of their COVID vaccine—that its injection was linked to a myriad of adverse events. Far from being "chills," "fever," "fatigue," as the CDC and other authorities claimed were the most worrying side effects, the actual side effects were catastrophically serious.

These side effects included: death (which Pfizer does list as a "serious adverse event"). Indeed, over 1,233 deaths in first three months of the drug being publicly available.

Severe COVID-19; liver injury; neurological adverse events; facial paralysis; kidney injury; autoimmune diseases; chilblains (a localized form of vasculitis that affects the fingers and toes); multiple organ dysfunction syndrome (when more than one organ system is failing at once); the activation of dormant herpes zoster infections; skin and mucus membrane lesions; respiratory issues; damaged lung structure; respiratory failure; acute respiratory distress syndrome (a lung injury in which fluid leaks from the blood vessels into the lung tissue, causing stiffness which makes it harder to breathe and causes a reduction of oxygen and carbon dioxide exchange), and SARS (or SARS-CoV-1, which had not been seen in the world since 2004, but appears in the Pfizer documents as a side effect of the injections).

Thousands of people with arthritis-type joint pain, the one of most common side effect, were recorded. Other thousands with muscle pain, the second most common. Then, industrial-scale blood diseases: blood clots, lung clots, leg clots; thrombotic thrombocytopenia, a clotting disease of the blood vessels; vasculitis (the destruction of blood vessels via inflammation); astronomical rates of neurological disorders—dementias, tremors, Parkinson's, Alzheimer's, epilepsies. Horrific skin conditions. A florid plethora of cardiac issues; myocarditis, pericarditis, tachycardia, arrhythmia, and so on. Half of the serious adverse events related to the liver, including death, took place within seventy-two hours of the shot. Half of the strokes took place within forty-eight hours of injection.

But what really emerged from the first forty-six reports, was the fact that though COVID is ostensibly a respiratory disease, the papers did not focus on lungs or mucus membranes, but rather they center, creepily and consistently, on disrupting human reproduction.

By the time Pfizer's vaccine rolled out to the public, the pharmaceutical giant knew that they would be killing babies and significantly harming women and men's reproduction. The material in the documents makes it clear that damaging human's ability to reproduce and causing spontaneous abortions of babies is "not a bug, it is a feature."

Pfizer told vaccinated men to use two reliable forms of contraception or else to abstain from sex with childbearing-age women. In its protocol, the company defined "exposure" to the vaccine as including skin-to-skin contact, inhalation, and sexual contact. Pfizer mated vaccinated female rats and "untreated" male rats, and then examined those males, females, and their offspring for vaccine-related "toxicity." Based on just forty-four rats (and no humans), Pfizer declared no negative outcomes for ". . . mating performance, fertility, or any ovarian or uterine parameters . . . nor on embryo-fetal or postnatal survival, growth, or development," the implication being that its COVID vaccine was safe in pregnancy and did not harm babies. Pfizer knew that lipid nanoparticles have been known for years, to degrade sexual systems, and Amy Kelly in fact found nanoparticles, of which lipid nanoparticles are a subtype, pass through the blood-testis barrier and damage males' Sertoli cells, Leydig cells, and germ cells. Those are the factories of masculinity, affecting the hormones that turn boys at adolescence into men, with deep voices, broad shoulders, and the ability to father children. So, we have no idea if baby boys born to vaccinated moms, will turn into adults who are recognizably male and fertile. Pfizer enumerated the menstrual damages it knew it was causing to thousands of women, and the damage ranges from women bleeding every day, to having two periods a month, to no periods at all; to women hemorrhaging and passing tissue; to menopausal and post-menopausal women beginning to bleed again. Pfizer's scientists calmly observed and noted it all but did not tell women.

Babies suffered and died. In one section of the documents, over 80 percent of the pregnancies followed resulted in miscarriage or spontaneous abortion. In another section of the documents, two newborn babies died, and Pfizer described the cause of death as "maternal exposure" to the vaccine.

Pfizer knew that vaccine materials entered vaccinated moms' breast milk and poisoned babies. Four women's breast milk turned "blue-green." Pfizer produced a chart of sick babies, made ill from breastfeeding from vaccinated moms, with symptoms ranging from fever to edema (swollen flesh) to hives to vomiting. One poor baby had convulsions and was taken to the ER, where it died of multi-organ system failure.

I will now take you to the thirty-six reports you will find in this book. Some of the headlines from the reports that follow are:

On Feb 28, 2021, Pfizer produced a "Pregnancy and Lactation Cumulative Review" showing that after mothers' vaccination with its vaccine:

- Adverse events occurred in over 54 percent of cases of "maternal exposure" to vaccine and included 53 reports of spontaneous abortion (51)/ abortion (1)/ abortion missed (1) following vaccination.
- Premature labor and delivery cases occurred, as well as two newborn deaths.
- Some newborns suffered severe respiratory distress or "illness" after exposure via breast milk.
- "Substantial" birth rate drops happened across thirteen countries: countries in Europe, as well as Britain, Australia, and Taiwan, within nine months of public vaccine rollout.
- Approximately 70 percent of Pfizer vaccine-related adverse events occur in women.
- Spike protein and inflammation were still present in heart tissue one year after receipt of the mRNA COVID vaccine.
- In Pfizer's clinical trial, there were more deaths among the vaccinated than the placebo participants. However, Pfizer submitted inaccurate data, showing more deaths in the placebo group, to the FDA when seeking emergency use authorization.
- Infants and children under twelve received Pfizer's vaccine seven months before a pediatric vaccine approval resulting in:
  - Stroke.
  - Facial paralysis.
  - Kidney injury or failure.
- There was an over 3.7-fold increase in the number of deaths due to cardiovascular events in vaccinated clinical trial subjects compared to placebo subjects.
- The vaccine Pfizer rolled out to the public was different than the formulation used on the majority of clinical trial participants, and the public was not informed of this.
- Histopathologic analyses (the staining of tissues to show disease states) show clear evidence of vaccine-induced, autoimmune-like pathology in multiple organs; spike protein–caused erosion of the blood vessels, heart, and lymphatic vessels; amyloids in multiple tissues; unusual, aggressive cancers; and atypical "clot" formations.
- Following vaccination, younger patients began presenting with cancers; tumors were bigger and grew more aggressively and faster than cancers had prior to mass inoculation of populations; co-temporal onset (the onset more than one cancer at the same time) of cancers became more common—a situation that was typically very unusual before the mRNA vaccines' rollout. Benign tumors' growth accelerated.
- By March 12, 2021, Pfizer researchers vaccinated almost the entire placebo (non-vaccinated) cohort from the trial, though Pfizer had previously committed to following both the vaccinated and placebo cohorts for two years. Immediately after receiving the Emergency Use Authorization, Pfizer lobbied the FDA to allow them to vaccinate the unvaccinated cohort for "humanitarian" reasons. Vaccinating the placebo group ended the ability to pursue safety studies over time.

- Autoimmunity cases reported to the Vaccine Adverse Events Reporting System (VAERS) increased 24-fold from 2020 to 2021, and annual autoimmunity-related fatalities increased 37x in the same time period.
- In Pfizer's October 2021 emergency use authorization data and documents submission for children ages five to eleven, Pfizer investigators speculated in writing that subclinical damages would manifest in patients in the long term, implying that continued doses with subclinical damages would eventually manifest as clinical damages.
- In trial studies, Moderna mRNA COVID-19 vaccine damaged mammals' reproduction—resulting in 22 percent fewer pregnancies; skeletal malformations; and nursing problems.
- There were hundreds of possible vaccine-associated enhanced disease (VAED) cases in the first three months of Pfizer's mRNA COVID vaccine rollout. Public health spokespeople minimized their severity by calling them "breakthrough COVID cases."
- Pfizer concealed eight vaccinated deaths that occurred during the clinical trial in order to make its results look favorable for receiving its ages 16+ EUA.

The most powerful forces in the world—including the White House, the staffers of the United States president himself; Dr. Rochelle Walensky of the CDC; the head of the FDA, Dr. Robert M Califf; Dr. Anthony Fauci; Twitter and Facebook; legacy media, including the *New York Times*, the BBC, the *Guardian* and NPR; OfCom, the British media regulatory agency; professional organizations such as the American College of Obstetricians and Gynecology, and the European Medicines Agency, the European equivalent of the FDA, and the Therapeutics Goods Administration, Australia's equivalent of the FDA—all sought to suppress the information that Amy Kelly, the research volunteers, and I brought to the world starting in 2022, and that you are about to absorb in the following pages.

Nonetheless, in spite of the most powerful censorship and retribution campaign launched in human history—made more powerful than past such campaigns by the amplifying effects of social media and AI—these volunteers' findings were not suppressed at last, and survived on alternative media, and on our site DailyClout.io; to be shared from mouth to mouth, saving millions of lives.

Fast forward to more recent events. What has the role of this information been in stopping this greatest crime ever committed against humanity?

The worst has happened. Disabilities are up by a million a month in the United States, according to former BlackRock hedge fund manager Edward Dowd. Excess deaths are way up in the US and Western Europe. Birth rates have plummeted, according to the mathematician Igor Chudov (and WarRoom/DailyClout Volunteer Researcher Dr. Robert Chandler) by 13–20 percent since 2021, based on government databases. Athletes are dropping dead. Turbo-cancers are on the rise. Conventional doctors may be "baffled" by all of this, but sadly, we, thanks to Amy Kelly and the volunteers, understand exactly what is happening.

Our relentless effort to get this information to the world, in an unimpeachable form, has finally paid off with results. The uptake for boosters is now 4 percent. Very few people "boosted" their children. Most colleges in the United States withdrew their vaccine "mandates." Pfizer's net revenue dropped in Q1 of 2024 to pre-2016 levels. OfCom, which had targeted Mark Steyn for "platforming" on his show my description of the reproductive and other harms in the Pfizer documents, is being sued by Steyn. The BBC had to report that vaccine injuries are real, as did the *New York Times*. AstraZeneca, a somewhat differently configured COVID vaccine in Europe, was withdrawn from the market in May 2024, following lawsuits involving thrombotic thrombocytopenia (a side effect about which our research volunteer Dr. Carol Taccetta had informed the FDA by letter in 2022), and the European Medicines Agency notably withdrew its EUA for AstraZeneca. Three days after we published our report showing that the FDA and CDC had received the eight-page "Pregnancy and Lactation Cumulative Review" confirming that Dr. Walensky knew about the lethality of the vaccine when she held her press conference telling pregnant women to get the injection, Dr. Walensky resigned.

It is difficult indeed to face this material in the roles that Amy Kelly and I play. No doubt for the volunteers, unearthing this criminal evidence is painful indeed. It may be hard to read some of what follows. As I have said elsewhere, seeing this material is like being among the Allied soldiers who first opened the gates of Auschwitz.

But the truth must be told.

Among other important reasons to tell these truths, people were injured and killed with a novel technology not deployed before in medicine; and these pages hold important clues as to the mechanisms of these injuries, and thus, they provide many signposts for physicians and scientists in the future, for treating the many injuries that these new mRNA technologies, injected into people's bodies, have brought about.

We must share the truth, as the truth saves and sustains; and eventually, the truth will heal.

We thank Steve Bannon, and his wonderful team at *WarRoom*, for being the instigator of this entire project and for consistently bringing us onto his show so that we can tell the world what the volunteers find.

We thank Skyhorse Publishing, publisher Tony Lyons, and our editor Hector Carosso, for taking the critical step of publishing this material in a book that will be available everywhere. Books matter, and this publication will make a difference in bringing about accountability and an accurate history of this catastrophic set of events.

We thank the volunteers, 3,250 strangers around the world who banded together in the love of truth and of their fellow human beings. We thank our two hundred lawyers, who helped us to FOIA emails from the CDC and helped us to understand the crimes that we were seeing in the following pages.

Many of our volunteers themselves have suffered ostracism, job loss, marginalization, and other penalties, as a consequence of their commitment to real science, real medicine, and to bringing forth the truth to save their fellow human beings, and generations yet unborn.

The battle is ongoing. No one who committed this massive crime against humanity is in jail, or even facing civil or criminal charges. There are at least three lawsuits against Pfizer—two of ours, and one of Brook Jackson's—but, to date, none of the lawsuits have completely prevailed. The litigation drags on.

Nonetheless—nonetheless. The word is out.

Amy Kelly and I get hundreds of emails from grateful families, telling us about their healthy babies or grandchildren and thanking us for saving those babies, or sons and daughters and daughters-in-law, and we know this project has saved many lives; perhaps hundreds of thousands of lives and maybe saved millions from disabling injuries. Steve Bannon, who started it all, saved hundreds of thousands of lives and saved his listeners and ours from sustaining millions of injuries. God know how many babies will be born in the future, safe and well, because of our collective, arduous, much-targeted work.

The story of this project is not over.

Your own actions, upon your having read these reports, are part of the ongoing ripples of this work.

Whom will you tell?

How will you process the information?

What will you do to avenge the crimes of the past?

What will you do to save the future?

One: "Liver Adverse Events—Five Deaths Within 20 Days of Pfizer's mRNA COVID Injection. 50% of Adverse Events Occurred Within Three Days."

—Joseph Gehrett, MD; Barbara Gehrett, MD; Chris Flowers, MD; and Loree Britt

The WarRoom/DailyClout Pfizer Analysis Post-Marketing Group—Team 1 created the following Liver (i.e., Hepatic) System Organ Class (SOC) Review from data in Pfizer document *5.3.6 Cumulative Analysis of Post-Authorization Adverse Event Reports of PF-07302048 (BNT162B2) Received Through 28-FEB-2021* (a.k.a., "*5.3.6*"). (https://www.phmpt.org/wp-content/uploads/2022/04/reissue_5.3.6-postmarketing-experience.pdf) There were 70 patient cases with 94 adverse events reported in the hepatic SOC category.

The report shows that *five deaths*, unexplained by the broad adverse event (AE) descriptions, occurred within 20 days of injection, which suggests severe and rapid liver injury or failure. Also, *50% of adverse events* occurred *within three days* of receiving Pfizer's mRNA COVID drug. At the data collection cutoff date of February 28, 2021, more than *half of the outcomes were unknown* and remain unknown to this day.

This report is unique compared to other SOC category reports, because the data published by Pfizer largely consist of lab test abnormalities related to liver enzymes rather than clinical disease descriptions. No further categorization or classification under medically recognized diseases, such as hepatitis or hepatobiliary (gallbladder or bile duct) conditions, was done, though the lab tests cited often point to disease entities. Additionally, no justification is offered to explain this inconsistency in Pfizer's data collection and reporting.

Shockingly, for liver adverse events, Pfizer deviated from listing every adverse event, as it had for strokes and cardiovascular abnormalities, and set a threshold of three separate occurrences of an adverse event before the AE became "reportable." Therefore, when a liver abnormality occurred only once or twice during Pfizer's post-marketing analysis time frame, it did not reach the threshold of "reportability."

Why was a threshold of "three or more" used before Pfizer would list liver-specific diagnoses? What is potentially hidden in those diagnoses conveniently not reported because they did not reach that arbitrary threshold? One can only conclude that Pfizer deliberately underestimated the number of adverse events in its *5.3.6* document, which it knew would have to be submitted to the Food and Drug Administration (FDA) for safety signal monitoring.

It is important to note that the adverse events in the *5.3.6* document were reported to Pfizer for *only a 90-day period* starting on December 1, 2020, the date of the United Kingdom's public rollout of Pfizer's COVID-19 experimental mRNA "vaccine" product.

Please read this important report below.

War Room/DailyClout Pfizer Document Analysis

Post-Marketing Team Micro-Report 4:

Liver (Hepatic) System Organ Class (SOC) Review of 5.3.6

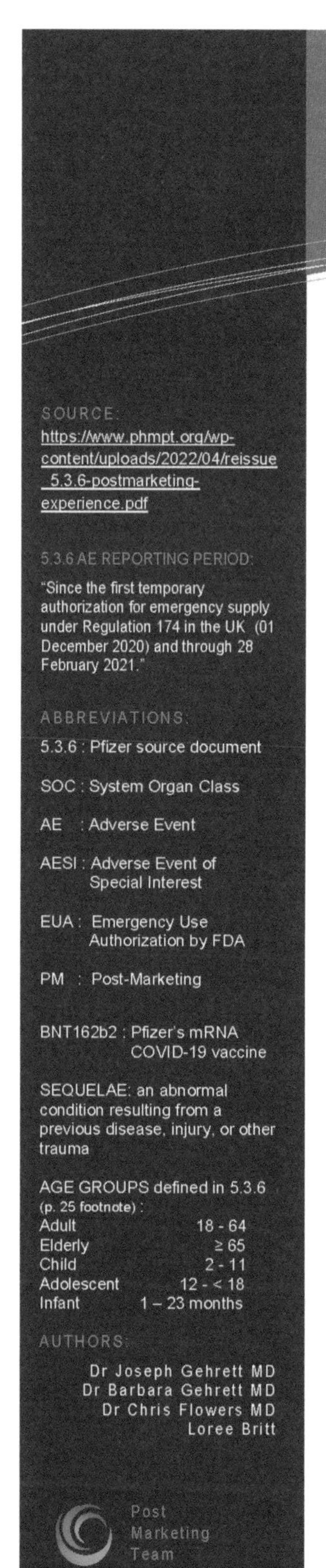

SOURCE:
https://www.phmpt.org/wp-content/uploads/2022/04/reissue_5.3.6-postmarketing-experience.pdf

5.3.6 AE REPORTING PERIOD:
"Since the first temporary authorization for emergency supply under Regulation 174 in the UK (01 December 2020) and through 28 February 2021."

ABBREVIATIONS:
5.3.6 : Pfizer source document
SOC : System Organ Class
AE : Adverse Event
AESI : Adverse Event of Special Interest
EUA : Emergency Use Authorization by FDA
PM : Post-Marketing
BNT162b2 : Pfizer's mRNA COVID-19 vaccine
SEQUELAE: an abnormal condition resulting from a previous disease, injury, or other trauma

AGE GROUPS defined in 5.3.6 (p. 25 footnote) :

Adult	18 - 64
Elderly	≥ 65
Child	2 - 11
Adolescent	12 - < 18
Infant	1 – 23 months

AUTHORS:
Dr Joseph Gehrett MD
Dr Barbara Gehrett MD
Dr Chris Flowers MD
Loree Britt

Post Marketing Team

08Jan22

Hepatic AESIs *Search criteria: Liver related investigations, signs and symptoms (SMQ) (Narrow and Broad) OR PT Liver injury*	• Number of cases: 70 cases (0.2% of the total PM dataset), of which 54 medically confirmed and 16 non-medically confirmed; • Country of incidence: UK (19), US (14), France (7), Italy (5), Germany (4), Belgium, Mexico and Spain (3 each), Austria, and Iceland (2 each); the remaining 8 cases originated from 8 different countries; • Subjects' gender: female (43), male (26) and unknown (1); • Subjects' age group (n=64): Adult (37), Elderly (27);

CONFIDENTIAL
Page 18
FDA-CBER-2021-5683-0000071

BNT162b2
5.3.6 Cumulative Analysis of Post-authorization Adverse Event Reports

Table 7. AESIs Evaluation for BNT162b2

AESIs[a] Category	Post-Marketing Cases Evaluation[b] Total Number of Cases (N=42086)
	• Number of relevant events: 94, of which 53 serious, 41 non-serious; • Most frequently reported relevant PTs (≥3 occurrences) include: Alanine aminotransferase increased (16), Transaminases increased and Hepatic pain (9 each), Liver function test increased (8), Aspartate aminotransferase increased and Liver function test abnormal (7 each), Gamma-glutamyltransferase increased and Hepatic enzyme increased (6 each), Blood alkaline phosphatase increased and Liver injury (5 each), Ascites, Blood bilirubin increased and Hypertransaminasaemia (3 each); • Relevant event onset latency (n = 57): Range from <24 hours to 20 days, median 3 days; • Relevant event outcome: fatal (5), resolved/resolving (27), resolved with sequelae (1), not resolved (14) and unknown (47). Conclusion: This cumulative case review does not raise new safety issues. Surveillance will continue

https://www.phmpt.org/wp-content/uploads/2022/04/reissue_5.3.6-postmarketing-experience.pdf

This event category was comprised of abnormal laboratory tests and not defined under any specific disease designations. There was no further categorization or classification under medically recognized diseases such as hepatitis or hepatobiliary (gallbladder or bile duct) conditions, though the lab tests cited often point to different disease entities. The common terms normally used, such as hepatitis, gallstones, and others, were not included in the search terms for patient cases in this document.

There were nine reports of "hepatic pain," three reports of ascites (fluid free within the abdominal cavity) and three cases of high bilirubin, which is the chemical that causes jaundice. **Pfizer chose to specify only those events with three or more occurrences.** All of the specific reported adverse events were elevated levels of proteins reflective of hepatocyte (the major type of liver cell) injury, bile processing system cell injury, symptoms, or physical findings.

Of those patients with age reported, 37 were categorized as adult and 27 as elderly. There were reports from 18 countries.

- **Adverse Events were reported to Pfizer during a 90-day period, following the December 1, 2020, public rollout of its COVID-19 experimental "vaccine" product.**
- **In the Pfizer 5.3.6 document, these AEs were categorized by System Organ Classes (SOC) – in other words, by systems in the body.**
- **There were 70 cases with 94 adverse events reported in the hepatic SOC category.**
- **The hepatic adverse events were defined as "liver-related investigations, signs and symptoms" or reported as "liver injury."**

The time reported from vaccine injection to adverse event ranged from within 24 hours to 20 days, **with half occurring within three days.**

There were **five deaths (7% of the patients)**. Of the reported events that were not fatal, 27 (30%) were resolved or resolving, although the figures in these two outcome categories were not independently provided. One (1%) was "resolved with sequelae," 14 (15%) were unresolved, and 47 (50%) were unknown. Given the imprecise method of outcome reporting, combined with the lack of long-term follow-up, the stated fatality rate is questionable and may be much higher.

This report is unique compared to other SOC categories under review by the Post-Marketing Team, in that the data presented by Pfizer largely consists of laboratory abnormalities, rather than clinical disease descriptions. No justification is offered to explain this inconsistency in data collection and reporting.

Pfizer presents this SOC data as single, abnormal lab results rather than diseases or conditions. Liver injury, short of catastrophic acute liver failure, requires complex assessment of multiple lab and other diagnostic studies performed over time. For example, fluid in the abdomen (ascites), which was found in three patients with abnormal liver enzymes, suggests a potentially severe liver condition. One test or a panel of enzyme tests at a single point in time are not sufficient data to evaluate safety or predict future liver health. Pfizer's own data show they have no follow-up information on over 50% of the patients in this SOC. Given the biochemical mixture in this novel genetic product, which includes mRNA plus lipid nanoparticles and other chemicals, the impact on the liver should have been of the highest safety monitoring priority.

Five deaths, and numerous other patients demonstrating serious enzyme elevations, demand further evaluation beyond Pfizer's dismissive conclusion published at the end of the hepatic SOC section, which reads:

"This cumulative case review does not raise new safety issues. Surveillance will continue."

Pfizer's inadequate assessment of the hepatic AEs, in terms of complete disregard of the safety signals revealed in this data set, foretells the overarching summary conclusion in Pfizer's 5.3.6 Post Marketing Adverse Events document which proclaims a "favorable benefit risk profile" of their investigational product:

> 4. DISCUSSION
> Pfizer performs frequent and rigorous signal detection on BNT162b2 cases. The findings of these signal detection analyses are consistent with the known safety profile of the vaccine. This cumulative analysis to support the Biologics License Application for BNT162b2, is an integrated analysis of post-authorization safety data, from U.S. and foreign experience, focused on Important Identified Risks, Important Potential Risks, and areas of Important Missing Information identified in the Pharmacovigilance Plan, as well as adverse events of special interest and vaccine administration errors (whether or not associated with an adverse event). The data do not reveal any novel safety concerns or risks requiring label changes and support a favorable benefit risk profile of to the BNT162b2 vaccine.
>
> 5. SUMMARY AND CONCLUSION
> Review of the available data for this cumulative PM experience, confirms a favorable benefit: risk balance for BNT162b2.
>
> https://www.phmpt.org/wp-content/uploads/2022/04/reissue_5.3.6-postmarketing-experience.pdf

It is important to note, since the end of the 5.3.6 reporting period (February 28, 2021), Pfizer has not voluntarily released any publicly available data to support their pledge to continue safety surveillance on the company's novel COVID-19 experimental product. Ongoing surveillance by Pfizer was a condition of the vaccine approval granted Pfizer by the FDA (August 2021). Furthermore, as of the date of this Post-Marketing Team report (January 8, 2023), there is no indication that the FDA has enforced this surveillance mandate.

The FDA's dereliction of regulatory duty is stunning given the fact that Pfizer's own 5.3.6 data reveal no definitive outcome or follow-up information on over half of the patients in the hepatic adverse events category.

Post-Marketing Team's CONCLUSION:
WHAT DOES IT TAKE?

How many serious **ADVERSE EVENTS** does it take?
How many **UNRESOLVED and UNKNOWN** outcomes does it take?
How many **DEATHS** does it take?

What does it take to **RECALL PFIZER'S UNSAFE "VACCINE"?**

Two: "Nine Months Post-COVID mRNA 'Vaccine' Rollout, Substantial Birth Rate Drops in 13 European Countries, England/Wales, Australia, and Taiwan."
—Robert W. Chandler, MD, MBA

Robert W. Chandler, MD, completed extensive research to write the article below. Some of the highlights of this important piece include:

- Nine months following the rollout of the COVID-19 mRNA "vaccines," substantial birth rate drops were seen in 13 of 19 European countries, England and Wales (one entity based on how data is published), Australia, and Taiwan.
- The decline in births in Switzerland was the largest in 150 years—more than during two World Wars, the Great Depression, and the advent of widely available birth control.
- There was an 8.3% drop in the birth rate in Germany through three quarters of 2022.
- England and Wales had a 12% birth rate drop through June 2022, which is when their government stopped publishing data related to this.
- Taiwan reported an alarming birth rate drop, but its data are incomplete.
- Australian birth rates fell 21% from October to November 2021, followed by a 63% decrease from November to December 2021.
- On August 25, 2022, the Swiss Hagemann group published a statement regarding the decline of live births in Europe: "My analysis puts the monthly birth figures in relation to the average of the last three years. In advance it should be noted that every single examined European country shows a monthly decline in birth rates of up to more than 10% compared to the last three years. I can be shown that this very alarming signal cannot be explained by infections with Covid-19. However, one can establish a clear temporal correlation to Covid vaccinations incidence in the age group of men and women between 18 and 49 years. Therefore, in-depth statistical and medical analyses have to be demanded." (https://www.initiative-corona.info/fileadmin/dokumente/Geburtenrueckgang-Europe-EN.pdf)

I. Background

Pfizer's Preclinical Studies, 2.4 Nonclinical Overview, revealed concentration of lipid nanoparticles containing experimental mRNA in ovaries of Wistar rats. (https://www.phmpt.org/wp-content/uploads/2022/03/125742_S1_M2_24_nonclinical-overview.pdf) The study was completed in 48 hours. Unfortunately, the tissue levels of lipid nanoparticle and mRNA were rising sharply at the time the animals were sacrificed, and the biodistribution time course of LNP and mRNA remains largely unknown. (https://robertchandler.substack.com/p/tissue-distribution-of-bnt162b2-pre) and (https://dailyclout.io/pfizer-used-dangerous-assumptions-rather-than-research-to-guess-at-outcomes/)

There has been no evidence located to date in the Pfizer records that necropsy examinations with special staining of ovarian tissues for spike proteins under light and electron microscopy were performed, which is an important omission. Additional animal studies were indicated but were not performed. Deficiencies were reviewed previously. (https://robertchandler.substack.com/p/pfizer-pre-clinical-studies-review)

Sasha Latypova in a January 1, 2023, review of Pfizer's Pre-Clinical (Pfizer document 2.4) testing concluded:

> The cursory nature of the entire preclinical program for mRNA injections conducted by Pfizer can be briefly summarized as "*we did not find any safety signals because we did not look for them.*" The omissions of standard safety studies and glaring scientific dishonesty in the studies that were performed are so obvious that they cannot be attributed to the incompetence of the manufacturers and regulators. Rather, the questions of fraud and willful negligence should be raised.
>
> (https://sashalatypova.substack.com/p/did-pfizer-perform-safety-testing)

Additional omissions occurred in Pfizer's clinical trials:

1. Critically, the Phase ⅔ Clinical Trial (Polack, et al.) involving over 40,000 subjects did not include pregnant women, at least not by design. A small number of pregnant women were injected, but no follow-up reporting on these women was provided.

> This report does not address the prevention of Covid-19 in other populations, such as younger adolescents, children, and pregnant women. Safety and immune response data from this trial after immunization of adolescents 12 to 15 years of age will be reported subsequently, and additional studies are planned to evaluate BNT162b2 in pregnant women, children younger than 12 years, and those in special risk groups, such as immunocompromised persons.
>
> (https://www.nejm.org/doi/pdf/10.1056/NEJMoa2034577?articleTools=true)

2. A 12/22/2022 paper by Irrarang, et al. identified dose-related effects in the distribution of IgG profile in humans after the second and third doses of Pfizer's SARS-CoV-2 mRNA drug (BNT162b2):

> Here, we report that several months after the second vaccination, SARS-CoV-2 specific antibodies were increasingly composed of non-inflammatory IgG2, which were further boosted by a third mRNA vaccination and/or SARS-CoV-2 variant breakthrough infections.
>
> (https://www.science.org/doi/10.1126/sciimmunol.ade2798)

Figure 1: Dose-Related Shift in IgG Immunoglobulins.

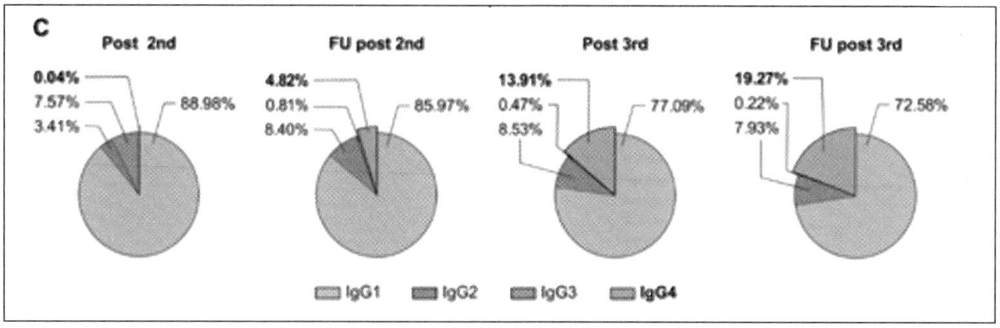

C
Post 2nd: 0.04%, 7.57%, 3.41%, 88.98%
FU post 2nd: 4.82%, 0.81%, 8.40%, 85.97%
Post 3rd: 13.91%, 0.47%, 8.53%, 77.09%
FU post 3rd: 19.27%, 0.22%, 7.93%, 72.58%
IgG1 IgG2 IgG3 IgG4

Dose-related shift in IgG profile with decline in IgG 1 and 3 and rise in IgG 2 and 4 with increasing doses of LNP/mRNA.

Figure 2: Gain in IgG 4 as the Number of LNP/mRNA Doses Increase.

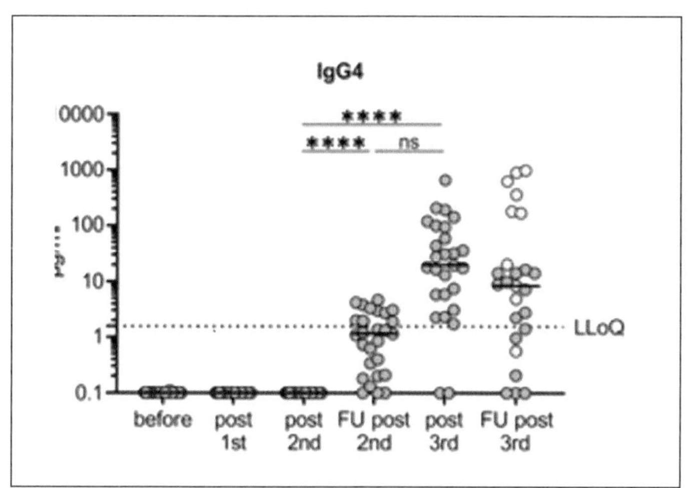

Figure 2 is a plot of the rise in IgG4 with successive doses of BNT162b2 after Dose 2. The significance of this IgG shift is only beginning to be explored. What is certain is that alteration of the IgG profile was not anticipated and therefore not studied.

Jessica Rose discusses these findings in the context of pregnancy noting:

> IgG can be passed to the foetus via the placental barrier via endosomes "within syncytiotrophoblasts of the placenta, through a pH dependent mechanism involving FcRn receptors, with a possible role for other IgG Fc receptors, yet to be fully elucidated." They also showed a preferential transfer of IgG4 (and IgG1 and IgG3). Right. *So (sic) what is the effect, therefore, on the foetus when there is a dramatic shift in IgG subclass ratio to subclass IgG4? I cannot imagine that the effects would be nil.* (Italics added)
>
> (https://jessicar.substack.com/p/igg4-and-pregnancy)

3. Röltgen, et al. dispelled the notion that the LNP/mRNA products briefly remain at the injection site and in local lymph nodes when they identified mRNA in local lymph nodes for two months after injection, at which point the study ended. So, it is unknown how long mRNA persists, where it is located, what it does to the host genome, and for how long it produces a largely unidentified artificial protein(s). (https://www.cell.com/cell/fulltext/S0092–8674(22)00076–9)

4. Long-term data is limited at present but is accruing. The studies promised by Pfizer and the Centers for Disease Control and Prevention (CDC) have not appeared.

5. The control group that was meant to be followed for two years was unblinded after a few months, which contaminated the group. Eliminating the control group was tragic decision.

II. Recommendations During Pregnancy

Remarkably, in spite of concentration of both lipid nanoparticles and mRNA in the ovaries of experimental animals, dose-related effects in animals and humans, and no testing in pregnant women during clinical trials or surveillance following the Experimental Use Authorization (EUA) granted by the Food and Drug Administration (FDA) December 14, 2020, the Centers for Disease Control and Prevention (CDC) and the American College of Obstetrics and Gynecology (ACOG) have recommended use of LNP/mRNA products in pregnant women without knowledge of either the short-term or long-term effects of what is contained in the vials of LNP/mRNA and their effects on the human reproductive system.

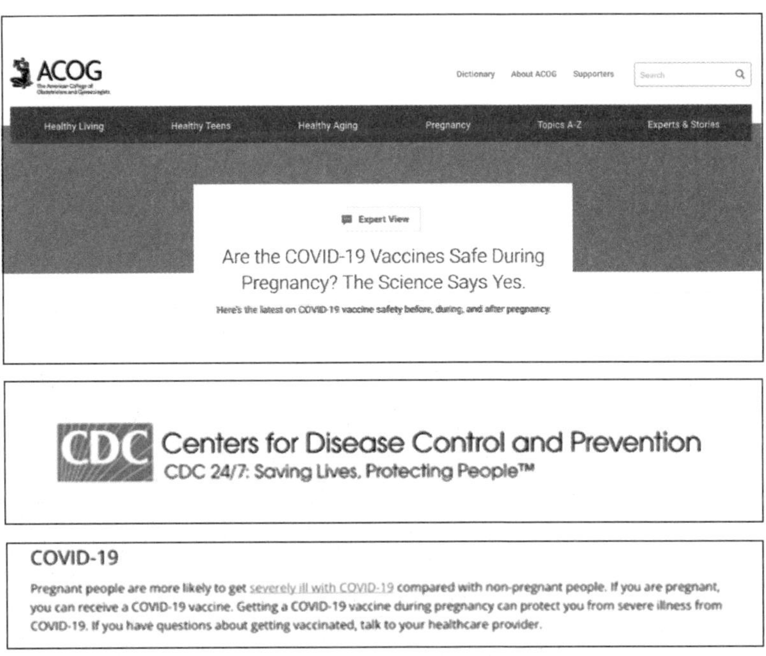

(Accessed 12/29/2022.)

III. Post Emergency Use Authorization (EUA) and Pregnancy

Widespread injection of the American population began in mid-December of 2020. Almost immediately Pfizer had to initially hire 600, then an additional 1400, extra personnel to document the tsunami of adverse event reports following administration of BNT162b2, the Pfizer LNP/mRNA therapeutic. (https://thetruedefender.com/just-in-pfizer-hires-additional-1800-employees-to-process-adverse-reactions-to-the-c-19-vaccine/)

Over 42,000 individuals experienced adverse events in the first 10 weeks after the December 2020 EUA. Cumulatively, through February 28, 2021, there were "**42,086 case reports** (**25,379 medically confirmed** and 16,707 non-medically confirmed) containing **158,893 events.**" (https://drjessesantiano.com/pfizer-bnt162b2-adverse-events-as-of-february-2021-after-the-roll-out/)

Of those, 72% involved women. (https://robertchandler.substack.com/p/cdcfda-safety-evaluation-in-pregnant and https://dailyclout.io/data-do-not-support-safety-of-mrna-covid-vaccination-for-pregnant-women/) Furthermore, 16% of the adverse events involved the reproductive organs and functions. (https://robertchandler.substack.com/p/why-do-females-have-more-adverse and https://dailyclout.io/women-have-three-times-the-risk-of-adverse-events-than-men-risk-to-the-reproductive-organs-is-even-greater-report/)

In 2021, no definitive data was collected, particularly properly powered prospective cohort studies of pregnant women and their babies. The CDC and FDA made an attempt to use a call-in reporting system, called v-safe, to determine an outcome for these women but the effort was a failure. (https://robertchandler.substack.com/p/cdcfda-safety-evaluation-in-pregnant and https://dailyclout.io/data-do-not-support-safety-of-mrna-covid-vaccination-for-pregnant-women/)

Josh Guetzkow presented a graphical representation of data released by the CDC after a Freedom of Information Act (FOIA) request from *The Epoch Times* and reported by Zachary Stieber: (https://childrenshealthdefense.org/defender/cdc-safety-signals-pfizer-moderna-covid-vaccines-et/ andhttps://jackanapes.substack.com/p/cdc-finally-released-its-vaers-safety)

Chart 1: VAERS Report 12/14/2020—7/29/2022.

I grouped these 758 safety signals into different categories. The figure below shows the total number of AEs reported for each of the major categories of safety signals:

Major Categories of Adverse Event Safety Signals from CDC's Analysis of VAERS Reports for mRNA COVID Vaccines

Cardiovascular
Neurological
Thrombo-embolic
Pulmonary
Menstrual
Death
Hemorrhagic
Gastrointestinal
Hematological
Immunological
Cancer

- 10,000 20,000 30,000 40,000 50,000 60,000 70,000

Number of Adverse Events Reported

There were almost 20,000 reported cases of menstrual problems after LNP/mRNA injections. It is not known for certain how many cases go unreported. The Vaccine Adverse Events Reporting System (VAERS) reporting is arduous and unfamiliar to some or many health-care providers. So, 20,000 could easily represent exponentially more.

Chart 2: Menstrual Irregularities after LNP/mRNA.

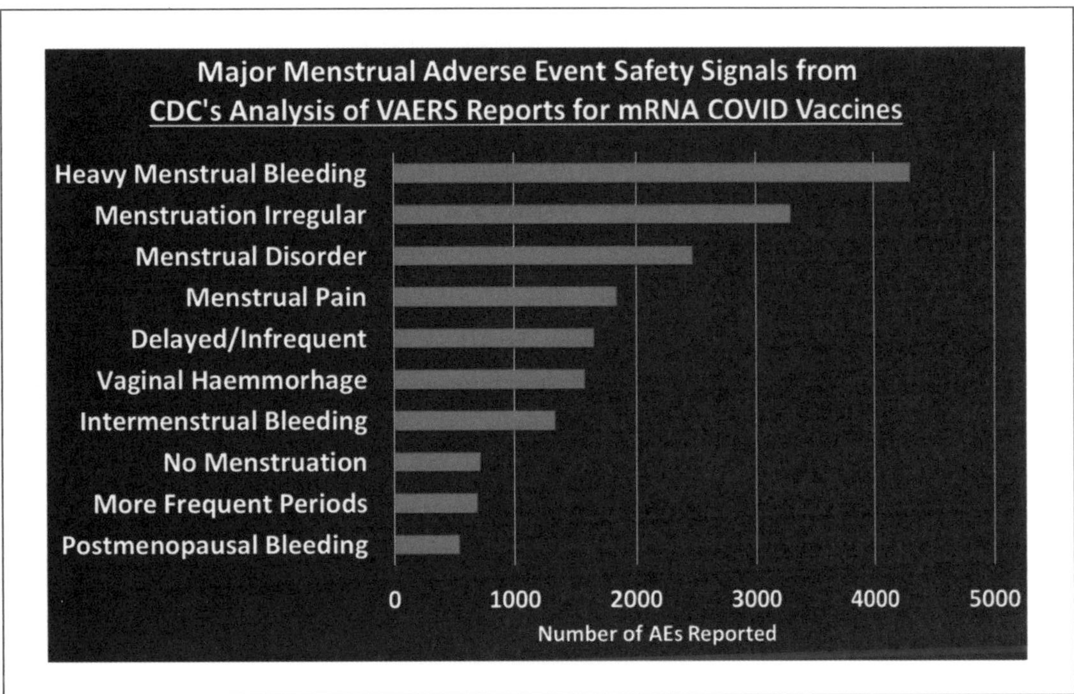

These disturbances might show up as a decline in births nine months later. We will not know for some time whether there are permanent alterations in fertility from LNP/mRNA.

Although data are accruing slowly, there remains a substantial information void concerning spontaneous abortion, stillbirth, preterm birth, small size for gestational age, congenital anomalies, and neonatal adverse events. (https://robertchandler.substack.com/p/misinformation-cdcfda-style-retroactive and https://dailyclout.io/report-40–2021-cdc-and-fda-misinformation-retroactive-editing-erroneous-spontaneous-abortion-rate-calculation-obfuscation-in-the-new-england-journal-of-medicine/)

IV. Birth Rates Following Rollout of the LNP/mRNA Products in Twenty-Two Countries

Time has brought forth data pertinent to the question of whether the LNP/mRNA products, and specifically Pfizer's BNT162b2, impair fertility. It now appears that LNP/mRNA products are associated with a decline in live births, the subject of this article.

Time series data will be presented making generous use of graphical presentation of data from Australia, Taiwan, and England/Wales (considered one country as the data are combined). Statistical analysis of data from 19 European countries will then be examined.

Caution here is advisable. Population studies have many technical issues, some identifiable and some not. There are numerous challenges to diligent and accurate capture and distribution of data from hundreds or thousands of primary sources in a timely and consistent manner.

Published data often must be revised later as the flow of data matures. Data collecting details vary widely in different countries. Sometimes numbers will vary according to when data collection periods are closed.

Sadly, the possible role of governments in falsifying data or obstructing the flow of data for political purposes must be kept in mind.

V. Live Birth Patterns in Twenty-Two Countries

The following sections will examine changes in birth rates relative to the rollout of LNP/mRNA using simple, descriptive statistics and visual representations of data for England/Wales, Australia, and Taiwan. These data are examined in the context of short-term changes in birth rates to get a sense of change in birth rate following widespread use of LNP/mRNA.

Data from European countries will be examined using the Spearman Rank Order statistic to measure correlation between the administration of LNP/mRNA and changes in birth rates.

(https://fbf.one/wp-content/uploads/2022/09/Geburtenrueckgang-Europe-EN.pdf)

Data from Switzerland (Group 1, high correlation and very high statistical significance) will be examined using both the Spearman Rank Order statistic by Hagemann, et al. as well as the Difference in Differences method as reported by Beck and Vernazza.

(https://transition-news.org/IMG/pdf/geburtenrueckgang-in-den-schweizer-kantonen_13082022.pdf and https://www.aletheia-scimed.ch/wp-content/uploads/2022/08/Geburtenrueckgang-in-den-Schweizer-Kantonen_13082022.pdf)

This section will then conclude with examination of outliers.

A. England and Wales

Chart 3: Declining Birth Rate 2008–2021.

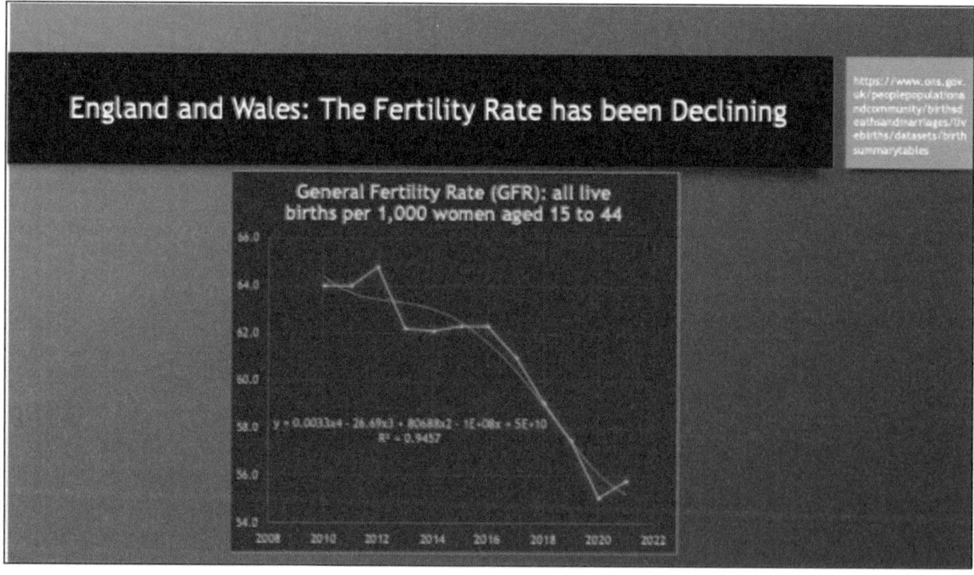

The 13-year pattern of declining birth rate in England/Wales is illustrated above in Chart 3. Note the uptick at the end of 2020 going into 2021 that may indicate the effect of lockdowns and then release from lockdowns.

The task, using population data showing this pattern of long-term decline, will be to identify acceleration (second derivative) in this decline.

Chart 4 examines this uptick in 2021 followed by a reversal of the 2021 uptick in 2022.

Chart 4: Close-Up Detail of Births from January 2021 through June 2022.

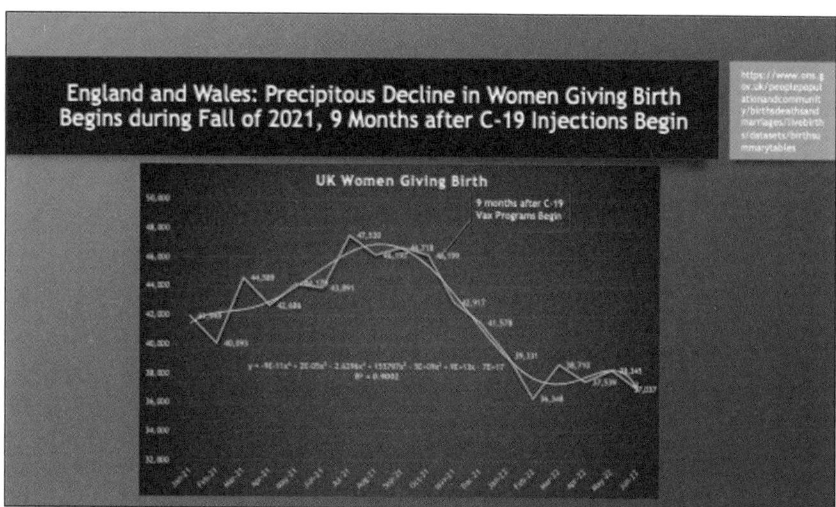

The uptick in births in 2021 was rapidly reversed during the months of September through November of 2021, approximately nine months after the widespread release of COVID-19 vaccines.

Chart 5: The Decline in Births Month-to-Month Compared with 2021.

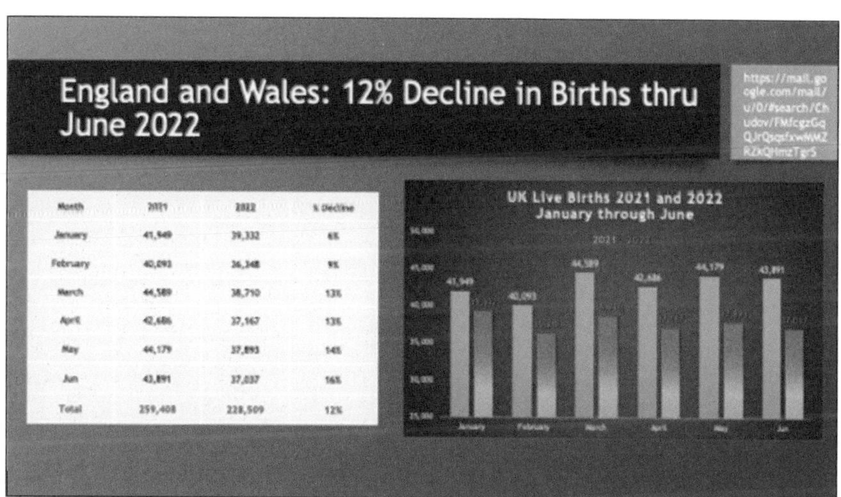

Month	2021	2022	% Decline
January	41,949	39,332	6%
February	40,093	36,348	9%
March	44,589	38,710	13%
April	42,686	37,167	13%
May	44,179	37,893	14%
Jun	43,891	37,037	16%
Total	259,408	228,509	12%

The decline in births appears to be accelerating according to these data.

The United Kingdom Security Agency publishes live birth data. The report is issued monthly, but the live birth data **has not been updated since June 2022**.

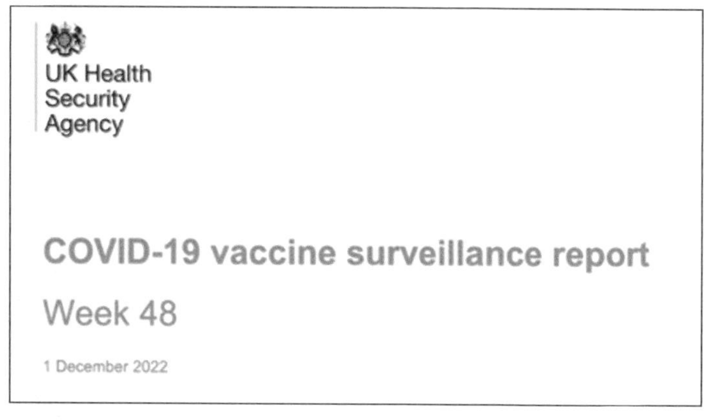
UK Health Security Agency

COVID-19 vaccine surveillance report

Week 48

1 December 2022

B. Australia

Chart 6: Australia Pattern of Births 2000–2021.

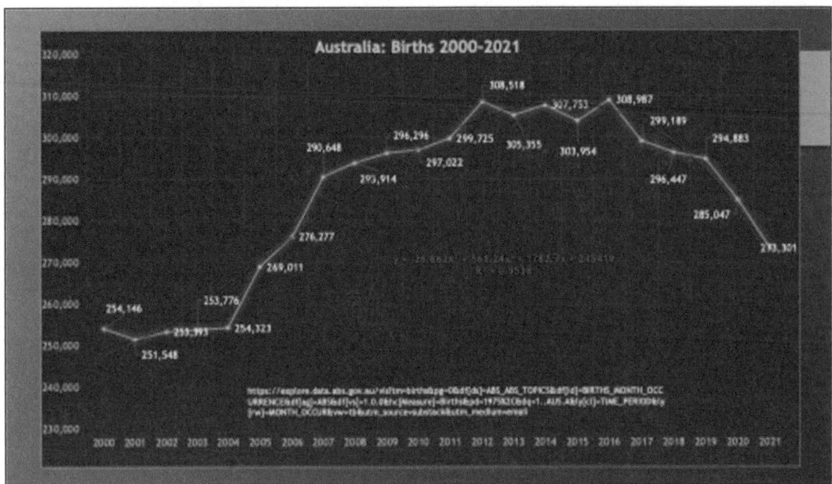

Birth rates in Australia were increasing until hitting a plateau during 2012–2016. Then there was a period of decline leading into the COVID-19 era.

Chart 7: Births in Australia from 2016–2021.

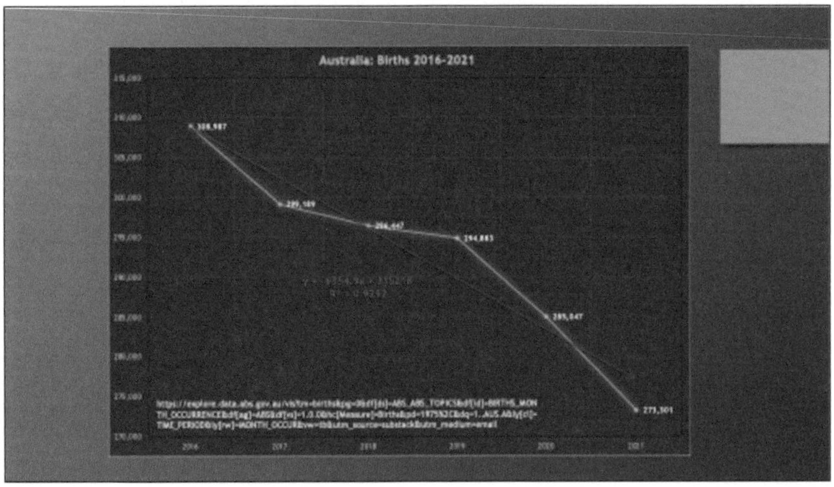

A closer look at the decline from 2016 to 2021 shows decline with a high degree of linearity.

Chart 8: Australia: Acceleration in Decline of Births.

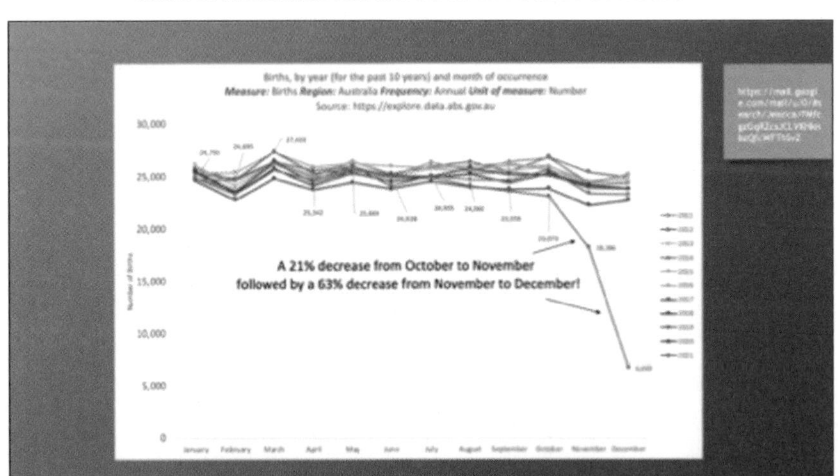

Chart 8 above, prepared by Jessica Rose, shows a nosedive in births like that seen in the England/Wales data, accelerating decline in births beginning approximately 9 months after the implementation of C-19 mRNA gene products. (https://jessicar.substack.com/p/whats-going-on-with-births-down-under)

The decline in October and November of 2021 is dramatic. Anomalies in the data collection and reporting process should be ruled out as the cause of this large drop-off in births. These figures need to be revisited from time to time to see if what began as menstrual irregularity is the first sign of infertility.

C. Taiwan

Chart 9: Births/1,000 in Taiwan 1958–2020.

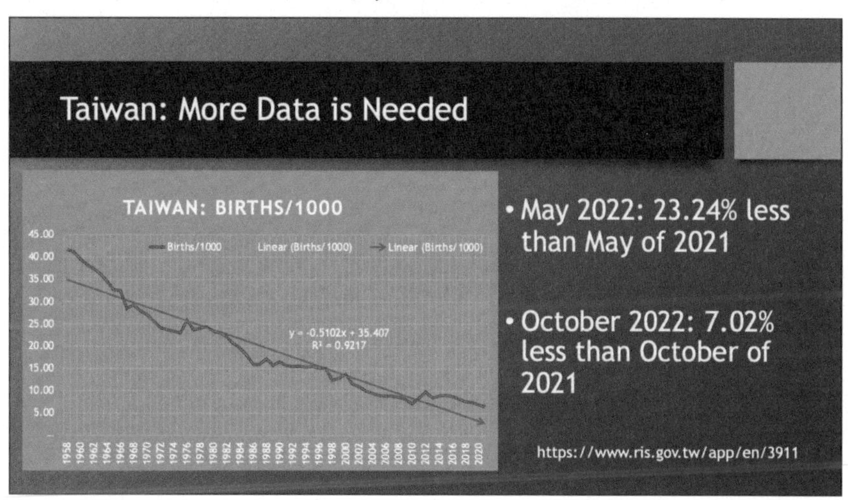

A government report indicated a drop-off in births per thousand in 2022, but a full data set needs to be located. The long-term trend in declining births is again illustrated but with possible acceleration in 2022 following widespread COVID immunization in 2021.

After this chart was made, the following data appeared on Wikipedia cited with reference [52] (in Mandarin). Through November 2022, there was a decline in live births of 9.22%; 12,885 expected babies did not arrive. The birth rate has been substantially below what is necessary for replacement. Now there are decreasing births and increasing deaths exacerbating the decline.

Current vital statistics [edit]

[52]

Period	Live births	Deaths	Natural increase
January–November 2021	139,693	167,993	-28,300
January–November 2022	126,808	189,545	-62,737
Difference	▼ -12,885 (-9.22%)	▲ +21,552 (+12.83%)	▼ -34,437

(https://en.wikipedia.org/wiki/Demographics_of_Taiwan#Fertility_rate)

The accelerating decline in births in England/Wales, Australia, and Taiwan following widespread administration of LNP/mRNA gene therapy products is disconcerting and points to an association between LNP

/mRNA gene therapy and acceleration in the declining rate of birth. Taiwan has more to worry about than the People's Liberation Army.

D. Europe

Chart 10 below is a visual representation of 60 years of births/1,000 in the European Union. The long-term declining trendline has a high degree of linearity.

Sixty years of declining births needs to be taken into consideration when looking at short-term changes.

Chart 10: Long-Term Births/1,000 in Europe.

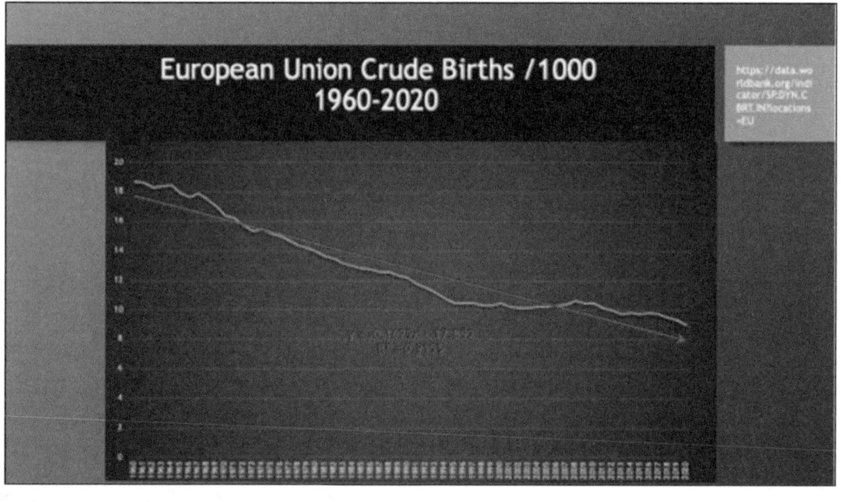

Large and abrupt changes in short-term births rates is one indicator of a causative relationship between LNP/mRNA and declining birth rates.

Hagemann, Lorré, and Kremer Analysis

On August 25, 2022, Raimund Hagemann, Ulf Lorré, and Dr. Hans-Joachim Kremer published a detailed, 91-page report on their analysis of live birth data from 19 European countries. Their report contains detailed data and analysis and is well worth reading. (https://www.aletheia-scimed.ch/wp-content/uploads/2022/08/Geburtenrueckgang-Europe-DE_25082022_2.pdf)

Most of the following section, Europe, will present data from Hagemann, et al., and the citation above applies unless otherwise specified.

Hagemann et al. used the Spearman Rank Order statistic to compare the correlation between the rank of birth rate change with the rate of injection of LNP/mRNA in persons ages 18 to 49. The correlation coefficient is represented by rho (ρ).

Discussion of methods

Spearman's rho versus Pearson correlation: With p-values of 4.9E-14, 2.2E-16 and 0.0003, an examination of the normal distribution of the total data (Shapiro-Wilk test) revealed only negligible probabilities for the existence of normal distributions of the vaccination frequency, the vaccination rate and the birth changes,, respectively. This circumstance could not be remedied by log transformation.

For all correlation calculations, Spearman's rho (rank correlation) was therefore used, where normally distributed data do not have to be assumed.

In evaluating the hypothesis tests with the help of the p-values, I apply Jürgen Bortz's suggestions for prospective studies. In April 2018, Prof. Ioannidis criticized the practice of choosing a threshold value of 0.05 to determine significance and suggested reducing it to 0.005.[4]

This statistic converts continuous data into ranks or categories before calculating the correlation between the two ranks.

The authors offer the following guides to the interpretation of the results.

Interpretation Guidelines:

Correlation: The degree of association between the rate of LNP/mRNA injection and the rate of change in births represented by rho. From Hagemann, et al.:

Interpretation of Spearman's ρ (rho) according to Cohen[2]

The interpretation of the calculated rank correlation coefficients is carried out according to Cohen (1988) in the levels:

weak correlation: 0.1≤ |ρ| < 0.3 [note the difference between Greek ρ (rho) and English p]
medium correlation: 0.3≤ |ρ| < 0.5
strong correlation: |ρ| > 0.5

Statistical Significance: The degree of probability that the correlation deviates from random chance is expressed by the p-value. In medicine p-values less than 0.05 or 95% chance that the correlation is non-random are usually considered to be statistically significant. Hagemann, et al. considered a p-value <0.001 to be highly significant.

Statistical significance

In the hypothesis test performed, the p-value indicates the probability of drawing the present random sample from a basic population whose true correlation is zero or positive (null hypothesis). If the p-value is small, the null hypothesis is very unlikely and one decides in favour of the alternative hypothesis (true correlation is less than zero). The smaller the p-value, the better confirmed is the decision in favour of the tested alternative hypothesis that increasing vaccination frequencies cause decreasing birth rates.

Interpretation according to Jürgen Bortz:[3]

- With a p-value of ≤ 0.05, Jürgen Bortz, for example, speaks of a significant,
- a value of ≤ 0.01 (2.3 standard deviations) is called very significant and
- a value of ≤ 0.001 (3.1 standard deviations) is a highly significant result.

Table 1 below presents the summary data from Hagemann, et al. for 19 European countries.

Hagemann, et al. kindly highlighted a key statistic with orange coloring. Note that all 19 countries registered declining birth rates through June 2022. Continuing to be helpful, the authors ranked the countries according to the degree of correlation and statistical significance of the association between the percent of the population ages 18 to 49 who received LNP/mRNA products and the rate of live births.

The comments below the data table in Chart 9 note the decline in birth rate in all 19 countries ranging from -1.3% in France to -18.8% in Romania. Note also that 68% (13/16) of the countries, Romania and above, had Spearman rho values of -0.527 or higher with p-values less than 0.05, which indicates very strong support for causation attributable to LNP/mRNA injections.

The association between LNP/mRNA inoculation rate and the rate of decline was closely examined in each country with the overall conclusion that there was a statistically significant association between the two rates but with a negative association, as the vaccination rate increased as the birth rate declined. Furthermore, the decline followed nine months after the rollout of the LNP/mRNA inoculation program. The decline was not associated with COVID-19.

Table 1: Births and Vaccination Rates in 19 European Countries.

Analysis: Live births in Europe

Evaluation: Europe

Region	Country	Births 2022	Ø 2019-21	Change Ø → 2022	Spearman ρ	p-value	1st vacc. 9 month prior 18-49 *	Population 18-49 years	Vacc. rate* total 18-49
North	Finland	22,180	23,266	-4.7%	-0.918	0.000033	1,712,463	2,164,149	79.1%
West	Switzerland	39,326	43,079	-8.7%	-0.873	0.00023	2,344,443	3,653,573	64.2%
West	Netherlands	81,125	83,339	-2.7%	-0.802	0.0015	5,056,399	7,019,309	72.0%
North	Latvia	8,026	8,859	-9.4%	-0.800	0.0016	416,436	745,854	55.8%
West	Austria	39,635	41,448	-4.4%	-0.773	0.0027	2,489,729	3,682,383	67.6%
West	Germany	285,753	313,543	-8.9%	-0.770	0.0046	30,725,410	45,321,314	67.8%
North	Lithuania	12,392	14,988	-17.3%	-0.741	0.0029	839,806	1,123,367	74.8%
East	Hungary	41,902	43,504	-3.7%	-0.682	0.0104	2,550,513	4,231,659	60.3%
East	Poland	126,400	146,145	-13.5%	-0.673	0.0165	8,535,540	16,639,191	51.3%
North	Sweden	54,560	58,457	-6.7%	-0.664	0.0130	3,239,628	4,222,335	76.7%
East	Slovenia	8,426	9,211	-8.5%	-0.627	0.0194	457,167	852,427	53.6%
North	Estonia	5,810	6,534	-11.1%	-0.582	0.0302	330,014	544,258	60.6%
East	Romania	78,792	97,022	-18.8%	-0.527	0.0478	2,674,679	8,029,346	33.3%
East	Czech Republic	24,232	27,146	-10.7%	-0.524	0.0914	1,975,874	4,538,565	43.5%
North	Denmark	28,828	30,049	-4.1%	-0.427	0.0949	1,908,007	2,361,498	80.8%
West	France	357,900	362,541	-1.3%	-0.355	0.1423	23,913,873	26,186,117	91.3%
South	Portugal	32,048	34,743	-7.8%	-0.297	0.2024	3,748,115	4,112,736	91.1%
South	Spain	159,705	172,399	-7.4%	-0.209	0.2686	16,490,325	19,638,928	84.0%
West	Belgium	56,604	57,430	-1.4%	-0.145	0.3348	3,780,494	4,678,439	80.8%
	Σ Europe - Selection	1,463,644	1,573,703	-7.0%	-0.522	3.014E-14	113,188,915	159,745,448	70.9%

* Number of persons with first vaccination in the age group 18-49 years, 9 months before the last reporting month of births.

The table was sorted in ascending order according to Spearman's rho.

- At present, all countries show a decline in births between −1.3% and −18.8% compared to the same period last year.
- All countries show a negative correlation between vaccination frequency and birth rate decline, whereby CZ, DK, PT, FR, BE, ES are not classified as significant – France, Belgium, Portugal, and Spain are also three statistically unusable countries with a proven lockdown effect.
- he significance of the negative correlation lies in 7 countries below the limit of 0.005 as demanded by Prof. Ioannidis.
- The decline in births in the analysed european country sample compared to the previous year's average, amounts to a total of −110,059 births or −7.0%.

Chart 11 below is a combination plot of the change in births in **yellow** and the rate of inoculation with LNP/mRNA products in **orange** for all 19 countries.

Chart 11: Percent Immunized (Orange) and Percent Decline in Births (Yellow).

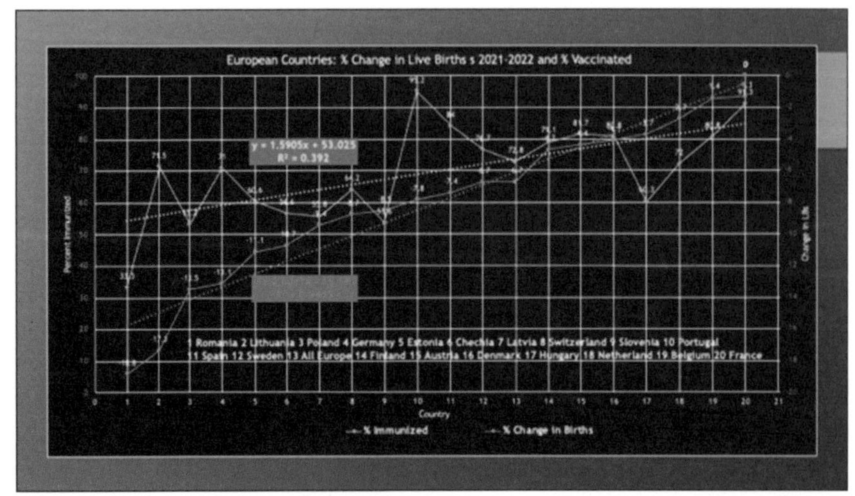

Countries are numbered and listed below the plots (previous chart). The birth rate decline data indicates a high degree of linearity (R^2 = 0.9415), while the rate of inoculation has far less fit with its linear trend line (R^2 = 0.392).

For the purposes of this analysis, the birth data from the 19 European countries listed in Table 1 will be examined in groups:

Group 1: Strong to very strong negative correlation $|\rho|$[1] >= 0.741 between LNP/mRNA vaccination rates and birth rates and very strong statistical significance with p-values <0.005. [[1]The Spearman Coefficients is negative in the study thus the designation, $|\rho|$ represents the absolute value of the Correlation Coefficient rho.]

Group 2: Moderate to strong correlation 0.527 > = $|\rho|$ < = 0. 682 with statistical significance with p-values <0.05.

Group 3: Moderate to weak correlation > 0.1 $|\rho|$ < 0.5 low statistical significance with p-values > 0.05.

D. 1. Europe Group 1

Table 2 gives the Spearman Rank Order Correlation Coefficient and p-values indicating strong correlation between the rate of injection of LNP/mRNA and acceleration in the decline of births nine months later.

Table 2 Group 1: Very Strong Correlation and Very Significant p-value < or << 0.005.

Country	Rho	p-value
Finland	-0.918	0.000033
Switzerland	-0.873	0.00023
Netherlands	-0.802	0.0015
Latvia	-0.800	0.00156
Austria	-0.773	0.00265
Germany	-0.770	0.00461
Lithuania	-0.741	0.0029

Switzerland will be used as an example of Group 1, because two independent statistical analyses of the birth rates and rates of LNP/mRNA injection were performed by Hagemann, et al. and Beck and Vernazza. The Swiss data will be reviewed in detail.

Example Switzerland Hagemann, et al. Analysis.

- 8.7% Decline in Births through May 2022.
- Inoculation Rate = 64.2%.
- Very Strong Negative Correlation Between Rate of LNP/mRNA Vaccination and Decline in Birth Rate Spearman rho = -0.873.
- Highly Statistically Significant p-value = 0.00023.

Chart 12: Monthly Birth Rates.

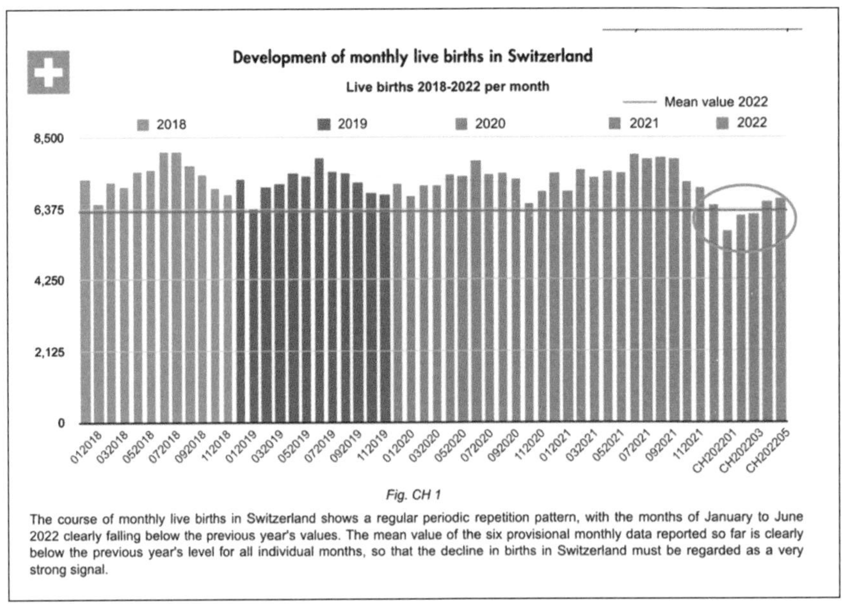

Fig. CH 1

The course of monthly live births in Switzerland shows a regular periodic repetition pattern, with the months of January to June 2022 clearly falling below the previous year's values. The mean value of the six provisional monthly data reported so far is clearly below the previous year's level for all individual months, so that the decline in births in Switzerland must be regarded as a very strong signal.

There is a visible drop-off in births in 2022 compared with the previous four years, consistent with an acceleration in the rate of birth decline.

Chart 13: Swiss Birth Decline Temporally Related to LNP/mRNA Not COVID-19.

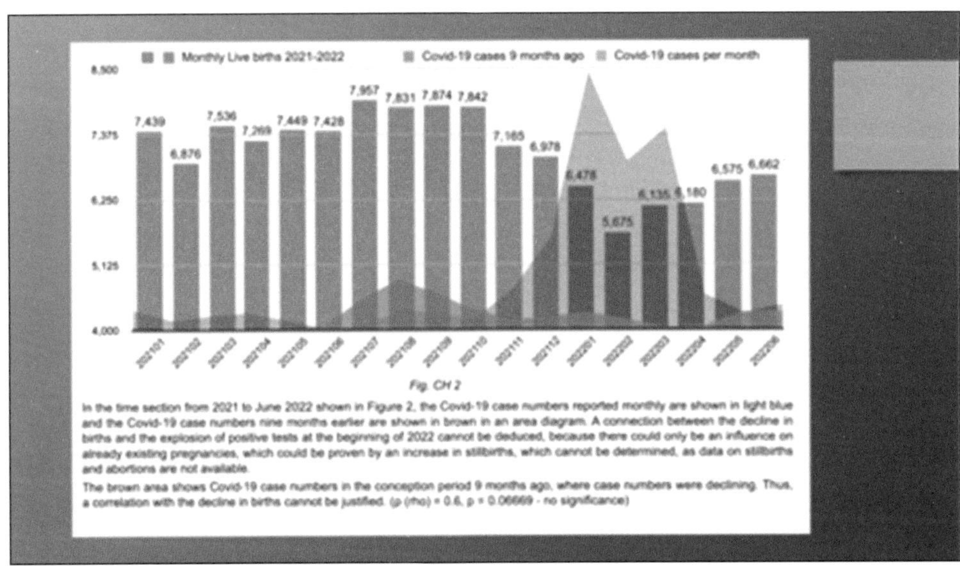

Fig. CH 2

In the time section from 2021 to June 2022 shown in Figure 2, the Covid-19 case numbers reported monthly are shown in light blue and the Covid-19 case numbers nine months earlier are shown in brown in an area diagram. A connection between the decline in births and the explosion of positive tests at the beginning of 2022 cannot be deduced, because there could only be an influence on already existing pregnancies, which could be proven by an increase in stillbirths, which cannot be determined, as data on stillbirths and abortions are not available.

The brown area shows Covid-19 case numbers in the conception period 9 months ago, where case numbers were declining. Thus, a correlation with the decline in births cannot be justified. (ρ (rho) = 0.6, p = 0.06669 - no significance)

COVID-19 peaked in August 2021, as represented by the brown curve, but the birth rate decline began in November of the same year. Hagemann et al. concluded that peak COVID-19 has little to no association with the decline in births.

Chart 14: Monthly Change in Births 2019–2022.

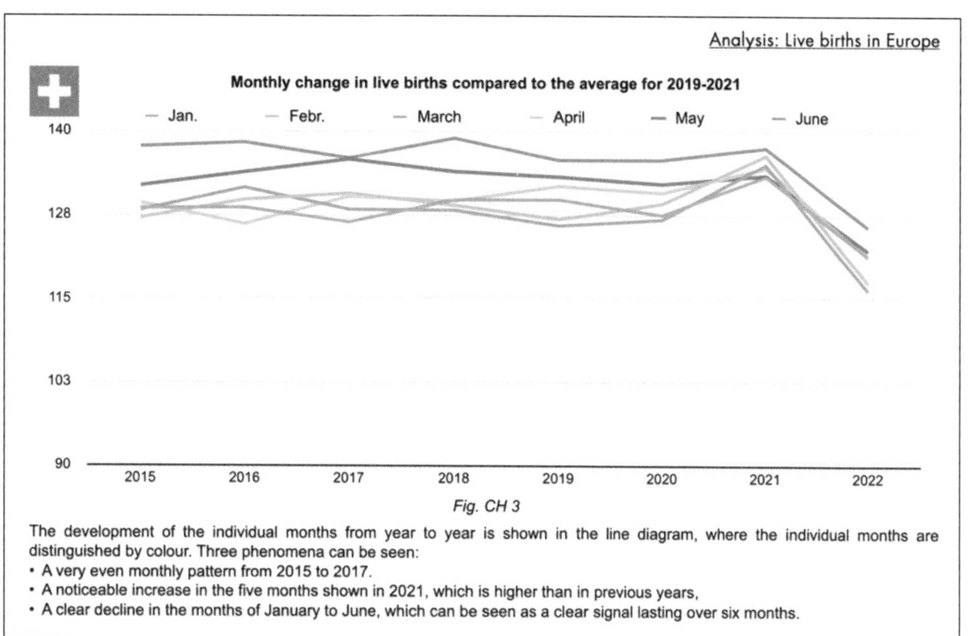

Fig. CH 3

The development of the individual months from year to year is shown in the line diagram, where the individual months are distinguished by colour. Three phenomena can be seen:

- A very even monthly pattern from 2015 to 2017.
- A noticeable increase in the five months shown in 2021, which is higher than in previous years,
- A clear decline in the months of January to June, which can be seen as a clear signal lasting over six months.

The sharp drop-off in monthly births for the first six months of 2022 compared with the prior four years is striking.

Table 3: Monthly Average Births Per Million Women Ages 20 to 49.

CH – Average monthly live births / day per million women, 20-49 years

Birth month	Oct.	Nov.	Dec.	Jan.	Febr.	March	April	May	June	July
2014-15	132.6	126.8	123.1	128.5	127.0	128.2	129.3	131.8	137.7	136.1
2015-16	130.8	127.0	124.9	128.5	129.7	131.6	126.1	133.9	138.3	142.6
2016-17	131.4	125.5	124.5	126.4	130.7	128.3	130.3	135.8	136.0	139.3
2017-18	133.0	129.2	126.1	129.7	128.9	128.2	129.6	134.0	139.0	144.3
2018-19	131.7	128.6	121.4	129.8	126.8	125.9	131.8	133.2	135.7	141.5
2019-20	128.4	126.3	121.8	127.5	129.2	126.8	130.8	132.1	135.7	140.1
2020-21	130.0	120.3	123.3	133.3	136.4	135.0	134.5	133.4	137.5	142.5
2021-22	140.5	132.6	125.0	121.3	117.3	116.1	121.2	122.1	125.7	
Difference to Ø 19-21	10.46	7.53	2.82	-8.81	-13.50	-13.10	-11.21	-10.78	-10.63	
Difference [%]	**8.0%**	**6.0%**	**2.3%**	**-6.8%**	**-10.3%**	**-10.1%**	**-8.5%**	**-8.1%**	**-7.8%**	
month of 1st. vaccination	**Jan. 2021**	**Febr. 2021**	**March 2021**	**April 2021**	**May 2021**	**June 2021**	**July 2021**	**Aug. 2021**	**Sept. 2021**	
Vacc./month	**1.0%**	**0.9%**	**1.5%**	**3.7%**	**16.0%**	**23.0%**	**5.5%**	**4.6%**	**7.9%**	

Statistical analysis	Spearman's ρ (rho)	-0.8727	strong negative relationship
Interpretation (Cohen)	p-value	0.00023	large effect

Table CH 1

* the vaccination age group reported by the FSO includes the cohorts 20-49 years.
The statistical analysis examines the correlation between percentage birth decline and vaccination frequency nine months before: There is a highly statistically significant strong negative correlation between the level of vaccination frequency and the decline in births nine months later!

Version 3 – 2022-08-31 14

Table 3 above contains the average monthly live births/day per million ages 20 to 49. Comparing 2022 to the prior seven years, a precipitous decline in births began in January 2022.

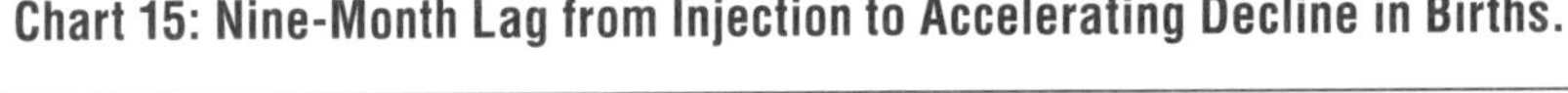

Chart 15: Nine-Month Lag from Injection to Accelerating Decline in Births.

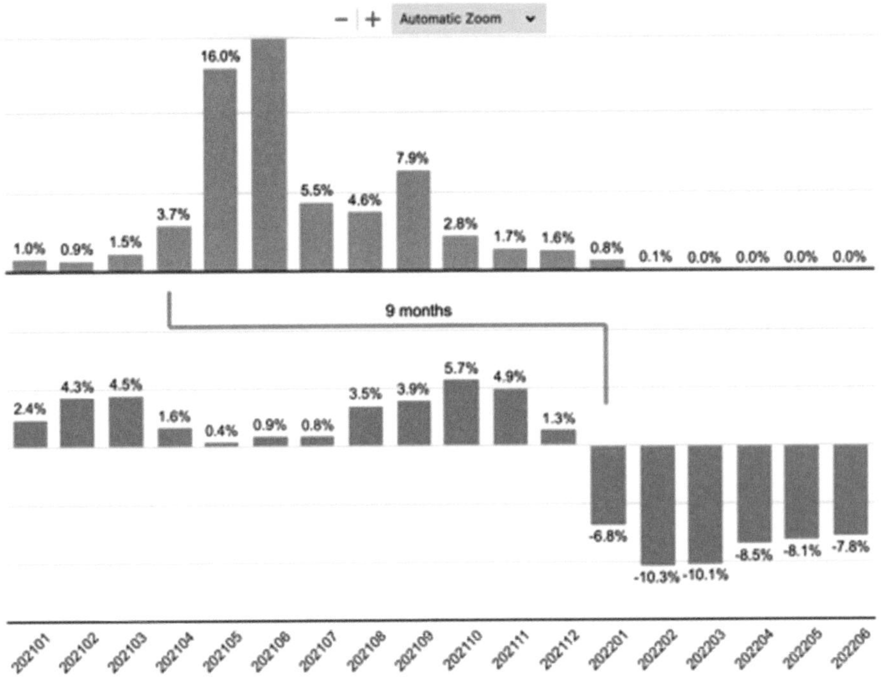

Fig. CH 4, 5

From January 2021 to May 2022, the two diagrams show the statistically significant correlation between monthly vaccination frequency of the age group 20-49 and the decline in births (normalised to the change in the number of childbearing women) between 10 and 15 per cent that started nine months later. Such an abrupt decline in the number of births cannot be explained as a "saturation effect" of completed family planning due to the somewhat increased number of births in 2021. Moreover, in 2021 there was only a plus of 2,400 births compared to the previous year's average for the whole year, whereas in the months of January to June alone the decline in births compared to the previous year's average is -6,292 births according to the current reporting status. The following aspects speak for a possible causal connection between the decline in births and the vaccination campaign:

- Temporal coincidence of birth rate decline and high vaccination activity nine months earlier in the age group of women and men aged 20-49.
- Strong negative correlation between vaccination activity and birth rate decline,
- the statistically very high significance of the strong negative correlation,
- Exclusion of a correlation with Covid-19 infection incidence.
- Historic decline in the number of births in the period from January to May compared to the previous year, which has never occurred to this extent in Switzerland since 1969 (see graph on the right).

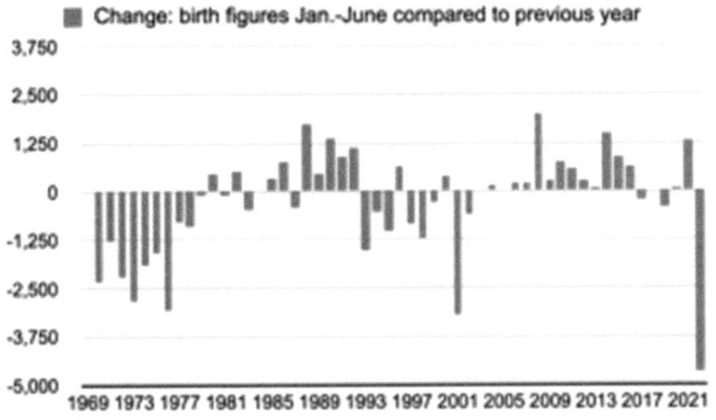

Chart 15 above illustrates the nine-month time lag between the LNP/mRNA gene therapy with LNP/mRNA and the acceleration in the decline of births in Switzerland.

Chart 16: Visual Display of January–June Birth Data from 2012–2022.

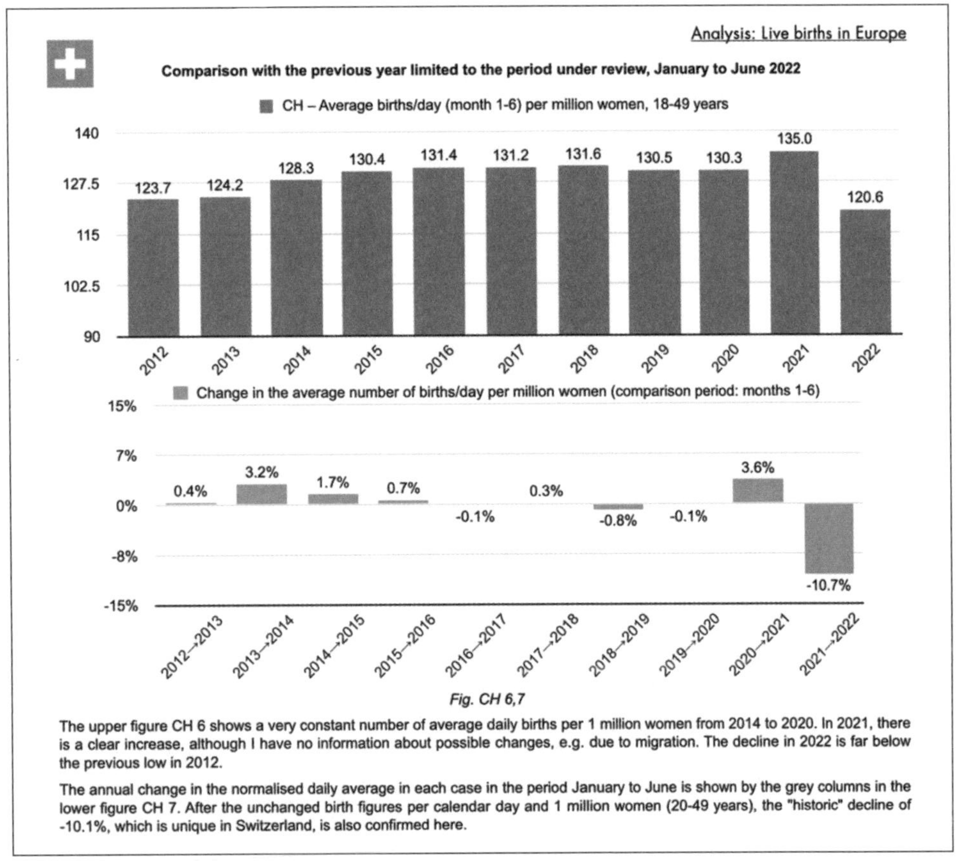

A drop of 10.7% occurred during the first six months of 2022, compared with much smaller changes in the years from 2012 through 2021, and was another indication of substantial acceleration in the decline of birth rate in 2022.

Chart 17: Swiss Birth Decline Monthly Comparison 2019–2022.

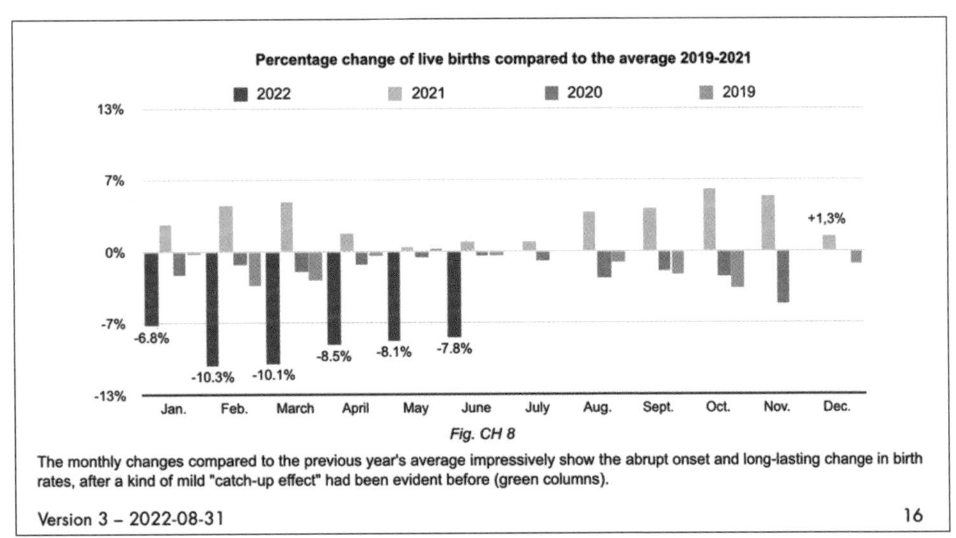

Chart 17 above presents monthly birth data from four years to illustrate the large drop in birth rates in 2022 compared with those in 2019, 2020, and 2021.

Hagemann et al. examined all 26 Swiss states, known as Cantons, using this methodology and found:

> In sixteen cantons, this decline is over 10%, in eight cantons over 15% and in three cantons close to or well over 20%.
>
> A strong negative correlation between the decline in the birth rate and vaccination frequency can be seen in Switzerland as a whole in the cantons of Zurich, Bern, Lucerne, Schwyz, Solothurn, Basel-Stadt, Basel-Landschaft, Graubünden, Aargau, Ticino, Vaud, Geneva and Jura. The nine months vaccination frequencies and the current decline in births are with high probability not statistically independent due to the strong negative correlations, combined with low p-values, which indicate a significance or high significance. These are the largest cantons with 6.3 million inhabitants, which together make up 72.7% of the Swiss population.

Beck and Vernazza Analysis of Swiss Birth Data

Professor Konstantin Beck, economist at the University of Lucerne, and Professor Emeritus Dr. Pietro Vernazza, infectious disease specialist at the Kantonsspital St. Gallen, published a second look analysis of the Swiss data using a statistical tool called a Difference in Difference Analysis for Swissmedic, the Swiss authority responsible for the authorization and supervision of therapeutic products, on September 22, 2022.

Analysis of a possible connection between the Covid-19 vaccination and the Fall in the birth rate in Switzerland in 2022

Report for Swissmedic*

Prof. Dr. Konstantin Beck
University of Lucerne
Prof. Emeritus dr Pietro Vernazza St Gallen

An extensive analysis of the LNP/mRNA inoculation rates and birth rates in the 26 Swiss Cantons is described in their 39-page report.

They concluded:

> ***Conclusion:*** *The reviews of the significance of the birthrate decline in cantons with high Vaccination rates compared to cantons with low vaccination rates leads in each individual case to the same result. The decline in births in cantons with high vaccination rates is higher than in the cantons with a lower vaccination rate, even if there are differences in the size of the cantons and the general decline in births in 2022 will be corrected.*
> *This difference was significant in the various calculations with half-yearly data the 99.8%, 98.2%, 93.7%, 92.4% and 90.6% level, respectively. In doing so, they confirm the tendency to overestimate the significance level in the regression, but show at the same time that the difference in the baby gap of the canton groups compared is* ***significant at least at the 90% level*** *. There was at all calculations presented here and also all other calculations not mentioned here never have a significance level below 90% or a wrong sign for the birth gap.*
>
> *The null hypothesis, 'there is no causal relationship between the Covid-19 vaccination campaign in 2021 and the drop in births in 2022', must be discarded,*

Beck and Vernazza prepared a graphic which presents a striking illustration of what happened to the birth rate in Switzerland after institution of the LNP/mRNA injection program.

Chart 18: 150 Years of Birth History in Switzerland.

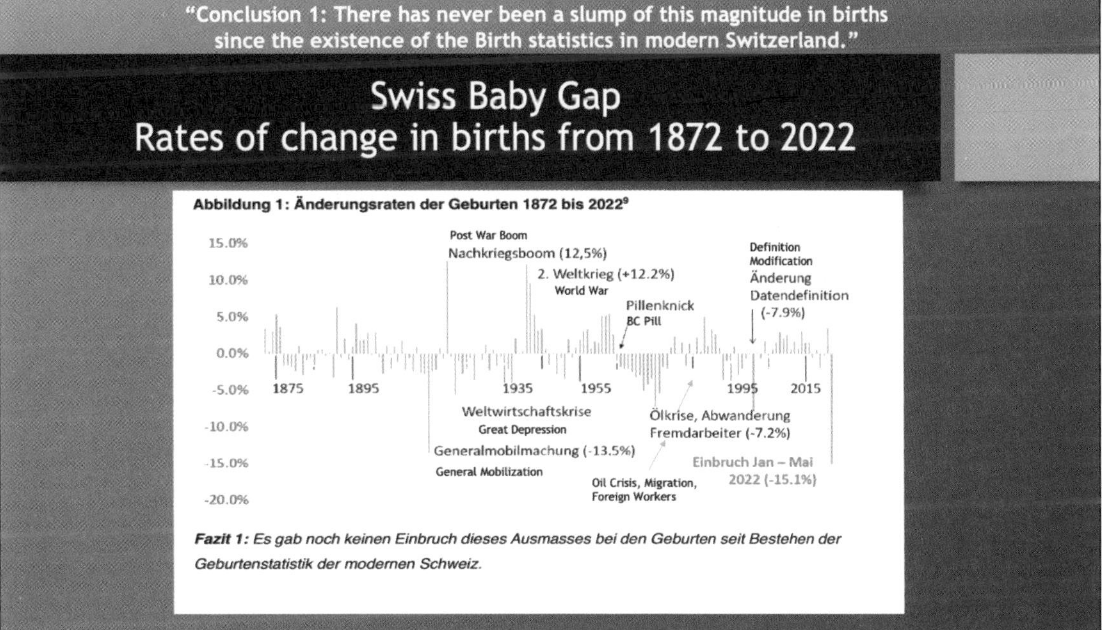

Chart 18 shows the drop in births year-to-date in May 2022 after LNP/mRNA injection program initiated in 2021 in comparison with other significant events such as World War I, the Great Depression, World War II, and the advent of the birth control pill. Beginning approximately nine months following the rollout of the LNP/mRNA program in Switzerland there was **the largest drop in birth rates in Switzerland in the last 150 years.**

The Swiss Federal Council is Alerted but Rejects the Appeal to Investigate

On September 9, 2022, a letter was sent to the Federal Council, the executive body of the Swiss Government in Bern, the capital of the Swiss Confederation, signed by Kullmann, Martin, Speiser-Niess, Rashidi, and Krähenühl who were alarmed by the large drop in birth rates identified in the Beck-Vernazza report. (https://docslib.org/doc/2062860/intervention-parlementaire-n-parlementaire)

The Council responded by noting that there was a recovery in the birth rates beginning in May and June of 2022, final figures were required, there had been past declines in births, the drop followed a rise in 2021, and countries like France had a very high vaccination rate but no substantial drop in births.

In defense of their position, they pointed out the findings in the Shimabukuro, et al. report of the CDC and FDA and two other sources. See https://robertchandler.substack.com/p/cdcfda-safety-evaluation-in-pregnant https://robertchandler.substack.com/p/misinformation-cdcfda-style-retroactive or https://dailyclout.io/report-40–2021-cdc-and-fda-misinformation-retroactive-editing-erroneous-spontaneous-abortion-rate-calculation-obfuscation-in-the-new-england-journal-of-medicine/ for a discussion of the highly flawed and misleading report from the CDC and FDA.

Not everyone was satisfied with this response:

Figure 3: Article in *Swiss Magazine*.

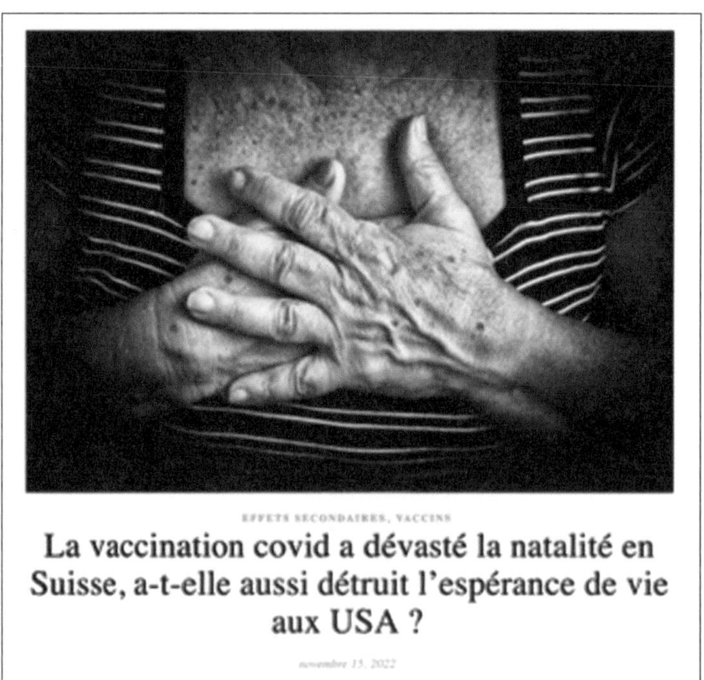

EFFETS SECONDAIRES, VACCINS

La vaccination covid a dévasté la natalité en Suisse, a-t-elle aussi détruit l'espérance de vie aux USA ?

novembre 15, 2022

The above article in "The Truth Will Set You Free" published November 15, 2022, asks "Covid vaccination has devastated the birth rate in Switzerland, has it destroyed life expectancy in the USA?"

Clearly, this matter is not settled in Switzerland as there is a case pending in the Swiss courts that includes criminal charges against Swissmedic and individual doctors at the Inselspital University Hospital in Bern including allegations of wrongful death among other allegations. (https://coronacomplaint.ch/criminal-complaint/ and https://www.youtube.com/watch?v=fufq_KdyuVo)

Figure 4: Summary by Kruse Law Firm Zurich, Switzerland.

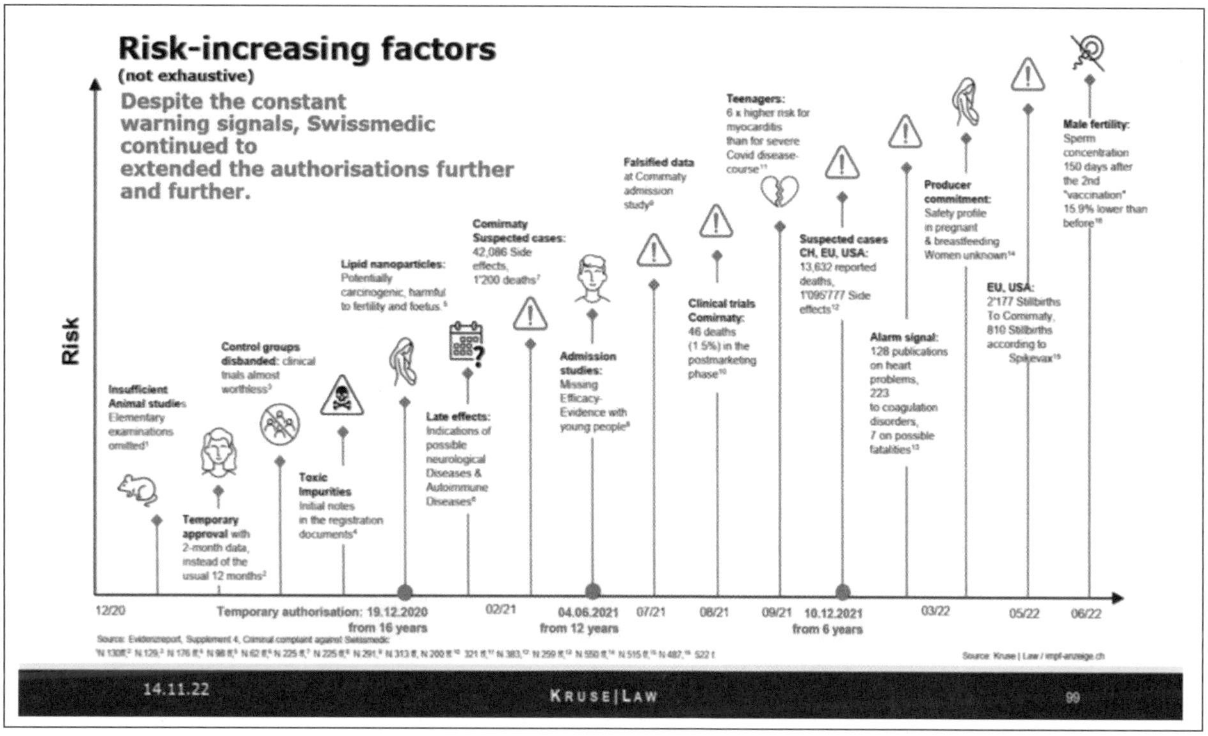

Mary Beth Pfeiffer published a worthwhile review of the Swiss/European decline in births related to the LNP/mRNA products:

(https://rescue.substack.com/p/the-missing-babies-of-europe)

The evidence report from the criminal case is available here:

(https://audio.solari.com//covid-law-suits/EN_Evidence-Report_v1.0_DEEPL.pdf)

Similar action may result from the investigation in Florida:

"Florida Governor Ron DeSantis said Tuesday he plans to ask the state Supreme Court to investigate 'any and all wrongdoing' connected to the COVID-19 vaccines, comparing the effort to recent rulings against manufacturers and distributors of opioids." (https://www.yahoo.com/entertainment/desantis-compares-potential-covid-vaccine-233549015.html)

D. 2. Europe Group II: Strong Correlation, Significant p-value < 0.05

Table 4: Europe Group II

Country	Rho	p-value
Hungary	-0.682	0.0104
Poland	-0.673	0.0165
Sweden	-0.664	0.0130
Slovenia	-0.627	0.0194
Estonia	-0.582	0.0302
Romania	-0.527	0.0478

Example: Sweden

- -6.7% in Hagemann, et. Al. Decline in Births through June, (-8.3% through October 2022).
- Inoculation Rate = 76.7%.
- Moderate Correlation Spearman rho = -0.664.
- Statistically Significant p-value = 0.0130.

Chart 19: Sweden: Monthly Live Births January–October 2019–2022.

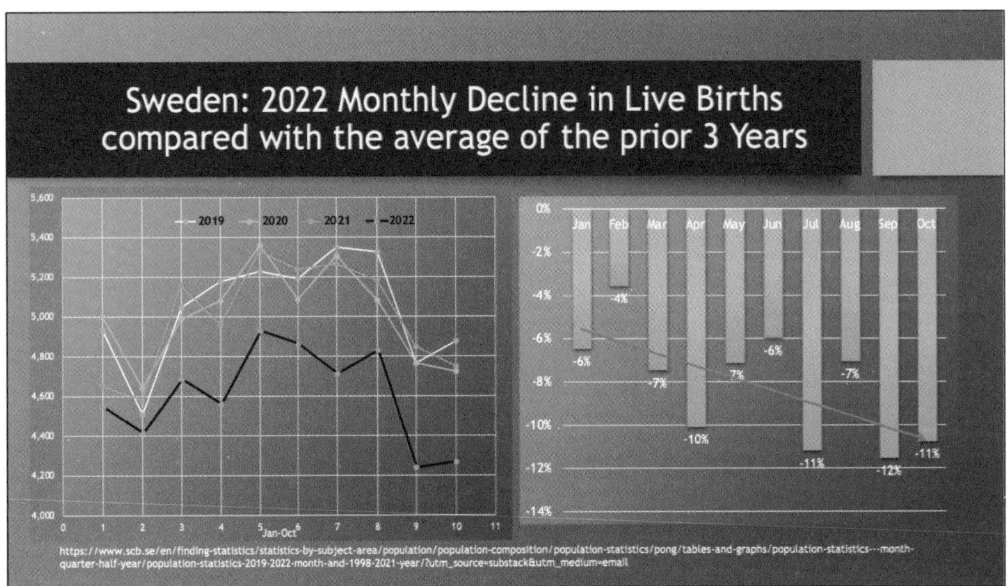

Monthly drops of 6% to 12% were recorded from January through October of 2022. There was poor fit with linear regression, y = -0.0057x—0.0498 R^2 = 0.432, but there was suggestion of accelerating decline in live births as three of the largest drops in births were in the last four months.

Chart 20: Sweden Annual Births January to October 2019–2022.

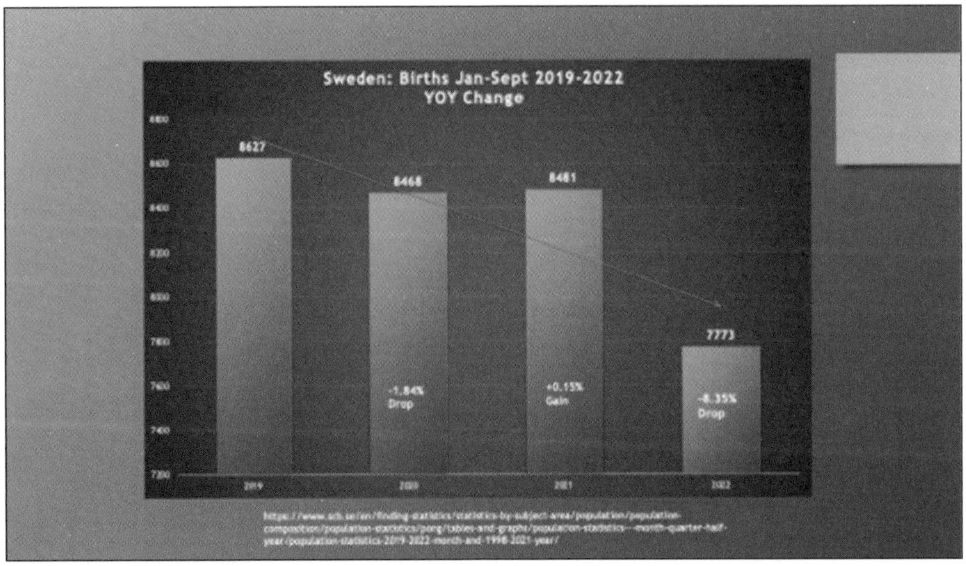

There was over 8% decline in births in 2022 compared with the three prior years through October 2022.

D. 3. Europe Group III: Weak Correlation, Insignificant p-value > 0.05 or >> 0.05

Table 5: Europe Group III

Country	Spearman's rho	p-value
Czechia	-0.524	0.0914
Denmark	-0.427	0.0949
France	-0.355	0.1423
Portugal	-0.297	0.2024
Spain	-0.209	0.2686
Belgium	-0.145	0.3348

Example: France (Group 3)

- -1.3% Decline in Births through May 2022.
- Inoculation Rate = 91.3%.
- Weak Correlation Spearman rho = -0.355.
- Not Statistically Significant p-value 0.1423.

Chart 21: Comparison of Monthly Births from 2019–2022 January through October.

Chart 22: France: No Significant Association Between LNP/mRNA Injection Rate and Birth Rate Decline.

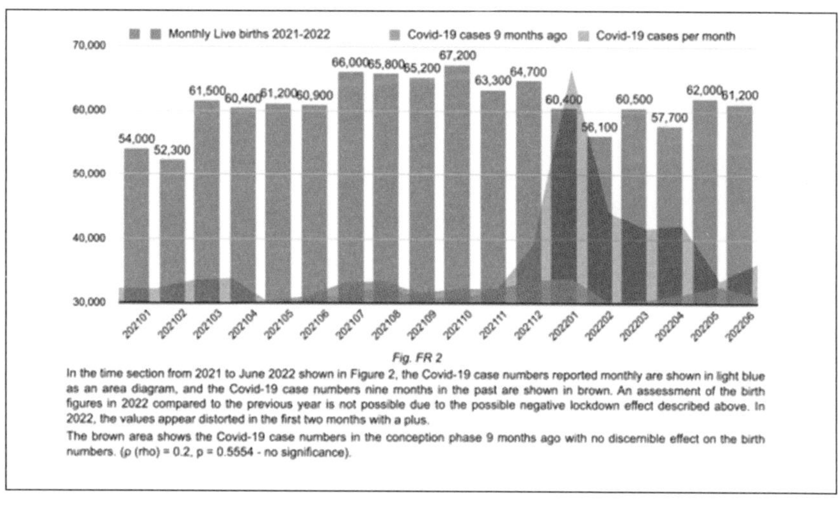

There is no pattern of acceleration in the decline in the live birth data from France (previous chart). A high rate of inoculation, 91%, was not found to be followed by an acceleration in the declining rate of birth.

This fact brings into question the integrity of this data from France as there is a very strong signal indicating an acceleration in the rate of decline in live births in 13 of the 19 countries (68%) following injection of LNP/mRNA products. France is an outlier.

E. Further Analysis

Hungary (Group II)

- 3.7% Decline in Births.
- Strong Correlation, Spearman's rho = -0.6818
- Statistically Significant p-value = 0.0104

The long-term fertility pattern in Hungary is similar to many other countries, gradually declining live births. See Chart 23 below. Once again, the key is to look for acceleration in the rate of declining live births, not just the velocity. In this instance, there is what looks like a paradox with a small rate of decline but with a strong correlation and a strong level of statistical significance.

Chart 23: Hungary Long Term Birth Rate/1000.

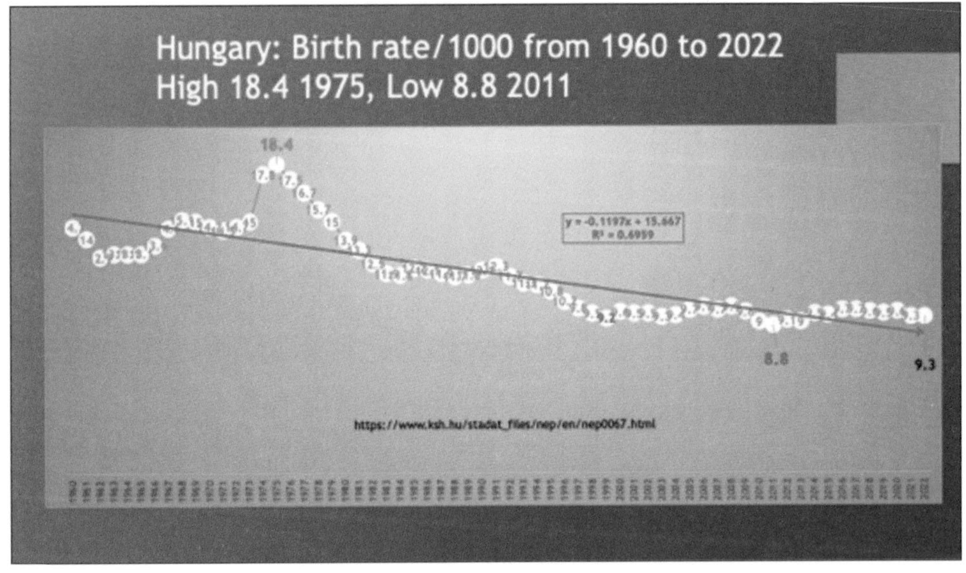

Attempting to explain these findings, Igor Chudov published an article on his Substack website on July 3, 2022, after looking at the vaccination rate in Hungary and the change in birth rate. He matched the birth declines with vaccination rates for the different counties in Hungary and placed the data on a color-coded map, Chart 24 below. (https://igorchudov.substack.com/p/hungary-most-vaccinated-counties)

Chart 24: Birth Rate Drop and Vaccination Rates.

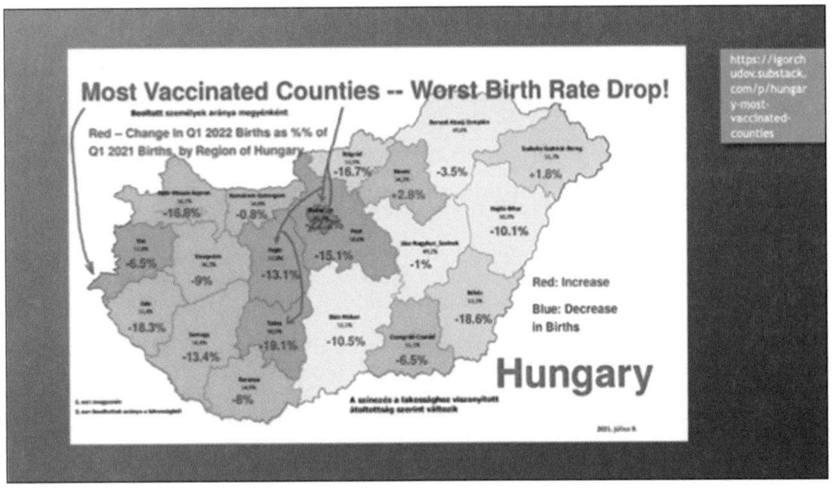

Chudov then observed:

> Unfortunately, **this data is noisy**, as it presents only a **single-moment snapshot of vaccination rates, and they are not super dissimilar**. To make the comparison less noisy, **I decided to pick five MOST vaccinated counties, and five LEAST vaccinated counties.**

Chart 25: Comparison of Top Counties with High Vaccination Rates.

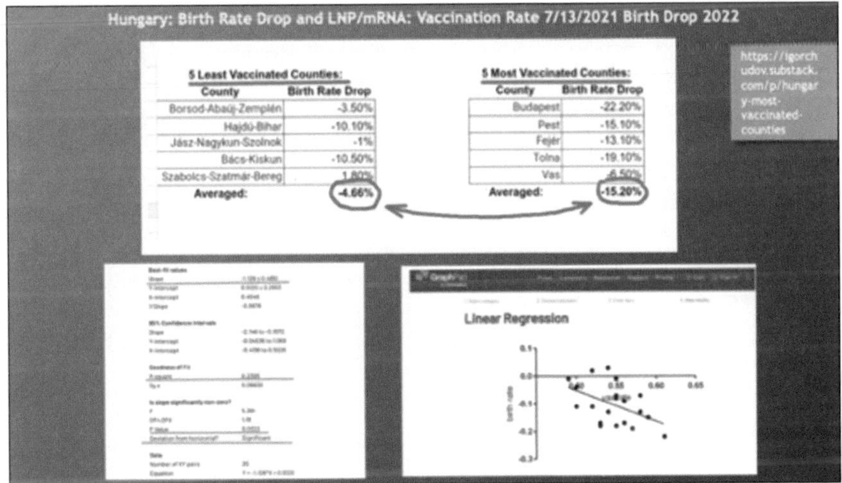

5 Least Vaccinated Counties:

County	Birth Rate Drop
Borsod-Abaúj-Zemplén	-3.50%
Hajdú-Bihar	-10.10%
Jász-Nagykun-Szolnok	-1%
Bács-Kiskun	-10.50%
Szabolcs-Szatmár-Bereg	1.80%
Averaged:	-4.66%

5 Most Vaccinated Counties:

County	Birth Rate Drop
Budapest	-22.20%
Pest	-15.10%
Fejér	-13.10%
Tolna	-19.10%
Vas	-6.50%
Averaged:	-15.20%

The birth rate drop was more than three times greater in the highest vaccinated counties compared with the five counties with the lowest vaccination rates. The trendline had negative slope supporting this statistic. The R^2 indicated a negative correlation between the vaccination rate and the birth rate with a p-value of 0.0322, significant statistically.

Chudov's conclusion:

> What is important is that statistical analysis shows the slope to be "**statistically significantly different from zero,**" in other words, **the effect of vaccination on birth rate is highly likely NOT a fluke**.

The evidence is piling up to support the conclusion that fertility impairment results from injection of LNP/mRNA products in humans.

F. Shenanigans?

In the context of the 22 countries discussed above where there was substantial evidence of significant acceleration in the declining birth rates nine months following the rollout of the LNP/mRNA inoculation campaign is the curious case of Victoria, Australia.

Victoria is the second smallest state in Australia by land area, location of the second largest city, Melbourne, and has a population of 6.5 million according to Wikipedia. (https://en.wikipedia.org/wiki/Victoria_(Australia)#cite_note-ABSPop-1)

Chart 26: Registered Births in Victoria, Australia Monthly from September 2019 through November 2022.

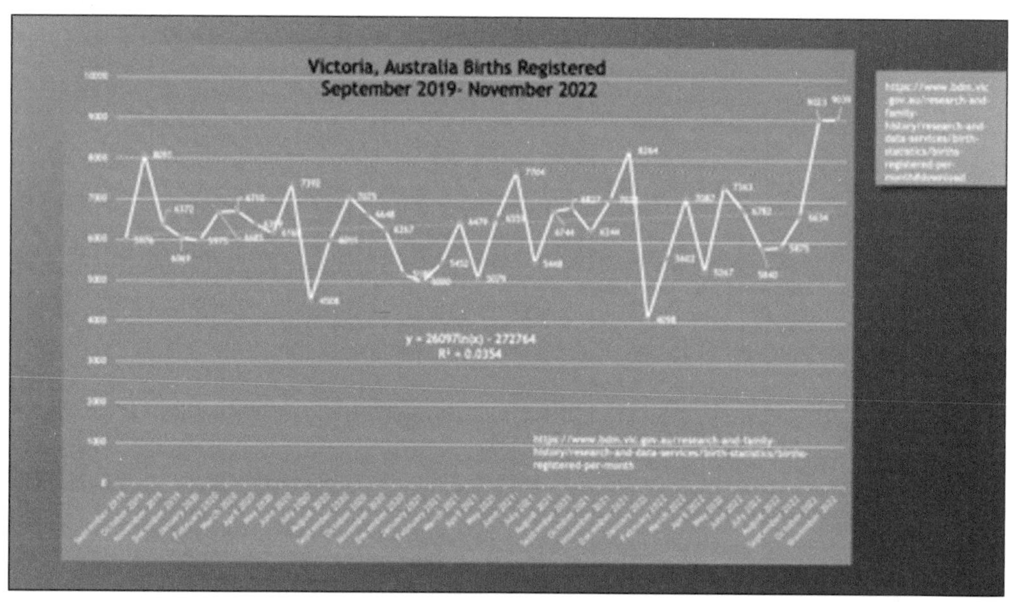

The birth rate oscillated between 5,267 and 7,363 until October when there was a gain of about 35%, a level that continued in November 2022. This trend goes against the long-term pattern of decline and is one of the only examples of a positive acceleration of birth rate seen to date.

Births tend to follow a circannual pattern with fewer winter births and more in the spring and summer.

We can all hope that our health-care authorities are diligent and honest in their data collection and reporting. Such may not be the case.

VI. Discussion

Analysis of birth data from 22 countries using a variety of analytic methods has identified strong correlation between the rate of injection with LNP/mRNA products and a subsequent acceleration of the baseline rate of decline in birth rates in 68% of European countries studied beginning approximately nine months following institution of LNP/mRNA gene therapy. For all 19 countries representing 1,59,745,448 individuals ages 18 to 49, there was a 7% decline in birth rate with a correlation coefficient of -0.522 and p-value of 3.014E-14 or 0.00000000000003014. Similar findings were found in England/Wales, Australia, and Taiwan.

Patient-level data is sorely needed. In the absence of double-blind, randomized clinical trials of at least two years' duration, prospective matched control studies, detailed and large-scale retrospective studies, detailed autopsy reports, and other forms of traditional medical research which has largely been banned by governments, researchers are left analyzing population data.

Population studies are complex in many respects although graphical presentation of data, as shown in the Swiss section, when backed up by two different statistical studies of population data by two different teams are providing the best information currently concerning birth rates.

The Hagemann, et al. study identified accelerated birth declines in 13 of 19 European countries (68%) in August 2022 and tied the decline to LNP/mRNA injections received nine months previously. They had previous found the same strong statistical correlation with low p-values in their August 12, 2022, study of the 26 Swiss Cantons. Beck and Vernazza questioned the methodology of these findings and used different statistical techniques that supported the findings of Hagemann, et al.

US, UK, and Australian data may not meet high standards. American data has yet to appear and may be suspect when it does, given the control exerted by the Department of Defense and the US intelligence agencies. The UK data has ceased to flow. (https://natyliesbaldwin.com/2022/11/debbie-lerman-governments-national-security-arm-took-charge-during-the-covid-response/, https://jdfor2020.com/2023/01/on-american-state-level-prosecution-for-federal-government-chemical-and-biological-wmd-crimes/,and(https://sashalatypova.substack.com/p/the-role-of-the-us-dod-and-their)

Further studies like those of Hagemann, et al. and Beck et al. are needed. Organizations should be established to perform statistical analyses on macro data; however, it must be complemented with patient-level studies, supported by laboratory study of tissue samples from autopsy and surgically removed tissues, to strengthen the case for a causal link between LNP/mRNA gene products and impaired fertility.

The evidence is growing that both male and female reproductive functions and organs are adversely affected by LNP/mRNA products with lowered sperm motility and counts, menstrual irregularities, and reproduction organ dysfunction.

It now appears that this adverse impact on reproductive organs and functions has become manifest as an acceleration in the decline of birth rates in the UK, Oceania, Asia, and 13 of 19 countries in Europe. There is strong support of a causal link between LNP/mRNA injections and acceleration of declining births nine months later. (https://sashalatypova.substack.com/p/my-affidavit-on-modernas-nonclinical, https://rescue.substack.com/p/deep-in-the-wombs-of-women-the-hidden, https://www.preprints.org/manuscript/202209.0430/v2, https://lostintranslations.substack.com/p/menstrual-changes-and-very-early, and https://behindthefdacurtain.substack.com/p/pfizer-fda-cdc-hid-proven-harms-to?amp)

VII. Conclusion

The pattern that began years to decades ago of declining birth rates in the developed world appears to be accelerating after the introduction of LNP/mRNA gene products suggesting at least a temporary reduction in fertility as a result of interference with reproductive function in both men and women.

Appendix 1 contains the full list of conclusions from Hagemann, Lorré, and Kremer; and Appendix 2 contains those from Beck and Vernazza.

Epilogue

How long will it take people around the world to blame the USA for vaccine harms as this "street art" from the Bahnhoff Strasse in Zurich in October 2022 indicates.

Figure 5: Sidewalk in Zurich, Switzerland.

Will Americans face the scorn of the world community like innocent Germans did after World War II? Or will citizens of other countries blame their own health agencies and political leaders?

Appendix 1: Conclusion from Hagemann, Lorré, and Kremer

Conclusion

- The first half of 2022 was marked by a significant decline in births ranging from 1.3% in France to 19% in Romania.
- In 15 countries this decline exceeded 4%, in 7 countries it exceeded 10%.
- A significant negative correlation between birth rates decline and vaccination frequency is found in 13 of 18 countries. In Finland, Switzerland, the Netherlands, Latvia, Austria, Germany, and Lithuania, as well as for Europe as a whole, the correlation analysis even yielded p-values of 0.005 or less.
- No correlation was found between the decline in birth rates and the incidence of Covid-19 infections or hospitalizations assigned to Covid-19.
- Adverse reactions related to the female reproductive organs and study findings related to male fertility point to a causal interpretation of the association of birth declines and the Covid-19 vaccinations.
- Observations of fertility centres for corresponding signs should be collected.
- With reference to the Bradford-Hill criteria, a relationship temporality between the decline in births and the course of the initial vaccination campaign (doses 1 and 2) nine months earlier could be demonstrated. There was a very high analogy between the European countries. The uniformly observed decline in the number of births with a temporal connection to the start of the vaccination campaign is thus not an isolated national phenomenon. Some countries are still withholding their data. Norway has sent data for the second quarters, but is currently withholding data for the first quarters.
- Given the considerable individual and social relevance of the link between vaccination campaigns and declining birth rates, the immediate suspension of Covid-19 vaccination for all persons of childbearing and reproductive age should be called for.
- Data on stillbirths, spontaneous, and any other abortions must be provided in a timely manner.
- It remains to be explored:
  - How exactly does the Covid-19 vaccine exert its apparent deleterious effect on female reproductive capacity?
  - Does the Covid-19 vaccine also affect male reproductive capacity?
  - How long do these effects last?
- My remarks and comments are not intended to exclude any interpretations, they are as factual as possible. Waiting for traditional scientific publications is unacceptable in view of the considerable individual and social threats posed by the emergency-approved vaccines.

Raimund Hagemann, Data analyst, technically supported by Ulf Lorré and Dr Hans-Joachim Kremer

Appendix 2: Conclusion from Beck and Vernazza

Conclusion

The present investigation of the demographic data of Switzerland in combination with the cantonal vaccination quotas, as well as the review of the relevant medical studies, allow the conclusion that the hypothesis that vaccination and the decline in birth rates are not causally related to one another must be rejected. The following arguments should be cited:

1. There is a striking temporal correlation between the peak of the first vaccination and the decline in births in Switzerland.
2. The fall in the birth rate in the first half of 2022 is assuming historic proportions. This also applies in the event that data is subsequently delivered at national level. It also applies in particular to areas with complete data collection (e.g., for the city of Zurich).
3. The argument that the decline is a consequence of the 2021 baby boom is unfounded because it has never happened in Switzerland that the baby boom years would have compensated for a subsequent decline in the birth rate. In addition, the baby boom is especially pronounced in those cantons where the decline in the birth rate is weaker (cf. Tab. 6).
4. The difference-in-difference analysis of which cantons with high vaccination rates have low vaccination rates shows a significant difference in declining birth rates in both groups. The group of cantons with a high vaccination rate shows a stronger decline compared to cantons with a lower vaccination rate. This provides robust evidence for the existence of a causal relationship.
5. The significance of this difference in the birth rate decline in the canton groups with high or low vaccination rate was calculated in different ways and checked for various possible distortions according to the rules of the art. The significance was in all examined cases more than 90%, usually even higher.
6. Finally, counter-evidence could also be provided. When canton groups are compared, which do not differ in terms of their mean vaccination rate, the decline in births was equally strong in both groups.
7. Difference-in-difference is a method that is currently being used in connection with Corona for countless proofs of the effectiveness or ineffectiveness of certain measures which have been taken. Unlike

most of this study, this can in our case prove the common trend in the period before. Vaccination impressively demonstrated the distorting effects of time series analysis be corrected and the canton data come from a very homogeneous Environment (vaccination campaign coordinated by the federal government, uniform national regulation Birth, abortion and reimbursement of medical expenses, central data collection by two federal offices, BIS and BAG).

8. To clarify whether the causality is through biological factors or through pure behavioral change of the vaccinated persons can be explained were younger medical studies used. It turns out that studies with of sufficient duration, there is definitely evidence of declining fertility can be provided to men.
9. It is also known from animal experiments by Pfizer/Medema that enrichment of the mRNA could be detected in the sex organs.

Because of this long list of statistically confirmed findings, we call on Swissmedic

- An explicit warning for people who do not want to have children use of an mRNA-based Covid-19 vaccine.
- Ask Pfizer/Medema to share their data on long-term mRNA accumulation in animal experiments and to publish them.

Three: "77% of Cardiovascular Adverse Events from Pfizer's mRNA COVID Shot Occurred in Women, as Well as in People Under Age 65. Two Minors Suffered Cardiac Events."

—Barbara Gehrett, MD; Joseph Gehrett, MD; Chris Flowers, MD; and Loree Britt

The WarRoom/DailyClout Pfizer Documents Analysis Project Post-Marketing Group (Team 1) produced a shocking review of the Cardiovascular System Organ Class (SOC) from data in Pfizer document *5.3.6 Cumulative Analysis of Post-Authorization Adverse Event Reports of PF-07302048 (BNT162B2) Received Through 28-FEB-2021* (a.k.a., "*5.3.6*"). (https://www.phmpt.org/wp-content/uploads/2022/04/reissue_5.3.6-post-marketing-experience.pdf) 1,403 patients, or 3.3% of the total patient population, reported cardiovascular adverse events, which did not include myocarditis and pericarditis. These reports came from 38 countries.

Highlights from this report include:

- 50% of the cardiovascular adverse events were reported in the **first 24 hours** post-injection.
- There were **136 deaths, which equates to nearly 10% of affected patients**.
- Of the total of 1,403 patients, 946, or **66%, had severe adverse events**.
- **Pfizer excluded myocarditis and pericarditis from the Cardiovascular category** and instead reported those adverse events under the Immune-Mediated/Autoimmune category. While immune-mediated myocarditis and pericarditis can occur, it seems disingenuous that those adverse events were left out of the Cardiovascular category.
- **One child and one adolescent** suffered cardiovascular adverse events, but Pfizer did not provide any details of what happened to them. These children also were not included in Pfizer's Pediatric Report. Did they experience heart attacks? Does the Food and Drug Administration (FDA) know about these cases?
- A **much younger population**, ages 18–64 years, made up the bulk of adverse event cases in this category (77%). Cardiovascular disease is typically a hallmark of age. Why were there so many cardiovascular side effects in younger adults?
- Cardiovascular adverse events **occurred more than three times as often in women**—1,076 **(77%) were female,** 291 (21%) were male, and 36 (2.5%) were unreported.

It is important to note that the adverse events in the *5.3.6* document were reported to Pfizer for *only a 90-day period* starting on December 1, 2020, the date of the United Kingdom's public rollout of Pfizer's COVID-19 experimental mRNA "vaccine" product.

Please read the concerning report by the Post-Marketing Group (Team 1) below.

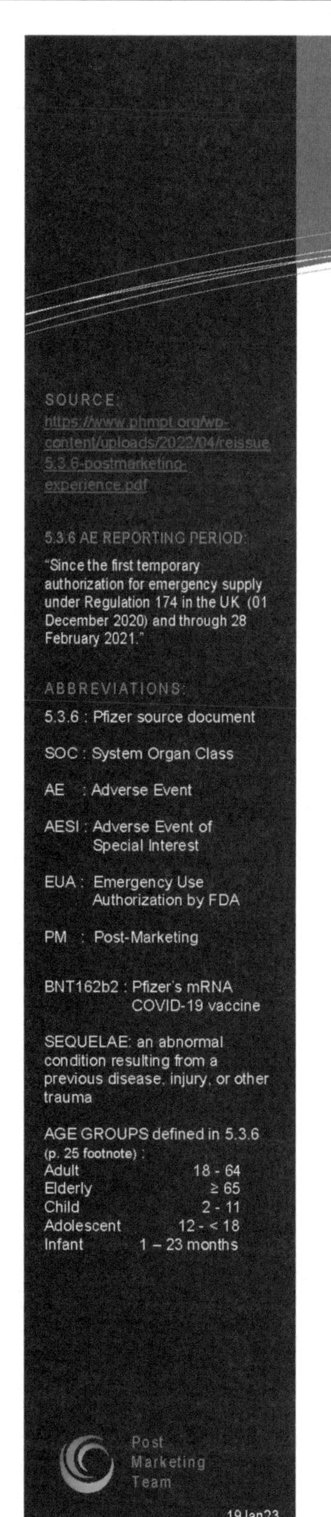

War Room/DailyClout Pfizer Document Analysis

Post-Marketing Team Micro-Report 5:

Cardiovascular System Organ Class (SOC) Review of 5.3.6

Cardiovascular AESIs *Search criteria: PTs Acute myocardial infarction; Arrhythmia; Cardiac failure; Cardiac failure acute; Cardiogenic shock; Coronary artery disease; Myocardial infarction; Postural orthostatic tachycardia syndrome; Stress cardiomyopathy; Tachycardia*	• Number of cases: 1403 (3.3% of the total PM dataset), of which 241 are medically confirmed and 1162 are non-medically confirmed; • Country of incidence: UK (268), US (233), Mexico (196), Italy (141), France (128), Germany (102), Spain (46), Greece (45), Portugal (37), Sweden (20), Ireland (17), Poland (16), Israel (13), Austria, Romania and Finland (12 each), Netherlands (11), Belgium and Norway (10 each), Czech Republic (9), Hungary and Canada (8 each), Croatia and Denmark (7 each), Iceland (5); the remaining 30 cases were distributed among 13 other countries; • Subjects' gender: female (1076), male (291) and unknown (36); • Subjects' age group (n = 1346): Adult (1078), Elderly (266) Child and Adolescent (1 each); • Number of relevant events: 1441, of which 946 serious, 495 non-serious; in the cases reporting relevant serious events; • Reported relevant PTs: Tachycardia (1098), Arrhythmia (102), Myocardial infarction (89), Cardiac failure (80), Acute myocardial infarction (41), Cardiac failure acute (11), Cardiogenic shock and Postural orthostatic tachycardia syndrome (7 each) and Coronary artery disease (6); • Relevant event onset latency (n = 1209): Range from <24 hours to 21 days, median <24 hours;

CONFIDENTIAL
Page 16

FDA-CBER-2021-5683-0000069

BNT162b2
5.3.6 Cumulative Analysis of Post-authorization Adverse Event Reports

Table 7. AESIs Evaluation for BNT162b2

AESIs Category	Post-Marketing Cases Evaluation Total Number of Cases (N=42086)
	• Relevant event outcome: fatal (136), resolved/resolving (767), resolved with sequelae (21), not resolved (140) and unknown (380); Conclusion: This cumulative case review does not raise new safety issues. Surveillance will continue

https://www.phmpt.org/wp-content/uploads/2022/04/reissue_5.3.6-postmarketing-experience.pdf

- **Adverse Events were reported to Pfizer during a 90-day period, following the December 1, 2020, public rollout of its COVID-19 experimental "vaccine" product.**
- **In the Pfizer 5.3.6 document, these AEs were categorized by System Organ Classes (SOC) – in other words, by systems in the body.**
- **Cardiovascular adverse events reports were received from 38 countries.**
- **In the cardiovascular category there were 1,403 patients, or 3.3% of the total patients reporting adverse events.**

The Cardiovascular AESI cases were a composite of searches made for heart failure, including shock (98 AEs), coronary artery disease including heart attacks (136 AEs) and disturbances of the heart rhythm under various specific diagnoses (1,200 AEs.) An additional syndrome of rapid heartbeat and low blood pressure when standing, termed postural orthostatic tachycardia syndrome (POTS), (7 AEs) was included in the search criteria.

The time from vaccination to the adverse event extended from one day to 21 days, though **half were reported within the first 24 hours**. Of the 1,441 diagnosed conditions, **946 (66%) were classified as serious.**

There were 136 deaths (9.7%). The report lacks further definition of the characteristics of the patients who died within this narrow window of time after vaccination. 767 conditions **(53%) were classified as resolved or resolving though there is no further information on the ultimate outcomes.** 21 (1.5%) resolved with ongoing consequences, 140 **(9.7%) were not resolved**, and 380 **(26%) had unknown outcome status.**

Tachycardia (rapid heartbeat) includes numerous specific fast heart rate syndromes that vary from normal (exercise-related) to deadly (ventricular tachycardia, fibrillation). **Arrhythmia** refers to any irregularity in the heartbeat. Again, this can range from a normal variation in the heart rate with breathing to a life-threatening problem. *It appears these arrhythmias are not related to myocarditis.*

A remarkable observation, from a medical point of view, is that a **number of diagnoses in the original search criteria seem to have been excluded**. Bradycardia (slow heartbeat), atrial fibrillation, atrial flutter, ventricular tachycardia, and ventricular fibrillation, among others, are not specifically listed. Specifics on which arrhythmias or tachycardias were serious or non-serious are not provided. **Yet Pfizer classifies 66% of the total AEs in this SOC as serious, which means many of the arrhythmias were serious.**

Is this general category of "arrhythmia" adequate in a search if only these limited conditions are specified? If these other diagnoses were not collected, the number of adverse events could be significantly higher.

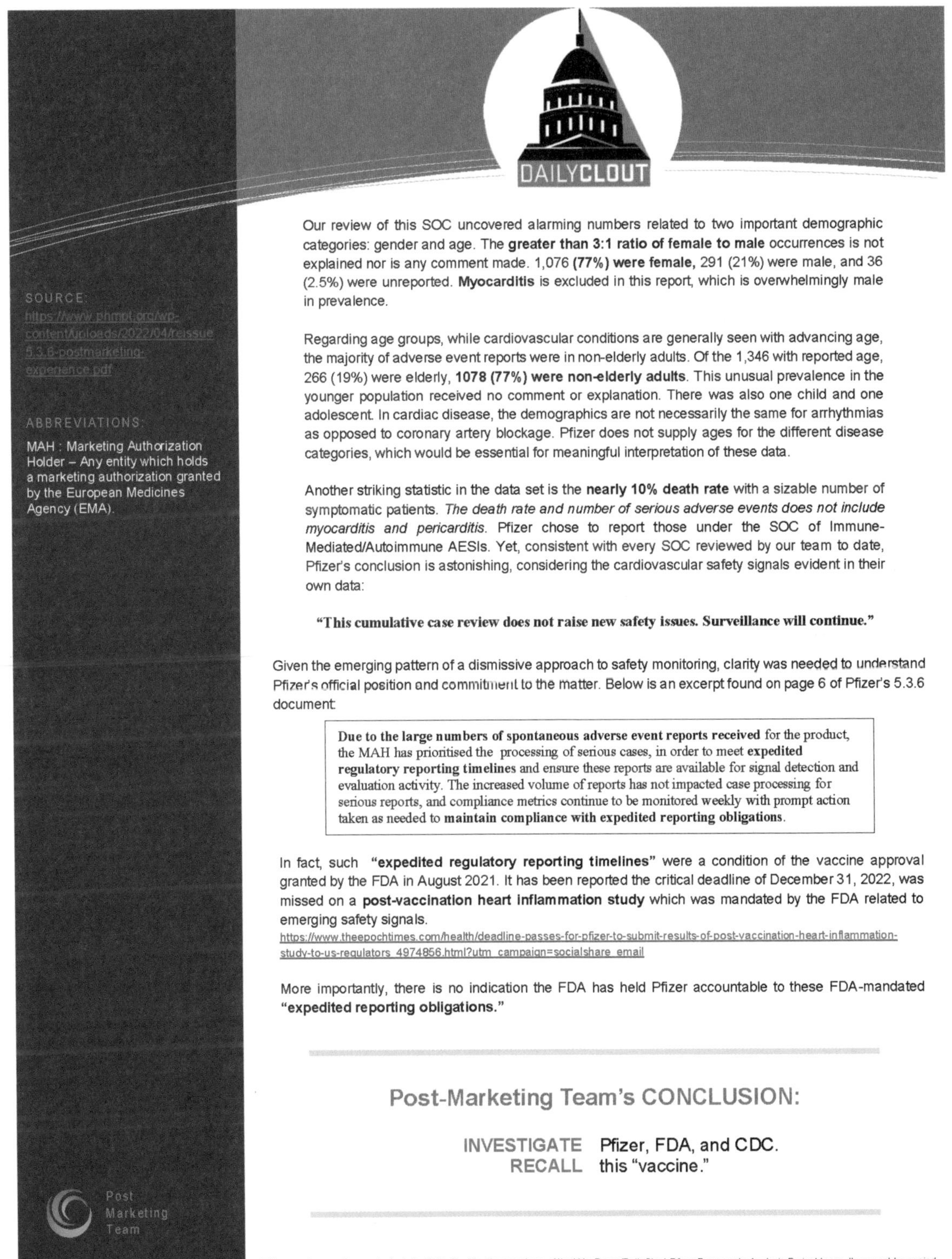

SOURCE:
https://www.phmpt.org/wp-content/uploads/2022/04/reissue_5.3.6-postmarketing-experience.pdf

ABBREVIATIONS:
MAH : Marketing Authorization Holder – Any entity which holds a marketing authorization granted by the European Medicines Agency (EMA).

Our review of this SOC uncovered alarming numbers related to two important demographic categories: gender and age. The **greater than 3:1 ratio of female to male** occurrences is not explained nor is any comment made. 1,076 **(77%) were female,** 291 (21%) were male, and 36 (2.5%) were unreported. **Myocarditis** is excluded in this report, which is overwhelmingly male in prevalence.

Regarding age groups, while cardiovascular conditions are generally seen with advancing age, the majority of adverse event reports were in non-elderly adults. Of the 1,346 with reported age, 266 (19%) were elderly, **1078 (77%) were non-elderly adults**. This unusual prevalence in the younger population received no comment or explanation. There was also one child and one adolescent. In cardiac disease, the demographics are not necessarily the same for arrhythmias as opposed to coronary artery blockage. Pfizer does not supply ages for the different disease categories, which would be essential for meaningful interpretation of these data.

Another striking statistic in the data set is the **nearly 10% death rate** with a sizable number of symptomatic patients. *The death rate and number of serious adverse events does not include myocarditis and pericarditis.* Pfizer chose to report those under the SOC of Immune-Mediated/Autoimmune AESIs. Yet, consistent with every SOC reviewed by our team to date, Pfizer's conclusion is astonishing, considering the cardiovascular safety signals evident in their own data:

"This cumulative case review does not raise new safety issues. Surveillance will continue."

Given the emerging pattern of a dismissive approach to safety monitoring, clarity was needed to understand Pfizer's official position and commitment to the matter. Below is an excerpt found on page 6 of Pfizer's 5.3.6 document:

> **Due to the large numbers of spontaneous adverse event reports received** for the product, the MAH has prioritised the processing of serious cases, in order to meet **expedited regulatory reporting timelines** and ensure these reports are available for signal detection and evaluation activity. The increased volume of reports has not impacted case processing for serious reports, and compliance metrics continue to be monitored weekly with prompt action taken as needed to **maintain compliance with expedited reporting obligations**.

In fact, such **"expedited regulatory reporting timelines"** were a condition of the vaccine approval granted by the FDA in August 2021. It has been reported the critical deadline of December 31, 2022, was missed on a **post-vaccination heart inflammation study** which was mandated by the FDA related to emerging safety signals.
https://www.theepochtimes.com/health/deadline-passes-for-pfizer-to-submit-results-of-post-vaccination-heart-inflammation-study-to-us-regulators_4974856.html?utm_campaign=socialshare_email

More importantly, there is no indication the FDA has held Pfizer accountable to these FDA-mandated **"expedited reporting obligations."**

Post-Marketing Team's CONCLUSION:

INVESTIGATE Pfizer, FDA, and CDC.
RECALL this "vaccine."

Post Marketing Team

Four: "Infants and Children Under 12 Given the Pfizer mRNA COVID 'Vaccine' Seven Months BEFORE Pediatric Approval. 71% of Adverse Event Cases Classified as Serious."

—Barbara Gehrett, MD; Joseph Gehrett, MD; Chris Flowers, MD; and Loree Britt

The WarRoom/DailyClout Pfizer Documents Analysis Project Post-Marketing Group (Team 1)—Barbara Gehrett, MD; Joseph Gehrett, MD; Chris Flowers, MD; and Loree Britt—produced a shocking review of the pediatric data found in Pfizer document *5.3.6 Cumulative Analysis of Post-Authorization Adverse Event Reports of PF-07302048 (BNT162B2) Received Through 28-FEB-2021* (a.k.a., "*5.3.6*"). (https://www.phmpt.org/wp-content/uploads/2022/04/reissue_5.3.6-postmarketing-experience.pdf)

It is important to note 1) that the adverse events (AEs) in the *5.3.6* document were reported to Pfizer for only a 90-day period starting on December 1, 2020, the date of the United Kingdom's public rollout of Pfizer's COVID-19 experimental mRNA "vaccine" product and 2) **no pediatric dose of the Pfizer product was approved for use during that time frame.**

What dose(s) of Pfizer's mRNA "vaccine" was given to these children since no approved dose existed?

Important points from this report include:

- **A seven-year-old experienced a stroke.**
- **One child and one infant suffered facial paralysis.**
- **One infant had a kidney adverse event, either kidney injury or failure.**
- Of the 34 adverse event cases, **24 (71%) were classified as serious.**
- **Predominantly female patients were affected**—at least 25 of 34 (73.5%) patients.
- Table 6 reports **34 cases of use in pediatric individuals**. However, **28 additional cases were excluded** because details such as height and weight were "not consistent with pediatric subjects."
- Ages ranged from **two months to nine years**, with **median 4.0 years**, which means **half the children were under four years of age.**
- 132 adverse events were reported in the 34 children—i.e., **an average of 3.88 AEs per child.**

Shockingly, Pfizer concluded:

> ***No new significant safety information was identified based on a review of these cases compared with the non-paediatric population.***

Please read the disturbing report by the Post-Marketing Group (Team 1) below.

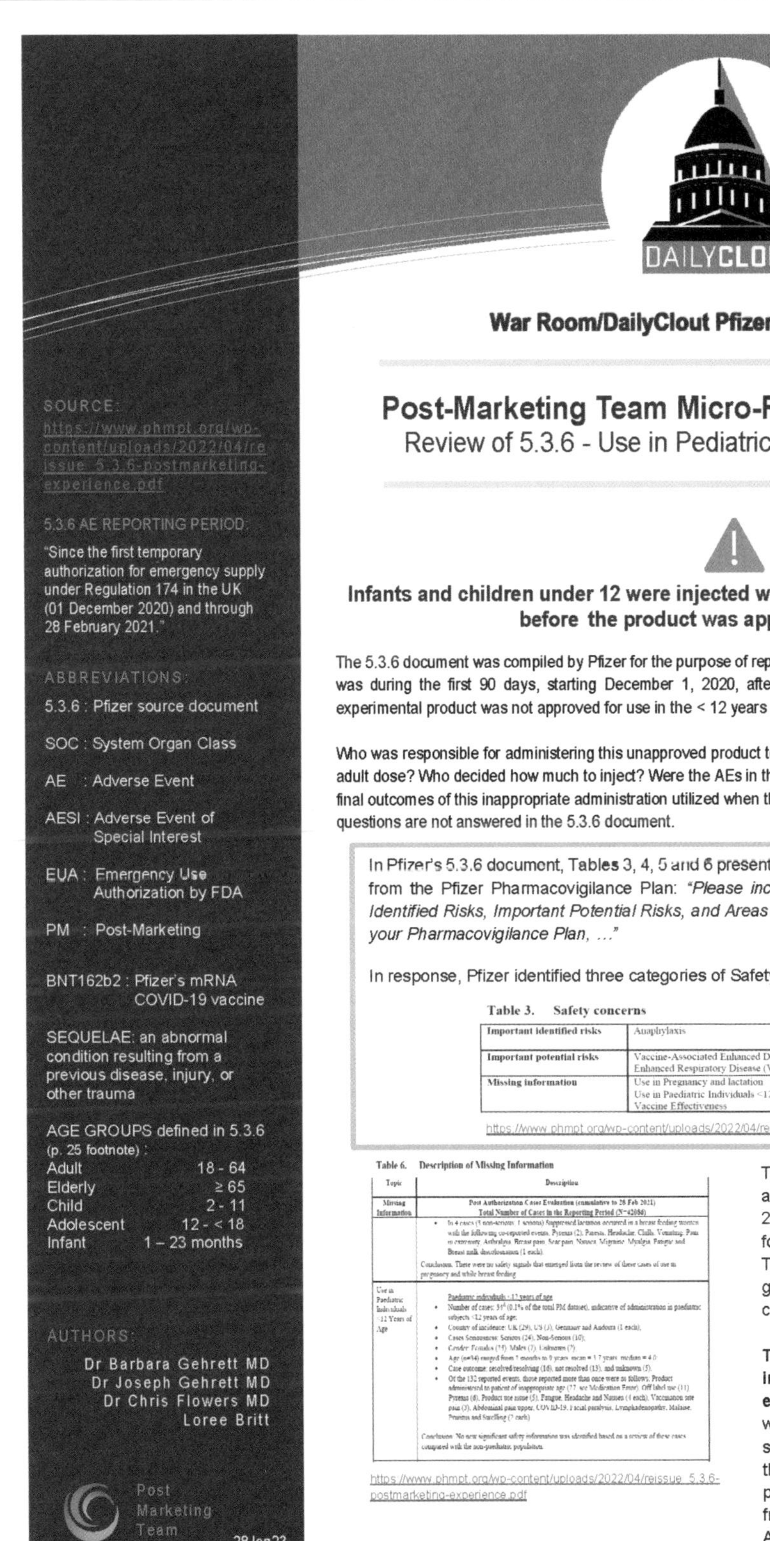

War Room/DailyClout Pfizer Document Analysis

SOURCE: https://www.phmpt.org/wp-content/uploads/2022/04/reissue_5.3.6-postmarketing-experience.pdf

5.3.6 AE REPORTING PERIOD:

"Since the first temporary authorization for emergency supply under Regulation 174 in the UK (01 December 2020) and through 28 February 2021."

ABBREVIATIONS:

5.3.6 : Pfizer source document

SOC : System Organ Class

AE : Adverse Event

AESI : Adverse Event of Special Interest

EUA : Emergency Use Authorization by FDA

PM : Post-Marketing

BNT162b2 : Pfizer's mRNA COVID-19 vaccine

SEQUELAE: an abnormal condition resulting from a previous disease, injury, or other trauma

AGE GROUPS defined in 5.3.6 (p. 25 footnote) :

Adult	18 - 64
Elderly	≥ 65
Child	2 - 11
Adolescent	12 - < 18
Infant	1 – 23 months

AUTHORS:

Dr Barbara Gehrett MD
Dr Joseph Gehrett MD
Dr Chris Flowers MD
Loree Britt

Post Marketing Team

28Jan23

Post-Marketing Team Micro-Report 6:
Review of 5.3.6 - Use in Pediatric Individuals < 12 years of age

Infants and children under 12 were injected with the Pfizer "vaccine" seven months before the product was approved for children.

The 5.3.6 document was compiled by Pfizer for the purpose of reporting adverse events to the FDA. The reporting period was during the first 90 days, starting December 1, 2020, after the COVID-19 "vaccine" rollout to the public. This experimental product was not approved for use in the < 12 years of age group at that time.

Who was responsible for administering this unapproved product to children? Did these children receive a full or a partial adult dose? Who decided how much to inject? Were the AEs in these children monitored long term? Were the results of final outcomes of this inappropriate administration utilized when the approval for this age group was considered? These questions are not answered in the 5.3.6 document.

In Pfizer's 5.3.6 document, Tables 3, 4, 5 and 6 present a summary of information the FDA requested from the Pfizer Pharmacovigilance Plan: *"Please include a cumulative analysis of the Important Identified Risks, Important Potential Risks, and Areas of Important Missing Information identified in your Pharmacovigilance Plan, ..."*

In response, Pfizer identified three categories of Safety Concerns in Table 3:

Table 3. Safety concerns

Important identified risks	Anaphylaxis
Important potential risks	Vaccine-Associated Enhanced Disease (VAED), Including Vaccine-associated Enhanced Respiratory Disease (VAERD)
Missing information	Use in Pregnancy and lactation Use in Paediatric Individuals <12 Years of Age Vaccine Effectiveness

https://www.phmpt.org/wp-content/uploads/2022/04/reissue_5.3.6-postmarketing-experience.pdf

Table 6. Description of Missing Information

Topic	Description
Missing Information	Post Authorization Cases Evaluation (cumulative to 28 Feb 2021) Total Number of Cases in the Reporting Period (N=42086)
	• In 4 cases (3 non-serious, 1 serious) Suppressed lactation occurred in a breast feeding women with the following co-reported events: Pyrexia (2), Paresis, Headache, Chills, Vomiting, Pain in extremity, Arthralgia, Breast pain, Scar pain, Nausea, Migraine, Myalgia, Fatigue and Breast milk discolouration (1 each). Conclusion: There were no safety signals that emerged from the review of these cases of use in pregnancy and while breast feeding.
Use in Paediatric Individuals <12 Years of Age	Paediatric individuals <12 years of age • Number of cases: 34[d] (0.1% of the total PM dataset), indicative of administration in paediatric subjects <12 years of age. • Country of incidence: UK (29), US (3), Germany and Andorra (1 each); • Cases Seriousness: Serious (24), Non-Serious (10); • Gender: Females (25), Males (7), Unknown (2); • Age (n=34) ranged from 2 months to 9 years, mean = 3.7 years, median = 4.0 • Case outcome: resolved/resolving (16), not resolved (13), and unknown (5). • Of the 132 reported events, those reported more than once were as follows: Product administered to patient of inappropriate age (27, see Medication Error), Off label use (11), Pyrexia (6), Product use issue (5), Fatigue, Headache and Nausea (4 each), Vaccination site pain (3), Abdominal pain upper, COVID-19, Facial paralysis, Lymphadenopathy, Malaise, Pruritus and Swelling (2 each) Conclusion: No new significant safety information was identified based on a review of these cases compared with the non-paediatric population.

https://www.phmpt.org/wp-content/uploads/2022/04/reissue_5.3.6-postmarketing-experience.pdf

The Pfizer COVID-19 "vaccine" was not approved for use in this age group until October 2021, when the FDA granted EUA authorization for (only) children aged 5 through 11 years. In Table 6, titled "Safety Concerns," the <12 age group appears under the "Missing Information" category.

Table 6 reports 34 cases of use in pediatric individuals. 28 additional cases were excluded because details such as height and weight were "not consistent with pediatric subjects." Pfizer did not supply further details of their pediatric parameters. Country of origin was predominantly UK (29), with an additional three from the U.S. and one each from Germany and Andorra.

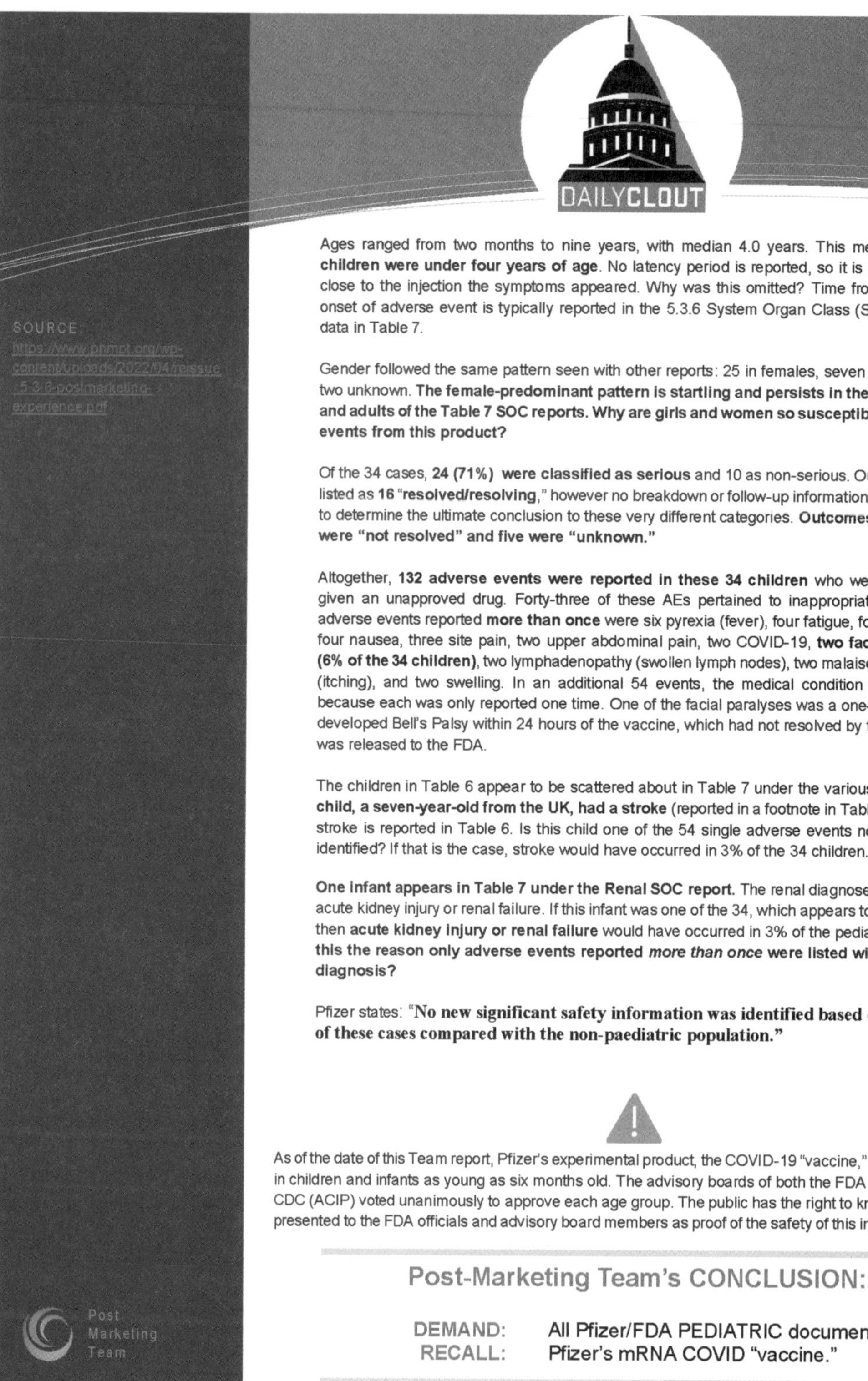

SOURCE:
https://www.phmpt.org/wp-content/uploads/2022/04/reissue_5.3.6-postmarketing-experience.pdf

Ages ranged from two months to nine years, with median 4.0 years. This means **half the children were under four years of age**. No latency period is reported, so it is unknown how close to the injection the symptoms appeared. Why was this omitted? Time from injection to onset of adverse event is typically reported in the 5.3.6 System Organ Class (SOC) category data in Table 7.

Gender followed the same pattern seen with other reports: 25 in females, seven in males, and two unknown. **The female-predominant pattern is startling and persists in the adolescents and adults of the Table 7 SOC reports. Why are girls and women so susceptible to adverse events from this product?**

Of the 34 cases, **24 (71%) were classified as serious** and 10 as non-serious. Outcomes were listed as **16 "resolved/resolving,"** however no breakdown or follow-up information was provided to determine the ultimate conclusion to these very different categories. **Outcomes of 13 cases were "not resolved" and five were "unknown."**

Altogether, **132 adverse events were reported in these 34 children** who were improperly given an unapproved drug. Forty-three of these AEs pertained to inappropriate use. Other adverse events reported **more than once** were six pyrexia (fever), four fatigue, four headache, four nausea, three site pain, two upper abdominal pain, two COVID-19, **two facial paralysis (6% of the 34 children)**, two lymphadenopathy (swollen lymph nodes), two malaise, two pruritus (itching), and two swelling. In an additional 54 events, the medical condition was omitted, because each was only reported one time. One of the facial paralyses was a one-year-old who developed Bell's Palsy within 24 hours of the vaccine, which had not resolved by the time 5.3.6 was released to the FDA.

The children in Table 6 appear to be scattered about in Table 7 under the various SOCs. **One child, a seven-year-old from the UK, had a stroke** (reported in a footnote in Table 7). No child stroke is reported in Table 6. Is this child one of the 54 single adverse events not specifically identified? If that is the case, stroke would have occurred in 3% of the 34 children.

One infant appears in Table 7 under the Renal SOC report. The renal diagnoses were either acute kidney injury or renal failure. If this infant was one of the 34, which appears to be the case, then **acute kidney injury or renal failure** would have occurred in 3% of the pediatric group. **Is this the reason only adverse events reported *more than once* were listed with a specific diagnosis?**

Pfizer states: **"No new significant safety information was identified based on a review of these cases compared with the non-paediatric population."**

As of the date of this Team report, Pfizer's experimental product, the COVID-19 "vaccine," has been approved in children and infants as young as six months old. The advisory boards of both the FDA (VRBPAC) and the CDC (ACIP) voted unanimously to approve each age group. The public has the right to know what data were presented to the FDA officials and advisory board members as proof of the safety of this injection for children.

Post-Marketing Team's CONCLUSION:

DEMAND: All Pfizer/FDA PEDIATRIC documents.
RECALL: Pfizer's mRNA COVID "vaccine."

Post Marketing Team

Five: "Autopsies Reveal Medical Atrocities of Genetic Therapies Being Used Against a Respiratory Virus"
—Robert W. Chandler, MD, MBA; Michael Palmer, MD

Summary

Dr. Arne Burkhardt is one of eight international pathologists, physicians, and scientists who were asked to perform a second autopsy, requested by friends and family of the deceased who were not satisfied with the results of the first autopsy.

Thirty autopsies and three biopsies were evaluated; 15 cases with routine histopathology (Step 1), three with advanced methods (Step 2), and some of the remaining 15 are included as illustrative cases.

The Step 1 group included eight women and seven men aged 28–95 (average 69).

Death occurred seven days to 180 days following the first or the second Spike-Mediated Gene Therapy (SMGT) with COMIRNATY in eight, Moderna in two, AstraZeneca in two, Janssen in one and Unknown in two.

Place of death was known in 17 cases:

- Nine Non-hospital: five at home, one on the street, one in a car, one at work, one in an elder care facility
- Eight Hospital: four ICU, four died having been in hospital less than two days

Special stains were used to identify Spike and Nucleocapsid Proteins, with the following differential:

- COVID-19 (C-19) = + Spike + Nucleocapsid.
- SMGT = + Spike—Nucleocapsid.

Causation by SMGT: Very probable in five cases, probable in seven, unclear in two, and no connection in one.

Lesions were on multiple organs including Brain, Heart, Kidney, Liver, Lungs, Lymph Node, Salivary Gland, Skin, Spleen, Testis, Thyroid, and Vascular.

Lymphocyte Infiltration, present in 14 of 20 cases (70%), was a common feature and involved multiple organs. Case 19 had at least five different organs involved. CD3+ Lymphocytes were dominant.

The Vascular System was targeted by Lymphocyte Infiltration in seven (35%) of the cases and included sloughing endothelium, destruction of the vessel wall, hemorrhage, and thrombosis.

A condition called Lymphocyte Amok was described by Dr. Burkhardt: Lymphocyte accumulation in non-lymphatic organs and tissues that might develop into lymphoma.

Five cases of unknown foreign material in blood vessels were identified. The favored explanation for origin of this material was aggregated Lipid Nanoparticles (LNPs).

Multiple pathologic processes were involved: Apoptosis, Coagulopathy, Clotting/Infarction, Infiltration/ Mass Formation, Inflammation, Lysis, Necrosis, and Neoplasia.

Röltgen, et al. (https://www.cell.com/cell/fulltext/S0092–8674(22)00076–9) found that COVID-19 depleted Lymphatic Germinal Centers (LGCs) whereas SMGT stimulated them, suggesting a possible origin of "Hunter/Killer" CD3+ Lymphocytes that are attracted to certain tissues, particularly the vascular system.

An expanded program of autopsy following SMGT is recommended in order to further understand the actions of SMGTs and to help formulate new treatments for the constellation of pathology associated with such drugs.

Burkhardt Group Conclusions

1. **Histopathologic analyses show clear evidence of vaccine-induced autoimmune-like pathology in multiple organs.**
2. **That myriad adverse events deriving from such auto-attack processes must be expected to very frequently occur in all individuals, particularly following booster injections.**
3. **Beyond any doubt, injection of gene-based COVID-19 vaccincs place lives under threat of illness and death.**
4. **We note that both mRNA and vector-based vaccines are represented among these cases, as are all four major manufacturers.**

Histopathology

This report is the first in a series in which harms from the Lipid Nanoparticle (LNP) Messenger Ribonucleic Acid (mRNA) therapeutics and other Spike-mediated products will be examined from the point of view of the pathologist, a medical doctor that studies specimens obtained from removal of tissue from living persons, bulk resection or biopsy, or after death. Such examinations make or confirm a diagnosis and provide a basis to determine causation of tissue mass or cause of death. Histopathology refers to the study of abnormal tissues.

Tissues are examined using careful inspection of specimens with the naked eye followed by examination by light microscopy employing a variety of different stains to highlight important features of cells, tissues, and organs. A common stain used is hematoxylin and eosin, H & E for short, which stains nuclei blue, cytoplasm pink or red, collagen fibers pink, and muscles red.

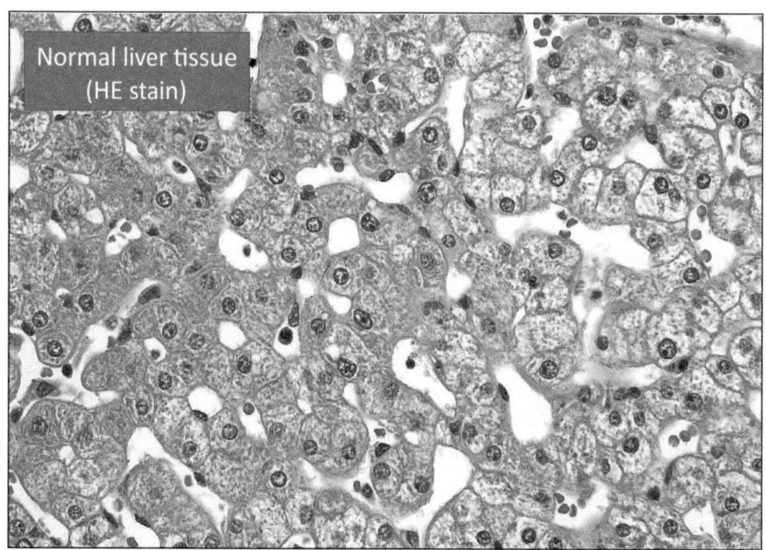

Many of the photomicrographs in this and subsequent articles will have been stained with H & E. Pathologists display sections prepared with H & E along with the magnification used, such as 40 times (40X) or 100 times (100X) magnification.

Immunohistochemical Stains for COVID-19 and Spike-Mediated Therapy

Special stains are vital to the identification of certain histopathology, such as cases involving Spike-mediated therapeutics. These examinations require special stains as outlined by Dr. Arne Burkhardt whose specimens and lecture notes will be the subject of this first report in a series.

Dr. Burkhardt discusses below the immunohistochemistry staining techniques necessary to differentially diagnose cell/tissue damage/organ from COVID-19, SARS-CoV-2 or something else, as well as specific cell types of interest such a T-lymphocytes and monocytes:

Immunohistochemistry to Detect Vaccine-Induced Spike Protein Expression

Prof. Dr. A. Burkhardt

(https://doctors4covidethics.org/notes-and-recommendations-for-conducting-post-mortem-examination-autopsy-of-persons-deceased-in-connection-with-covid-vaccination/)

- Use anti-SARS-COV-2 spike protein/S1 antibodies to test for presence of spike protein in tissue samples. Always include myocardium and spleen tissue samples.
- If spike protein is detected, use anti-nucleocapsid antibody to examine expression of SARS-COV-2 nucleocapsid: presence of nucleocapsid indicates viral "breakthrough" infection, absence of nucleocapsid supports vaccine-induced spike protein expression.
- Perform positive and negative controls using vaccine-transfected and non-transfected cell cultures.

Differential Staining to Identify CD3 and CD68 Cells and to Differentiate COVID-19 from Spike-Inducing Drugs

COVID-19 = Spike stain + Nucleocapsid stain +

LNP/mRNA = Spike stain + Nucleocapsid stain -:

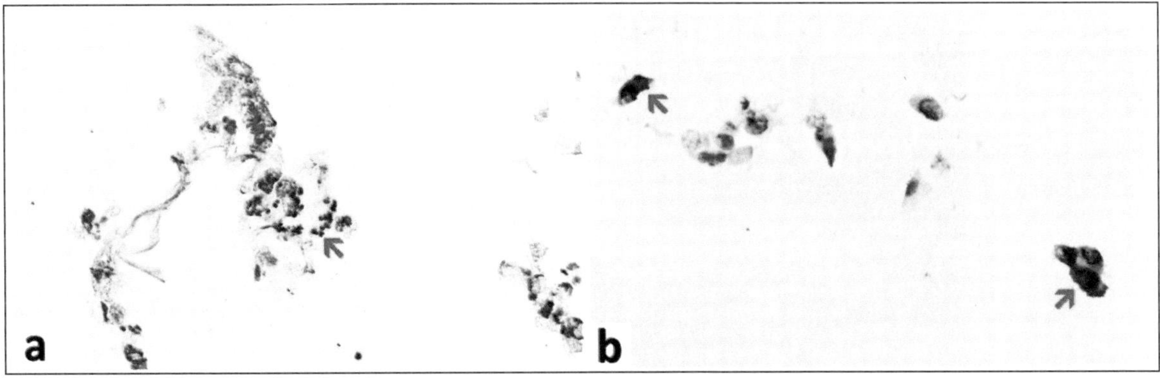

- Spike, red arrow. b. Nucleocapsid, red arrow.

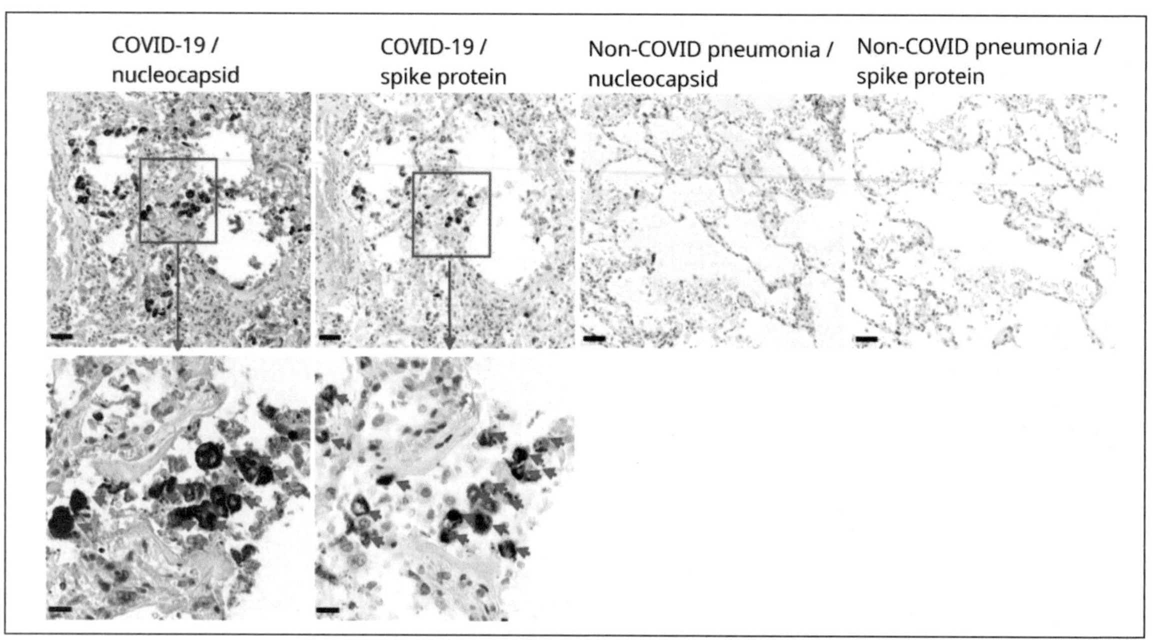

(https://www.prosci-inc.com)

Without these special stains and without an exhaustive search of the specimens, no autopsy should be considered complete.

The internet is an excellent source for examples of both normal and pathological cells, tissues and organs.

A useful guide to have available when looking at the photomicrographs to follow is the Histology Guide at: https://histologyguide.com/.

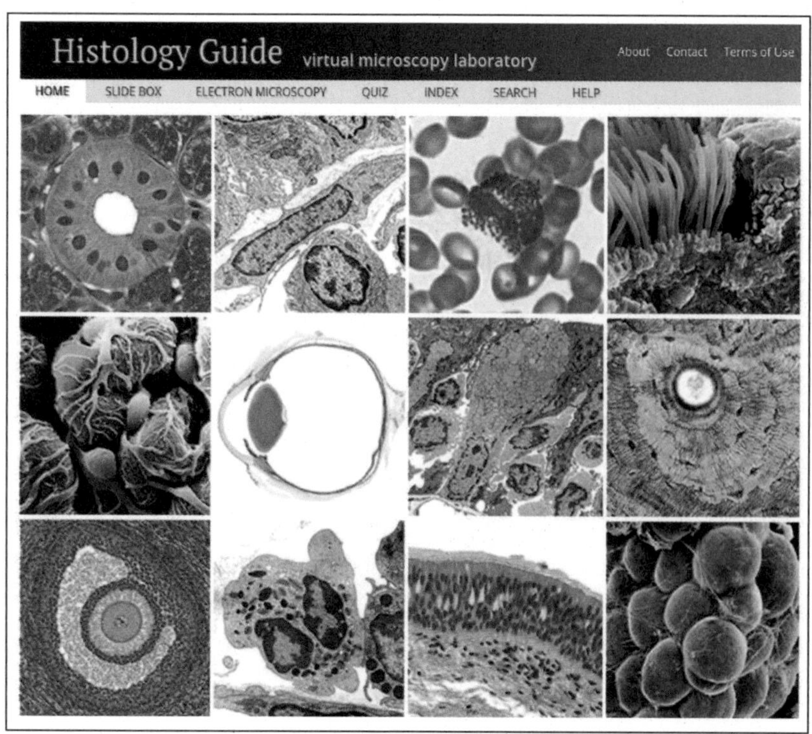

This internet tool can be used to examine normal histology and compare it to the histopathology seen in Dr. Burkhardt's slide deck, which has been integrated with the transcription from his lecture on February 5, 2022.

Autopsy-Histology-Study on Vaccination-Associated Complications and Deaths

(Dr. Burkhardt's slide deck was reproduced here and was integrated with the text derived from notes compiled from the voice recognition transcript with limited editing for readability with no intention to make substantive changes.)

Understanding Vaccine Causation Conference—February 5, 2022
World Council for Health
Dr. Arne Burkhardt
Pathologist
Reutlingen, Germany

Dr. Arne Burkhardt, born in 1944 in Germany, is a pathologist with more than 40 years diagnostic and teaching experience at the Universities of Hamburg, Bern, and Tübingen. He is the author of more than 150 original publications in international journals, currently engaged in autopsy studies of persons dying after taking the Covid vaccine.—**Shabhan Palesa Mohamed**

(https://worldcouncilforhealth.org/multimedia/uvc-arne-burkhardt/)

Dr. Burkhardt: I think it's very, very, important to have an international communication on this subject, because last year in May, April, I was confronted with some relatives of persons who died after vaccination. And I tried to establish a national registry of these persons dying.

And I tried to get autopsies done in these persons, but the national associations of pathology here didn't reply to this request by myself. So, when relatives continued to ask me, "Where can I get some solution to this problem?" finally, I said, "I can examine these organ probes that have been taken during autopsy, and we can try to get some other pathologists sent."

These are the most relevant data on our study. We have **eight cooperating pathologists, physicians, biologists;** and they are internationally from Germany and other European countries, and also some outside of Europe.

Autopsy-Histology-Study on Vaccination-associated Complications and Deaths - Reutlingen

- 8 Cooperating Pathologists, Physicians and Biologists / International
- 30 Autopsies / 3 Biopsies
- 15 Cases evaluated in Step 1: Routine Histology
- 3 Cases in Step 2: Advanced Methods
- 7 Men, 8 Women
- Age range: 28 to 95 Years
- Death occurred between 7 days and 6 month after last Injection
- Vaccines used: Comirnaty 8, Moderna 2, Janssen 1, AstraZeneca 2, unknown 2

So, by now we have 30 autopsies and three biopsies from vaccinated persons. Fifteen cases have been evaluated in the step one that has reached Routine Histology. Three cases are in step two, Advanced Methods. I will explain what I mean by this.

And, just to give you a rough impression, it's seven men, eight women, 28 years to 95 years, death seven days to six months after the last injection and vaccination, the typical vaccinations that are used in Germany.

Place of Death (20 Cases)

At Home	5
On the Street	1
In the Car	1
At Work	1
In Home for the Aged	1
Hospital less than 2 Days	4
Hospital Intensive Care	4
Unknown	2

So, one important fact is that most of these persons that we examined have not died in the hospital, but at home, on the street, in the car. And that is very important, because, in these cases, we can exclude that there is interference with therapeutical measures like artificial respiration and things like that. So, only four were in intensive care medicine before they died.

Death after Corona Vaccination

- 15 Cases with autopsies performed elsewhere (8 in forensic medicine institutes, 7 in pathology institutes)
- Cause of death according to autopsy „natural" / unclear
- Relatives insisted on a second Opinion
- Follow-up Examination of tissue specimens in Reutlingen:

Causal Connection with Vaccination very probable	5
Causal Connection with Vaccination probable	7
Causal Connection with Vaccination unclear/possible	2
No Connection	1

We had **15 cases** with autopsy elsewhere which we examined in step one. And all of these 15 cases were classified by the pathologist of legal medical persons who made the autopsy as **natural and unclear**. And the relatives insisted on a second opinion. And we, in Reutlingen, in our group, looked at the specimens of the organs that were taken.

Burkhardt Series: 15 Cases in Step 1

Average Age 69, 28-95 Time from Injection to Death Average 45 Days, 7-180

Case #	Sex	Age	Drug	Day from Last Injection Until Death
Case 1	Female	82	Moderna 1 & 2	37
Case 2	Male	72	Pfizer 1	31
Case 3	Female	95	Moderna 1 & 2	68
Case 4	Female	73	Pfizer 1	Unknown
Case 5	Male	54	Janssen 1	65
Case 6	Female	55	Pfizer 1 & 2	11
Case 7	Male	56	Pfizer 1 & 2	8
Case 8	Male	80	Pfizer 1 & 2	37
Case 9	Female	89	Unknown 1 & 2	180
Case 10	Female	81	Unknown 1 & 2	Unknown
Case 11	Male	64	AstraZenica 1 & 2	7
Case 12	Female	71	Pfizer 1 & 2	20
Case 13	Male	28	AstraZenica 1 Pfizer 2	28
Case 14	Male	78	Pfizer 1 & 2	65
Case 15	Female	60	Pfizer 1	23

Our follow-up gave **very probable correlation with the vaccination in five cases**, **probable in seven, unclear/possible in two,** and **no connections** with only minimal changes we saw in **one case.**

So, what were the organs where we saw lesions? The target organs and the main lesions in her (sic) space, vascular lesions. Not only to the small vessels, the endothelium, but also to large vessel walls, to the muscular and elastic wall components.

Target Organs and Main Lesions

- Vascular lesions - endothelium / vessel wall
- Unidentified intravascular material
- Spleen / lymph nodes
- Heart
- Lung
- Brain
- Lymphocyte „Amok" in many organs and tissues

1. In five cases, **we found unidentified, intervascular material that might stem from the vaccination material.** Then
2. Spleen and
3. Lymph nodes had changes.
4. Heart,
5. Lung,
6. Brain, and, finally, a phenomenon that we call
7. **Lymphocyte Amok**. That means that **we've found applications and nodular infiltration of lymphatic tissues and organs and tissues that are non-lymphatic.**

Methods

- Routine histological preparation
- Conventional stains

- Immunohistochemistry
  - Standard markers of lymphocytes / inflammatory cells
  - SARS-CoV2-Spike protein subunit 1

So, what are our methods?

1. First of all, routine histological preparation with conventional stain, and then
2. Immunohistochemistry first standard markers for lymphocytes/inflammatory cells.
3. But mostly this is one of the aims of this study, too, **we try to demonstrate the Spike protein in the organs and tissues that were damaged.**

And first of all, of course, we examine the **specificity of our antibody to a Spike protein**, and here a larger magnification. So, it seems to have a very high specificity for this Spike protein. And we did this in cell cultures.

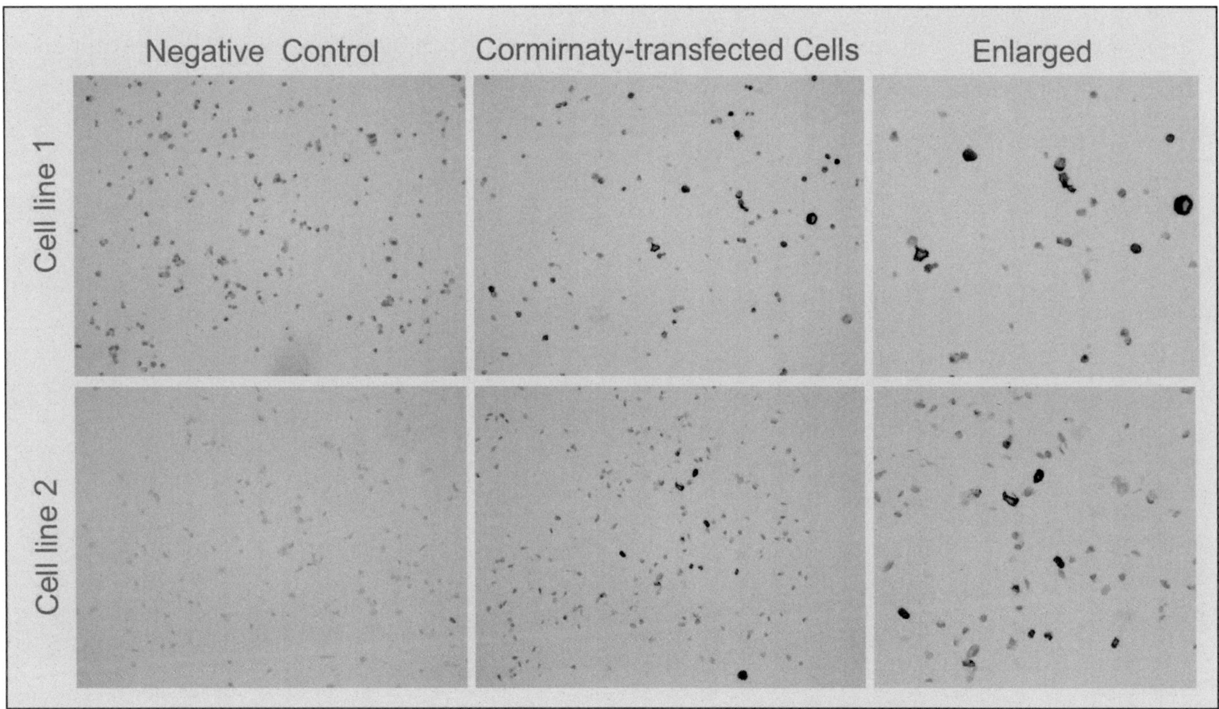

And you can see here net negative control, positive control.

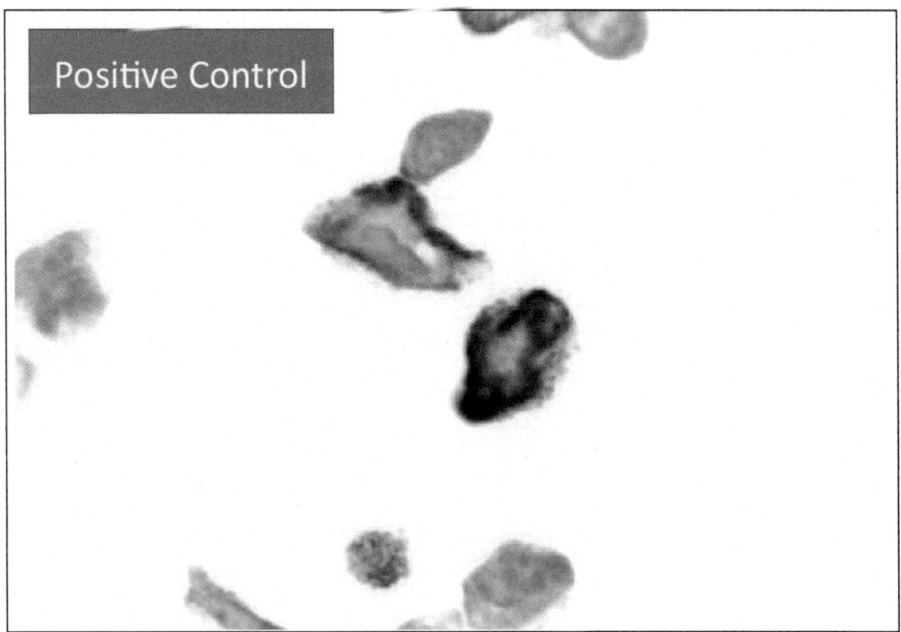

And I will just show you a few examples of the tissue damage that I have listed before.

Blood Vessels, Endovasculitis, Perivasculitis and Vasculitis

So here *(left)* you can see a normal, small vessel, and you can see the endothelium that is like a wallpaper and very small, elongated spindle cell nuclei.

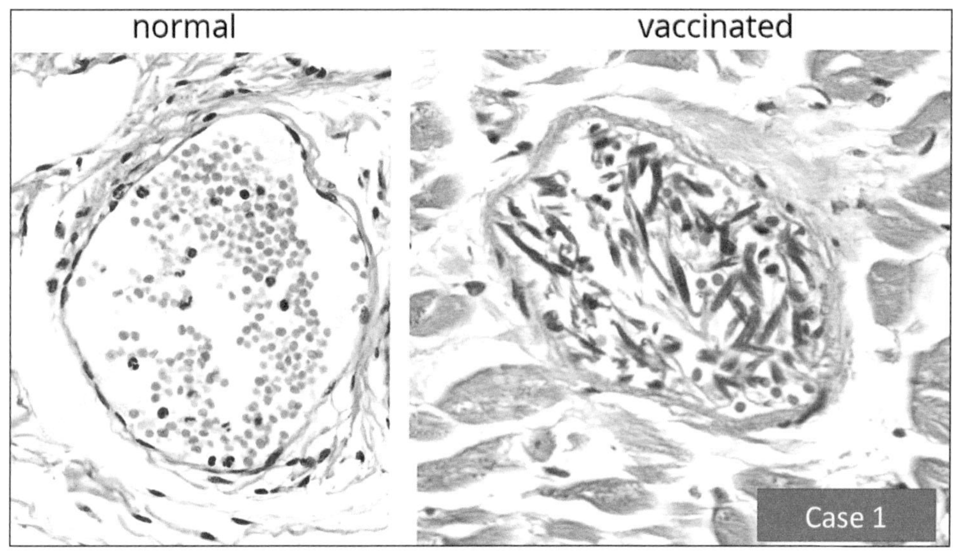

And here *(right)* in one of the cases, you can see that the

1. Endothelium is in the lumen. And it is in there mixed with
2. Lymphocytes and erythrocytes and the
3. Nuclei are swollen.

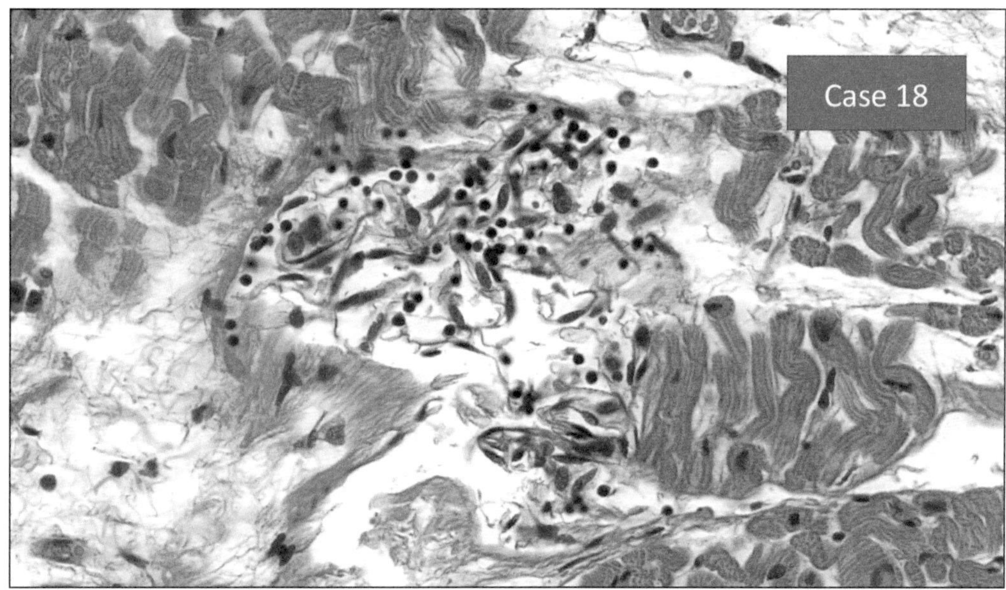

So, in some cases, the small vessels even are completely destroyed by inflammatory infiltrates, mostly lymphocytes. And this proves to me that it is an intravital reaction and not an autolytic phenomenon caused by degradation after death.

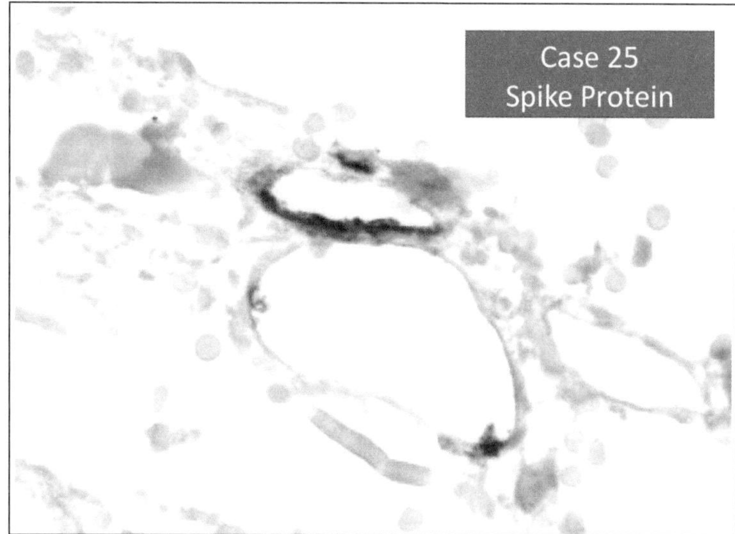

So, we get the Spike protein, immunohistochemistry on these cases, and we see, you see here a very marked and specific mark of the endothelium in these patients. *Rust-color stain more extensive in the smaller vessel wall.*

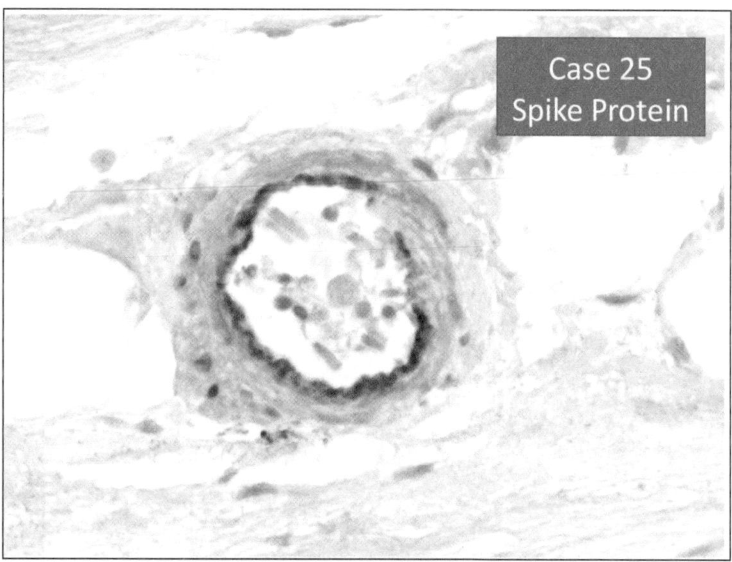

And not only in the small vessels, but also in the smaller arteries, you can see it in the inner part of the vessel. And you can see here there's decimated endothelial cells.

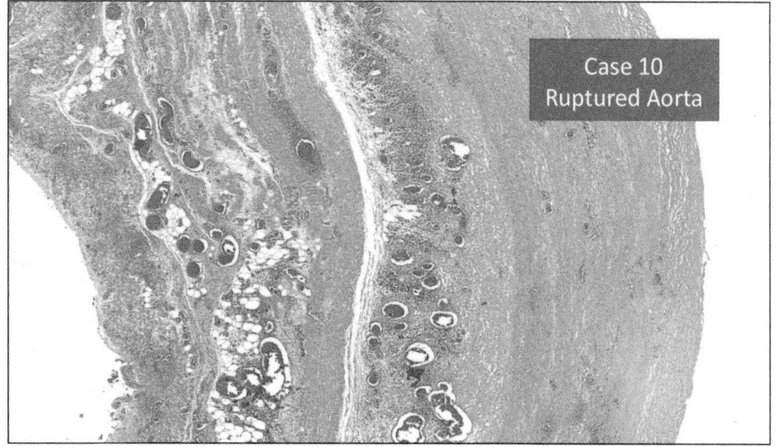

So, as I said, not only the smaller vessels were affected, but also the **aorta** and the larger arteries and **two cases have died of a ruptured artery**. And, actually, we found arteriosclerotic changes; but, as you see here (Case 10 Ruptured Aorta), it's not very pronounced. But you can see inflammatory changes around in the deep layers of this aorta, and also you can see some disturbance of structure of the smooth muscle and the elastic fibers.

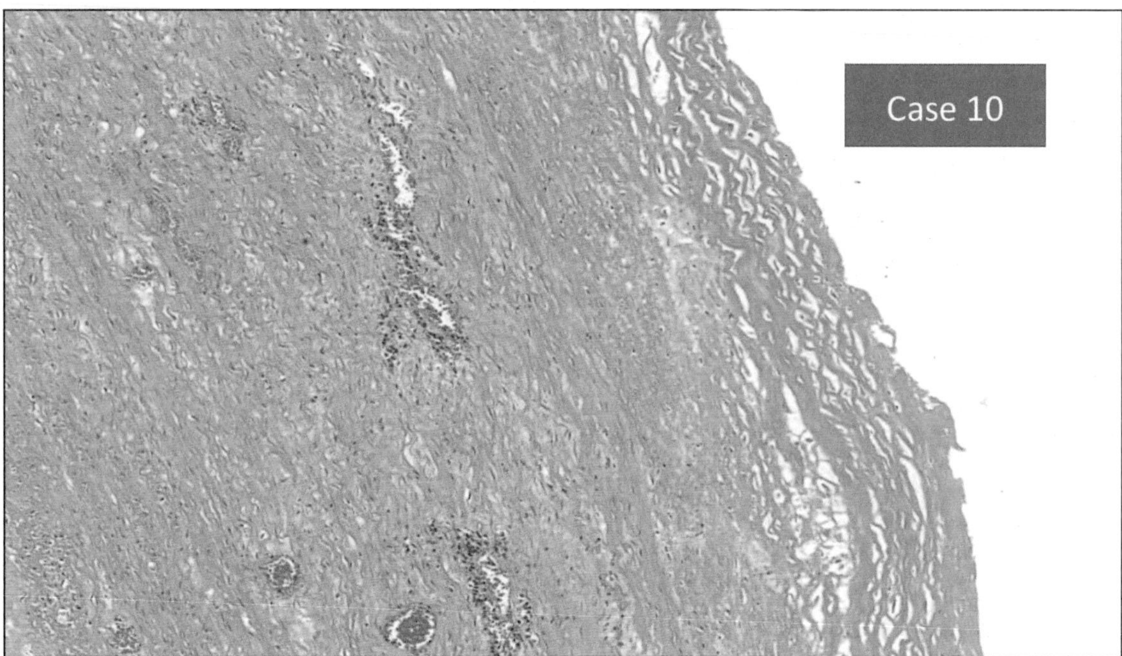

And if you have a higher magnification, you can see these small areas where the elastic fibers and smooth muscles are destroyed. And again, lymphocytic infiltration proving that it was an intravital process.

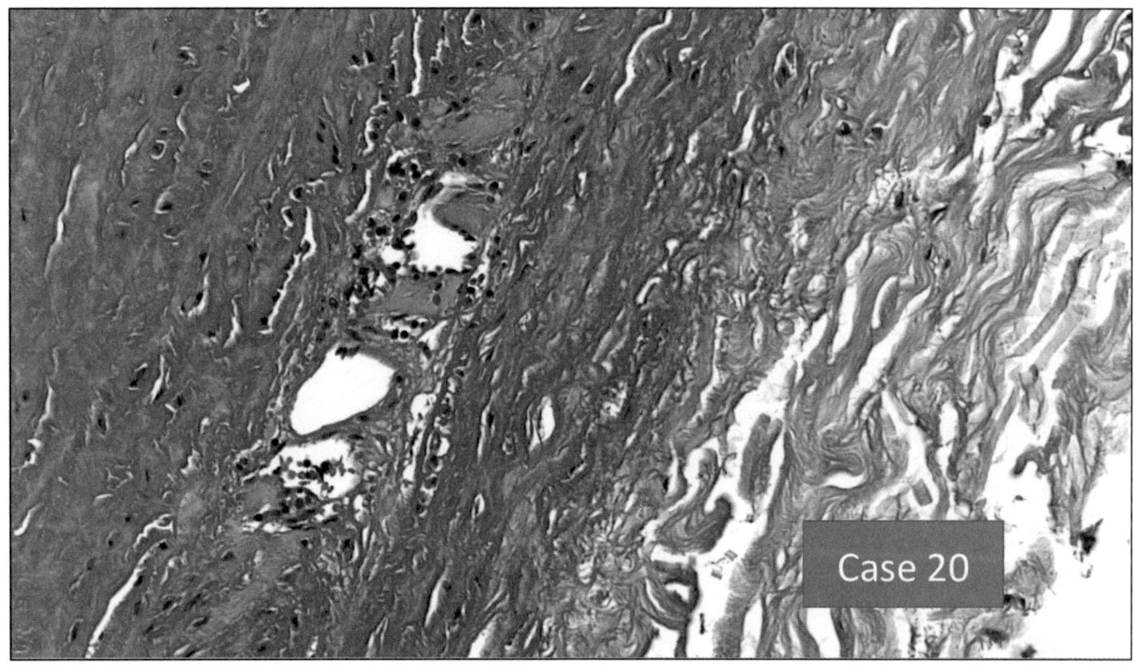

And here another case. We found it in five cases, so this cannot be a coincidence.

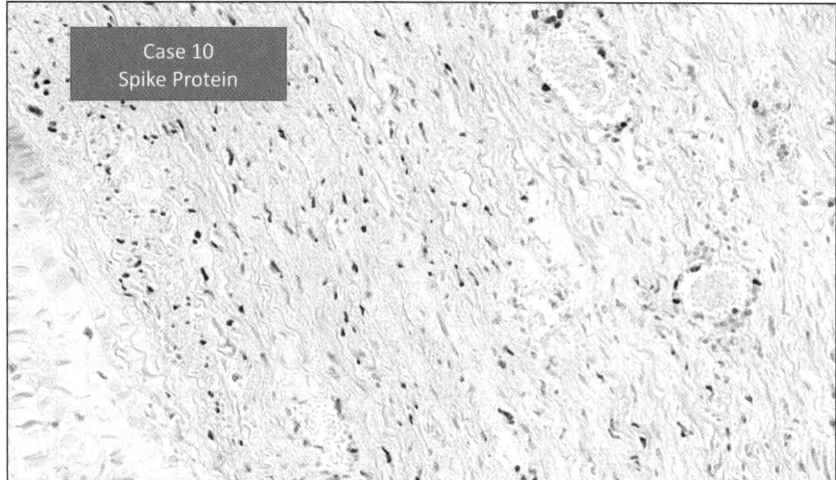

And we did the Spike protein *stain*, and you find a marked positive expermeation *(expression)* of Spike protein in the myofibroblasts of the arterial wall. This is the aorta.

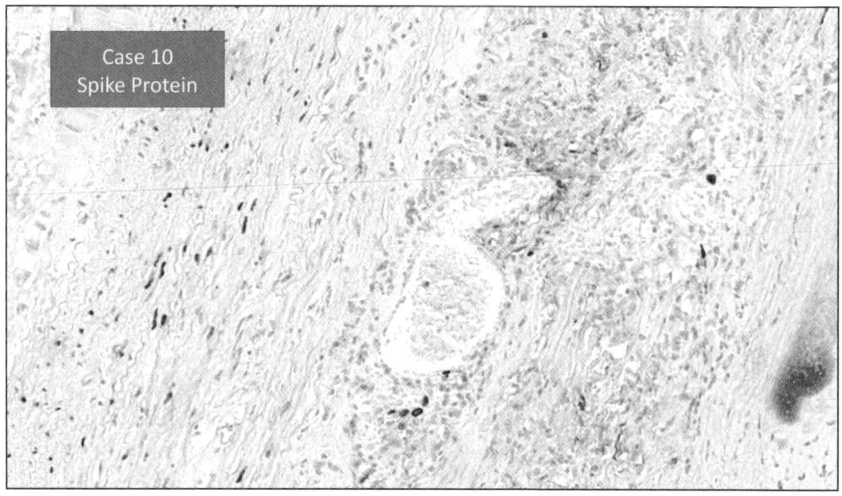

And also, in the vasa vasorum, you can see very strong expression of Spike protein in these areas. And this, I think, is a very important finding.

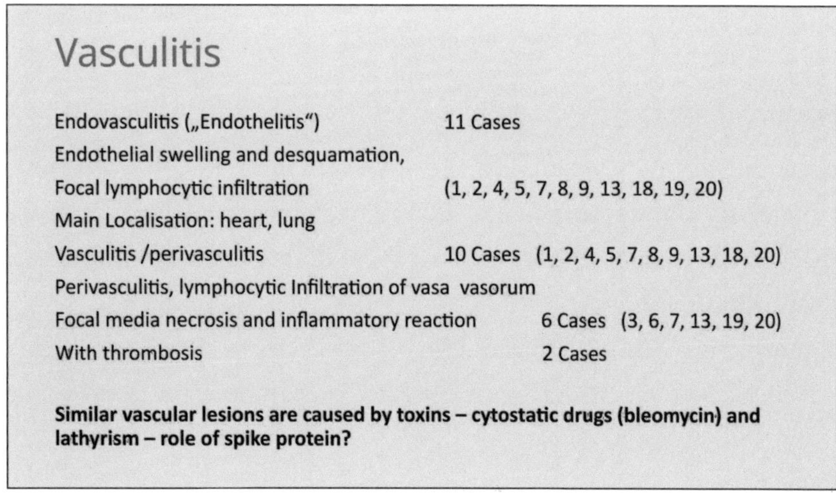

So how often did we see this vasculitis, this endovasculitis as some call it

1. **Endotheliitis, in 11 cases with focal lymphocytic infiltration,** then
2. **Vasculitis perivasculitis in 10 cases,**
3. **Focal media-necrosis in six cases,** and
4. **Thrombosis** caused on this area in **two cases.**

Spleen and Liver

So similar lesions are caused by toxins and drugs, cytostatic drugs, and in some food poisoning like lathyrism. And so, **we think also here a toxic element, the Spike protein might be the causative agent.**

Spleen

- „Onion-Skin" arteriolitis of the spleen arteries as seen in autoimmune disease (Lupus erythematodes)
- Focal destruction of follicular arteries in the spleen with prolapse of lymphatic follicular tissue into the vessel (never seen before?)

Now another, other lesions that we saw in the spleen.

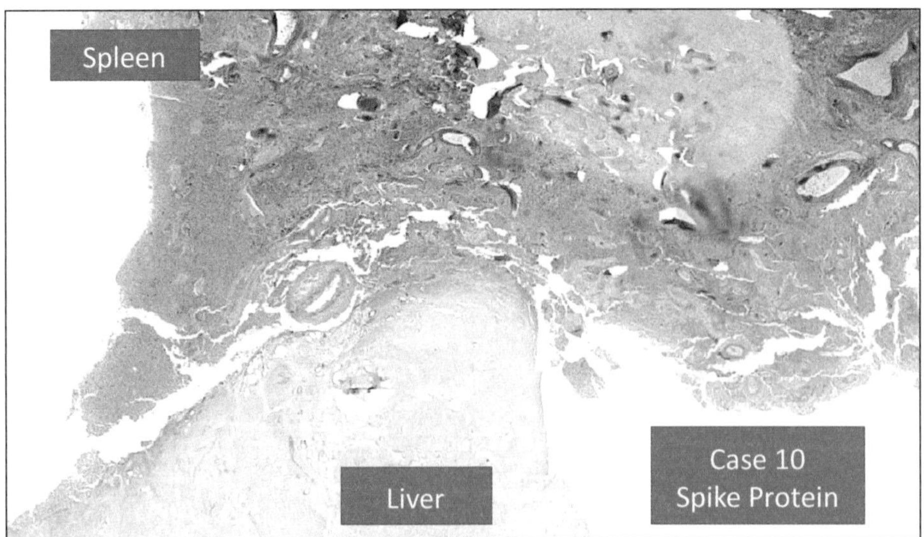

We first overlooked; but the more we looked, the more we found it. And this is one phenomenon that is known as **onionskin arteritis of the spleen,** which is **seen in some autoimmune diseases like lupus erythematosus.** And, in the course of these arteriolitis, we saw **focal destruction of follicular arteries in the spleen** and **products of the lymphatic follicles.**

Now, first of all, this picture is an overview of two organs in one paraffin dock. You can see here the liver and here the spleen. And so, both organs have had the same preparation, the same fixation and everything.

So, you can see easily that the **liver is practically non-reactive**. You can see **some small vessels *in the endothelium are* positive, but the liver itself is negative**.

And then, in contradiction, the spleen has a very pronounced mark around the vessels and small arterioles.

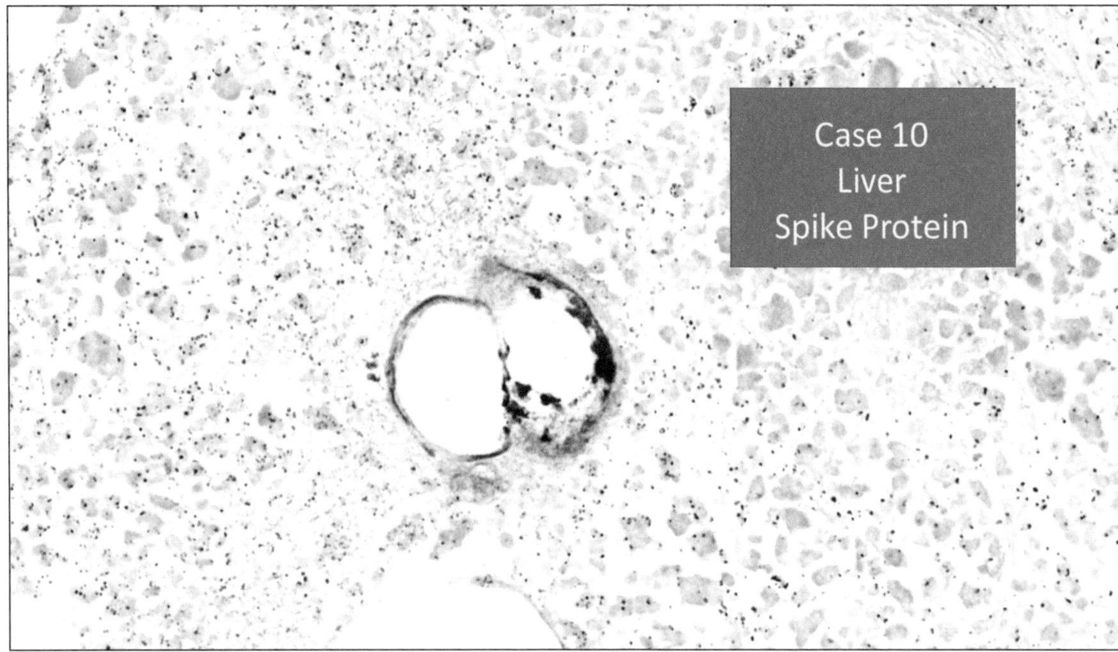

This is a liver now; and you can see the liver cells itself are negative, but the small vessels, the capillaries, have a strong, positive reaction of the endothelium *(indicative of Spike protein)*.

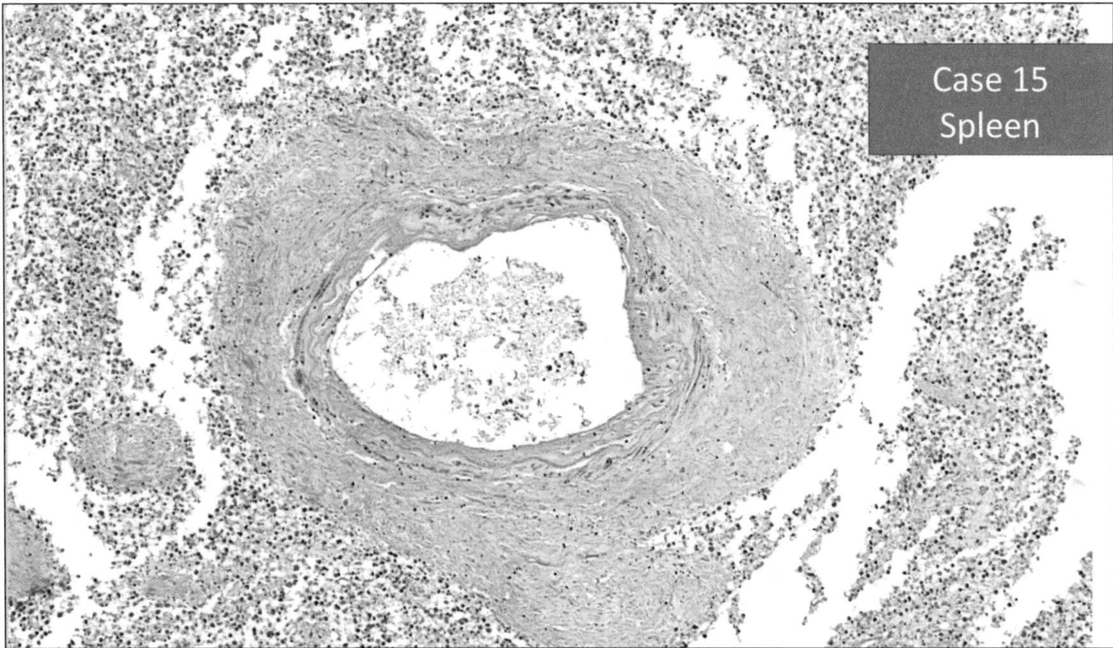

Spleen showing the "onion skin" phenomena (following image) that is seen in some autoimmune diseases. Concentric layering is seen in the thickened arterial wall. Surrounding the artery, hypercellularity is visible with intense nuclear material reaction (blue dots).

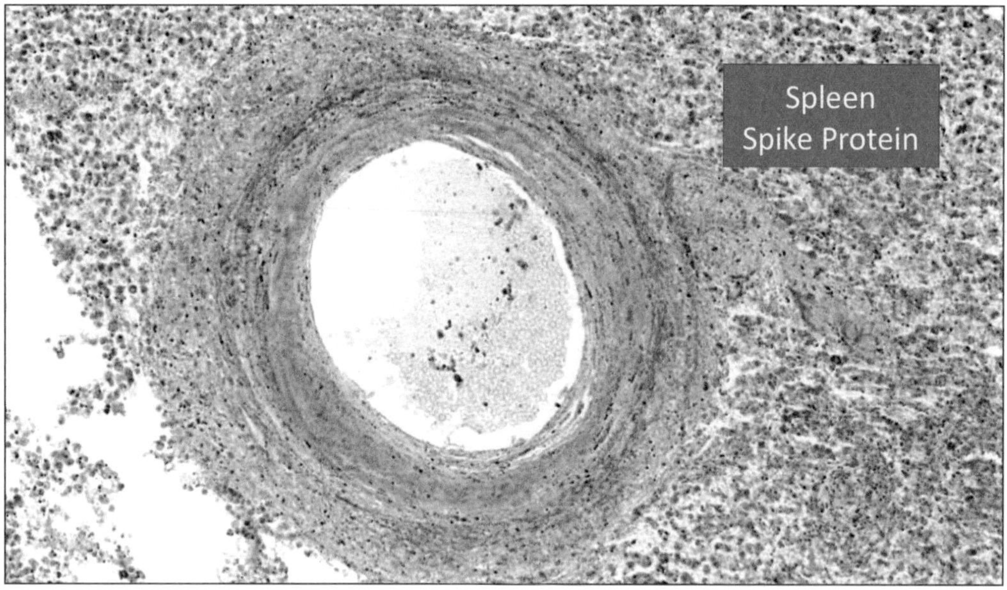

And this is this reaction that is called the onion skin phenomenon, which is seen in some autoimmune diseases. You can see the wall of the artery is split up in, in a way. And also, here we can show, as we saw on the overview that **there's a strong reaction for the Spike protein.**

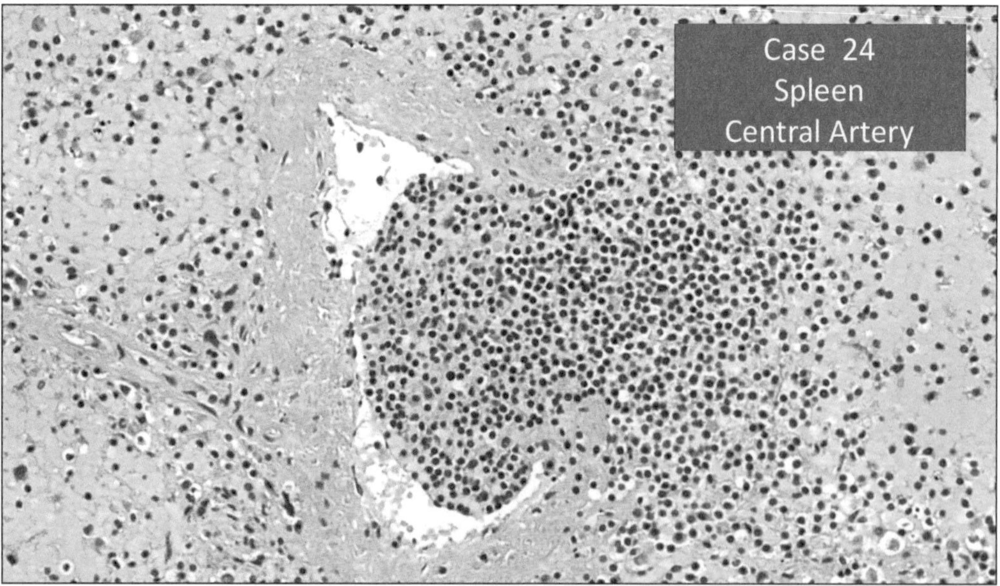

And **this is a phenomenon that none of the pathologists that I work together have ever seen.** This is a **small follicular artery in the spleen**; and you can see the **wall has a focal defect, and the lymphatic tissue is protruding into the vessel**.

Lymph Nodes, Pseudolymphoma, Neoplasia

So, it also changes in the lymph nodes. Densely packed with, possibly, **atypical lymphocytes that are locally invasive suggestive of neoplasia?**

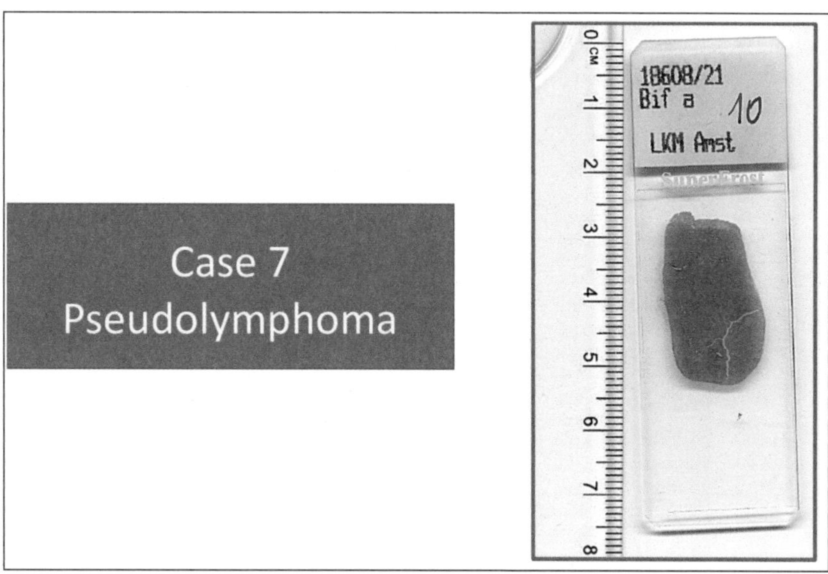

We have seen a case of a pseudolymphoma, as I can show you here. It's at least three centimeters large.

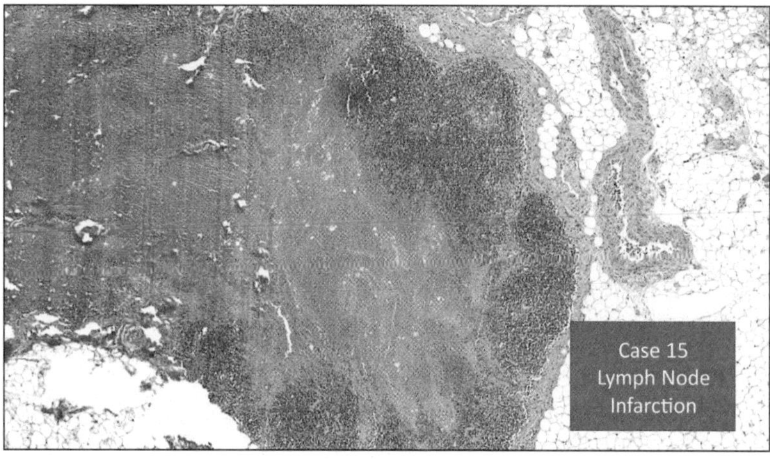

And then, in another case, we saw a focal central infarct of the lymph nodes. Infarction is suggestive of neoplasia, as cellular proliferation is so aggressive that it outstrips the local blood flow leading to infarction.

Heart

Now the **myocarditis** is now, I think, **it's internationally known that it is a side effect of vaccination.**

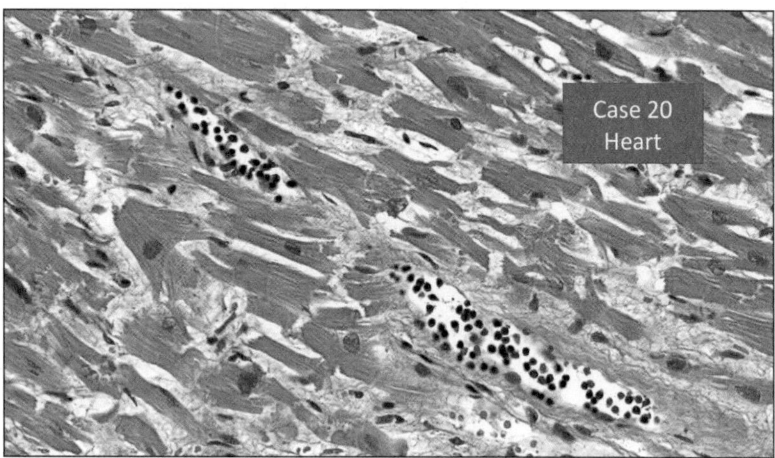

And you can see that (previous image), in our cases, we saw these **lymphocytes marching up in the small vessels here. You can see the intact muscular fibers.**

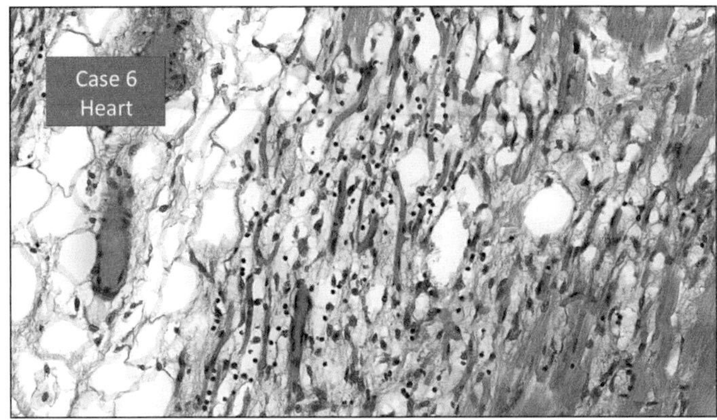

And, in this case, you can see that **they *(cardiac muscle fibers)* are destroyed by the lymphocytes that are infiltrating** in contradiction to true infarct of the cardiac muscles. Do not see granulocytes *(granulocytes are white blood cells also known as polymorphonuclear leukocytes that predominate in cases of non-vaccine-related myocardial infarction)* in these areas. Only very few with some macrophages, of course.

Lung

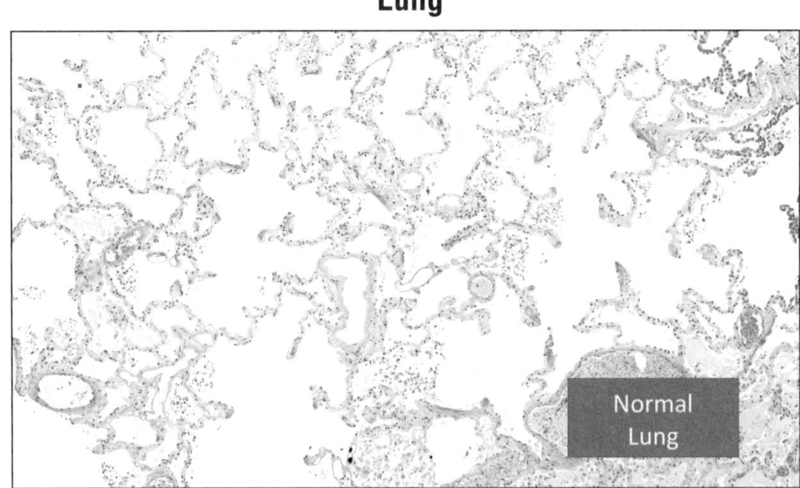

So, we come to the lung here. You see a normal lung. You can see here all the white areas of the lung alveoli.

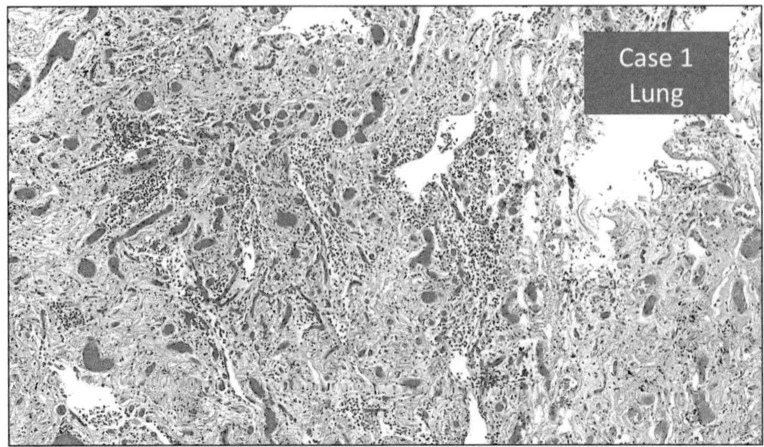

And what we found (previous image) are **very pronounced changes** in which you might call a **lymphocytic alveolitis, lymphocytic interstitial pneumonia**. And you can see that only **very few areas where there are still alveoli.**

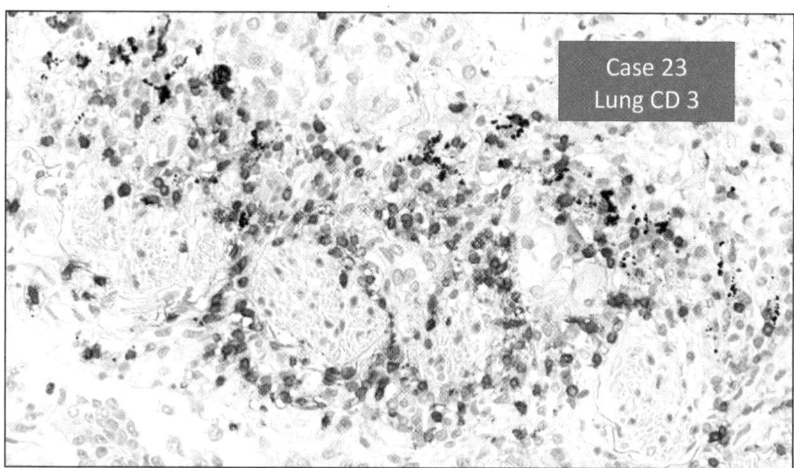

The infiltration is mostly T lymphocytes, CD3+.

Brain

Very important are the changes that we found in the brain.

Main Pathological Findings: Brain

- Transfection-associated encephalitis
- Lymphocytic Infiltration and focal destruction of intracerebral and arachnoidal blood vessels
- Subarachnoidal haemorrhage without aneurysm
- Focal lymphocytic infiltration of the dura mater
- Necrosis of the pituitary

We found a **transfection-associated encephalitis**, then **lymphocytic infiltration**, and focal **destruction of intracerebral and arachnoidal blood vessels**, subarachnoidal hemorrhage, without an aneurysm, in young people, focal **lymphocytic infiltration** is also **in the Dura mater**. In one case**, partial necrosis of the hypophysis** ***(pituitary gland)***.

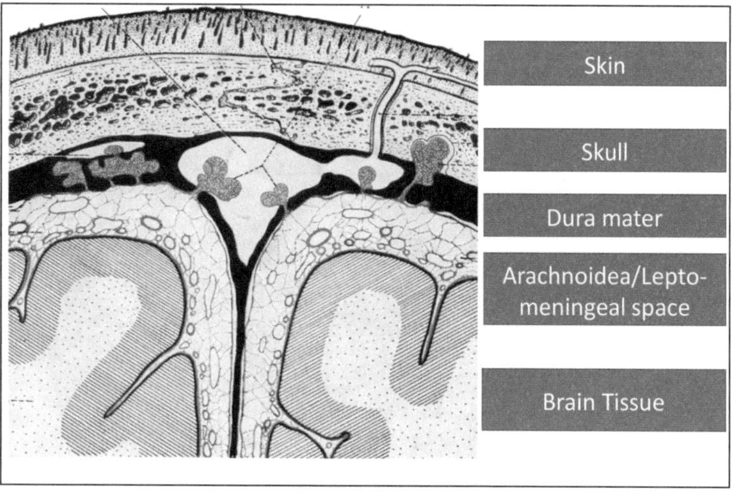

Now this, just for those that are not familiar, this is Dura mater. And here we found infiltration by lymphocytes. This is the **arachnoidea** where we found **perivascular inflammation**, and we **also** found it in the **brain**. This is Dura mater; you can see this focal infiltration by lymphocytes.

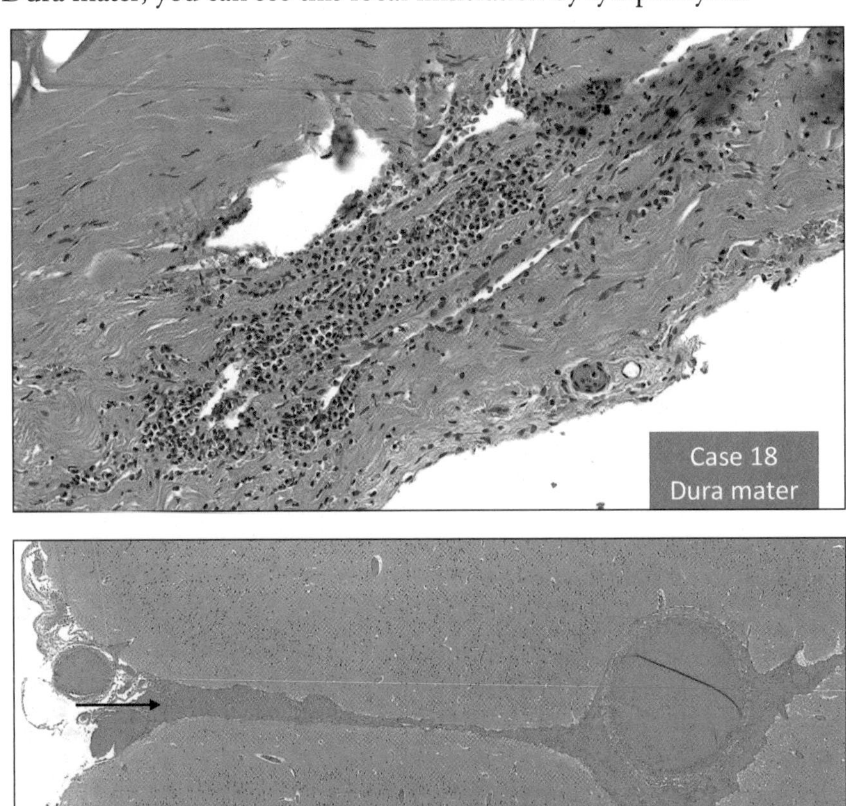

This young 26-year-old died of hemorrhage *(arrow at 9 o'clock).* No aneurysm.

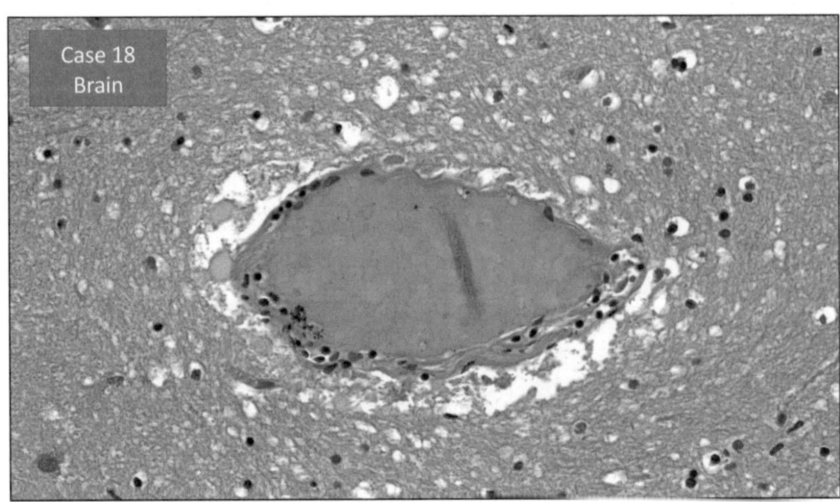

You can see (previous image) that the **vessels in the brain** and in the [inaudible] have a **focal lymphocytic infiltration, and probably that caused a rupture**. Also,

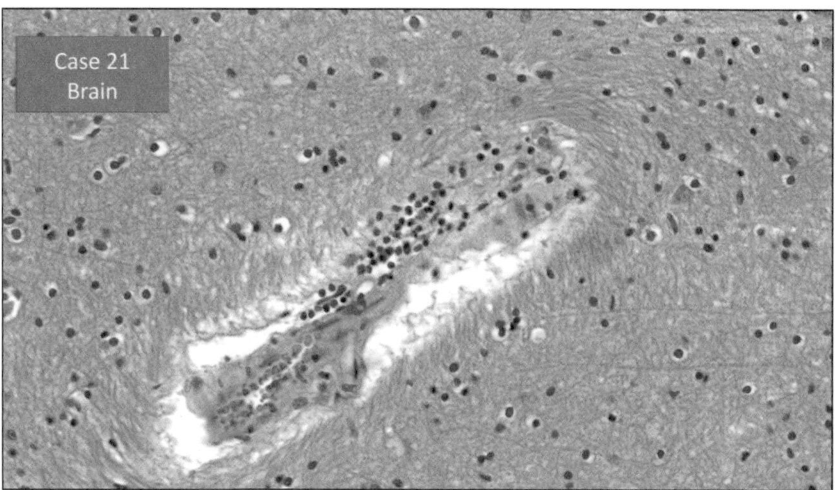

infiltration by lymphocytes.

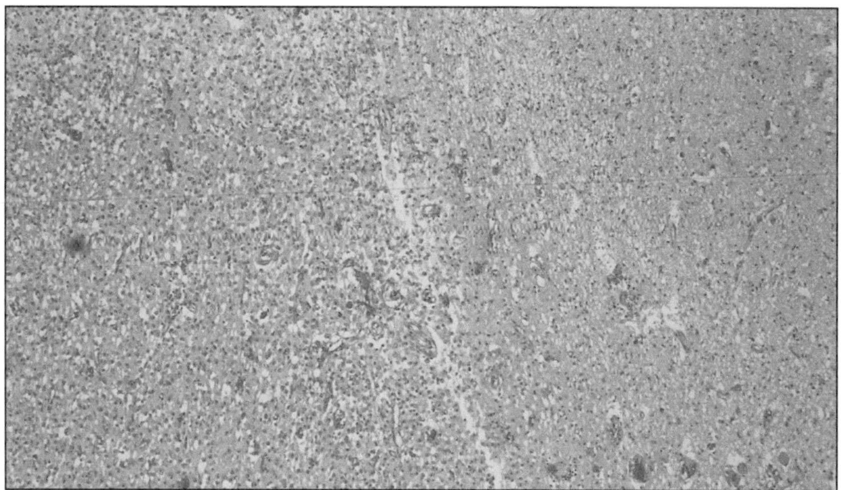

And this is a case of encephalitis which we observed.

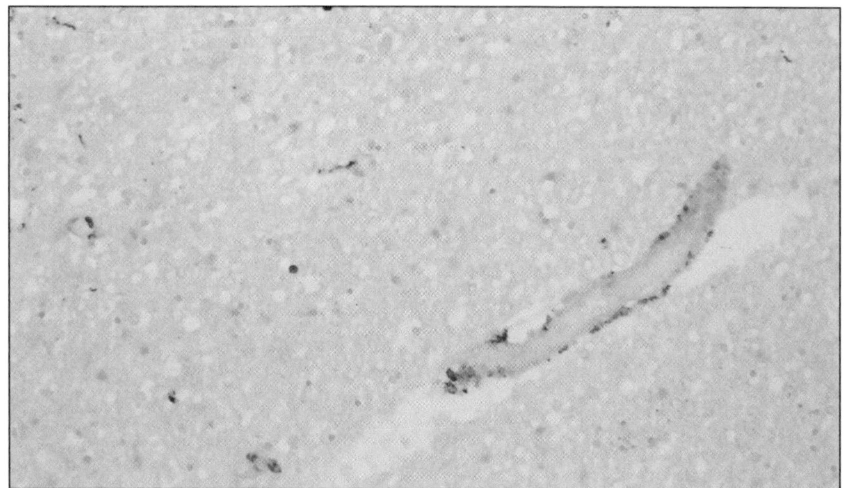

And in this case (previous image), we could demonstrate a **Spike protein**, again, **in the smaller vessels**. And here, a small artery; here, very pronounced, positive reaction.

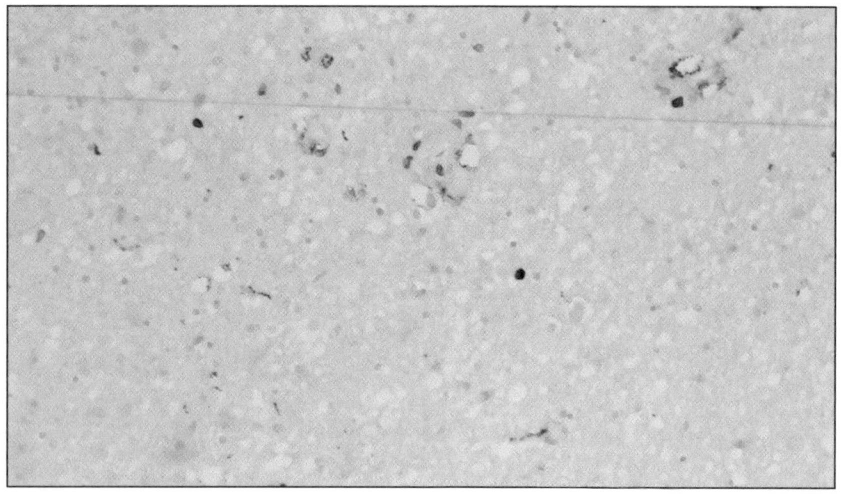

Another area where you can see this definitely and very clear positive cells. It's mostly in the small vessels, but **also in some neural cells**.

Lymphocyte "Bee-Hiving"

Lymphocyte Amok

- Lymphocyte accumulation (lymphocytosis)
- Lymphocyte-predominant tissue destruction („inflammation") with immanent prolonged auto-immune disease - in non-lymphatic organs and tissues
- Related entities with lymphocytosis:
  - Wegener's disease (granulomas, polyangiitis)
  - tropical splenomegaly
  - possible development of malignant lymphoma

And now in a phenomenon that we call the **Lymphocyte Amok, which is a lymphocyte accumulation and lymphocyte predominant tissue destruction outside of the myocardium and the lung** where I've already demonstrated this.

It's definitely the danger of a prolonged autoimmune disease. And this we found in non-lymphatic organs and tissues. There are some autoimmune diseases which are related to this phenomenon.

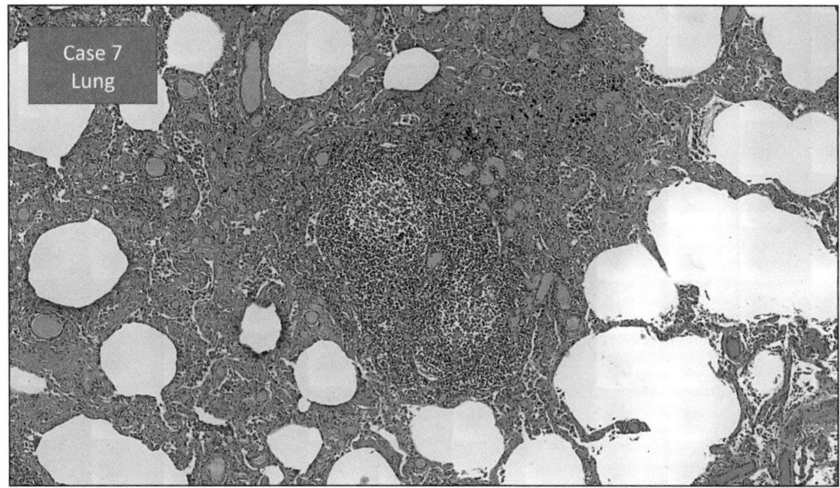

And this is in the lung. In the lung, sometimes you find small lymphocytic elements; but I have never seen before a quasi-lymph node with reaction centers and activation.

Lymphocytic infiltration outside the heart muscle and lung

Thyroid Gland	2 Cases (5, 13)
Salivary Glands	2 Cases (7, 15)
Aorta, larger Vessels	7 Cases (3, 6, 7, 9, 13, 19, 20)
Skin	2 Cases (4, 20)
Liver /NASH	1 Case (8)
Kidney:	3 Cases (12, 14, 19)
Lymphocytic Pyelonephritis/ Nephritis/Ureteritis	1 Case (19)
Testis	1 Case (10)
Dura	1 Case (19)

And this is the frequency that we found of these lymphocytic infiltrations, the thyroid glands, the salivary glands. And by the way, of course, we found this in two cases, but we only had these organ specimens in two cases. So, it was found in 100% of the cases that we examined.

So, in the aorta, I showed you before, skin liver, kidney, lymphocytic pyelonephritis, nephritis, in the testis, in one case, and in the Dura, I showed you this phenomenon.

Fourteen cases with lymphocytic infiltration outside of the lungs:

- Seven cases with aorta and large vessels;
- Three cases with kidney;
- Two cases each with salivary glands, skin, and thyroid;
- One case each with Dura, pyelonephritic/ureteritis, liver/NASH, testis;

One patient (#19) had five sites of lymphocytic infiltration: aorta/vessels, kidney, skin, Dura, lymphocytic pyelonephritis/Ureteritis.

So, these are the organ changes.

Unidentified Foreign Material in Vessels

And, just as the last lesions that I would like to show you, is unidentified foreign material in the vessels, especially in the spleen vessels.

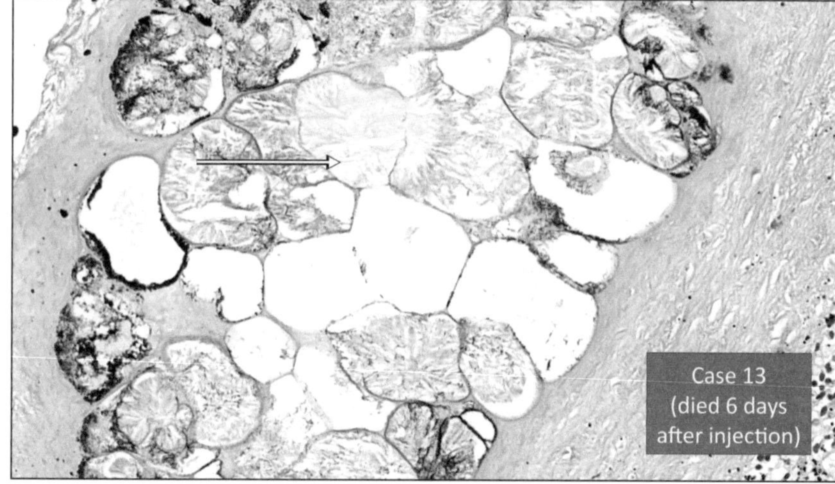

And this is something that we could not identify. *(White arrow at 9 o'clock.)*

No pathologists that have looked at it know what it is.

First, we saw these cells, but they have an inner structure like [inaudible]; and then we thought it might have been a contrast material, but this patient died at home and was not in the hospital for a long time.

And this was one case in the spleen, and this is another after some longer period of time.

So, **our theory is that these are the nanolipid particles, which when they come into the body and are warmed up, which coalesce and form larger particles that might, at one point, stick in the system, in the vessel system.**

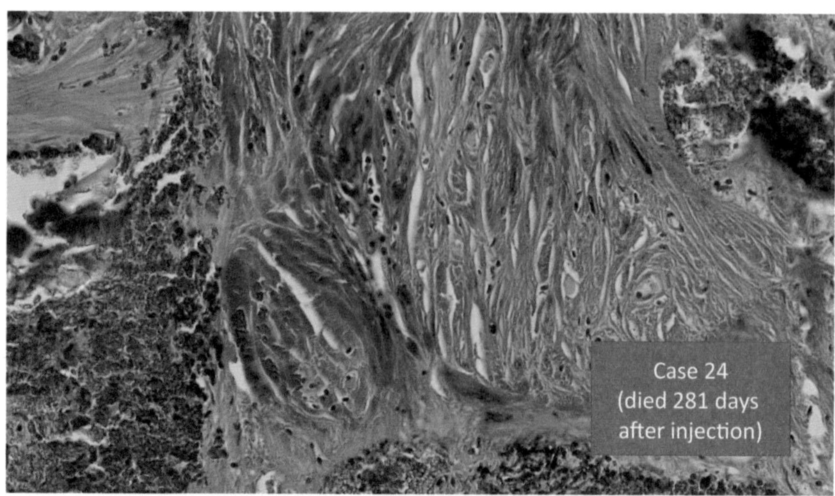

The last one was only a few days, and this is after some longer period of time. I might add that this is a coincidental finding because this was not macroscopically seen, but we had it by coincidence in our sections.

Case Report

I will show you a case report over natural death uncovered as caused by vaccination-induced vasculitis of the coronary artery.

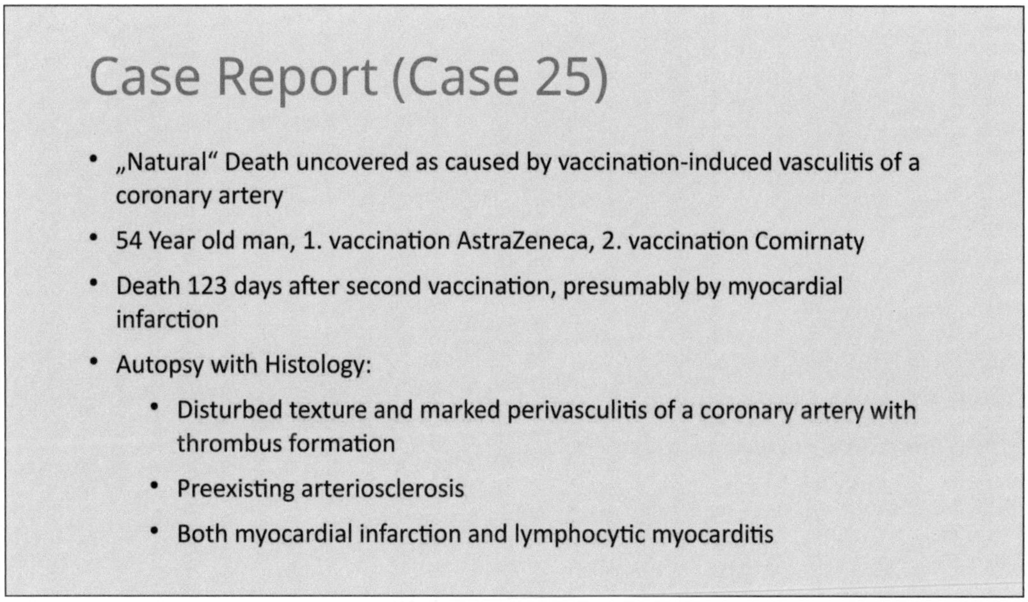

Case Report (Case 25)

- „Natural" Death uncovered as caused by vaccination-induced vasculitis of a coronary artery
- 54 Year old man, 1. vaccination AstraZeneca, 2. vaccination Comirnaty
- Death 123 days after second vaccination, presumably by myocardial infarction
- Autopsy with Histology:
    - Disturbed texture and marked perivasculitis of a coronary artery with thrombus formation
    - Preexisting arteriosclerosis
    - Both myocardial infarction and lymphocytic myocarditis

It was a 54-year old man, two vaccinations, and he died 123 days after the second vaccination.

And there was no doubt this was myocardial infarct. The primary autopsy was *contested.*

They saw a discoloration of the cardiac muscle and said, "Well, yes, he died of a myocardial infarct;" but then we did histology of the coronary artery, and we found these changes.

And, in addition, in the muscle we found the **myocarditis**, as we found typically in the other cases, as I have shown.

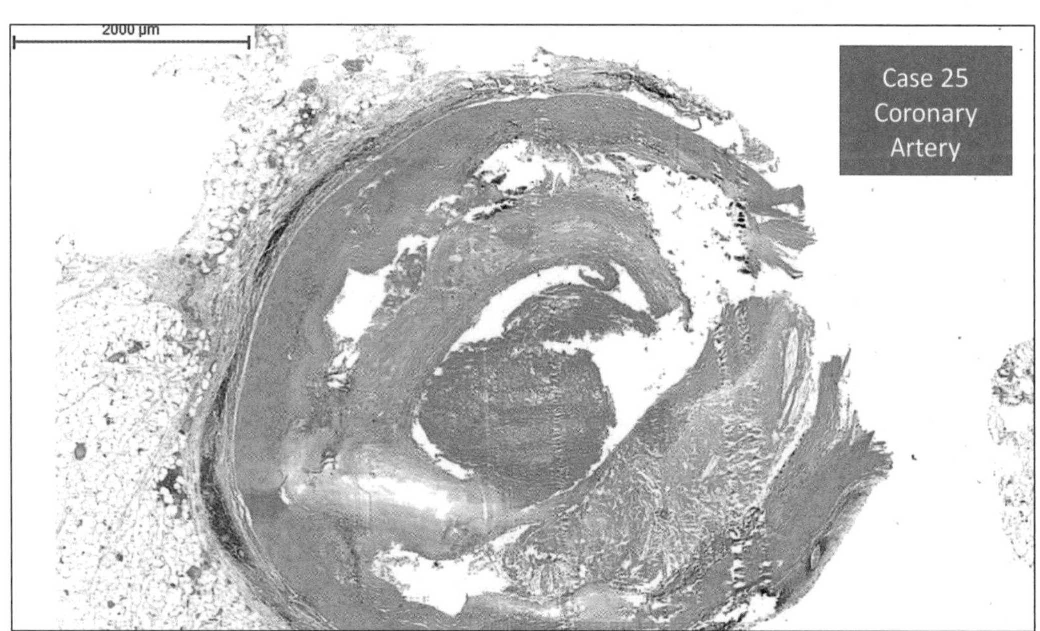

Now this is the coronary artery (previous image) and, yes, you see it's clearly there; *there is* thrombus formation inside. And yes, there are arteriosclerotic changes.

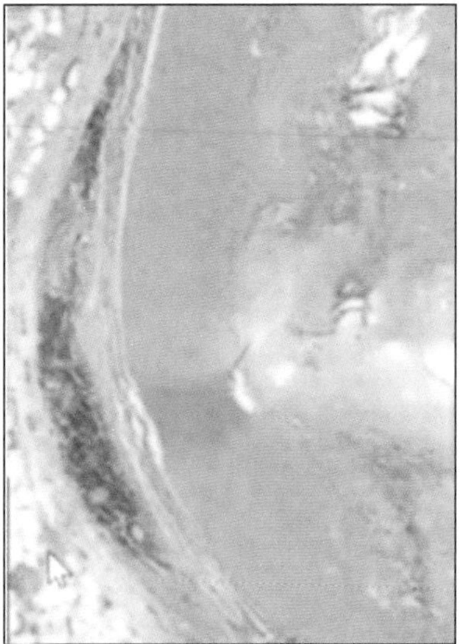

But look at these areas and here around the vessel. There's definitely inflammatory reaction. *(At 8 to 10 o'clock outside the pink thrombus in the coronary artery to the right of the clue staining tissue.)*

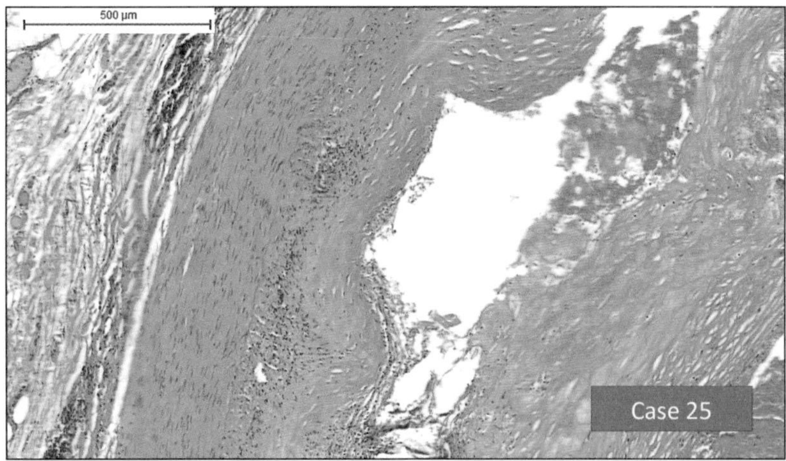

And you can see, this disturbance of texture of the smooth muscle and the myofibroblasts here, and you can see also some discrete lymphocytic infiltration.

And then, around the vessel, there's this dense lymphocytic infiltration.

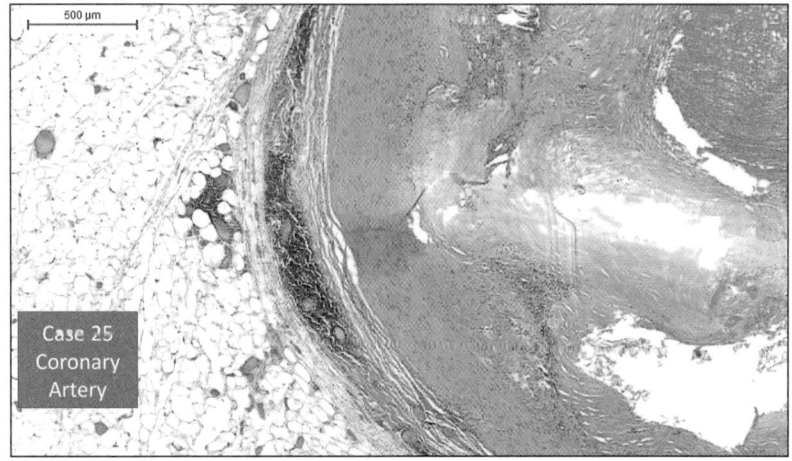

So, we concluded that the vessel, the coronary artery, had an inflammation induced by the vaccination, by the Spike protein (previous image).

And the thrombus was built on the ground of these inflammatory changes and led to the infarct, which definitely was present.

But also, the fact that there was a concomitant, myocarditis, very strong evidence that this secondary, because of the vaccination.

And we spent many, many hours looking at all these slides.

And for a long time, we were thinking, "Well, we are chasing a phantom;" and we looked at each other and ask each other, "Do you see this? Do you see this? Is this real?"

And we are now at a point, and especially after we could prove the presence of Spike protein for months after vaccination, we are come to the conclusion, no, we are not chasing a phantom.

Further studies are necessary, and I think it's a very exciting field that we are coming to, but also a very depressing and very scary phenomenon.

Thank you.

Dr. Mark Trozzi:

So, I'm going to pick two questions. One is very sweet, and it's an anonymous attendee who says, "Professor Burkhart, do I understand that you came out of retirement to help us, the people of the world in need?"

"Well, actually, I was just retiring a few months before I started to study. I gave back my license for the health insurance in April last year. And, in May, I was contacted, and they asked me if I would do this study. And I, by the way, I also have some consulting contracts with some other laboratories.

So, I'm not retired in the way that I am not active anymore, but I'm retired in the sense that I am not responsible to anybody, and that I cannot be thrown out of my job. I'm independent."

An underlying theme of the histopathology of LNP/mRNA harms is involvement of vascular structures, arteries, and veins. Table 1 below identifies vascular involvement in the 20 cases.

Patient	Endovasculitis: Endothelial swelling, Desquamation, Focal Lymphocytic Infiltration, Main Localization: Heart and Lung	Lymphocytic Infiltration of Vasa Vasorum	Focal Media-Necrosis and Inflammatory Reaction	With Thrombosis
1	X	X		
2	X	X		
3			X	
4	X	X		
5	X	X		
6			X	
7	X	X	X	
8	X	X		
9	X	X		
10				
11				
12				
13		X		
14	X		X	
15				
16				
17				
18	X	X		
19			X	
20	X	X	X	
	Vasculitis 11 Cases	Vasculitis/Perivasculitis 10 Cases	Vasculitis/Perivasculitis 6 Cases	2 Cases

Table 1: Histopathology from LNP/mRNA Involving the Vascular System (Burkhardt Series)

Lymphocytes of various types have been identified in proximity of cellular, tissue, and organ damage and may come from activated germinal centers as identified by the Boyd group at Stanford, https://doi.org/10.1016/j.cell.2022.01.018. (Röltgen, Katharina, et al. "Immune Imprinting, Breadth of Variant Recognition, and Germinal Center Response in Human SARS-COV-2 Infection and Vaccination." *Cell*, Elsevier, 24 Jan. 2022, https://www.cell.com/cell/fulltext/S0092–8674(22)00076–9.)

Table 2: Possible Origin of "Activated" Lymphocytes

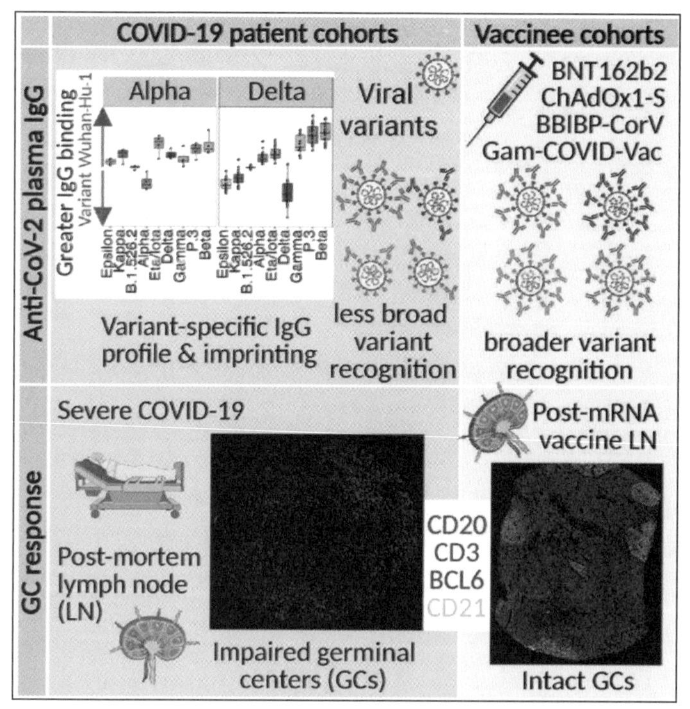

This study found that COVID-19 impairs the lymph node germinal centers but transfection with Spike-based therapeutics was found to stimulate the germinal centers:

> "The biodistribution, quantity, and persistence of vaccine mRNA and spike antigen after vaccination and viral antigens after SARS-CoV-2 infection are incompletely understood but are likely to be major determinants of immune responses."

Presumably, this also means that the activated lymphocytes produced by the ongoing activity of the transfected mRNA have not been studied as well including potential detrimental effects such as autoimmunity from protein mimicry and neoplasia from genome dysregulation.

> "The observed extended presence of vaccine mRNA and spike protein in vaccinee [lymph nodes (LNs)] [germinal centers (GCs)] for up to 2 months after vaccination was in contrast to rare foci of viral spike protein in COVID-19 patient LNs."

This is an important difference between COVID-19 and the Spike-based therapeutic products and at least partially explains the long-time course of harms from the Spike-producing products that range from seven to 180 days in the Burkhardt series (average 45 days).

Table 3: Lymphocyte-Associated Organ System Pathology from Spike-Producing Therapeutics

	Lymphocytic Infiltration Active Cases	14 Cases							
Case	Aorta and Larger Vessels	Kidney	Salivary Glands	Skin	Thyroid Gland	Dura Mater	Lymphocytic Pyelonephritis/Ureteritis	Liver/NASH	Testis
Case 1									
Case 2									
Case 3	X								
Case 4				X	X				
Case 5	X								
Case 6	X								
Case 7			X						
Case 8								X	
Case 9	X								
Case 10									X
Case 11									
Case 12		X							
Case 13	X								
Case 14		X			X				
Case 15			X						
Case 16									
Case 17									
Case 18				X					
Case 19	X	X				X	X		
Case 20	X								
Total	7	3	2	2	2	1	1	1	1

Consider the action of mRNA on the Lymphatic Germinal Centers with activation of lymphocytes while, at the same time, Spike-related foreign proteins are being produced in distant tissues and organs.

Are these activated lymphocytes released from the Germinal Centers and then hunt down and attack organs and tissues producing Spike proteins that are recognized as foreign? Do these activated lymphocytes produce tumors?

Table 4: Organ System Harms from Spike-Producing Therapeutics Burkhardt Series

Organs
Brain
Heart
Kidney
Liver
Lungs
Lymph Node
Salivary Gland
Skin
Spleen
Testis
Thyroid
Vascular

As cases accrue, the number of organs is expected to increase.

Table 5: Histopathology: Mechanisms of Injury

Histology	Definition
Apoptosis	Cell Death
Coagulopathy	Clotting
Infarction	Sudden Loss of Blood Supply
Infiltration	Abnormal Accumulation
Inflammation	Complex Chemical and Cellular Cascade that clears debris, microbes and begins Repair
Lysis	Dissolution
Necrosis	Death
Neoplasia	Pre Cancer or Cancer

Autopsy, Histopathology and Determination of Cause of Death
Case Report:
Hunter Brown, Age 21

"I'm heartbroken that I'll never see him on the football field again, graduate from the Academy, establish a career, get married, or have kids," she wrote. "He was my joy and I loved being his mom. I was so proud of the man he had become. We are so thankful for all the support and prayers we have received, they have definitely eased some of the pain. Please continue to pray for our family."

(https://taskandpurpose.com/news/air-force-academy-cadet-dies/)

The Gazette

Autopsy for Air Force football player Hunter Brown released

BRENT BRIGGEMAN brent.briggeman@gazette.com Feb 1, 2023 Updated May 19, 2023

"A coroner has determined that Air Force Academy Cadet 3rd Class Hunter Brown died of a blood clot in his lungs that was caused by an injury to his left foot that he sustained during football practice weeks earlier, according to a copy of Brown's autopsy report that was provided to Task & Purpose."

The final diagnosis also noted enlarged liver, heart, and spleen for Brown, a Louisiana native who was 6-foot-3 and 292 pounds at the time of his death at the at the age of 21." [https://gazette.com/sports/autopsy-for-air-force-football-player-hunter-brown-released/article_05da22d4-a242–11ed-88b1-b767cb0e502b.html]

Cadet Brown had a fracture dislocation of the base of his second toe that was repaired surgically. Blood clots (superficial or deep venous thrombosis) following surgery for an injury of this type are exceedingly rare and, when they occur, they seldom break free and travel to the lung (pulmonary embolus).

Selby, Rita MBBS, FRCPC, MSc[*,†]; Geerts, William H. MD[*]; Kreder, Hans J. MD, MSc[‡]; Crowther, Mark A. MD, MSc[§]; Kaus, Lisa[*]; Sealey, Faith RN[*]. A Double-Blind, Randomized Controlled Trial of the Prevention of Clinically Important Venous Thromboembolism After Isolated Lower Leg Fractures. Journal of Orthopaedic Trauma 29(5):p 224–230, May 2015. | DOI: 10.1097/BOT.000000000000025

ORIGINAL ARTICLE

A Double-Blind, Randomized Controlled Trial of the Prevention of Clinically Important Venous Thromboembolism After Isolated Lower Leg Fractures

Selby, Rita MBBS, FRCPC, MSc[*,†]; Geerts, William H. MD[*]; Kreder, Hans J. MD, MSc[‡]; Crowther, Mark A. MD, MSc[§]; Kaus, Lisa[*]; Sealey, Faith RN[*]

Author Information ⊙

Journal of Orthopaedic Trauma 29(5):p 224 230, May 2015. | ***DOI:*** 10.1097/BOT.0000000000000250

The primary effectiveness outcome was clinically important venous thromboembolism (CIVTE), defined as the composite of symptomatic venous thromboembolism within 3 months after surgery and asymptomatic proximal deep vein thrombosis on DUS. The primary safety outcome was major bleeding.

Two hundred fifty-eight patients (97%) were included in the primary outcome analysis for effectiveness (130: dalteparin; 128: placebo). ***Incidence of CIVTE in the dalteparin and placebo groups was 1.5% and 2.3%, respectively (absolute risk reduction, 0.8%; 95% confidence interval, -2.0 to 3.0).***

There were no fatal pulmonary emboli or major bleeding. (https://pubmed.ncbi.nlm.nih.gov/25900749/)

Assuming Hunter Brown had an extremely unlikely fatal pulmonary embolus, such an event is sudden and catastrophic with no time to develop associated pathology in the liver, heart, and spleen.

Hunter Brown's medical event and the associated autopsy findings do have an excellent match with the pattern of organ damage from Spike-producing therapeutic agents (SPTA). In a report (following) of what the authors believe was the first autopsy in a Spike-producing therapeutic associated fatality, nine different organs or tissues were damaged by the LNP/mRNA.

ELSEVIER

Contents lists available at ScienceDirect

International Journal of Infectious Diseases

journal homepage: www.elsevier.com/locate/ijid

INTERNATIONAL SOCIETY FOR INFECTIOUS DISEASES

Short Communication

First case of postmortem study in a patient vaccinated against SARS-CoV-2

Torsten Hansen[a,*], Ulf Titze[a], Nidhi Su Ann Kulamadayil-Heidenreich[b], Sabine Glombitza[c], Johannes Josef Tebbe[b], Christoph Röcken[d], Birte Schulz[a], Michael Weise[b], Ludwig Wilkens[c]

[a] *Institute of Pathology, University Hospital OWL of the University of Bielefeld, Campus Lippe, Detmold, Germany*

[b] *Department of Internal Medicine, Gastroenterology and Infectious Medicine, University Hospital OWL of the University of Bielefeld, Campus Lippe, Detmold, Germany*

[c] *Institute of Pathology, KRH Hospital Nordstadt, Hannover, Germany*

[d] *Institute of Pathology of the University of Schleswig-Holstein, Campus Kiel, Germany*

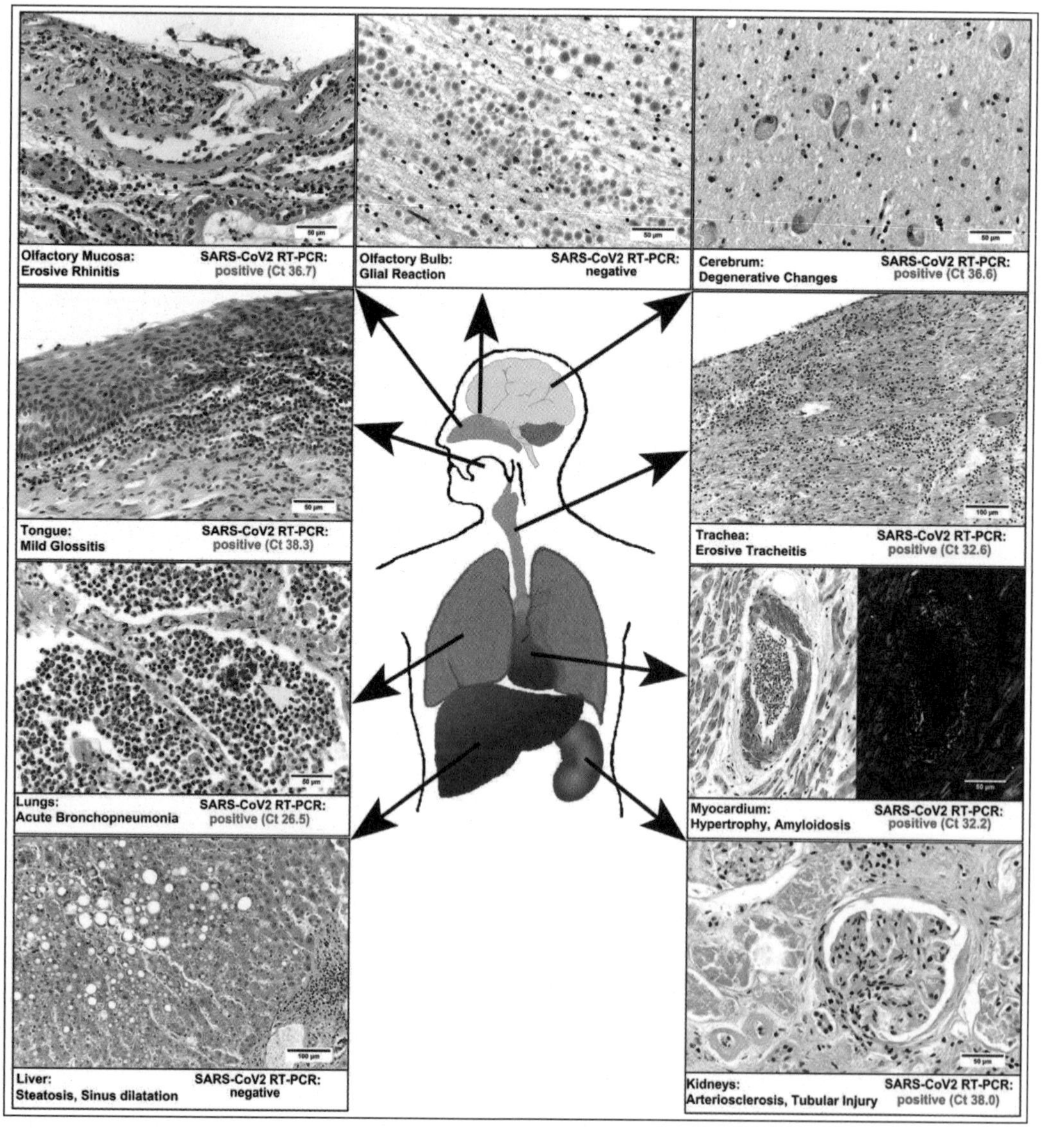

Olfactory Mucosa: Erosive Rhinitis — SARS-CoV2 RT-PCR: positive (Ct 36.7)

Olfactory Bulb: Glial Reaction — SARS-CoV2 RT-PCR: negative

Cerebrum: Degenerative Changes — SARS-CoV2 RT-PCR: positive (Ct 36.6)

Tongue: Mild Glossitis — SARS-CoV2 RT-PCR: positive (Ct 38.3)

Trachea: Erosive Tracheitis — SARS-CoV2 RT-PCR: positive (Ct 32.6)

Lungs: Acute Bronchopneumonia — SARS-CoV2 RT-PCR: positive (Ct 26.5)

Myocardium: Hypertrophy, Amyloidosis — SARS-CoV2 RT-PCR: positive (Ct 32.2)

Liver: Steatosis, Sinus dilatation — SARS-CoV2 RT-PCR: negative

Kidneys: Arteriosclerosis, Tubular Injury — SARS-CoV2 RT-PCR: positive (Ct 38.0)

https://www.ncbi.nlm.nih.gov/pmc/articles/PMC8051011/pdf/main.pdf

Conclusions from Drs. Bhakdi and Burkhardt

- **"Histopathologic analyses show clear evidence of vaccine-induced, autoimmune-like pathology in multiple organs.**
- **That myriad adverse events deriving from such auto-attack processes must be expected to very frequently occur in all [COVID-vaccinated] individuals, particularly following booster injections.**
- **Beyond any doubt, injection of gene-based COVID-19 vaccines place lives under threat of illness and death.**
- **We note that both mRNA and vector-based vaccines are represented among these cases, as are all four major manufacturers."**

(Bhakdi and Burkhardt, "On COVID vaccines: why they cannot work, and irrefutable evidence of their causative role in deaths after vaccination." Transcript from Live Streamed presentations at the Doctors for COVID Ethics Symposium December 10, 2021. https://doctors4covidethics.org/on-covid-vaccines-why-they-cannot-work-and-irrefutable-evidence-of-their-causative-role-in-deaths-after-vaccination/)

Six: "542 Neurological Adverse Events, 95% Serious, in First 90 Days of Pfizer mRNA Vaccine Rollout. 16 Deaths. Females Suffered AEs More Than Twice As Often As Males."

—Joseph Gehrett, MD; Barbara Gehrett, MD; Chris Flowers, MD; and Loree Britt

The WarRoom/DailyClout Pfizer Documents Analysis Project Post-Marketing Group (Team 1)—Barbara Gehrett, MD; Joseph Gehrett, MD; Chris Flowers, MD; and Loree Britt—produced an alarming review of the neurological System Organ Class (SOC) adverse events found in Pfizer document *5.3.6 Cumulative Analysis of Post-Authorization Adverse Event Reports of PF-07302048 (BNT162B2) Received Through 28-FEB-2021* (a.k.a., "*5.3.6*"). This SOC includes altered function of the brain, spinal cord, or peripheral nerves.

It is important to note that the adverse events (AEs) in the *5.3.6* document were reported to Pfizer for only a 90-day period starting on December 1, 2020, the date of the United Kingdom's public rollout of Pfizer's COVID-19 experimental mRNA "vaccine" product.

Key points in this report include:

- 542 neurological events, **95% of which were serious**, occurred in 501 patients.
- **16 patients died**.
- **50% of events occurred within the first 24 hours** after injection, equating to over 270 events in a single day.
- **69% of the neurological events affected females**, and 31% occurred in males.
- **376 seizures** were reported, twelve of which were **"status epilepticus,"** a rare condition of prolonged seizure or series of seizures that is **life-threatening**.
- **38 cases of multiple sclerosis**.
- **11 cases of transverse myelitis** (*a destructive inflammation of the spinal cord*).
- **10 cases of optic neuritis** (*inflammation of the optic nerve threatening blindness*).
- **24 cases of Guillain-Barré syndrome**, ascending paralysis from nerve inflammation.
- **Three cases of meningitis** (*infection and inflammation of the fluid and membranes surrounding the brain and spinal cord*).
- **Seven cases of encephalopathy** (*any disease of the brain that alters brain function or structure; hallmark is altered mental state*).
- Only adverse events that occurred *two or more times* are specifically reported in the diagnoses list. There were twenty events that happened once and, thus, were not included.

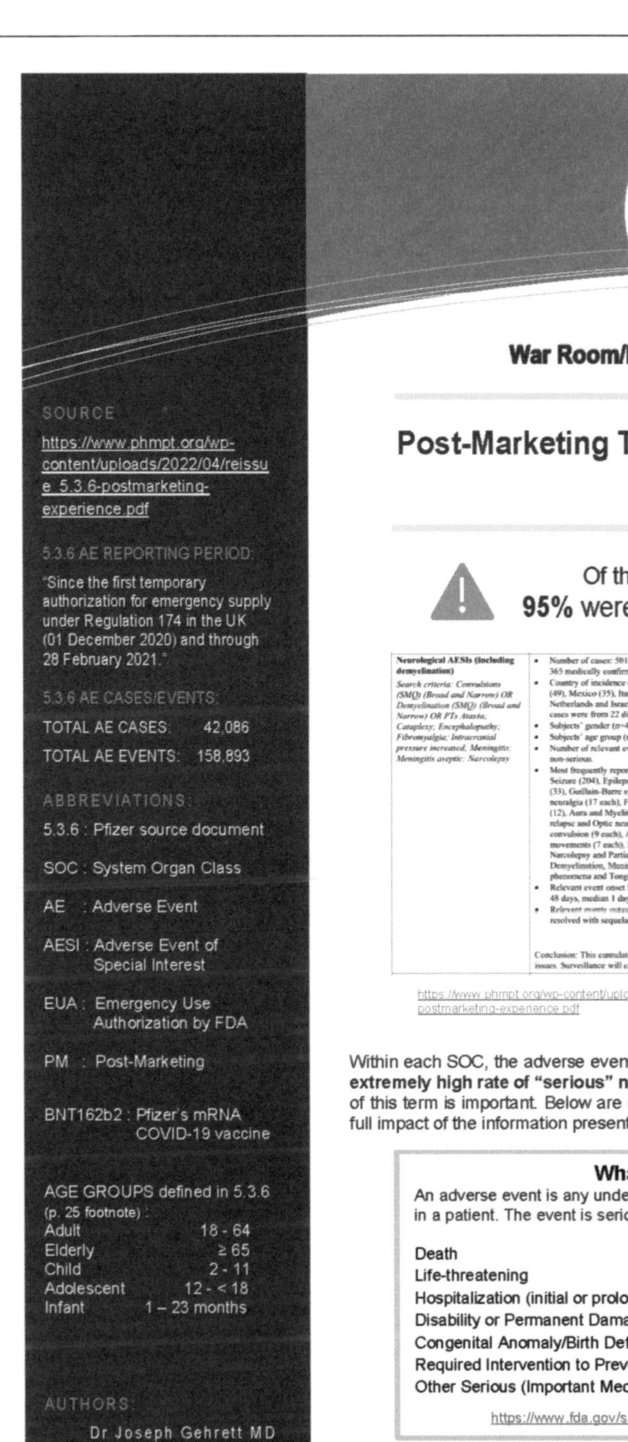

SOURCE

https://www.phmpt.org/wp-content/uploads/2022/04/reissue_5.3.6-postmarketing-experience.pdf

5.3.6 AE REPORTING PERIOD:

"Since the first temporary authorization for emergency supply under Regulation 174 in the UK (01 December 2020) and through 28 February 2021."

5.3.6 AE CASES/EVENTS:

TOTAL AE CASES: 42,086

TOTAL AE EVENTS: 158,893

ABBREVIATIONS:

5.3.6 : Pfizer source document

SOC : System Organ Class

AE : Adverse Event

AESI : Adverse Event of Special Interest

EUA : Emergency Use Authorization by FDA

PM : Post-Marketing

BNT162b2 : Pfizer's mRNA COVID-19 vaccine

AGE GROUPS defined in 5.3.6 (p. 25 footnote) :

Adult	18 - 64
Elderly	≥ 65
Child	2 - 11
Adolescent	12 - < 18
Infant	1 – 23 months

AUTHORS:

Dr Joseph Gehrett MD
Dr Barbara Gehrett MD
Dr Chris Flowers MD
Loree Britt

Post Marketing Team

19Feb23

War Room/DailyClout Pfizer Document Analysis

Post-Marketing Team Micro-Report 7:

Neurologic SOC Review of 5.3.6

Of the 542 neurological adverse events
95% were defined as serious. **16** were fatal.

Neurological AESIs (including demyelination)	
Search criteria: Convulsions (SMQ) (Broad and Narrow) OR Demyelination (SMQ) (Broad and Narrow) OR PTs Ataxia; Cataplexy; Encephalopathy; Fibromyalgia; Intracranial pressure increased; Meningitis; Meningitis aseptic; Narcolepsy	• Number of cases: 501 (1.2% of the total PM dataset), of which 365 medically confirmed and 136 non-medically confirmed. • Country of incidence (≥9 cases): UK (157), US (68), Germany (49), Mexico (35), Italy (31), France (25), Spain (18), Poland (17), Netherlands and Israel (15 each), Sweden (9). The remaining 71 cases were from 22 different countries. • Subjects' gender (n=478): female (328), male (150). • Subjects' age group (n=478): Adult (329), Elderly (149); • Number of relevant events: 542, of which 515 serious, 27 non-serious. • Most frequently reported relevant PTs (>2 occurrences) included: Seizure (204), Epilepsy (83), Generalised tonic-clonic seizure (33), Guillain-Barre syndrome (24), Fibromyalgia and Trigeminal neuralgia (17 each), Febrile convulsion, (15), Status epilepticus (12), Aura and Myelitis transverse (11 each), Multiple sclerosis relapse and Optic neuritis (10 each), Petit mal epilepsy and Tonic convulsion (9 each), Ataxia (8), Encephalopathy and Tonic clonic movements (7 each), Foaming at mouth (5), Multiple sclerosis, Narcolepsy and Partial seizures (4 each), Bad sensation, Demyelination, Meningitis, Postictal state, Seizure like phenomena and Tongue biting (3 each); • Relevant event onset latency (n = 423): Range from <24 hours to 48 days, median 1 day; • Relevant events outcome: fatal (16), resolved/resolving (265), resolved with sequelae (13), not resolved (89) and unknown (161); Conclusion: This cumulative case review does not raise new safety issues. Surveillance will continue

https://www.phmpt.org/wp-content/uploads/2022/04/reissue_5.3.6-postmarketing-experience.pdf

- **Adverse Events were reported to Pfizer during a 90-day period, following the December 1, 2020, public rollout of its COVID-19 experimental "vaccine" product.**
- **In the Pfizer 5.3.6 document, these AEs were categorized by System Organ Classes (SOC) – in other words, by systems in the body.**
- **In the neurologic SOC, of those patients whose sex was reported, 69% were female and 31% were male.**
- **Of the 478 subjects with age reported, 329 were adult and 149 were elderly.**

Within each SOC, the adverse events are further classified as either "serious" or "non-serious." **Given the extremely high rate of "serious" neurological adverse events**, an understanding of the FDA's definition of this term is important. Below are excerpts from the official FDA website. Provided with this context, the full impact of the information presented in this report can be realized.

What is a Serious Adverse Event?

An adverse event is any undesirable experience associated with the use of a medical product in a patient. The event is serious and should be reported to FDA when the patient outcome is:

Death
Life-threatening
Hospitalization (initial or prolonged)
Disability or Permanent Damage
Congenital Anomaly/Birth Defect
Required Intervention to Prevent Permanent Impairment or Damage (Devices)
Other Serious (Important Medical Events)

https://www.fda.gov/safety/reporting-serious-problems-fda/what-serious-adverse-event

This category includes conditions of altered function of the brain, spinal cord, or peripheral nerves (nerves that connect to the spinal cord and extend to the rest of the body). Also included are conditions resulting from direct damage to nerve tissue. Pfizer chose to report fibromyalgia in this category. However, those conditions categorized by Pfizer under the general term peripheral neuropathy (abnormal nerve function) are reported separately in the SOC of "Immune-related/Autoimmune" adverse events. "Polyneuropathy" (multiple nerve dysfunction) is categorized under the SOC "Musculoskeletal" adverse events. The distribution of these diagnoses into various other SOCs is medically debatable. Bell's palsy with facial nerve damage is summarized in its own report.

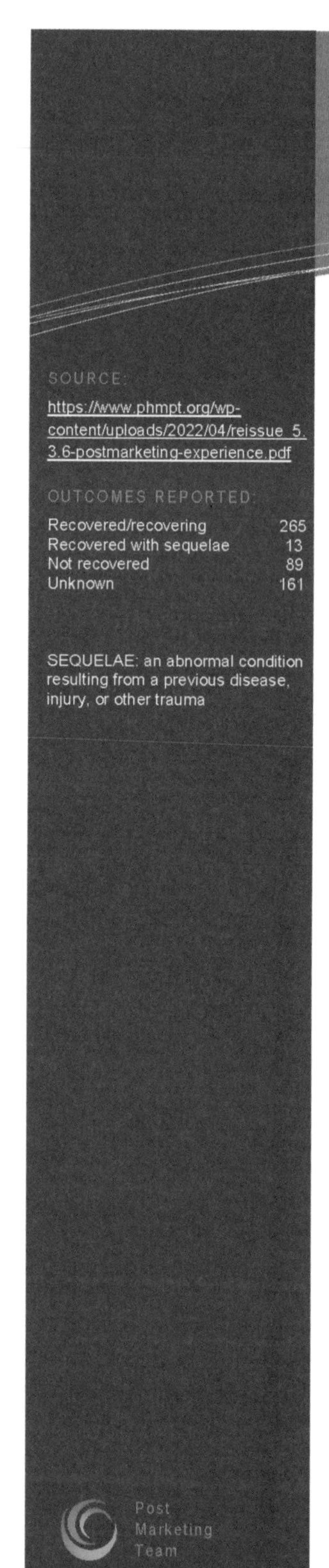

SOURCE:

https://www.phmpt.org/wp-content/uploads/2022/04/reissue_5.3.6-postmarketing-experience.pdf

OUTCOMES REPORTED:

Recovered/recovering	265
Recovered with sequelae	13
Not recovered	89
Unknown	161

SEQUELAE: an abnormal condition resulting from a previous disease, injury, or other trauma

To recap, **95%** of the adverse events were considered **serious**. **16 resulted in death**. Our review found the largest number of diagnoses were **seizures including generalized, petit mal and febrile.** There are **376** (69%) events of these types though notably **12** (3%) were **status epilepticus** (sustained seizures). This particular diagnosis deserves a full explanation here.

> **What is status epilepticus?**
> A seizure that lasts longer than 5 minutes, or having more than 1 seizure within a 5 minutes period, without returning to a normal level of consciousness between episodes is called status epilepticus. This is a medical emergency that may lead to permanent brain damage or death.
>
> https://www.hopkinsmedicine.org/health/conditions-and-diseases/status-epilepticus

Pfizer reports 15 febrile seizures. These are considered a pediatric event, but Pfizer reported no children in this SOC. Were these seizures erroneously listed as febrile seizures or did Pfizer not report ages properly?

Another 29 events (5%) include disturbances of sensation or function including aura, narcolepsy, and "bad sensation." **Three meningitis cases** and **seven cases of encephalopathy** (disordered brain function often resulting in confusion or disturbance of mental clarity) were diagnosed.

Guillain-Barré syndrome, often described as immune attack on nerves with paralysis starting in the legs and ascending at times to the chest with inability to breathe, was diagnosed in **24** (4%). There were eight (1.5%) with **disturbances of walking**. Seventeen (3%) had **trigeminal neuralgia**, an extremely painful facial condition. **Multiple sclerosis** and other cases of damage to the sheath surrounding some nerve cells, typically found to be from an immune system attack, comprised **38** (7%) of the events. Seventeen (3%) cases of fibromyalgia were reported. Clinical diagnoses were specified only if there were two or more reports, so 20 diagnoses or syndromes were not given to us. What were they?

The neurologic adverse events above are **in addition to the strokes** reported in a separate report. This set of patients has suffered an equally serious group of lethal or potentially disabling disorders. We see double-digit fatalities in the first 90 days with a litany of dreaded new diagnoses among which are multiple sclerosis, seizures, Guillain-Barré, meningitis, and encephalopathy. **And half of the total events occurred within 24 hours of receiving the injection. In addition, at least 250 of the non-fatal events have no documented recovery.**

Pfizer's conclusion?

"This cumulative case review does not raise new safety issues. Surveillance will continue."

Post-Marketing Team's CONCLUSION
RECALL **this unsafe "vaccine."**

Post Marketing Team

Seven: "Part 2—'Autopsies Reveal Medical Atrocities of Genetic Therapies Being Used Against a Respiratory Virus'"

—Robert W. Chandler, MD, MBA; Michael Palmer, MD

Histopathological reevaluation of serious adverse events and deaths following COVID-19 vaccination

Professor Arne Burkhardt
Pathologist

Reutlingen, Germany

Professor Dr. Arne Burkhardt gave an update on his series of autopsy and biopsy cases associated with Spike associated gene therapy entitled "Histopathological reevaluation of serious adverse events and deaths following COVID-19 vaccination" at the January 2023 Pandemic Strategies: Lessons and Strategies conference in Stockholm Sweden, presented by the Swedish physician group Läkaruppropet. *(The Doctors' Call)* https://lakaruppropet.se/

[Note: A rough draft was prepared using voice recognition then was edited for clarity with an effort to retain Professor Dr. Burkhardt's original meaning. The text was then added to the presentation graphics. Text in italics has been added by the current author.]

Arne Burkhardt, Professor and MD of Pathology studied Medicine at the Universities of Kiel, Munich and Heidelberg. He trained in Pathology at the Universities of Heidelberg and Hamburg (1970–1979) and became Professor of Pathology at the Universities of Hamburg (1979) and Tübingen (1991). He holds a position as an Extraordinarius Emeritus for General and Special Pathology at the University of Bern, Switzerland, and has been practicing as a pathologist in his own laboratory since 2008. Dr. Burkhardt has held guest professorships in numerous universities in the United States (Harvard, Brookhaven), Japan (Nihon), South Korea, and Europe. He has authored more than 150 original publications in international and German medical journals and contributed to textbooks in German, English, and Japanese.

Summary

The Burkhardt Group (TBG) now consists of a 10-member international research team of pathologists, coroners, biologists, chemists, and physicists.

TBG now has 100 autopsy and 20 biopsy cases in various stages of analysis, 51 of which are the subject of this report.

There were 26 men and 25 women; ages ranged from 21 to 94.

Death occurred from seven days to six months after the last injection of Spike-mediated gene therapy. The larger series had one case in which death occurred eight months after the last gene therapy injection.

The deceased received Spike-inducing drugs from four manufacturers: Janssen/Johnson and Johnson, Pfizer/BioNTech, Moderna, and AstraZeneca.

Initial autopsy reports listed cause of death as or "natural" or uncertain in 49/51 cases.

Evaluation consisted of Histology, Special Stains, Immunochemistry, and Advanced physicochemical methods.

Forensic autopsy disclosed that the cause of death at a highly likely or likely level of probability (to a reasonable degree of medical probability) was from Spike-inducing gene therapy products in 80% of cases.

Findings:

[SET I., II., III. lists, + numbered lists below each]

I. General Lesions affecting more than one organ were characterized by:

1. Presence of Spike protein and absence of nucleocapsid protein (SARS-CoV-2 only).
2. Both arterial and venous systems had inflammation of the inner lining of the blood vessel wall.
3. Larger vessels had evidence of inflammation in the elastic fibers of the aorta and larger vessels.
4. Crystals consistent with cholesterol were identified in remote tissues and were thought to have been released from atheromata that were unroofed after the inner arterial lining were disrupted by Spike-caused erosion of the endothelium releasing debris and cholesterol emboli.
5. Abnormal proteinaceous material consistent with amyloid was identified in multiple tissues.
6. Unusual and aggressive cancer was identified and labelled "Turbo Cancer."
7. Atypical "clot" formation was identified.
8. "True" foreign bodies from contaminated vaccine were identified.

II. Specific Organ and Tissue Lesions involving the vascular system were characterized by:

Small Vessels:

1. Heart, lung, and brain had evidence of inflammation of the inner wall of blood vessels (endotheliitis).
2. Evidence of bleeding (hemorrhage).
3. Unusual blood clot formation comprised of amyloid, spike protein, and fibrin.
4. Presence of small blood clots and clot-forming blood cells.
5. Obliteration of blood vessels.

Large Vessels:

1. Disrupted blood vessel wall of the aorta with associated lymphocytic vasculitis and perivasculitis.
2. Damage to the inner lining of blood vessels with "unroofing" of cholesterol filled plaque.
3. Disruption of the inner lining of blood vessels with dissection into the muscular middle layer of major arteries and subsequent dissection and aneurysm formation.

4. Full thickness disruption of the aorta with exsanguination.
5. Thrombotic casts.

III. Main Pathologic Findings (other organs) were characterized by:

1. Myocarditis— lymphocytic infiltration
with/without destruction of muscle fibers
scar formation
2. Alveolitis— diffuse alveolar damage (DAD)
lymphocytic interstitial pneumonia
endogen-allergic?

3. Lymphocytic nodules outside lymphatic organs
association with autoimmune diseases
Lymphocyte—Amok

Dr. Burkhardt: I have to tell you how it all started. . . .

Soon after the first vaccinations were done in Germany, I was approached by relatives whose loved ones had died suddenly after the vaccination. They were autopsied, and the pathologist said, "Well, it's all natural causes." The loved ones didn't believe it. So, they went to other pathologists. They declined to look at these slides. And I was approached if I could give a second opinion.

And I said, "Well, of course I've done this in 40 years of pathology practice, and I will do it." After the first five cases, I realized that this was not an easy task and that it had to change from a second opinion to a scientific project. First of all, I was alone, then joined by Professor Dr. Lang of the University of Hannover. He's also an experienced pathologist.

We started to look many times at these specimens that were sent to us, and which had come from the autopsies done by other pathologists. Now we are all in all 10 pathologists, coroners, biologists, chemists, and physicists that have joined to elucidate these cases.

Autopsy and histopathology studies on adverse events and deaths due to COVID-19 vaccinations, conducted at Reutlingen

- **International team of 10 pathologists, coroners, biologists, chemists, physicists**
- **Studies on 51 deceased and 4 living patients (08/2022) - 100/20 (01/2023)**
  - **26 men, 25 women**
  - **Age range: 21,2 – 94,7 years: median 65,7 years**
  - **Death occurred 7 days to 6 months after most recent injection**
  - **Vaccines: Pfizer/BioNTech 8, Moderna 2, Janssen 1, AstraZeneca 2, unknown 2**

I have to remind you that this is an ongoing examination. In the tables that will follow, I have different collectives because I cannot update every time I give a lecture.

In August 22 (previous studies' summary) we had:

- 51 deceased and four living persons that we examined. (As of January 2023, we have 100 autopsies and 20 biopsies.)
- Of the 51 deceased, we had 26 men and 25 women.
- Age range: 21 to 94.
- Death occurred seven days to six months after the recent injection.

The vaccines are the ones that are usually in Germany. The task was to see if the vaccination had anything to do with the death occurrence.

Death in connection with COVID-19 vaccination

Among the 51 cases that had previously been autopsied:

- **22 cases had been autopsied by a coroner**
- **28 cases had been autopsied by a pathologist**
- **1 case pathologist and coroner**
- **In all but 2 cases, cause of death reported as uncertain or "natural"**
- **1 case „probably caused by vaccination"**

The subsequent detailed examinations at Reutlingen found a causation of death by the vaccine to be

- **highly likely / likely** 80%
- **possible/unclear in 2 cases**
- **ruled out in 1 case**

Among 51 cases:

- 22 cases were autopsies by coroners and usually without histology.
- 20 cases were autopsies by pathologists.
- One case by a pathologist and a coroner.
- In all but two cases the cause of death was reported as "Uncertain" and mostly as "Natural." Whatever natural death is.
- It was stated the death could be possibly related to vaccination in only in one case.

After looking at all these specimens, histological slides in the microscope, we came to the conclusion that in 80% the vaccination had some influence on the death occurrence.

Death, of course, is a complicated occurrence, especially in an older person. But it may be influenced by vaccination, and there is usually a timely correlation. There's one other study (below) at Heidelberg University. They said in 30% that death after vaccination is correlated to the vaccination. (Schwab, et al.)

(*https://www.ncbi.nlm.nih.gov/pmc/articles/PMC9702955/*)

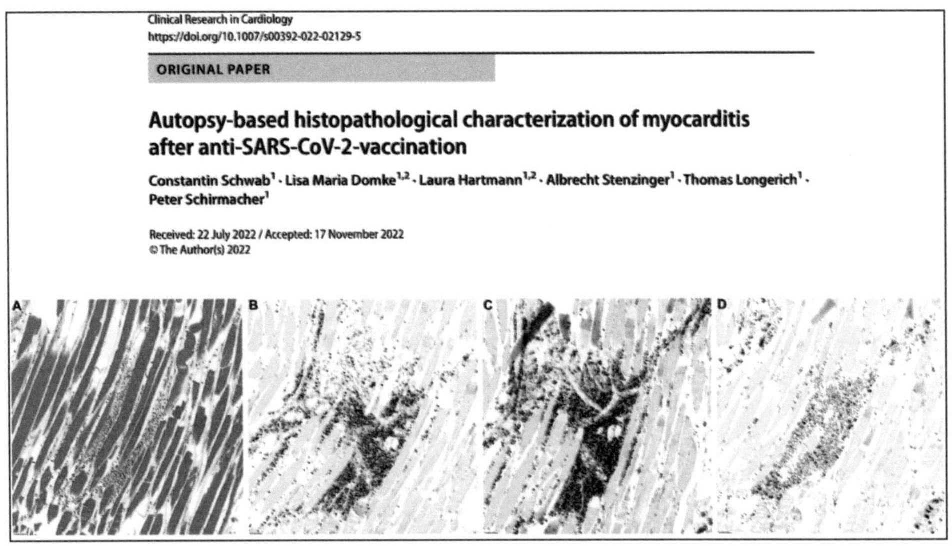
Clinical Research in Cardiology
https://doi.org/10.1007/s00392-022-02129-5

ORIGINAL PAPER

Autopsy-based histopathological characterization of myocarditis after anti-SARS-CoV-2-vaccination

Constantin Schwab[1] · Lisa Maria Domke[1,2] · Laura Hartmann[1,2] · Albrecht Stenzinger[1] · Thomas Longerich[1] · Peter Schirmacher[1]

Received: 22 July 2022 / Accepted: 17 November 2022
© The Author(s) 2022

Places of death (19 cases) **15 of 19 SADS**

Place	Number of cases	Individual cases no.
At home	5	1, 5, 15, 17, 20
In the street	1	11
In the car	1	2
At work	1	13
Assisted living	1	3
Hospital / ICU	4	4, 9, 14, 19
Hospital / short term (≤ 2 days)	4	6, 10, 12, 18
Not yet known	2	7, 8

Most of the patients that we examined had a sudden death. They were found at home, on the street, or in the car. So, we don't have changes caused by treatment like artificial respiration and so on. And of these, 15 of 19, were in the category of what we call now sudden adult death syndrome (SADS). And you may notice this is a new term that has not existed before the vaccination.

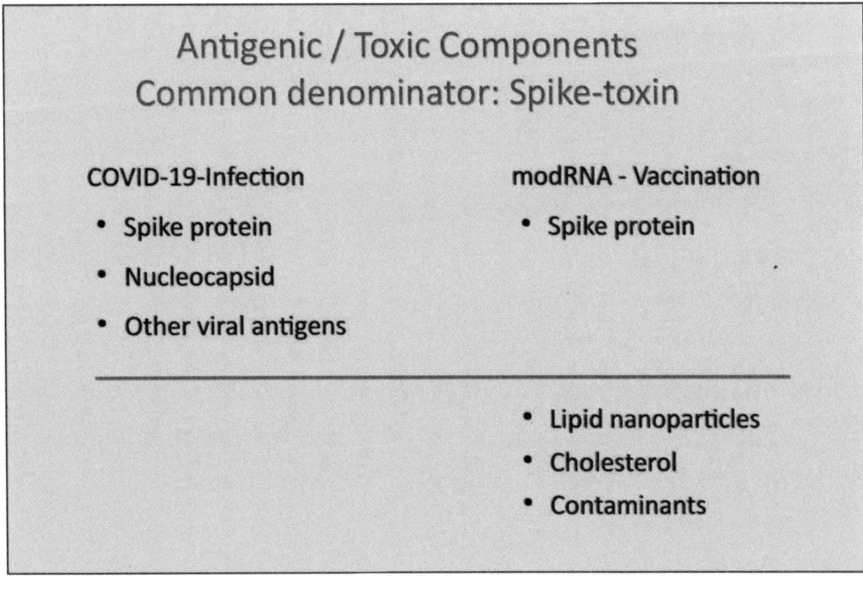

What did we do? First of all, we had to realize that there is a difference between the true Corona infection and the vaccination to the body. The Corona infection has protein and other viral antigens like nucleocapsid, while the vaccination only has a spike protein (previous graphic).

So, we have a common denominator. There is an overlap between the true infection and the vaccination; but, in the vaccination, we have other components that might be of relevance to pathological changes like the lipid nanoparticles, mRNA, cholesterol, and, in some cases, contamination, for example, by metals. The later has not been very closely examined until now.

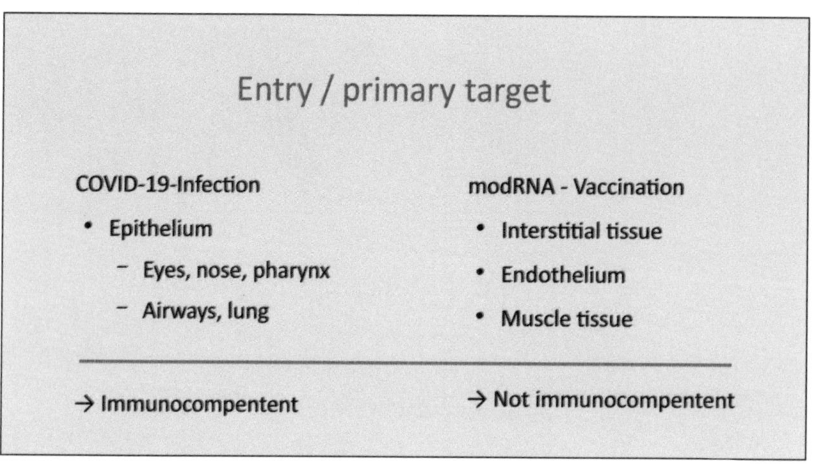

What is the difference between the entry and the primary target of the COVID-19 infection? The true infection goes to the epithelium: eyes, nose, pharynx, airways, and lung. This epithelium is immunocompetent. While, when we inject the vaccine, it goes into the interstitial tissue, into the muscle cells, the endothelium, and usually into the vessels. And these are not immunocompetent.

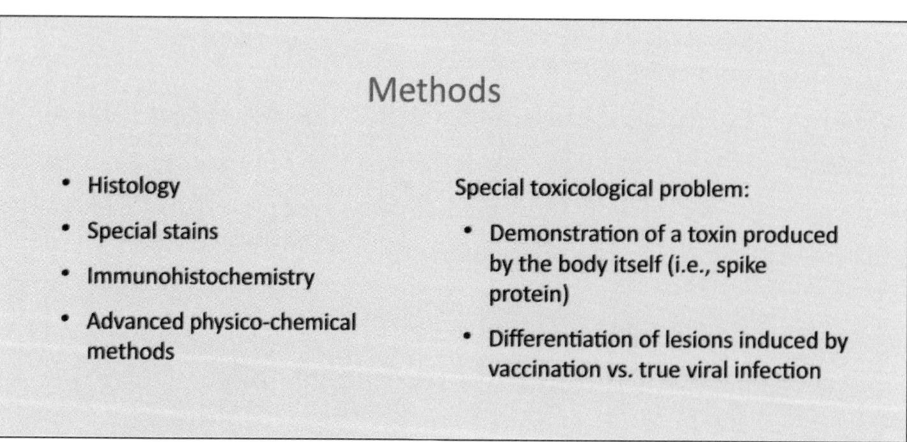

Now, what are the methods that we used? We used common histology, special stains, immunohistochemistry, and, in some cases, advanced physical chemical methods. From the beginning on, we were aware that this is a special challenge to us, because we don't have a toxin that is coming from the outside, either by ingestion or by injection.

For example, systemic or histologic demonstration of toxins would not be of any significance; but we had to demonstrate a toxin that the body itself produced, and this was a Spike protein. And that's why we very soon we developed a method to show the Spike protein in the tissues. *(Below)*

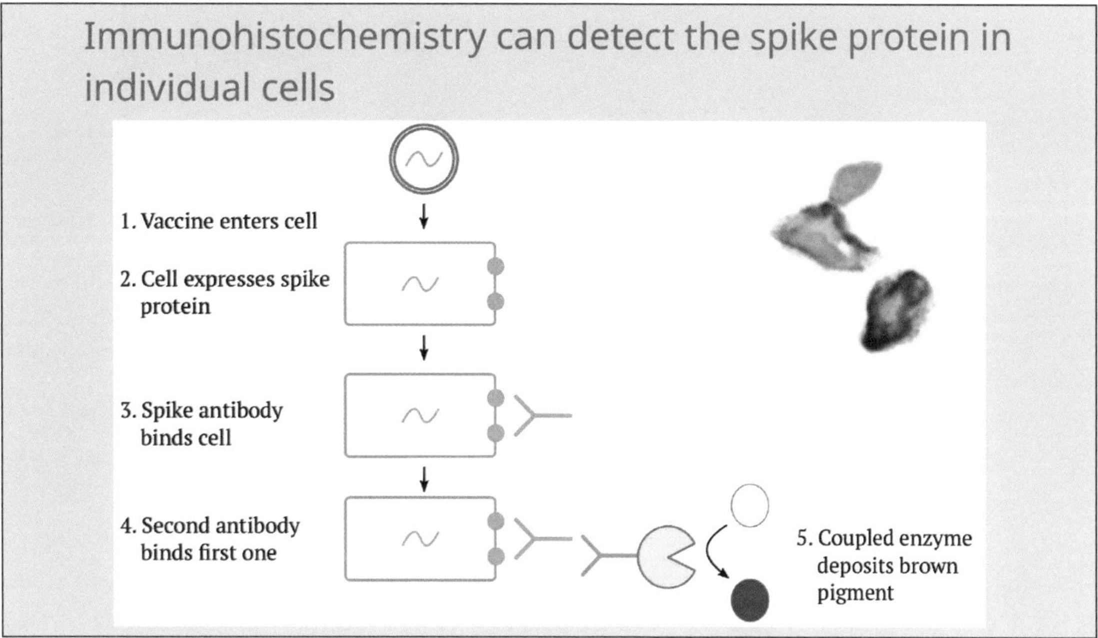

And we differentiated this from the true infection by demonstrating the nucleocapsid antigen.

General Findings

This is the first part of my presentation—general lesions expect affecting more than one organ.

Now, the first thing that we examined was the expression of Spike protein in the tissue. Then, we found that the endothelium is mostly affected, and there's a disturbance of vessels generally, especially the larger vessels.

General lesions affecting more than one organ

a. Expression of Spike protein S1
b. "Endothelialitis" /destruction and inflammation of endothelium
c. Disturbance of the texture with destruction of elastic fibers of the aorta and larger vessels

d. "Displaced" (unidentified (?)) vacuolar and crystalline particles (cholesterol?)
e. Proteinaceous deposits (functional amyloidosis, fibrin amyloid)
f. Unusual Cancer manifestations – „turbo cancer"

g. "Clot"-formation
h. True foreign bodies from contaminated vaccine

Then we have displaced unidentified vacuolar and crystalline particles, proteinaceous deposits, functional amyloidosis. We have unusual cancer manifestations, and we have clot formations in the blood and, in some cases, true foreign bodies. So, I will show you examples of these.

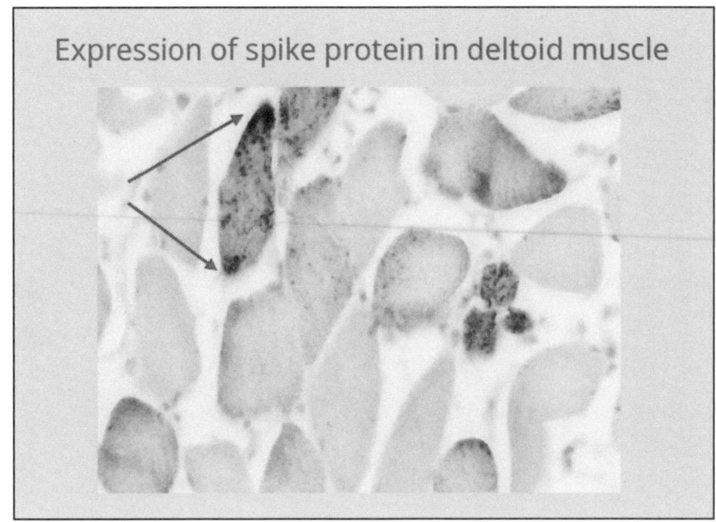

First of all, you've seen this already in the morning. This shows that we could confirm that the Spike protein is produced in the deltoid muscles where the vaccine is injected, but we could show it *(Spike protein)* in almost all organs, more or less explicitly.

And here *(Below)* you see a case where we show the testis. You can see that, in this 28-year-old man who had a healthy son and who died 140 days after injection, the Spike protein is strongly expressed in the testis.

Normal (https://histologyguide.com/) on the left. Compare the circled region in a normal subject with the corresponding section in the 28-year-old man. Look at the lack of cells, not the color, as the stains are different.

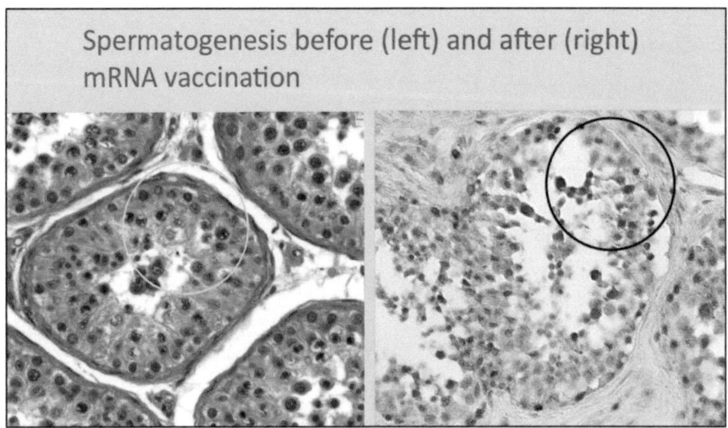

And you can see there are almost no spermatocytes *(above right)* in here, but there is strong expression of Spike protein in the spermatogonia.

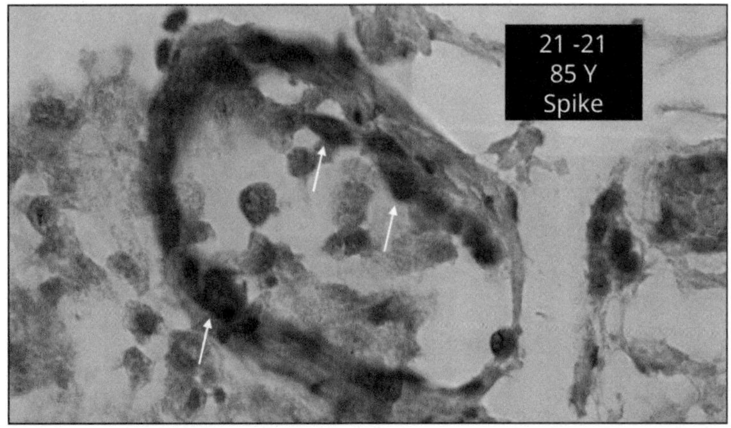

This is an old man. And you can see here (previous image), also, a strong expression of Spike in the spermatogonia. There's not one single spermatozoa in this.

So, if I may make a personal comment, this is not a scientific comment. **If I were a woman in fertile age, I would not plan a motherhood from a person, from a man, who has been vaccinated.** I think these pictures are very disturbing for me.

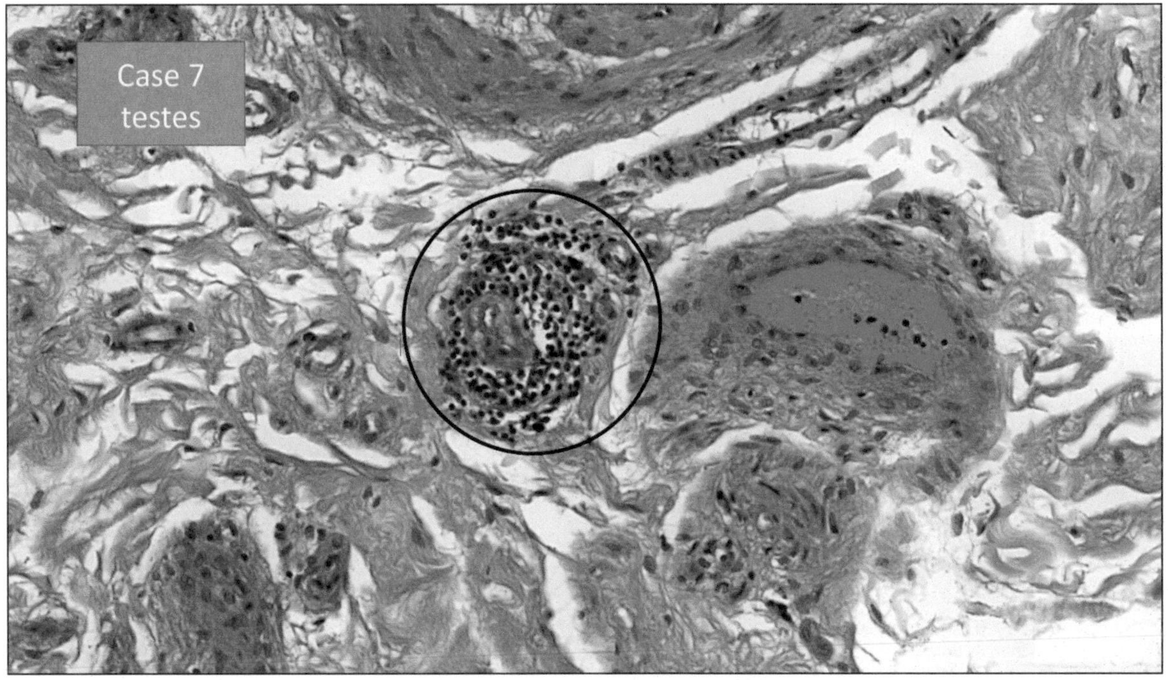

Lymphocyte Amok involving the testis.

In the testis, you can see this phenomenon which we called **Lymphocyte Amok**. We can see a lymphocytic infiltration and inflammation in the testes *(above)*.

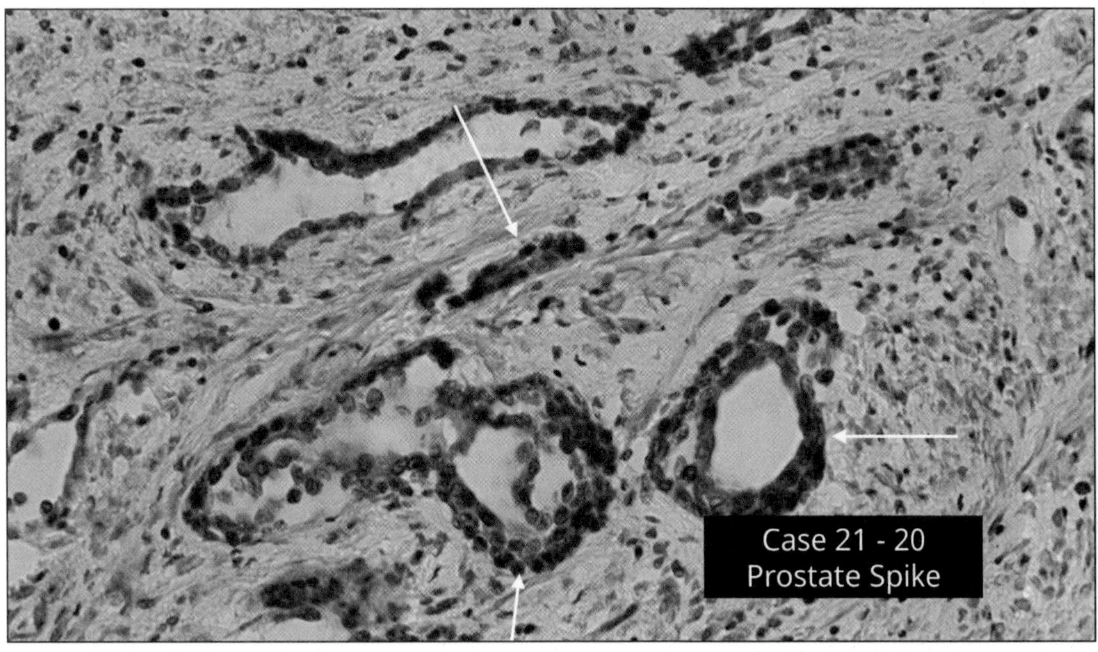

In the prostate gland, you can see a strong expression of Spike protein in this man (previous image).

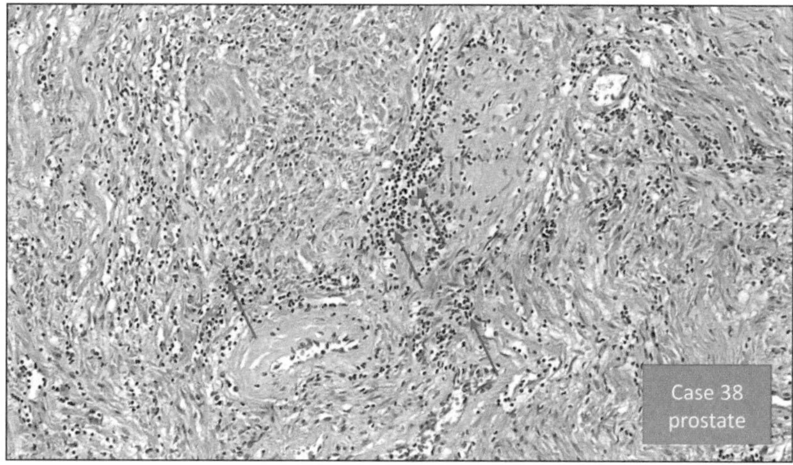

This is very disturbing, but it's meaning we don't know yet. This is inflammation in the prostate.

This is not a prostatitis. It's a **lymphocytic** infiltration in the prostate.

But I come now to the main changes, the **main damage that is done by the Spike protein which is induced by the vaccination.**

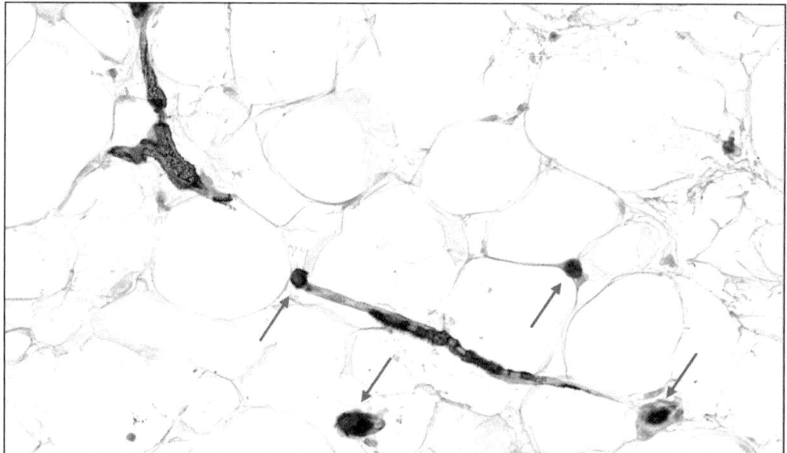

You can see here a Spike protein demonstration in a small capillary. You can see this is fat tissue. These are vascular structures, and you can see the Spike protein clearly and very distinctly marks capillary.

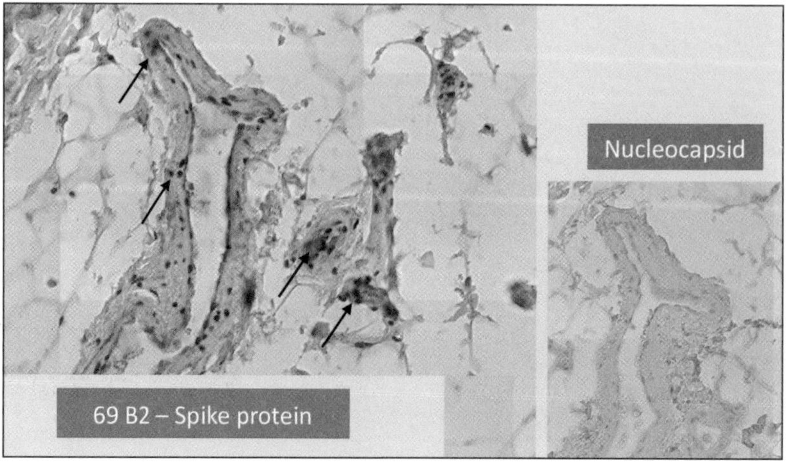

You can see here, on the left side, the spike protein in a small arteriole (previous image). *The right inset box shows tissue with NO nucleocapsid protein thus no evidence of Covid-19 Spike. Only Spike from drug therapy is present.*

Strong and distinct expression of Spike in the vascular endothelium of a small arteriole *(below).*

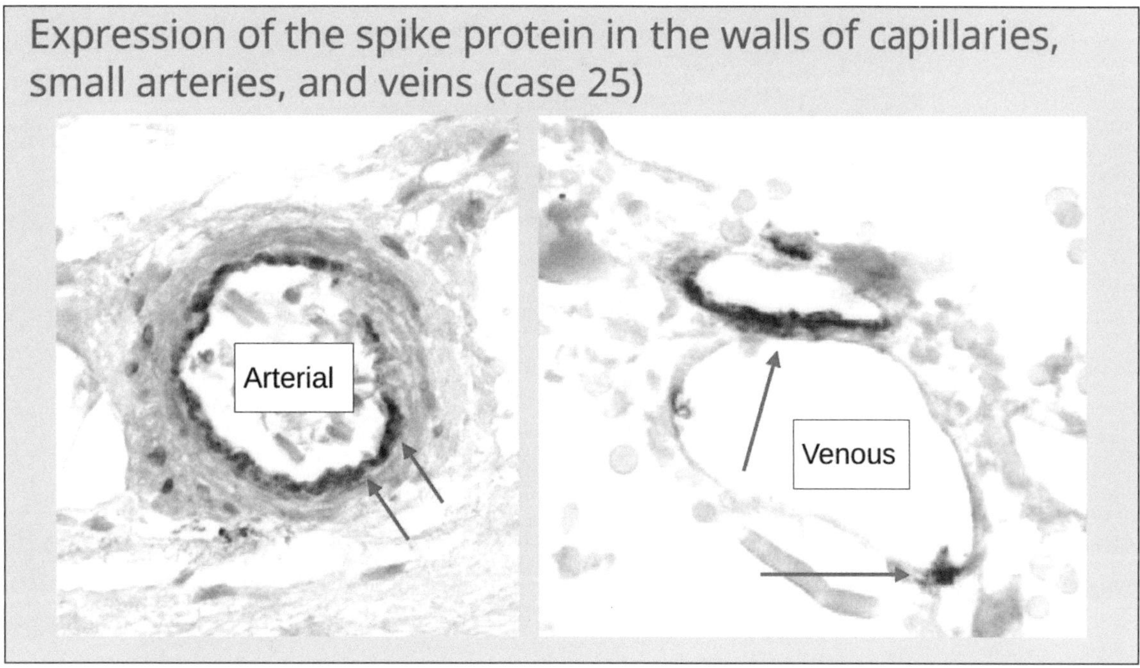

Inner part of these vessels has strong production of Spike protein that elicits a strong immunological reaction with destruction of endothelium. *(Autoimmunity).* Destruction of endothelium may be a major factor of the adverse effects of this vaccination.

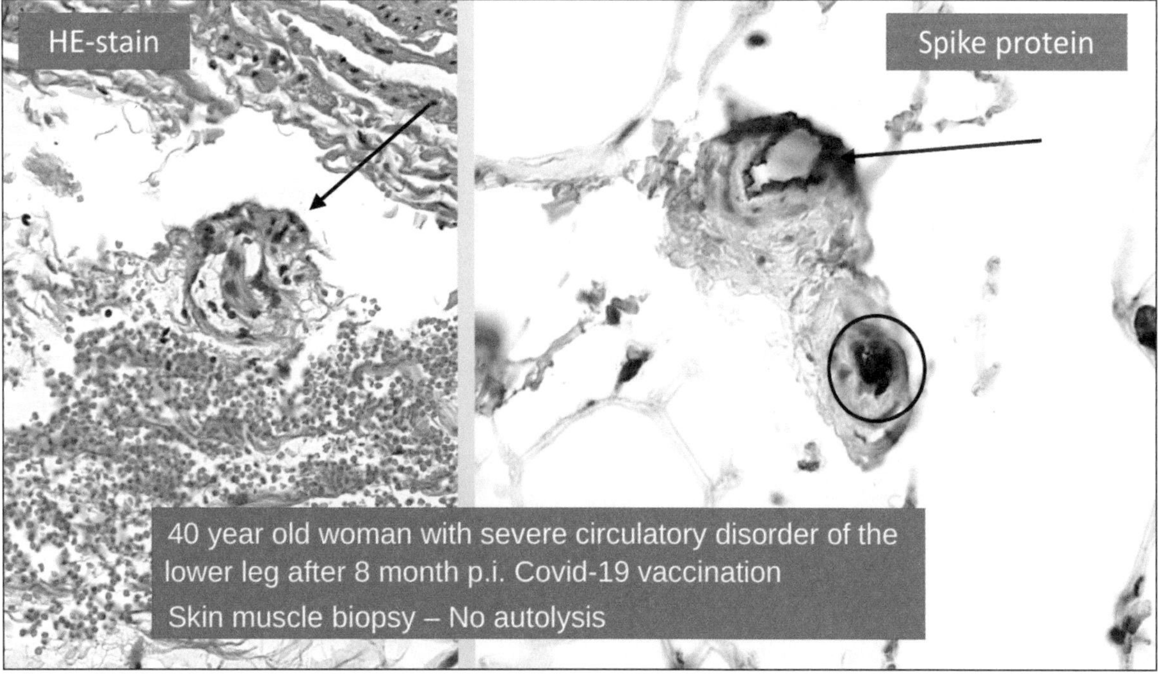

Swelling of the endothelium associated with impaired perfusion *(circulation)* in a lower leg eight months following vaccination. Endothelium clearly expresses Spike. Occluded vessel on the right.

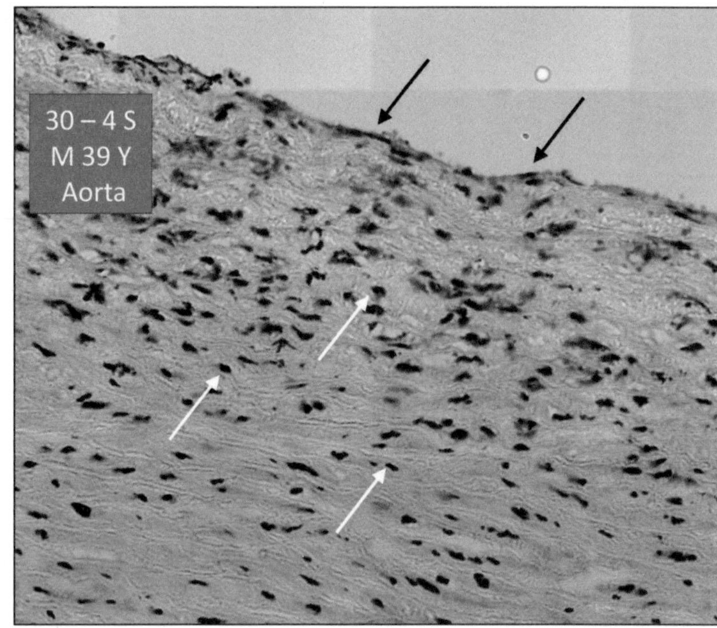

Aorta showing expression of Spike protein involving the endothelium *(inner cell lining)* and myofibroblastic *(damaged heart muscle is replaced by fibrous tissue made by these cells)* cells.

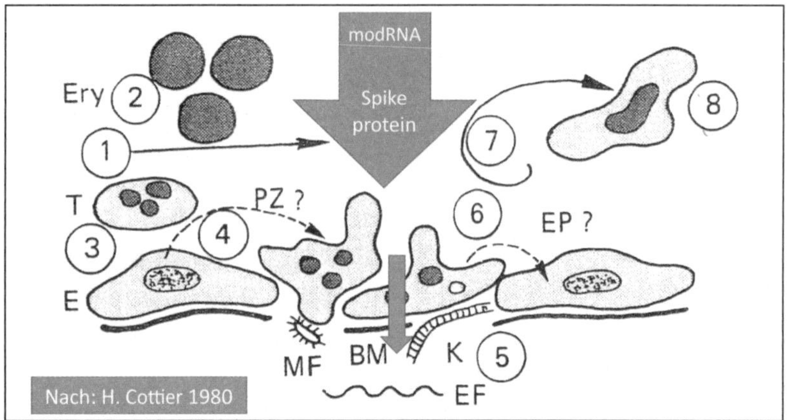

Foreign Body Reactions

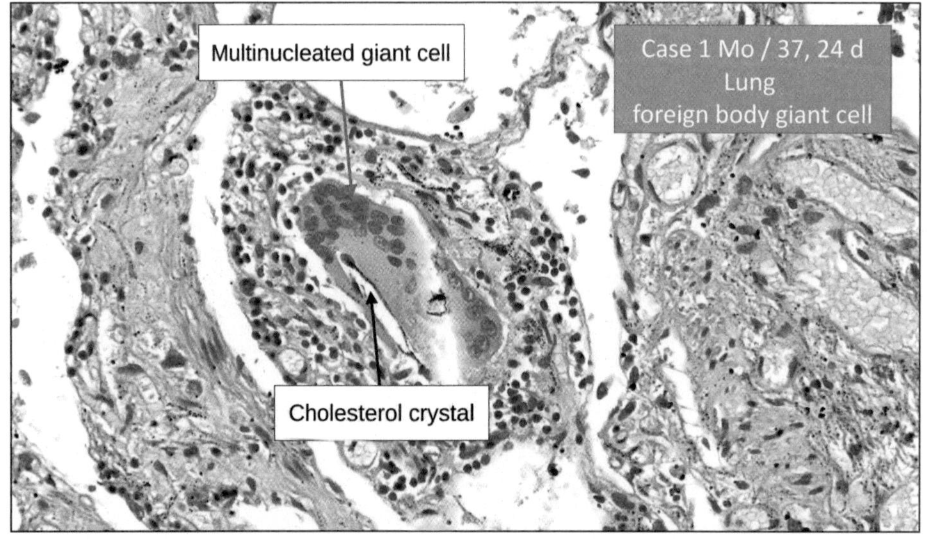

Giant cell formation with needle-like cholesterol crystals (previous image). (*Giant cells are formed from fusion of other white cells such as macrophages in areas of chronic inflammation involving a variety of agents including foreign material such as surgical implants, bacterial, viral, parasitic or fungal infections. (Amy K. McNally, James M. Anderson, Macrophage fusion and multinucleated giant cells of inflammation, in Cell Fusion in Health and Disease, Springer 2011, pp 97—111 Dittmar and Zänker (eds.)*

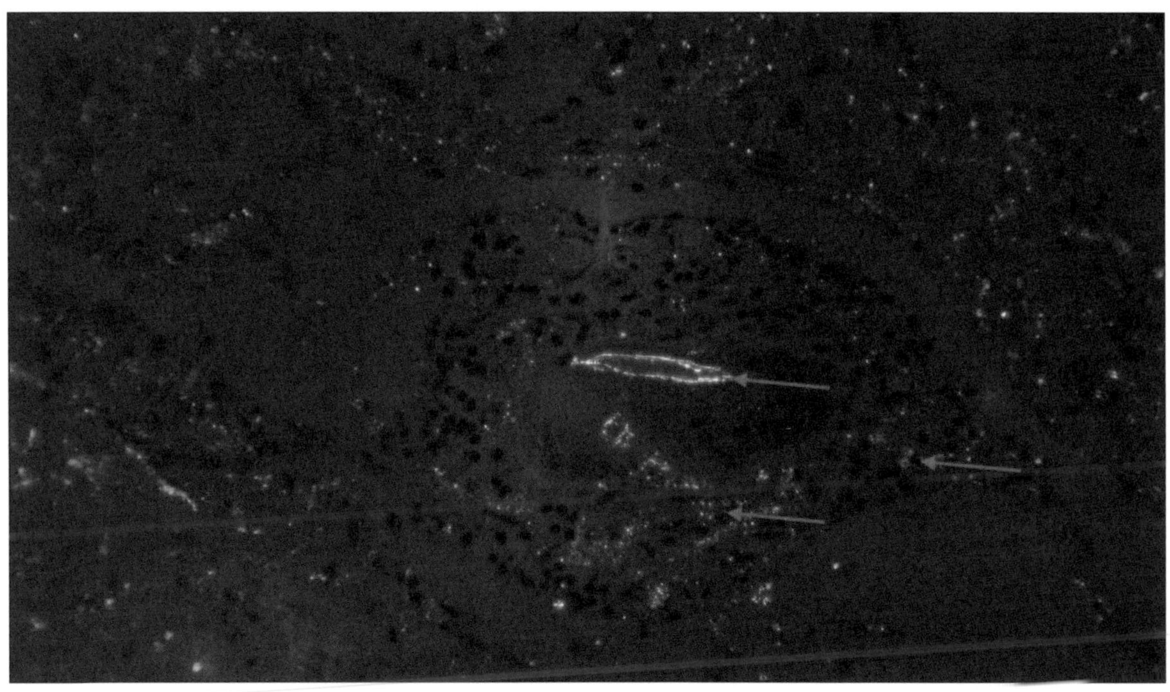

Birefringent *(https://www.youtube.com/watch?v=WdrYRJfiUv0)* microscopy shows small cholesterol crystals.

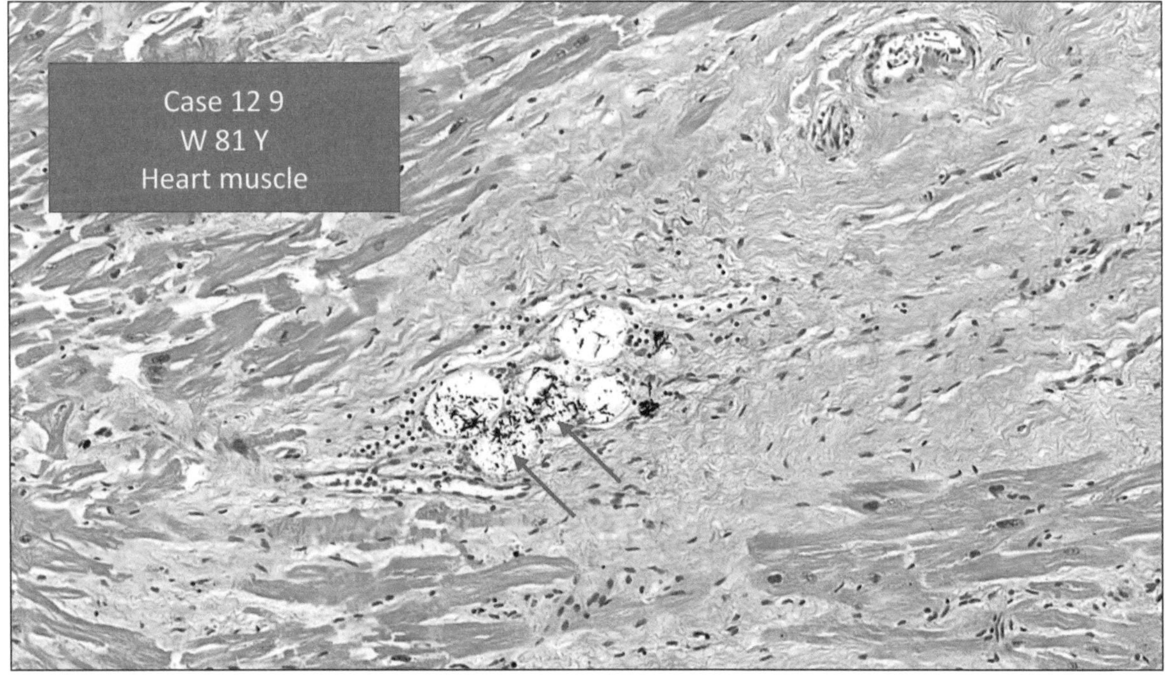

Heart muscle with rod-like structures in vacuoles made of cholesterol crystals (previous image).

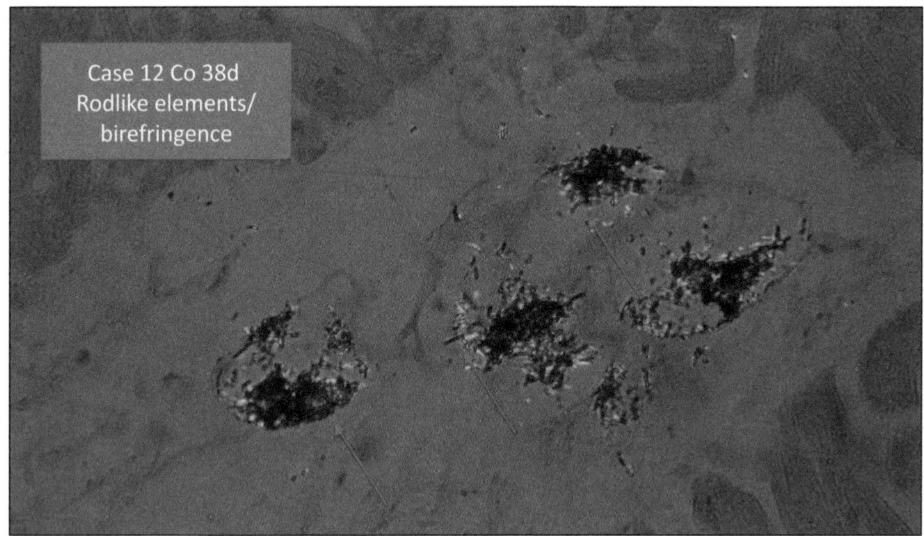

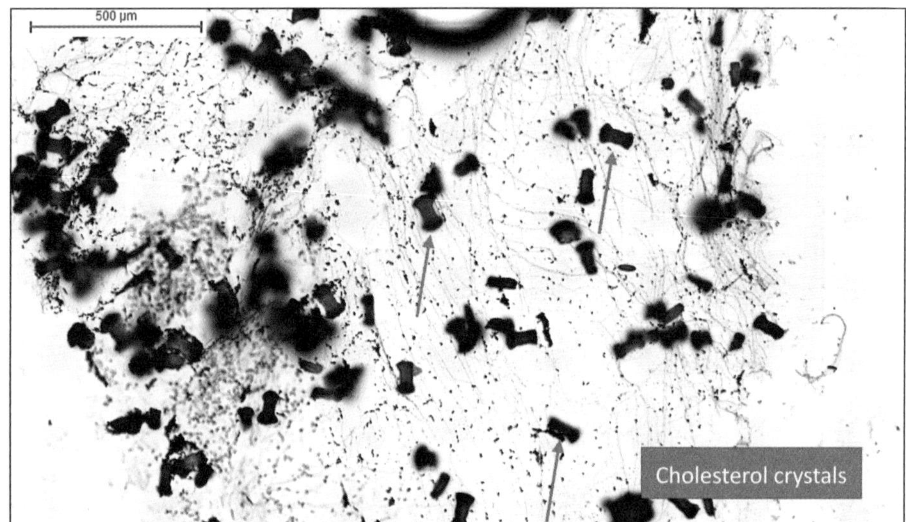

Tablet-shaped larger objects with small, granular ones thought to be cholesterol crystals.

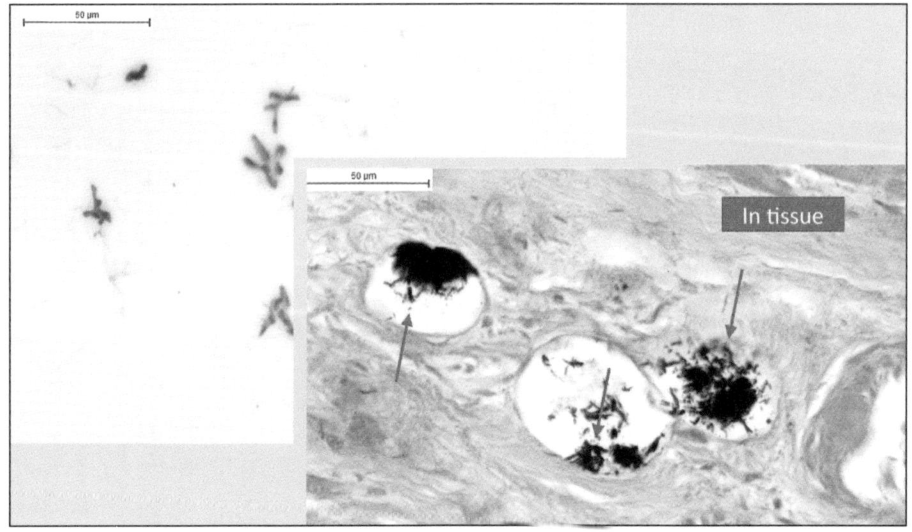

Left: Rod-like crystals in a cholesterol preparation (previous image). Right: Rod-like cholesterol crystals in heart muscle.

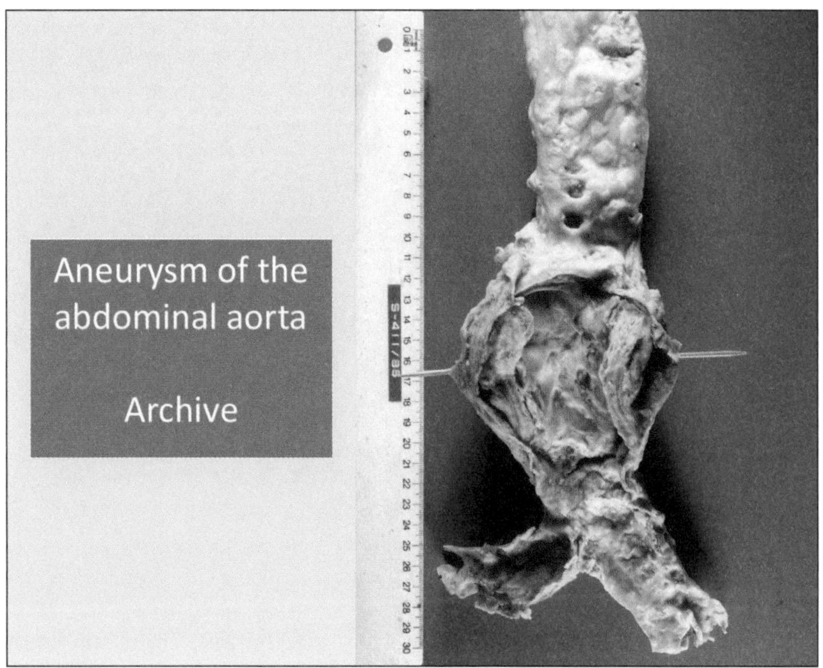

Archive photograph: Where does the cholesterol come from? Not from the injection, because the amount would not be sufficient. We now believe that the cholesterol comes from the wall of the aorta and atherosclerotic vessels. When the Spike attacks the endothelium, the cholesterol is released.

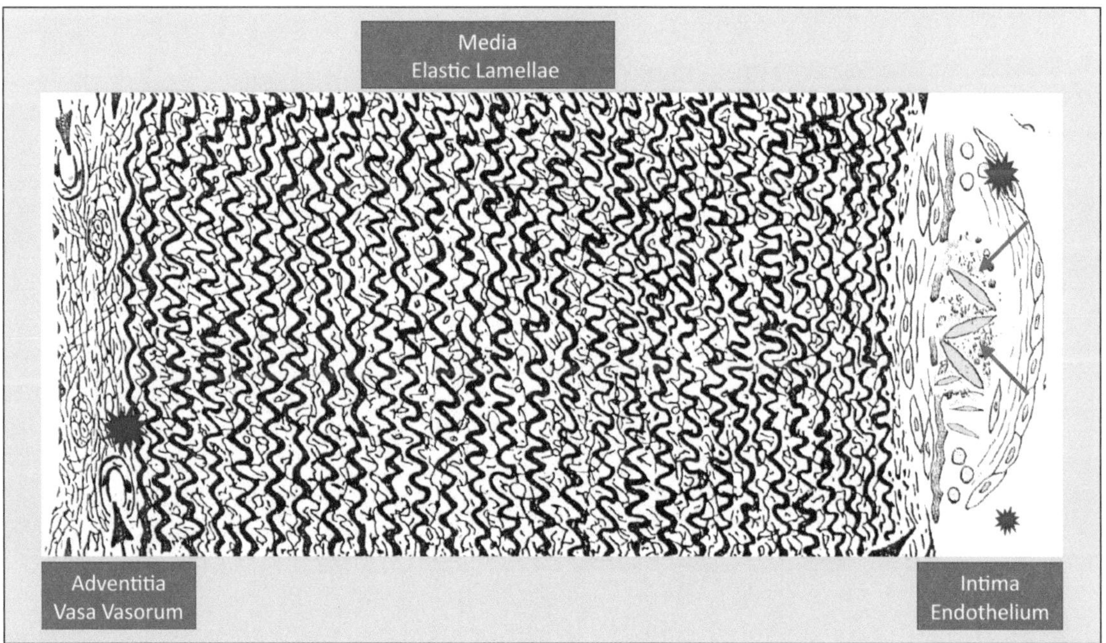

Schematic of arterial wall showing an **atheromatous plaque**, two red stars, containing spindle-like **cholesterol** needles on the right side, a thick muscular layer occupying about 80% of the full thickness of the vessel, and the external surface of the artery, the vasa vasorum. The vessels on the outside of the artery are also induced to form Spike proteins.

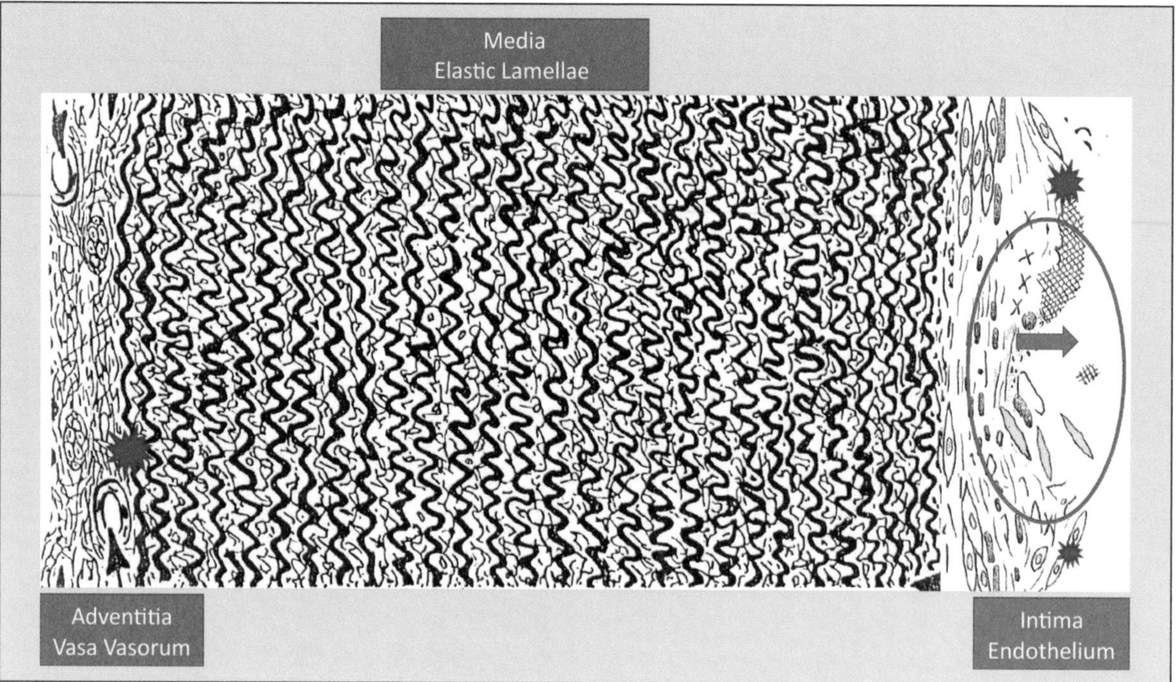

Atheromatous plaques could be set free into the body's circulation, right side showing cholesterol and debris being released into circulation.

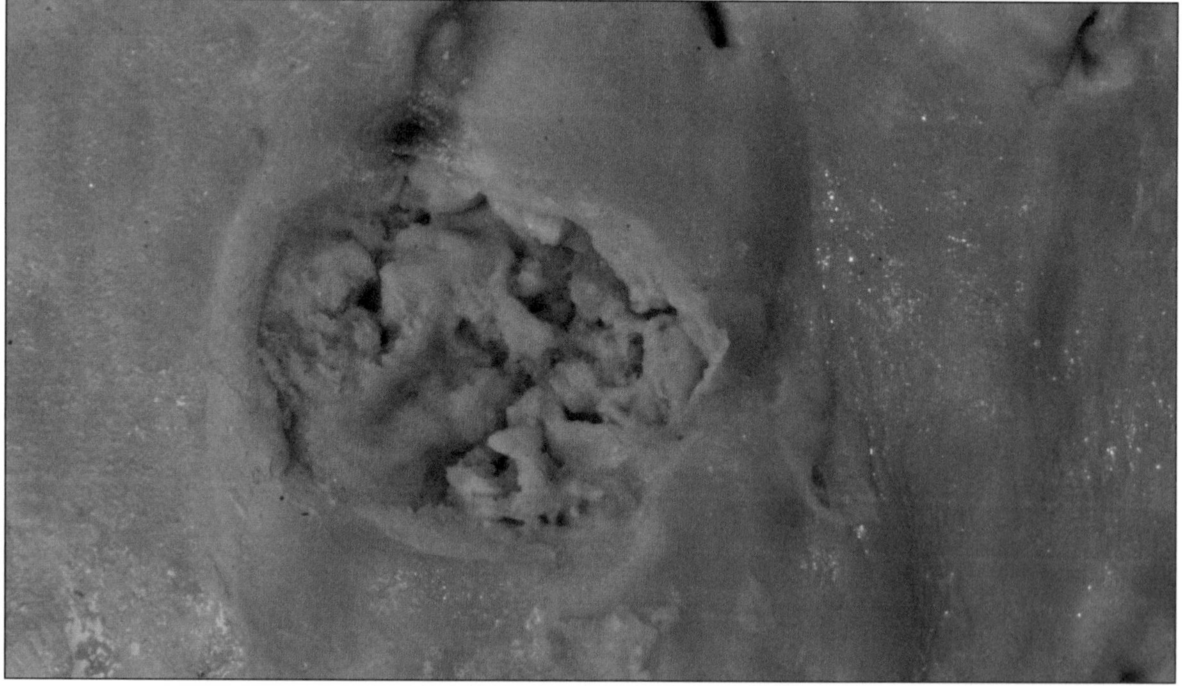

Vaccinated person with an atheroma laid open by the destruction of endothelium releasing the contents into the artery.

Same case:

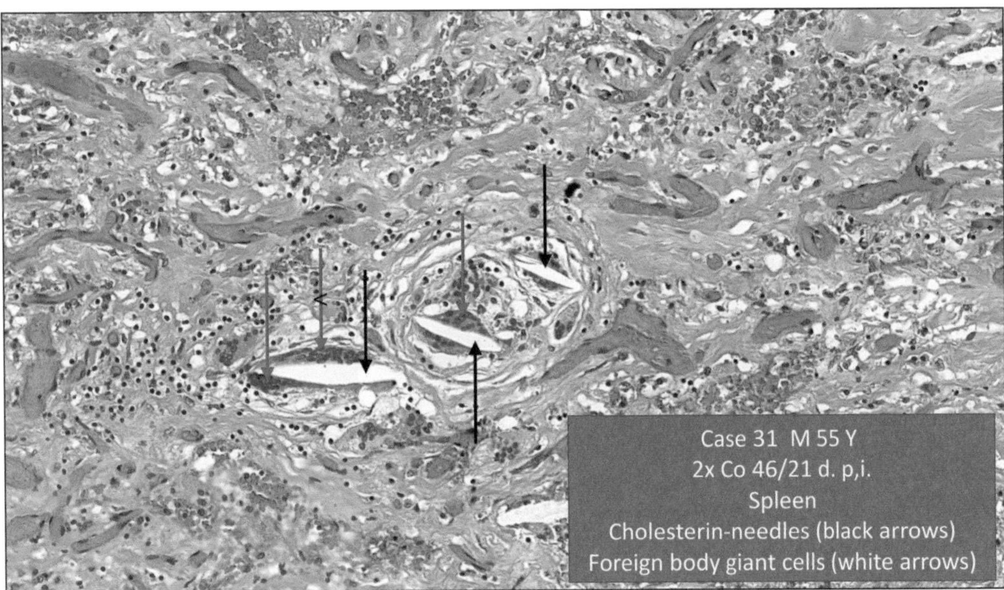

Cholesterol has come out of the atheromatous plaque of the aorta and lodged in a splenic vessel.

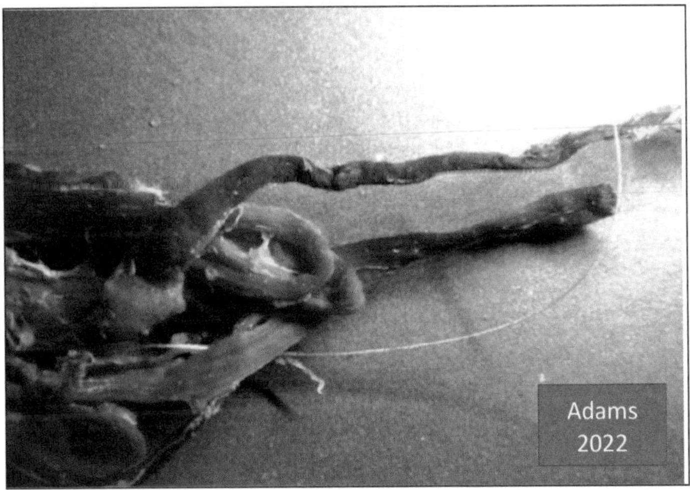

Strange clots have been associated with LNP/mRNA treatment.

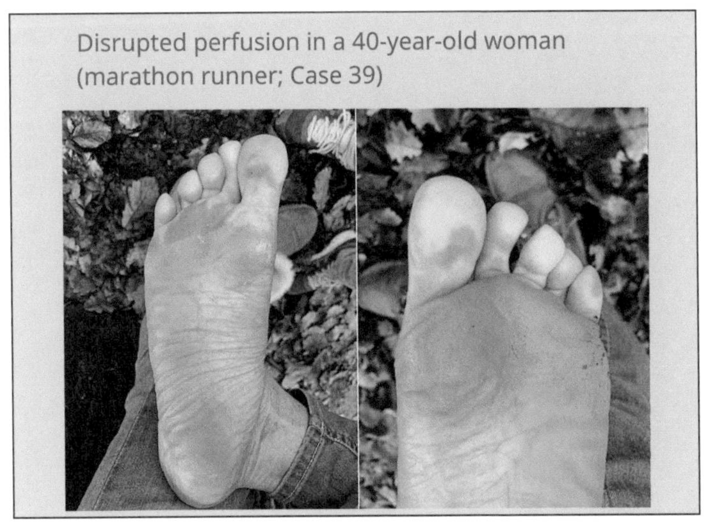

Lady with damaged capillary endothelium with severe disturbance to the circulation (previous image). Sometimes she is unable to walk. Pictured here, on the sole of the foot, is red discoloration signifying inflammation in blood vessels called vasculitis.

Radiological findings in the above patient:

- Double-barreled vessel wall of the femoral and popliteal artery
- Signs of increased flow resistance in the peripheral circulation

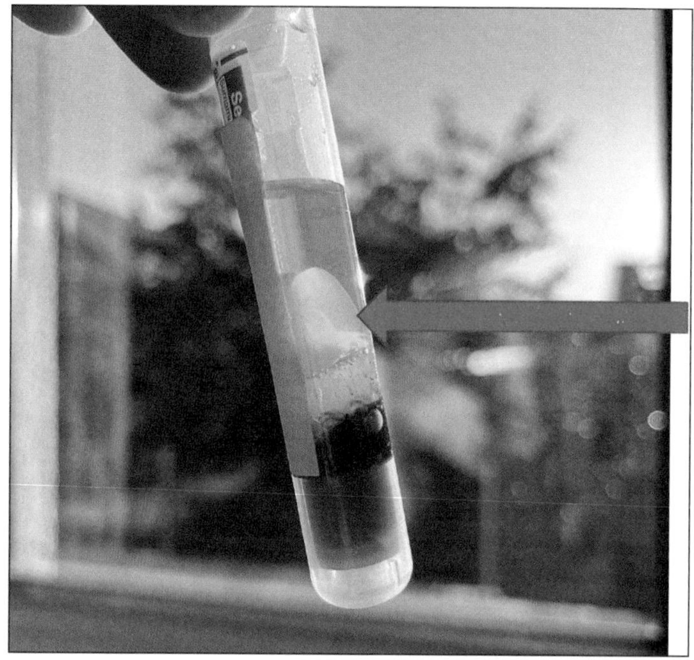

Blood was taken from this lady, and this clot formed in the serum after centrifugation and cooling in the refrigerator.

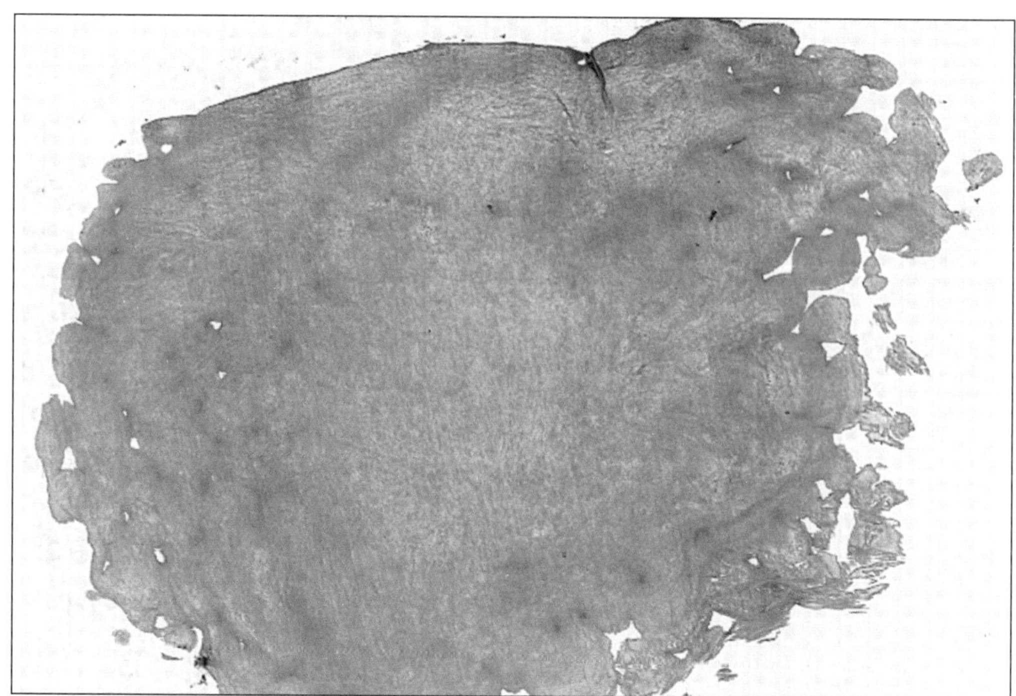

Clot stained and magnified (previous image). Proteinaceous practically acellular substance.

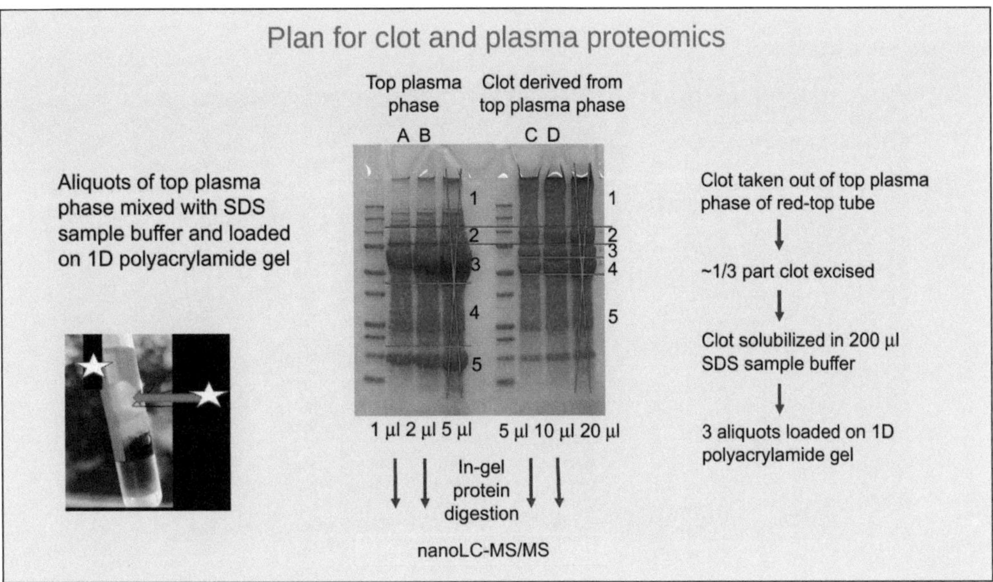

Examination of this clot was performed by specialized physical chemists, and they compared the contents of the plasma phase to the clot itself by mass spectroscopy.

137 clot-enriched proteins

GO 0005201	Extracellular matrix structural constituent
GO 0062023	Collagen-containing extracellular matrix
GO 0005518	Collagen binding
GO 0043236	Laminin binding
GO 0071953	Elastic fiber
GO 0050839	Cell adhesion molecule binding
GO 2001027	Negative regulation Endothelial cell chemotaxis
GO 1900024	Regulation of substrate adhesion-dependent cell spreading
GO 2000352	Negative regulation of endothelial cell apoptotic process
GO 0045907	Positive regulation of vasoconstriction
GO 0036002	pre-mRNA binding
GO 0035198	miRNA binding

137 proteins were present in the clot and not in the serum. In **red** are substances related to the **endothelium, a sign of continuous damage.**

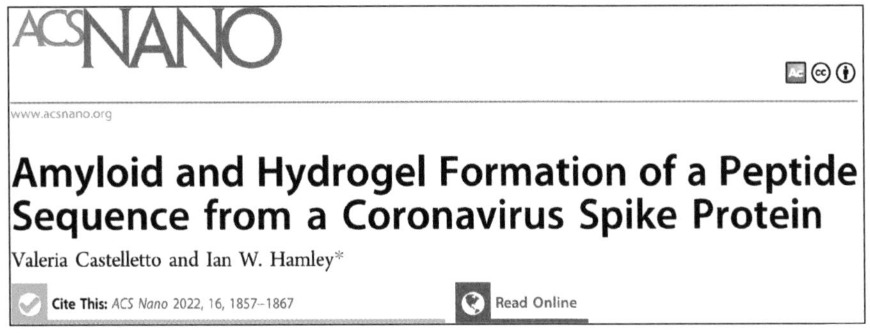
ACSNANO

www.acsnano.org

Amyloid and Hydrogel Formation of a Peptide Sequence from a Coronavirus Spike Protein

Valeria Castelletto and Ian W. Hamley*

Cite This: *ACS Nano* 2022, 16, 1857–1867 Read Online

There are some hints that **amyloid, a waxy translucent substance consisting primarily of protein that is deposited in some animal organs and tissues under abnormal conditions,** could be formed of or by Spike proteins (previous study).

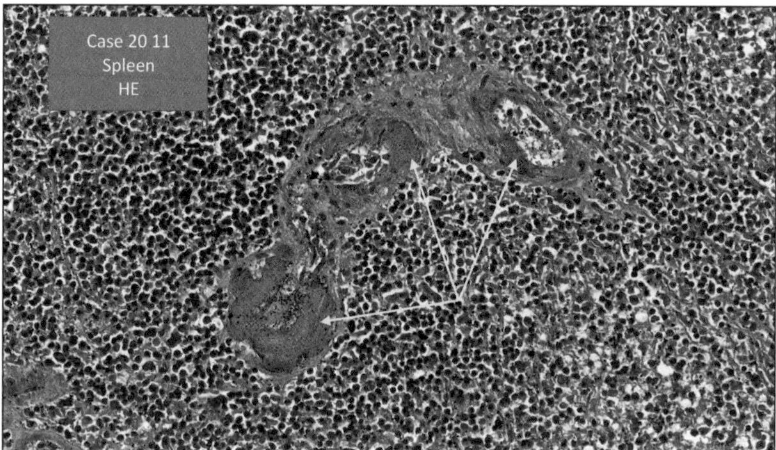

We very early found these deposits especially in vessel walls. You can see they are acellular (red stain) and compressed the vessel walls. Very early we had the suspicion this could be **amyloid**. *(https://www.karger.com/Article/FullText/506696)*

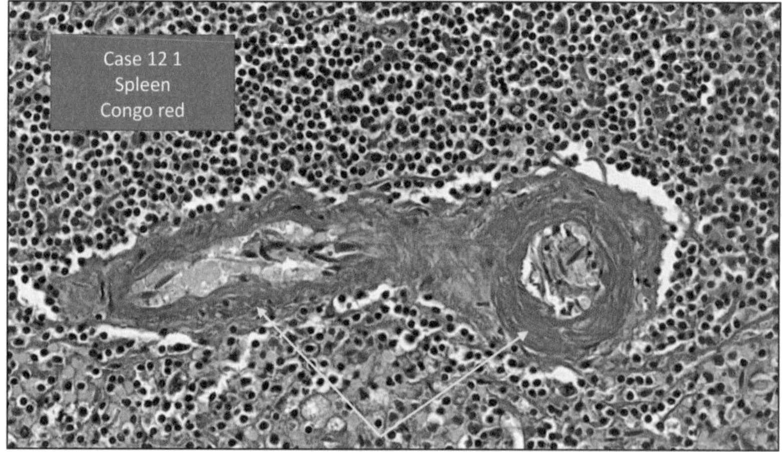

This could be proven by the special stain of Congo Red. *(https://www.pathologyoutlines.com/topic/stainscongored.html)*

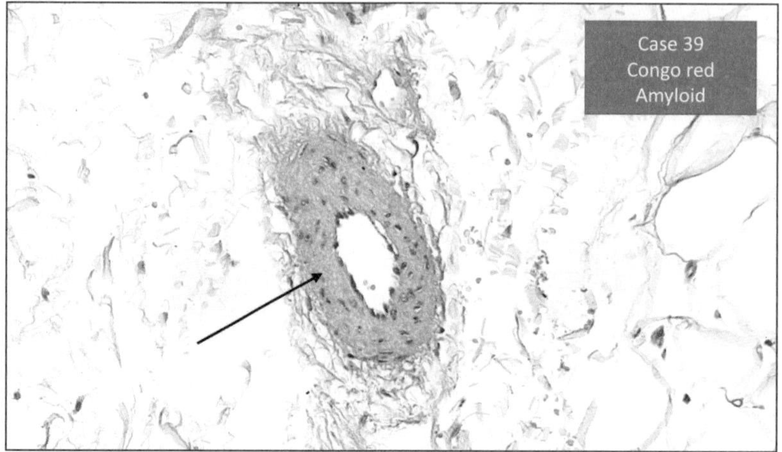

This is spleen and was found in the biopsy specimen of this lady that I showed you before (Case 39). So, she has some amyloid and, certainly, this has some meaning for the function of the vascular tissue. Also, we have these deposits in the brain.

Specific Organ and Tissue Lesions

Specific organ- and tissue lesions

Main pathological findings on blood vessels

Small vessels

- **Endothelialitis, most prominently in heart, lungs and brain**
- **Aggregation of erythrocytes, bleeding, hemosiderosis within vessel wall**
- **Complex formation of amyloid-spike protein-fibrin in vessels - amyloidosis**
- **Thrombocyte aggregates and microthrombi**
- **Obliteration**

We come to the specific on organ lesions. We have the small vessels showing destruction of the endothelium.

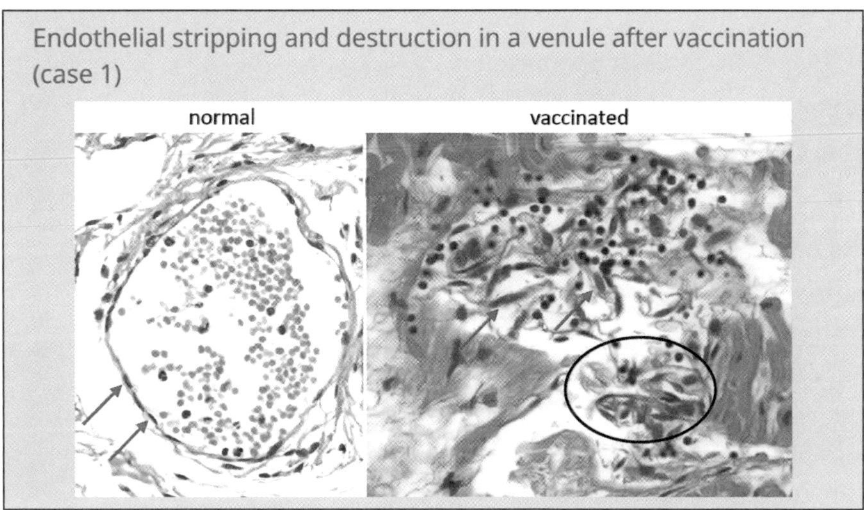

Here you can see the normal on the left side and the destructive capillary endothelium in heart muscle on the right side.

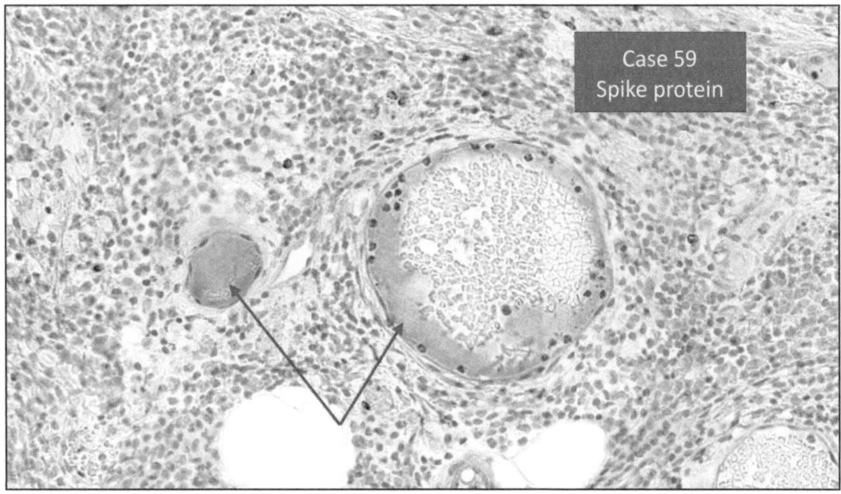

You can see that we can show Spike protein in these lesions (Case 59 above and below).

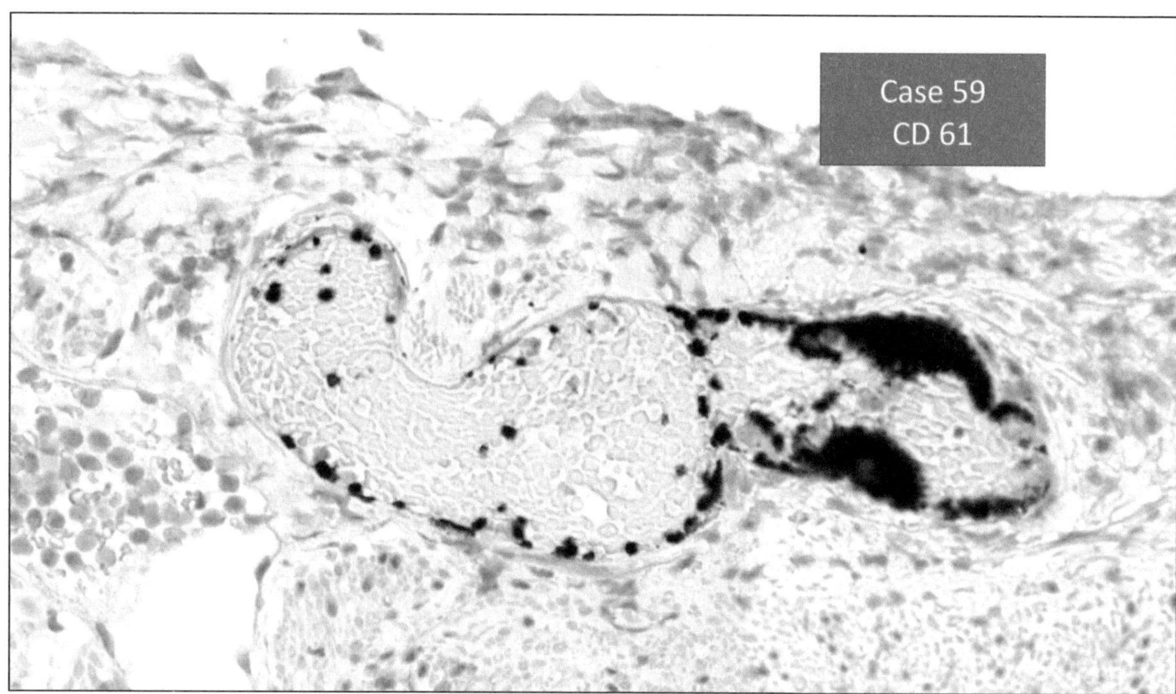

We can find a CD61+ thrombocyte apposition in these lesions. *The CD (Cluster of Differentiation) designation refers to a convention of nomenclature for molecules of the cell surface of certain white cells. (https://www.hcdm.org/index.php/component/molecule/?Itemid=132) These white cell surface molecules are important to the function of the immune system. (https://www.immunopaedia.org.za/immunology/basics/cd-nomenclature/) Combined with CD41, the CD61 cell plays a role in platelet aggregation and clotting. (Principles of Immunophenotyping Faramarz Naeim, in Hematopathology, 2008, https://www.sciencedirect.com/science/article/pii/B9780123706072000028.)*

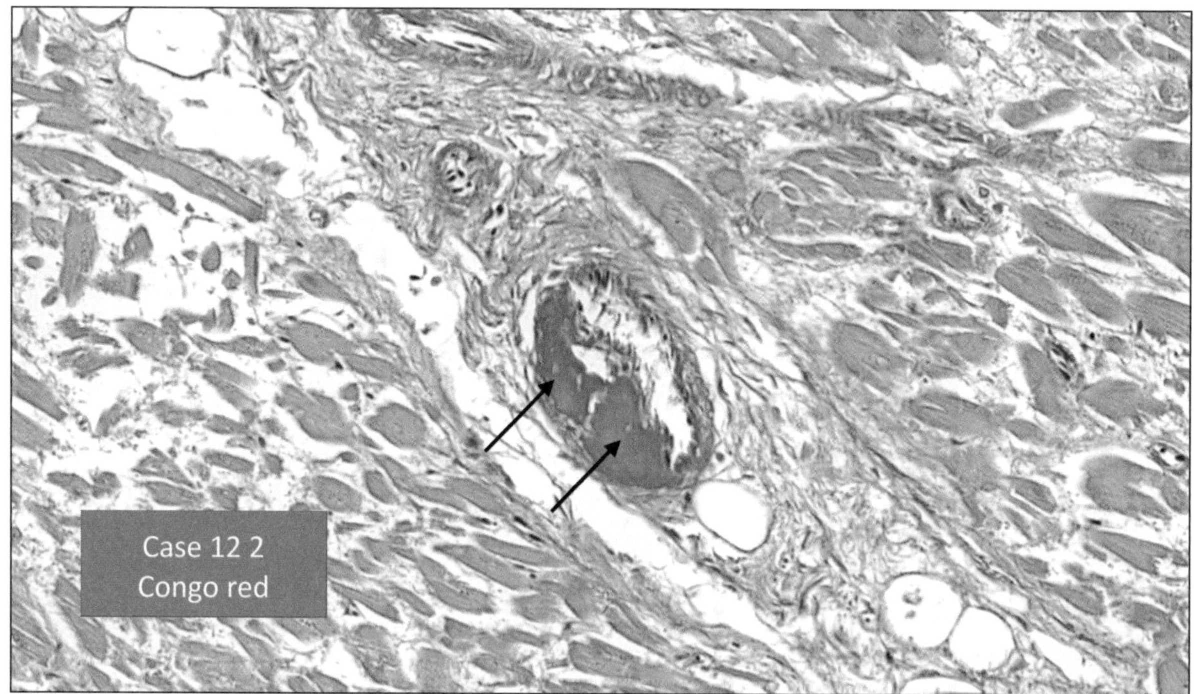

This is the amyloid deposition in a small vessel in the heart muscle (previous image).

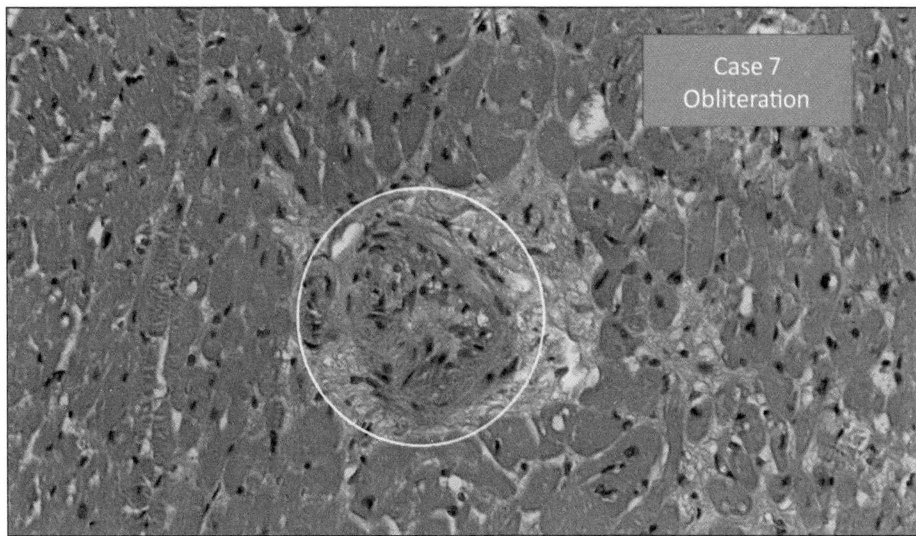

Occlusion *(blockage or closing)* of this vessel.

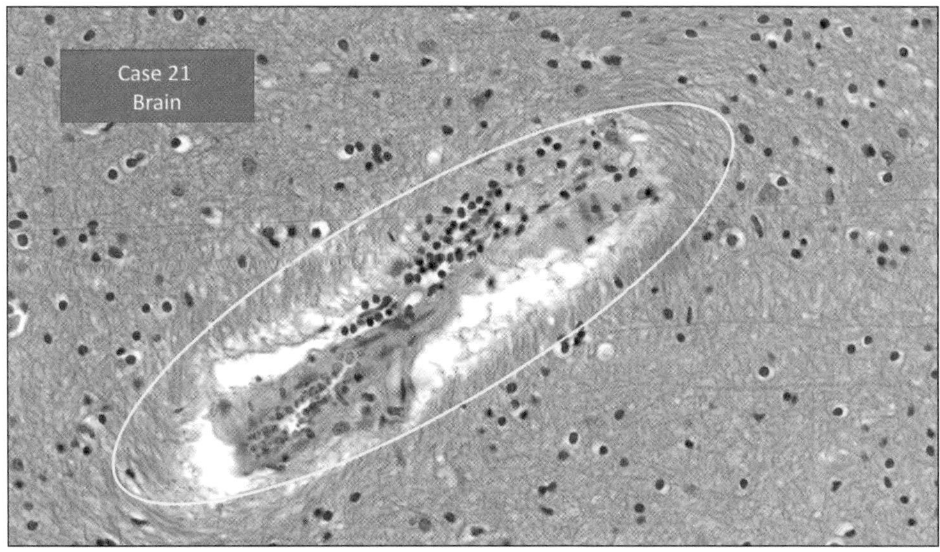

This is in the brain with inflammation of the small vessels.

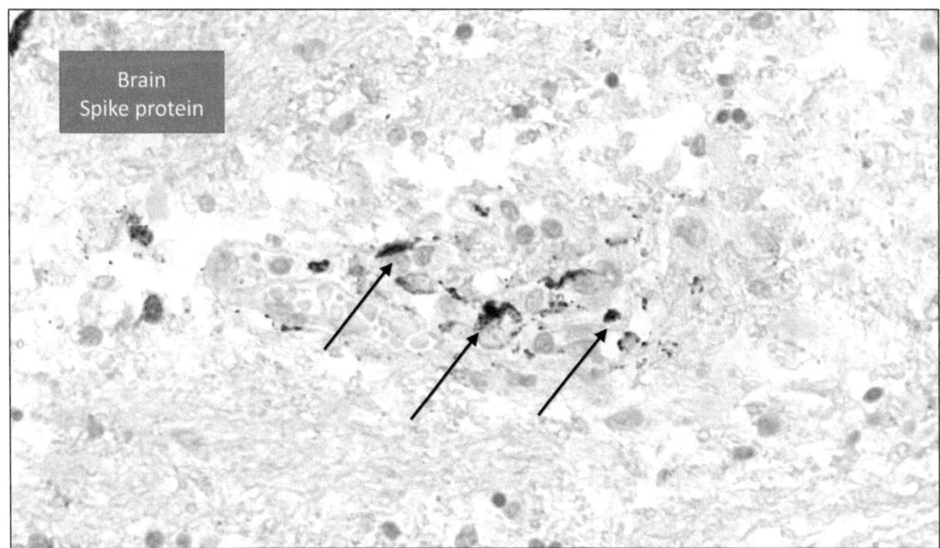

Here is Spike protein demonstration (previous image).

Main pathological findings on blood vessels

Large vessels

- Disrupted wall structure of aorta with lymphocytic vasculitis and perivasculitis
- Endotlhelial damage - Break-up of atheromatous plaques
- Medianecrosis / Dissection
- Perforation (5 cases)
- Thrombotic casts without erosion of atherosclerotic lesions

One very important finding, in our view, is the destruction and disruption of larger vessel walls.

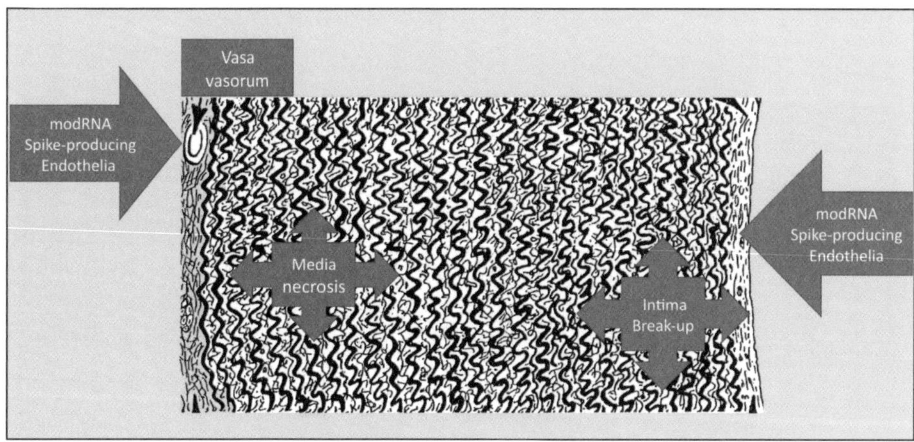

We found media necrosis and breakup in these larger vessels.

Where do we find the media necrosis? We found it in idiopathic arteriosclerosis as shown here, and infection-toxic syphilis, Lathyrism and, apparently, Spike-induced.

This is an historic specimen from before World War I (previous image) showing syphilitic destruction of the media *(the middle, muscle layer of the artery).*

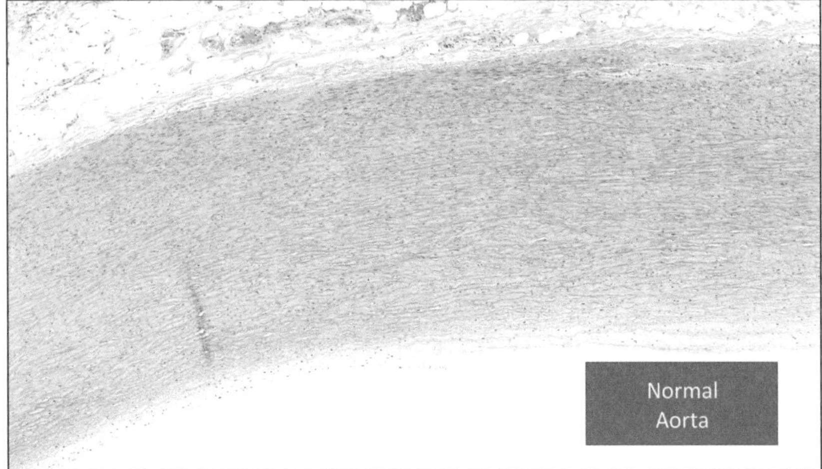

You can see what a normal aorta looks like this. It has a very regular, organized situation.

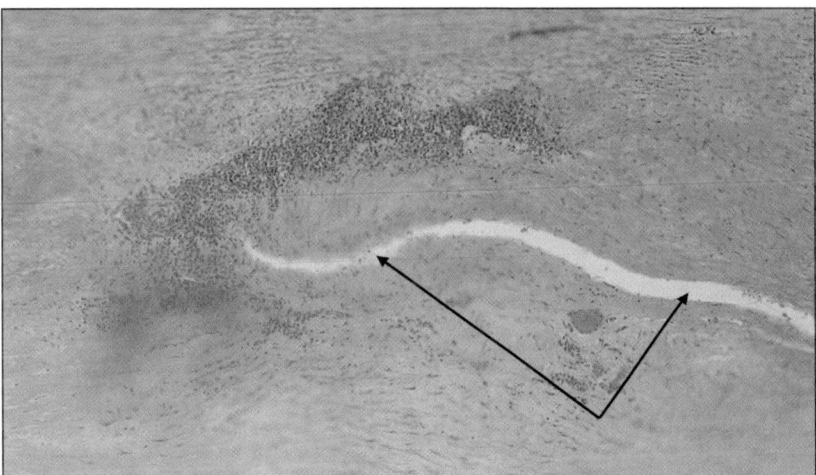

You can see here what we found in our deceased persons. You can see that the aorta is split.

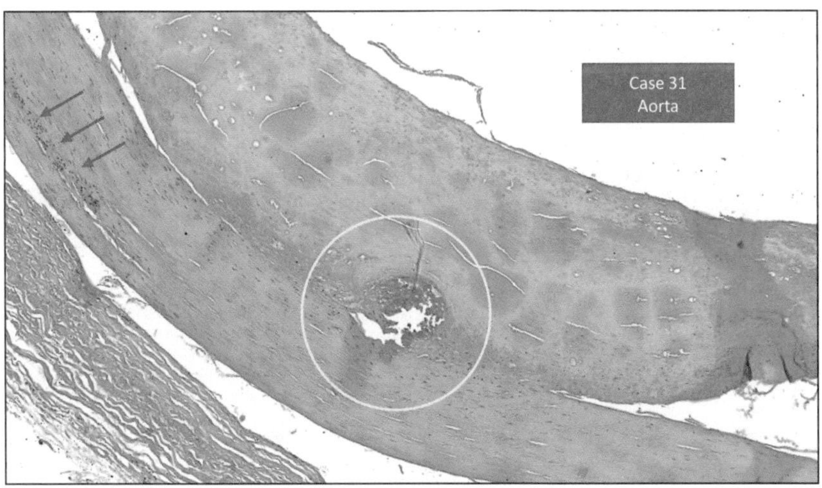

The media is necrotic, and there are **inflammatory infiltrates**, which in idiopathic form is not present. We can see that the media is largely destructed (previous image).

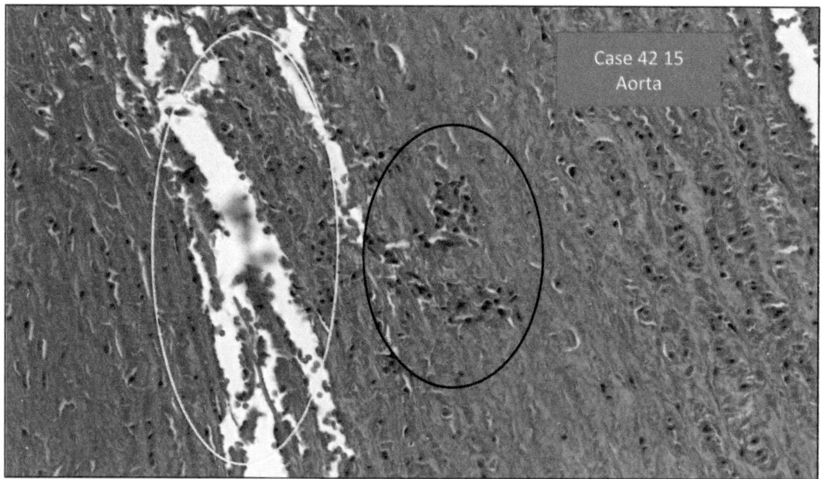

There are **inflammatory infiltrates** and destruction of the elastic lamella *(layer of tissue)*. You can see that the elastic lamellae are discontinuing. There are inflammatory infiltrates, which in the idiopathic form is not present.

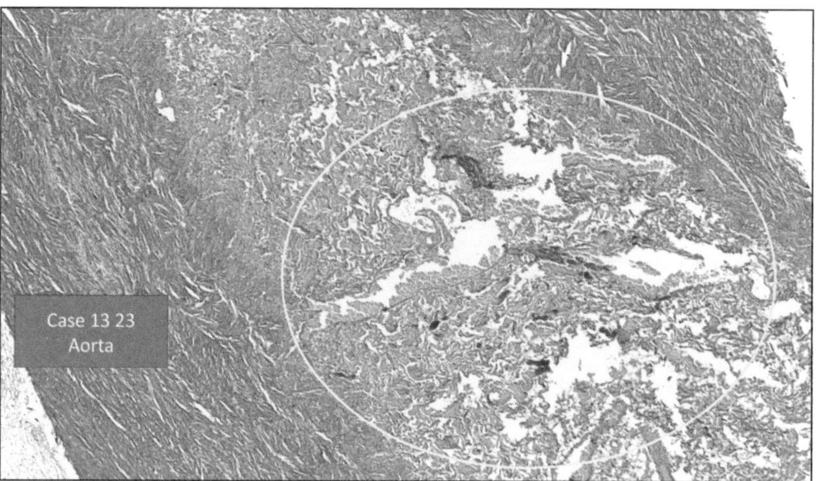

We can see media largely destroyed.

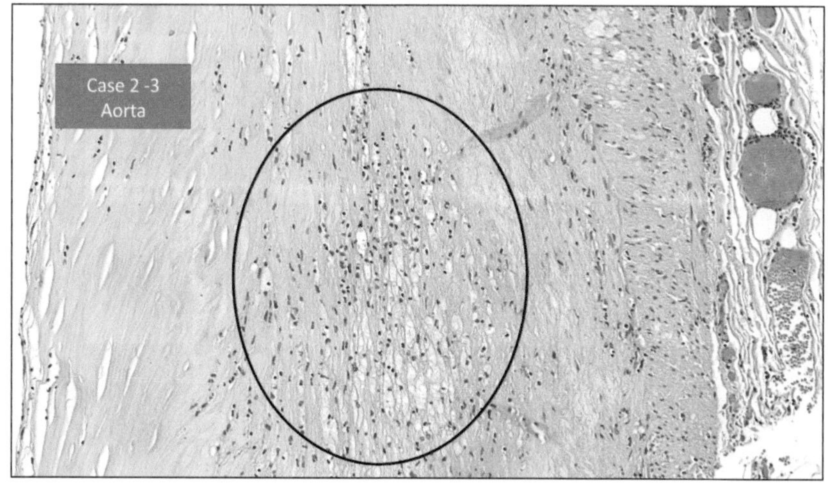

The inflammatory infiltrates (previous image)

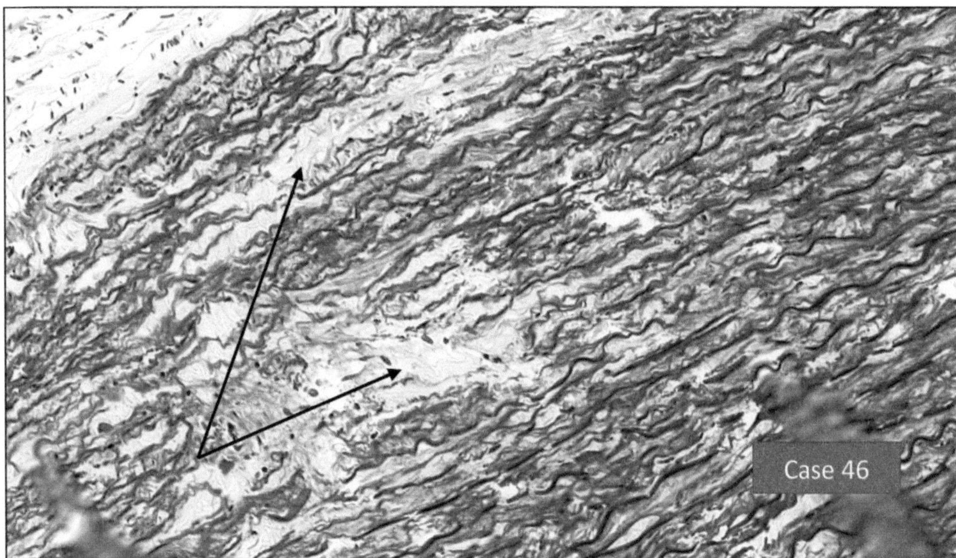

with destruction of the elastic lamella. You can see that the elastic lamellae are discontinuous.

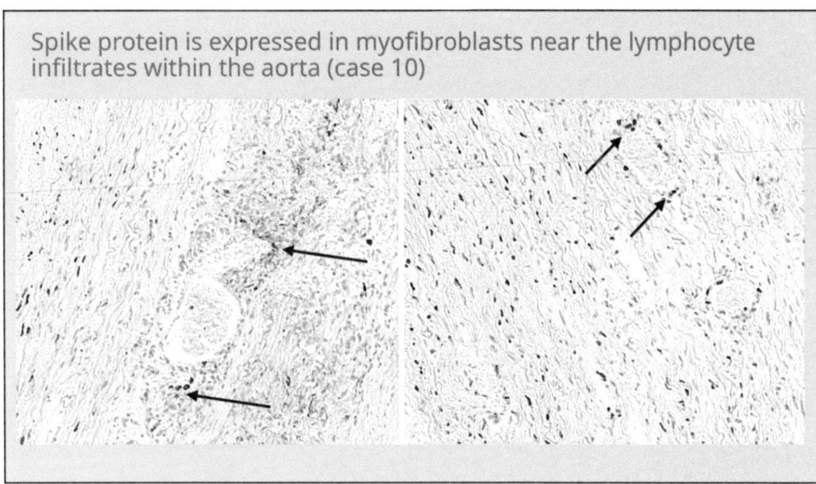

We can show here that the Spike protein expression in these hyperplastic myofibrils and also in the inflammatory infiltrate.

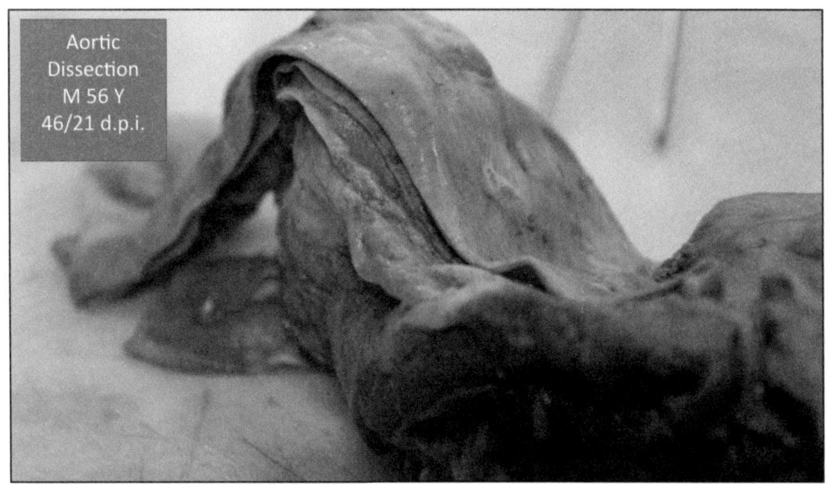

This is one of the autopsies that we did in a 56-year-old man (previous image).

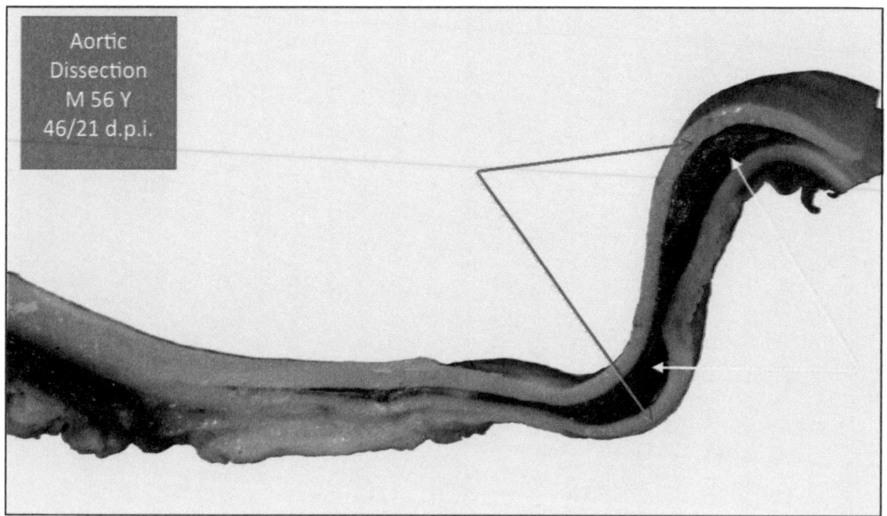

He had this, media[l] necrosis of the aorta. You can see here that the wall of the aorta is split (Red) into two parts, and in the middle there's black blood. (Yellow)

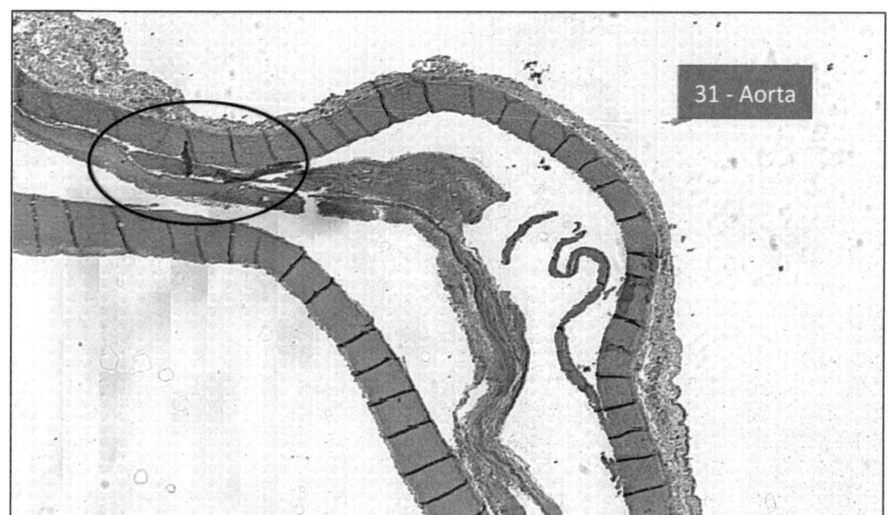

And this is a histological preparation. You can see very clear that there is medial necrosis,

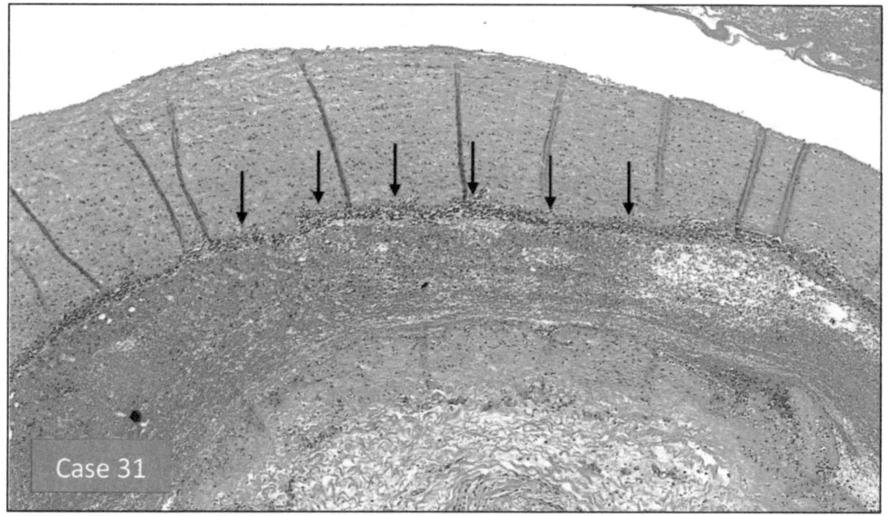

and you can see here (previous image) that there's a dense inflammatory infiltrate

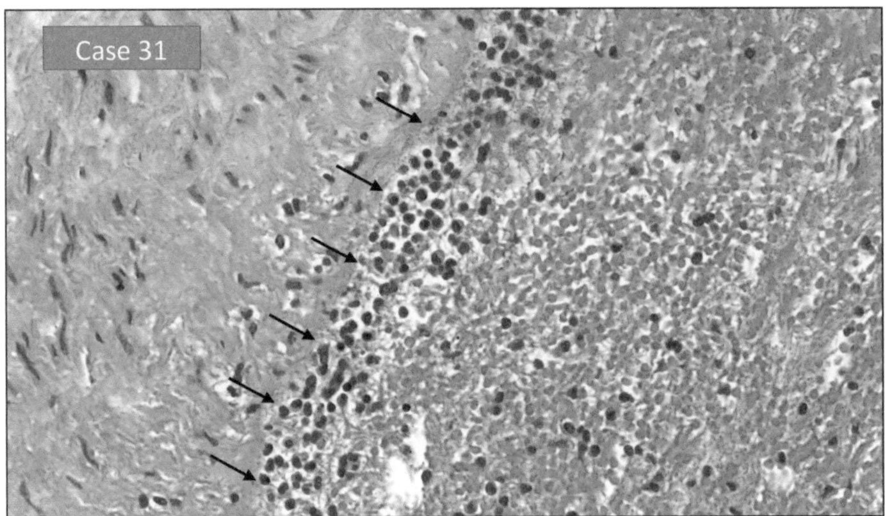

and a mass of histiocytes and macrophages.

Legal Medicine 59 (2022) 102154

Contents lists available at ScienceDirect

Legal Medicine

journal homepage: www.elsevier.com/locate/legalmed

ELSEVIER

Case Report

An autopsy case report of aortic dissection complicated with histiolymphocytic pericarditis and aortic inflammation after mRNA COVID-19 vaccination

Motonori Takahashi[a,b,*], Takeshi Kondo[a,b], Gentaro Yamasaki[a,b], Marie Sugimoto[a,b], Migiwa Asano[c], Yasuhiro Ueno[a,b], Yasushi Nagasaki[a]

[a] Medical Examiner's Office of Hyogo Prefecture, 2-1-31 Arata-cho, Hyogo-ku, Kobe, Hyogo, Japan
[b] Division of Legal Medicine, Department of Community Medicine and Social Health Science, Kobe University Graduate School of Medicine, 7-5-1 Kusunoki-cho, Chuo-ku, Kobe, Hyogo, Japan
[c] Department of Legal Medicine, Ehime University Graduate School of Medicine, 454 Shizugawa, Toon, Ehime, Japan

First of all, we thought, we may be looking at a phantom; but, in Japan, they saw the same thing—aorta dissection, complicated by pericarditis and inflammation.

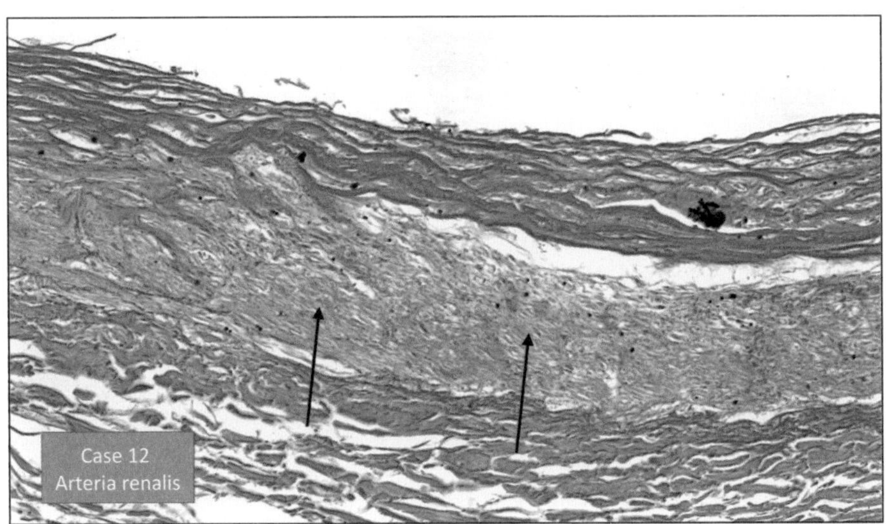

So, we did not only see this in the aorta, but also in the renal artery (previous image),

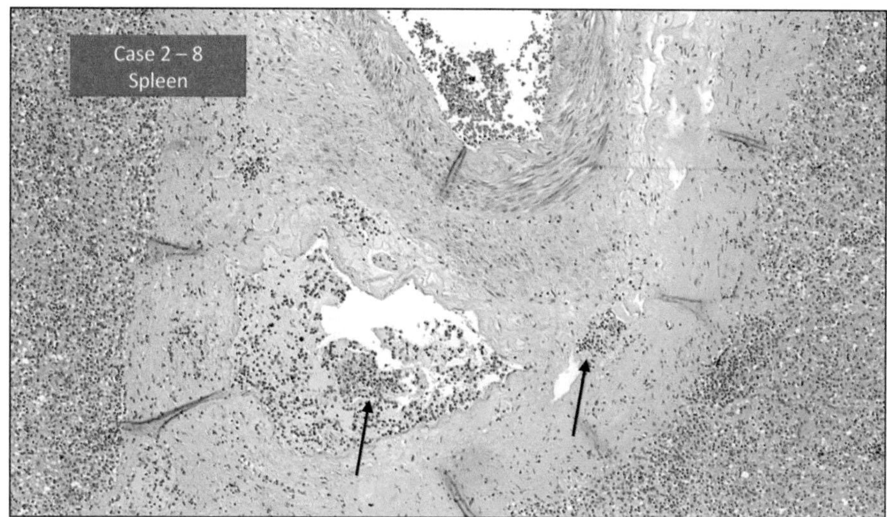

in the splenic artery,

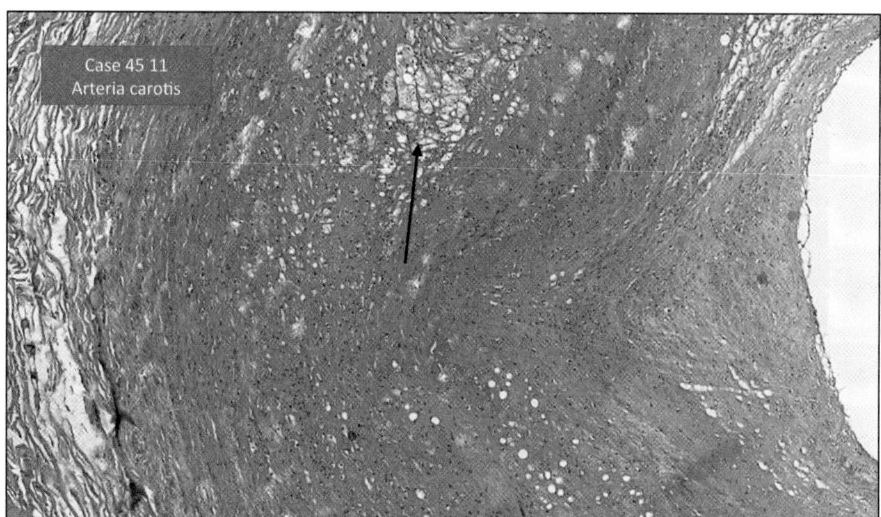

the carotid artery.

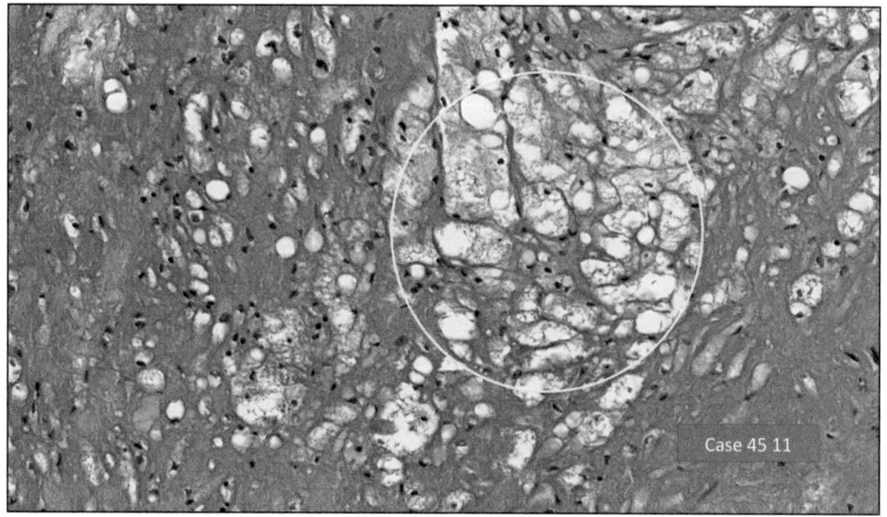

Here you can see vacuolar degeneration of the media (previous image).

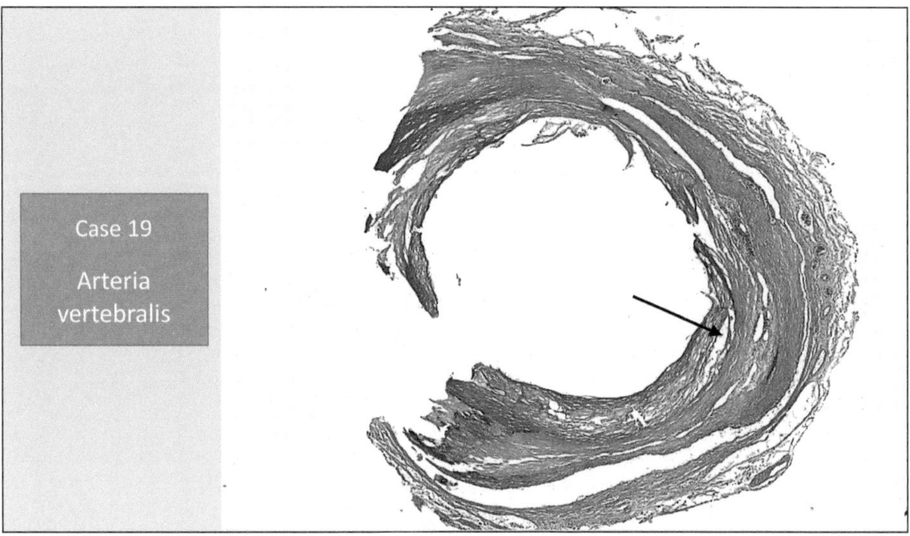

Also seen in the vertebral artery.

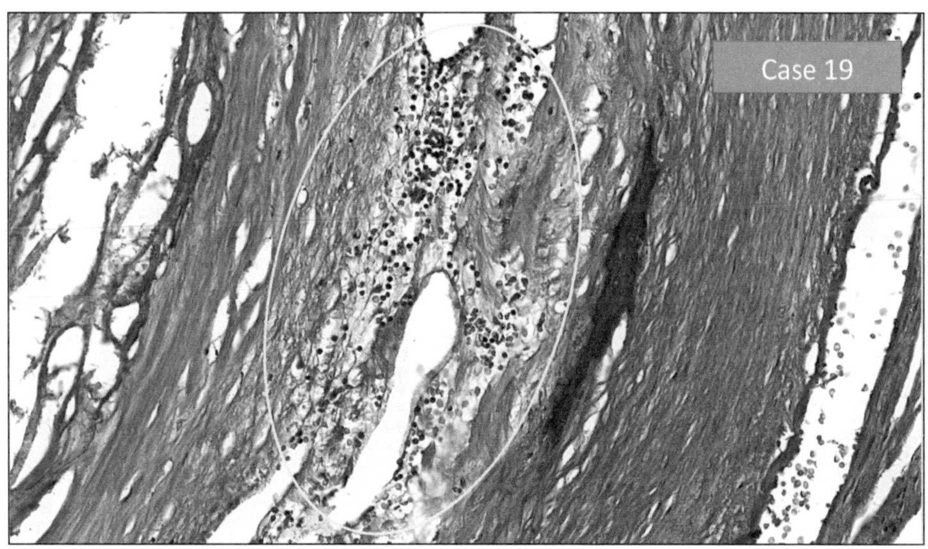

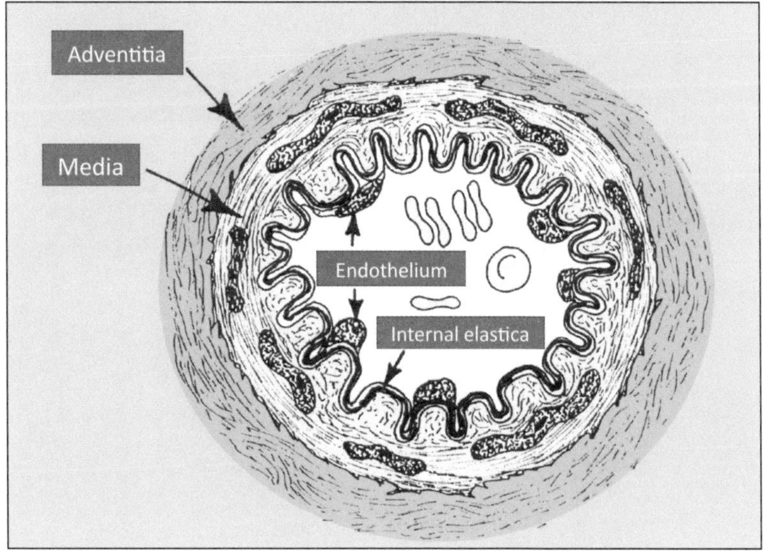

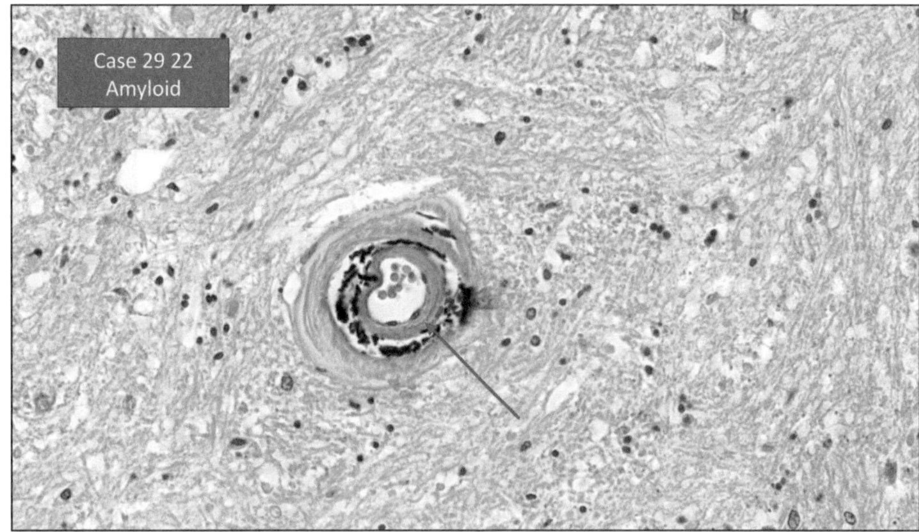

In the brain, there are residuals of bleeding. *(Congo Red stain)*

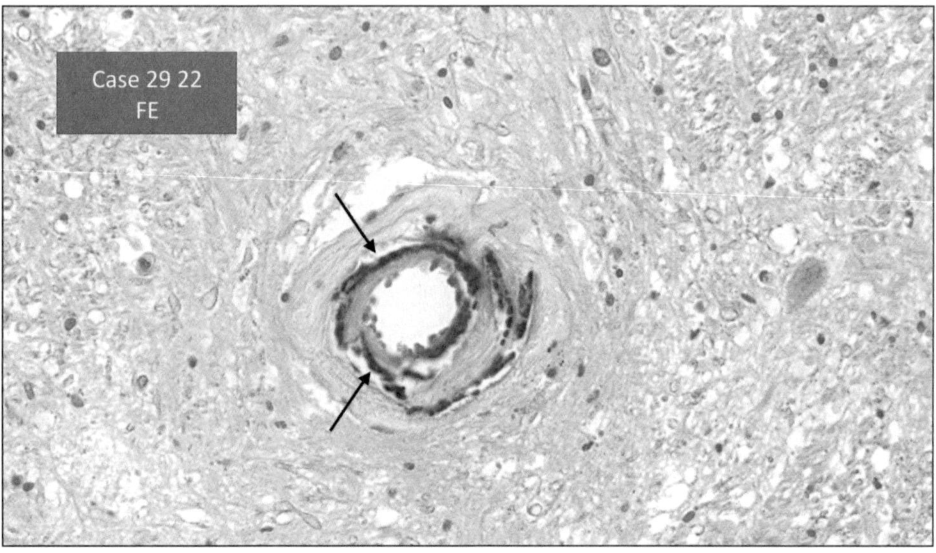

Blue stain represents Iron (Fe), so there has been bleeding in the small arteries in the brain.

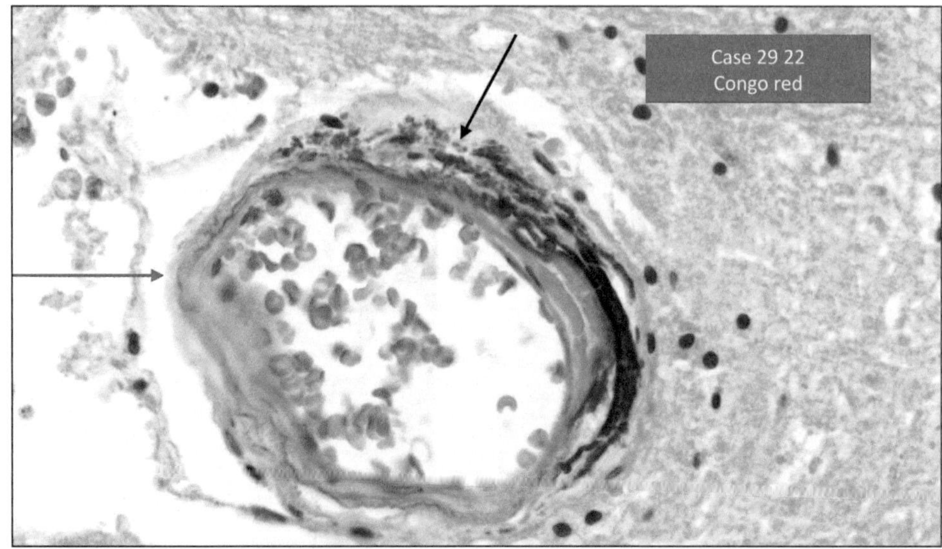

You can see that the elastic lamellae are disrupted (previous image), and there are small, what we call aneurysms,

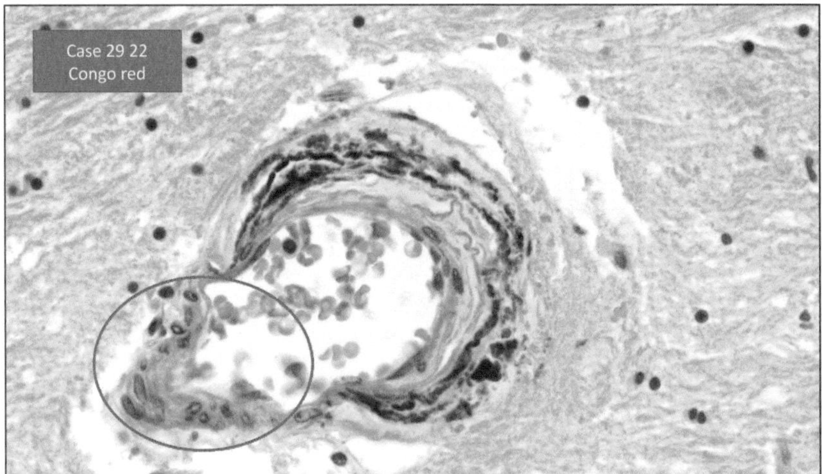

which might be ruptured at any time.

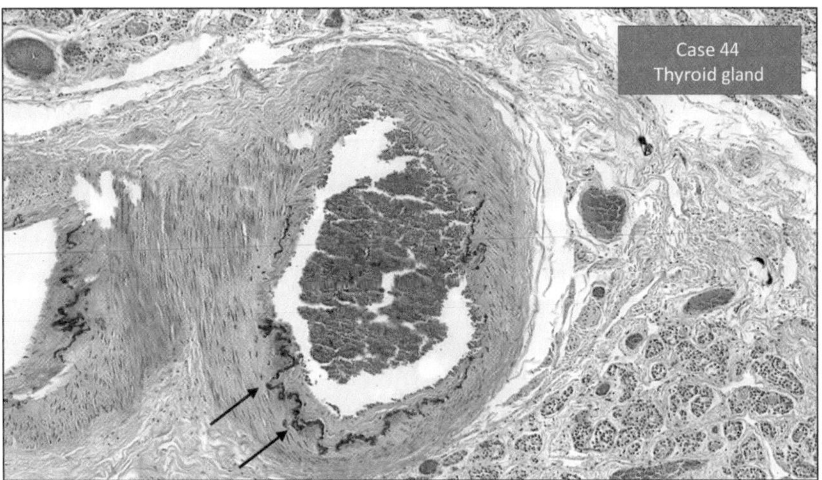

This occurred not only in the brain but also in the thyroid gland arteries. We saw these residuals of the destruction of the elastic lamella and iron deposition.

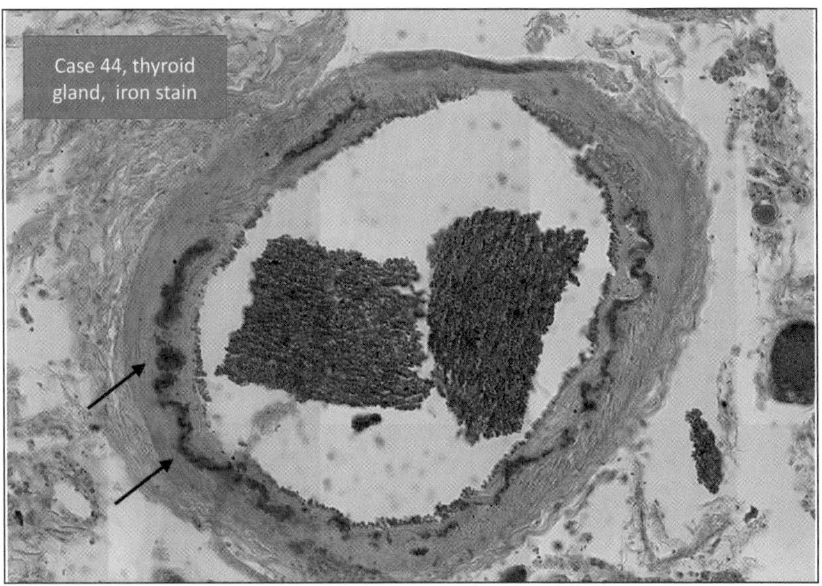

Here you can see the iron in the vessel wall (previous image).

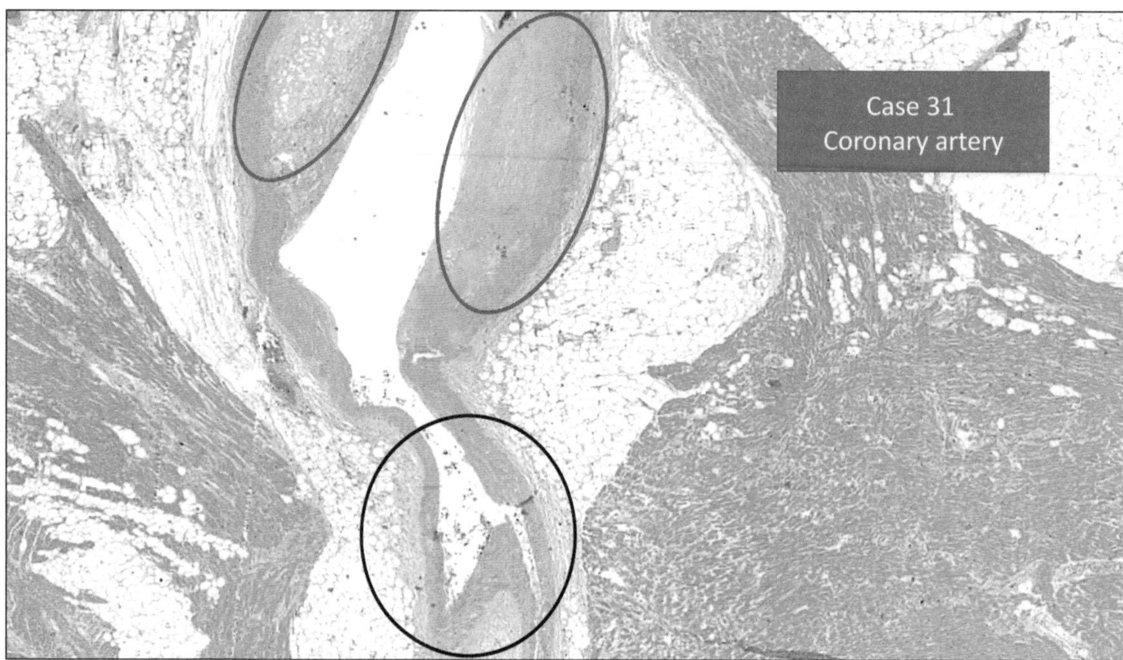

And now I come to the very delicate point because this might be the reason for some of the cases of a sudden death syndrome, because we see this in the coronary artery. At the bottom, you see a normal artery without arteriosclerosis. And in the upper part, you can see there's a cushion-like expansion and occlusion in the upper part. We saw this in many of the cases where the coronary artery was examined.

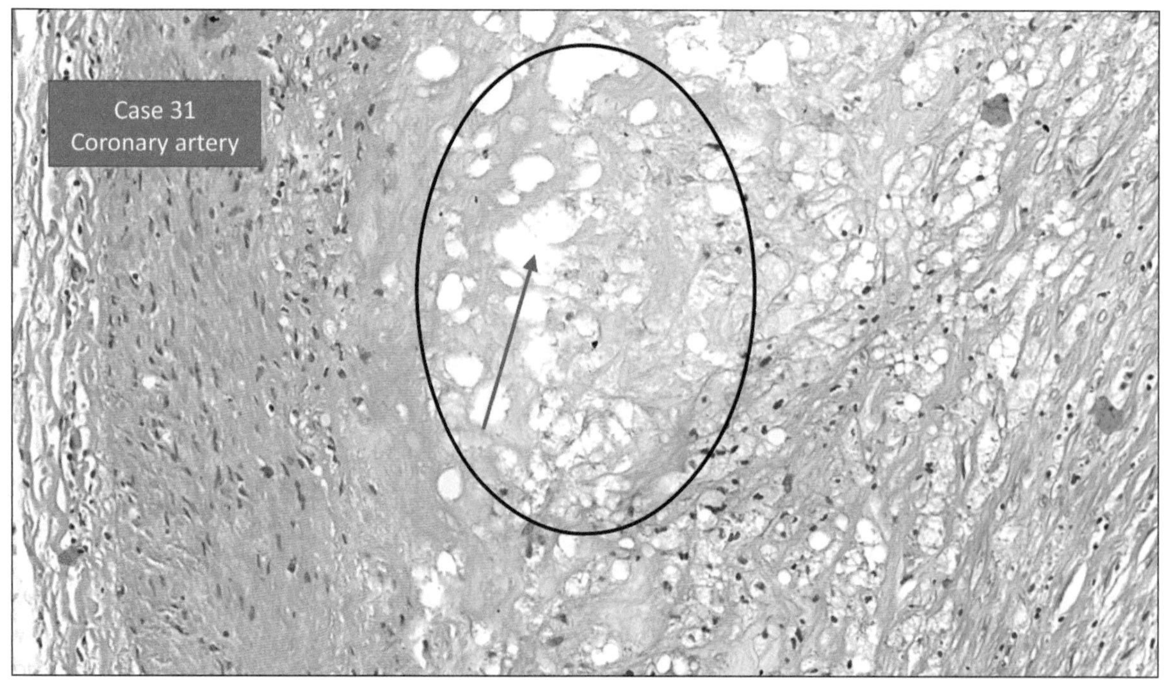

You can see here (previous image) the **medial necrosis** with **vacuolar degeneration**.

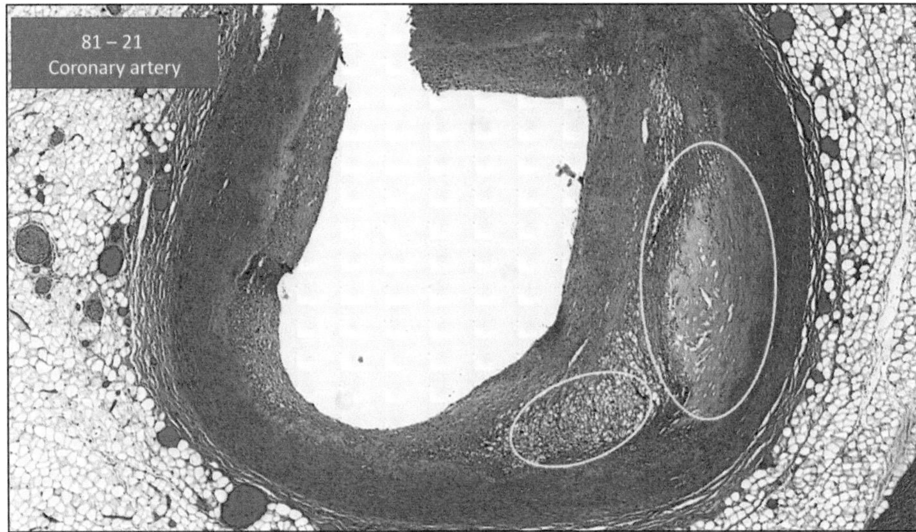

This is not arteriosclerosis (above). This is medial necrosis . . .

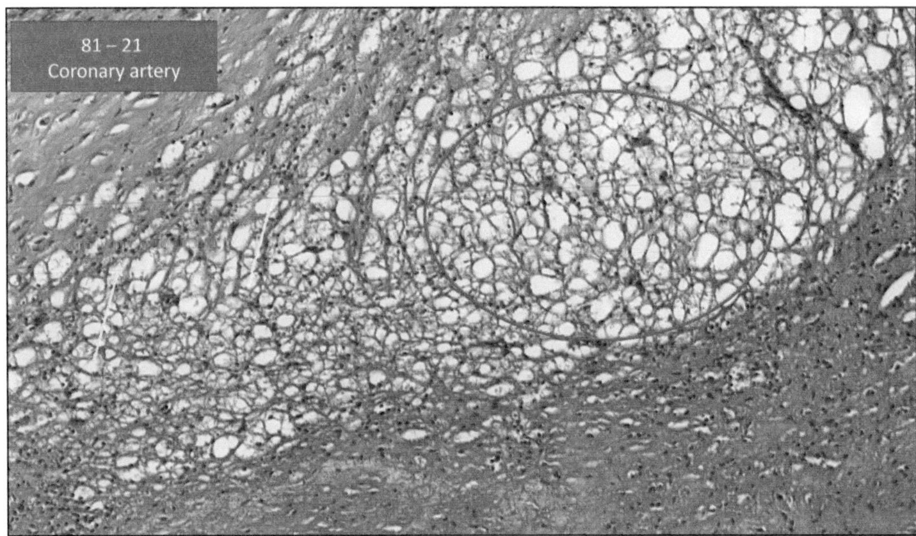

with some infiltration and **vacuolar degeneration.**

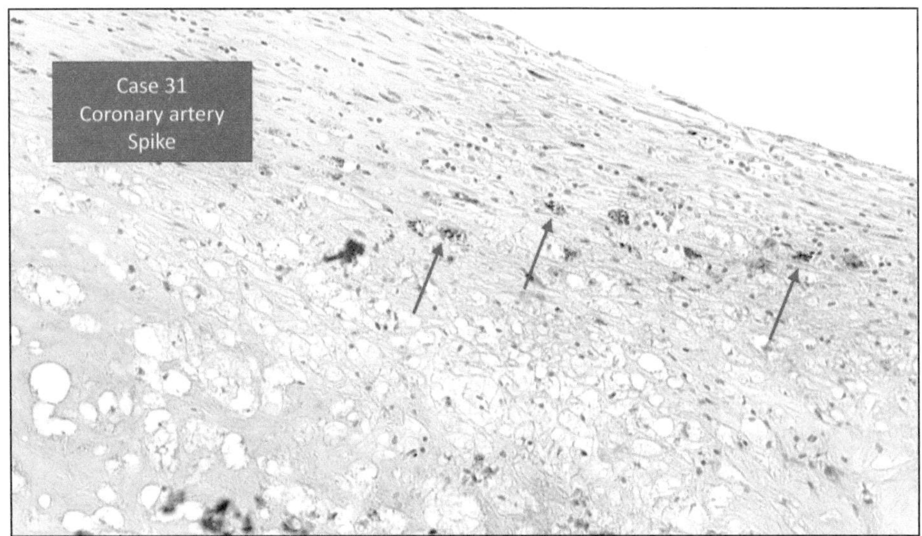

Here we demonstrate Spike protein in the inflammatory and myofibroblastic *(https://www.ahajournals.org/doi/10.1161/CIRCRESAHA.120.316958)* cells (previous image).

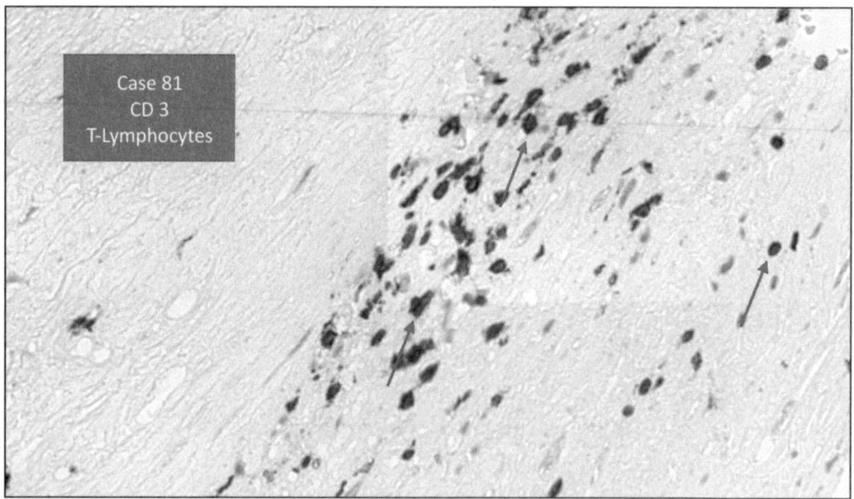

There are many T lymphocytes which, of course, cause immunological auto-aggression *(autoimmunity).*

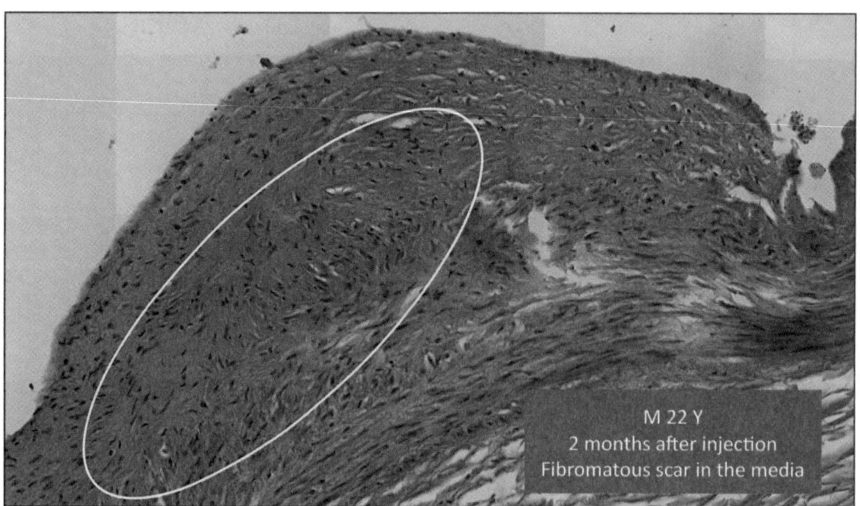

This is a case of a **22-year-old** man who has a scar in the coronary artery. This is **not an arteriosclerotic** plaque.

I come to the meaning of these findings.

„SADS – Sudden Adult Death Syndrome" - death without conventionally detectable cause

- Focal media necrosis of coronary arteries with swelling luminal constriction, with or without thrombosis
- Spike expression / T-lymphocyte – macrophage - myofibroblast reaction
- Lymphocytic perivasculitis of the vasa vasorum
- Acute heart failure without microscopic manifestation of necrosis

„acute coronary syndrome", „rhythmogenic heart failure"

I think that the Sudden Adult Death Syndrome, means a death without conventionally detectable cause. Pathologists term this as "arrhythmogenic heart failure," which I don't really know what they mean by that; but we have a focal medial necrosis of the coronary artery with swelling and luminal construction, with and without thrombosis.

We have Spike protein expression, T lymphocyte, macrophage, and myofibroblastic reaction. We have lymphocytic perivascular inflammation, and this may be the underlying cause of what we call an acute coronary artery.

There's no time that true necrosis of the muscle is manifest. So, there's no pathologic findings of myonecrosis *(dead heart muscle)*. We have no drugs outside the lymphatic organs showing an association of autoimmune disease.

Main pathologic findings (other organs)

3. Myocarditis
    - lymphocytic
    - with/without destruction of muscle fibers
    - scar formation

4. Alveolitis
    - diffuse alveolar damage (DAD)
    - lymphocytic interstitial pneumonia
    - extrinsic-allergic ?

5. Lymphocytic follicles outside lymphatic organs
    - associated with autoimmune diseases
    - „lymphocyte amok"

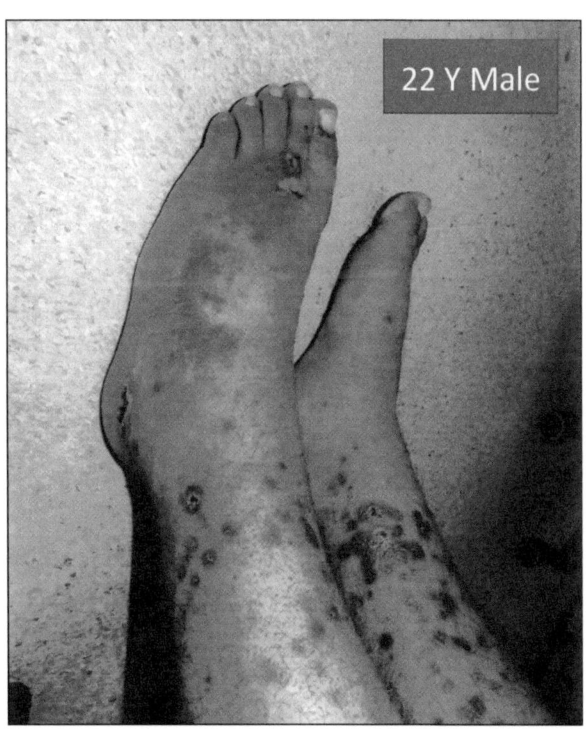

We are increasingly seeing younger persons with skin lesions (previous image). This is a **20-year-old man** with vasculitis and atypical lichen of the skin. You can see here the vasculitis.

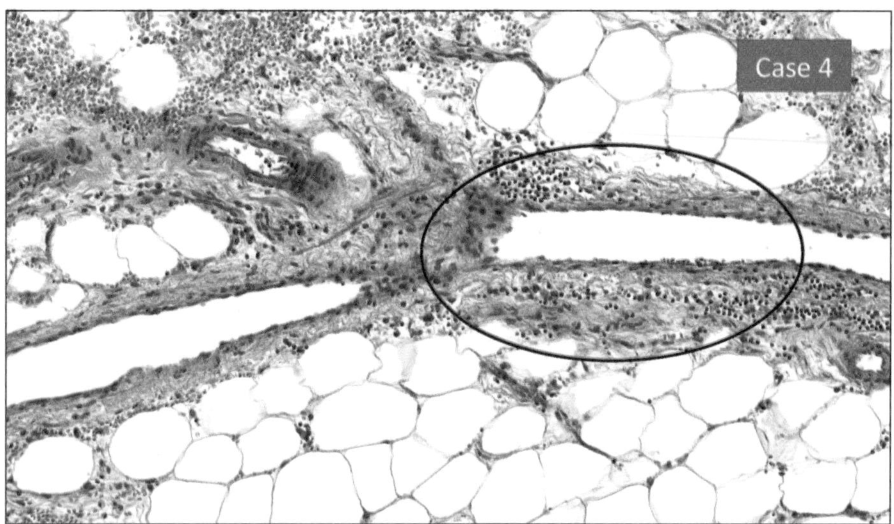

We have another case (image above), a very impressive case of a 30-year-old woman two weeks after the second injection. She had zero general side effects but had massive skin lesions, rash, *(inaudible),* and bullae. She corresponded with me. She told me ". . . my beautiful skin full of stains, my sexual life because I do not undress anymore."

Atypical lichen planus

- Case 68 35 Y W: 2 x Comirnaty
- Two weeks after the 2. injection serious general side effects with massive skin lesions: rash, whees, bullae
- Quote: „My beautiful skin full of stains, my sexual live ... because I do not undress anymore. I love the sea, it was taken from me, because I cannot wear a bikini. Before I felt fine, now I get panic attacks without reason."

I left out some details. "I love the sea. It was taken from me because I cannot wear a bikini. Before I have felt fine. **Now I get panic attacks without reasons**."

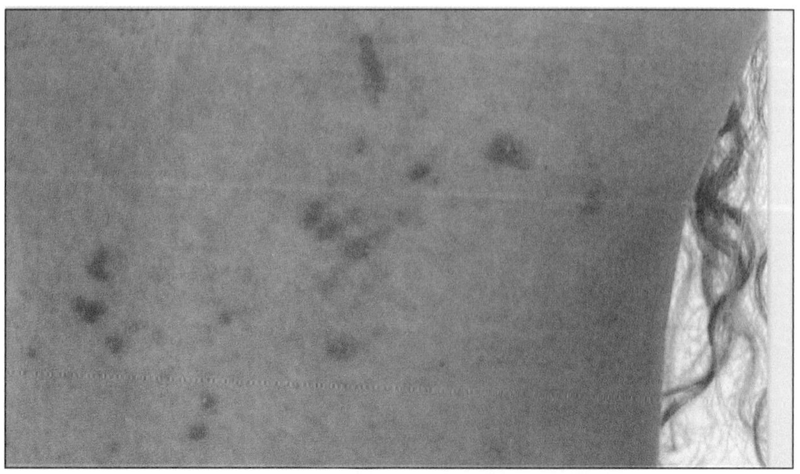

And these are the skin lesions that she never had before (previous image).

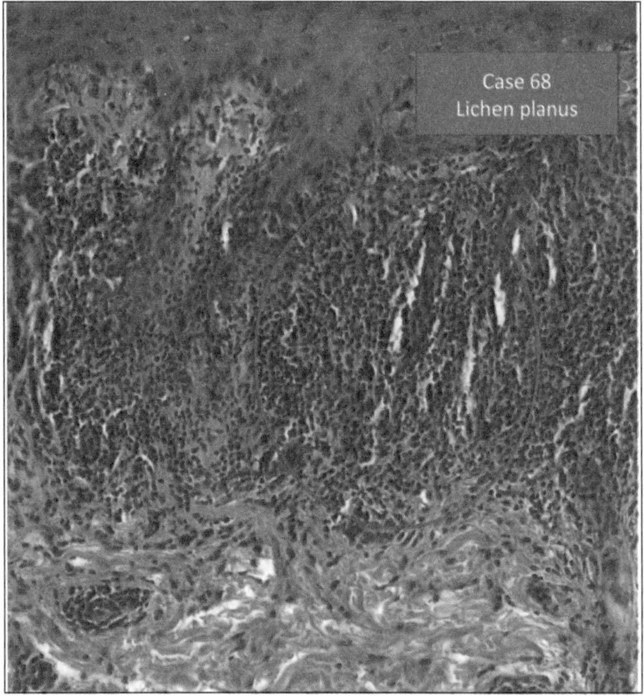

We see what we call, what is typical of an autoimmune disease, destruction of the basal cells with a **band like lymphocytic infiltration**.

But in the lower part, you can see a small vessel, and this is atypical. It's a vasculitis.

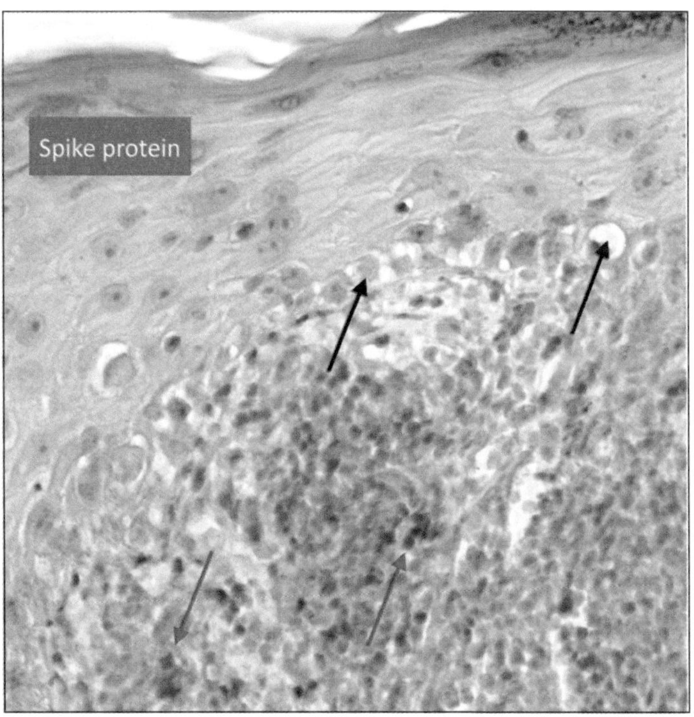

You can see the **vacuolated basal** cells are disrupted (previous image). And you can see again **Spike** protein can be demonstrated in the cell.

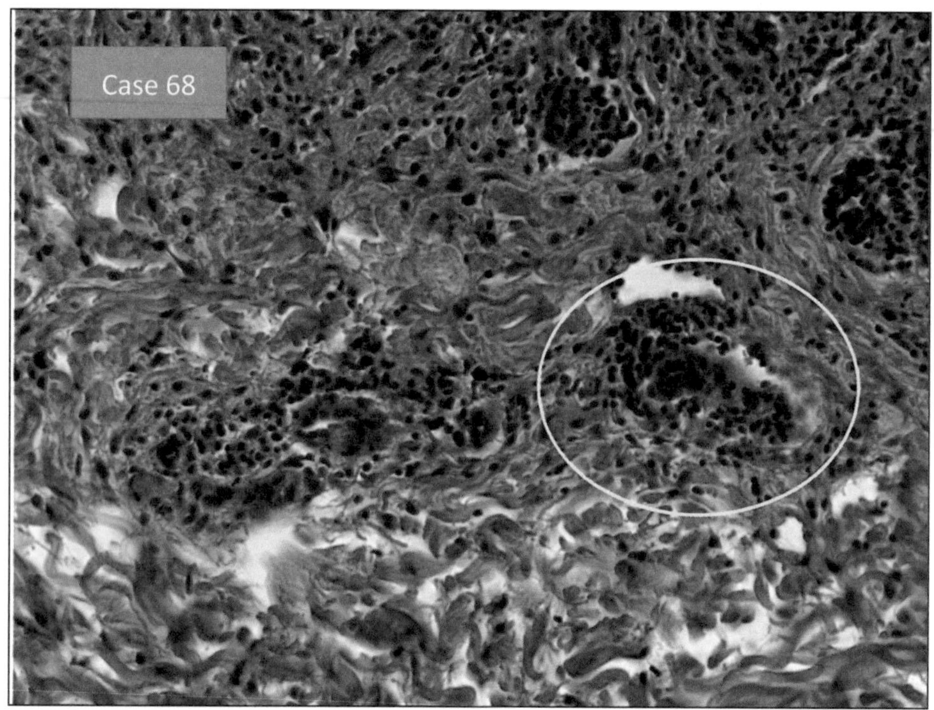

This is a **vasculitis** (above) and also in the small vessel are Spike protein.

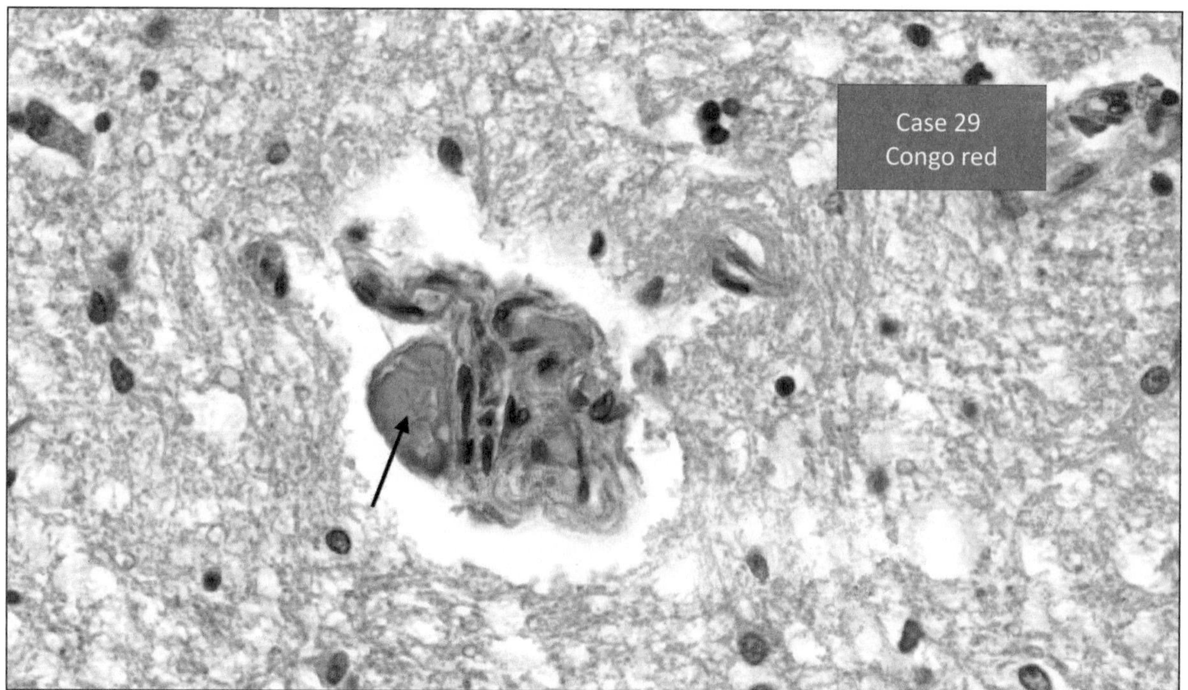

The last picture I will show you (previous image): the brain has Congo Red *(stain)* deposits which are very much like what we find in Alzheimer's disease and . . .

MITTWOCH, 21. SEPTEMBER 2022 – REUTLINGER GENERAL-ANZEIGER

Gesundheit – Doppelt so viele Diagnosen wie 2000

Alzheimer-Fälle nehmen zu

Cases of Alzheimer's disease increase – twice as many diagnoses as in 2000

WIESBADEN. Die Zahl der Alzheimer-Patienten steigt rapide. Wie das Statistische Bundesamt zum Welt-Alzheimertag am 21. September mitteilte, mussten bundesweit 19 356 Menschen im Jahr 2020 mit dieser Diagnose ins Krankenhaus. Damit hat sich die Zahl der stationären Behandlungen binnen 20 Jahren mehr als verdoppelt: Im Jahr 2000 hatte es 8 116 Behandlungen gegeben.

Im Jahr 2020 starben in Deutschland insgesamt 9 450 Menschen an Alzheimer – laut Statistischem Bundesamt so viele wie nie zuvor. Die Zahl der Todesfälle war mehr als doppelt so hoch wie im Jahr 2000 mit 4 535 Todesfällen.

»Das Risiko einer Erkrankung steigt mit zunehmendem Alter«, so die Statistiker: Rund 95 Prozent der im Jahr 2020 betroffenen Patienten waren 65 Jahre und älter. Dabei wurde die Altersgruppe der Hochbetagten ab 80 Jahren besonders häufig wegen Alzheimer im Krankenhaus behandelt: Mehr als die Hälfte aller Betroffenen gehörten dieser Altersgruppe

PORT LOUIS/FRANKFURT. Jeden Morgen, nachdem Direktor Giandev Moteea die Tür des Post-Museums von Mauritius aufschließt, geht sein erster Weg zu dem Kabinett, in dem die Rote Mauritius ausgestellt ist. Es ist zwar nur eine Replik der weltberühmten Briefmarke, die erstmals am 21. September 1847 herausgegeben wurde, doch Moteea kann sich an ihr nicht sattsehen. »Ich lebe den Traum eines jeden Briefmarkensammlers«, sagt er schmunzelnd.

Die Rote und auch die Blaue Mauritius symbolisieren einen wichtigen Teil der Geschichte des tropischen Inselparadieses, das im Indischen Ozean vor der Südostküste Afrikas liegt. Nachdem 1840 in England die erste vorausgezahlte Briefmarke der Welt gedruckt wurde, folgte die damalige britische Kolonie Mauritius dem Beispiel sieben Jahre später. Briefmarken waren damals ein neues Konzept, und nur eine Handvoll Länder verwendeten sie. Mauritius war das fünfte Land der Welt,

Alzheimer's disease is increasing.

Journal of Alzheimer's Disease 89 (2022) 411–414 411
DOI 10.3233/JAD-220717
IOS Press

Short Communication

Association of COVID-19 with New-Onset Alzheimer's Disease

Lindsey Wang[a], Pamela B. Davis[b], Nora D. Volkow[c], Nathan A. Berger[a], David C. Kaelber[d] and Rong Xu[e,*]

[a]*Center for Science, Health, and Society, Case Western Reserve University School of Medicine, Cleveland, OH, USA*
[b]*Center for Community Health Integration, Case Western Reserve University School of Medicine, Cleveland, OH, USA*
[c]*National Institute on Drug Abuse, National Institutes of Health, Bethesda, MD, USA*
[d]*The Center for Clinical Informatics Research and Education, The MetroHealth System, Cleveland, OH, USA*
[e]*Center for Artificial Intelligence in Drug Discovery, Case Western Reserve University School of Medicine, Cleveland, OH, USA*

There is an association of COVID-19, but as we see also with the vaccination.

I get about 20 to 35 telephone calls, because my name is now well-known in Germany and also in some adjourning country state, call me and they ask for help and they ask if I can do an examination and evaluation of their deceased relatives. All of themselves or the relatives with severe side effects.

I think I can close here. I could show you more cases. I have a collection, as I said, of 100 cases now, and it's very hard to select what I want to show you. But I think I have been able to show you that there are very disturbing, very alarming findings that we have in the autopsy and also in the biopsy cases.

I think this is a reason to stop this vaccination at once.

Eight: "The Flawed Trial of Pfizer's mRNA 'Vaccine'" —Linnea Wahl, MS—Team 5

The clinical trial of Pfizer's mRNA "vaccine" did not prove the mRNA injection is safe and effective, despite Pfizer's claims to the contrary. In fact, Pfizer stopped collecting useful data long before the planned end date of the clinical trial. Based on an inaccurate diagnostic test confirming COVID-19 in a tiny fraction of the study population, Pfizer researchers unblinded the control group and injected them with Pfizer's mRNA "vaccine." Because of these flaws in Pfizer's clinical trial, no valid conclusions can be drawn about safety and effectiveness of Pfizer's mRNA injection.

Pfizer researchers originally designed the clinical trial using the gold standard for drug testing: a double-blind, randomized, controlled trial. In this type of trial, neither patients nor researchers know who receives the drug (or intervention) being tested and who receives a placebo (double-blind). Researchers randomly choose trial participants to receive either the drug or placebo (randomized). Finally, researchers compare results of those who received the drug being tested (the experimental group) to those who received the placebo (the control group).

The Phase 3 randomized, controlled trial is the accepted standard method for assessing whether there is likely a beneficial effect due to an intervention (efficacy of the intervention). Randomized controlled trials "are the most stringent way of determining whether a cause-effect relation exists between the intervention and the outcome" (Kendall 2003).

Pfizer initiated a Phase 3 trial on July 27, 2020, which enrolled more than 40,000 adults (age 16 and older). About half of the participants received two doses of the Pfizer mRNA injection, and the other half received two doses of a saline placebo. According to the trial protocol, Pfizer expected participants in both groups to continue for up to 26 months (p. 15, Clinical Protocol), regardless of when various predetermined endpoints were reached.

Clinical trials typically involve both safety and efficacy endpoints. The safety of an intervention refers to the absence or low incidence rates of harmful side effects (adverse events). Efficacy endpoints refer to measurable outcomes, such as the rate of disease incidence (efficacy in the research world is generally equivalent to effectiveness in the real world). For a vaccine trial, researchers hypothesize that the disease rate in the experimental group will be significantly lower than the rate in the control group, based on a predefined expected difference.

As an endpoint, vaccine efficacy measures the reduction in disease rates (unvaccinated rate minus vaccinated rate) relative to the rate in the unvaccinated group. The Pfizer trial was designed to observe enough cases to provide a sufficient chance of detecting a minimal vaccine efficacy rate. Researchers can sometimes stop a trial early if the safety endpoint is clearly not being achieved and is thereby subjecting trial participants to high risks, or if the relative efficacy of the intervention is much greater than expected and can be established sooner than anticipated (that is, with fewer subjects). If researchers stop a vaccine trial early for reasons of efficacy, they may not have gathered sufficient data to establish the safety of the vaccine across various safety endpoints.

Pfizer designed its clinical trial to continue until 164 cases of COVID-19 were confirmed in trial participants after their second dose of Pfizer's mRNA injection. These 164 cases would "be sufficient to provide 90% power to conclude true VE [vaccine efficacy] >30% with high probability" (p. 38, Clinical Protocol). The US Food and Drug Administration (FDA) accepted this endpoint when it approved Pfizer's original protocol.

Pfizer declared Phase 3 of the trial a success on November 18, 2020, after 170 confirmed cases of COVID-19 and just about four months after the trial began. In a press release, Pfizer announced, "Pfizer and BioNTech Conclude Phase 3 Study of COVID-19 Vaccine Candidate, *Meeting All Primary Efficacy Endpoints*" (emphasis added). After just 170 cases of COVID-19 in 41,135 participants who had received their second doses by November 13, 2020, Pfizer called the efficacy test a success based on a COVID-19 incidence rate of only 0.4% (170/41,135).

To make matters worse, Pfizer researchers diagnosed these COVID-19 cases using the faulty polymerase chain reaction (PCR) test (p. 55, Clinical Protocol). Researchers knew at the time that this test was inaccurate and had high rates of both false negative and false positive results. Yet the FDA approved PCR use in this clinical trial. Because most of these questionable COVID-19 cases were in placebo recipients, Pfizer declared the mRNA injection effective in preventing COVID-19, and the FDA approved the Pfizer injection for emergency use (EUA) on December 11, 2020.

At the same time, Pfizer and the FDA knew that some clinical trial participants had reported serious side effects from the mRNA injection. On November 24, 2020, Pfizer informed the FDA of the trial results in an interim report. Those results included 285 serious adverse events such as heart, liver, and neurological disease; cancers; and deaths. At that point, all trial participants were still blinded, and Pfizer could have continued the gold standard conditions for drug testing. Perhaps Pfizer and the FDA preferred to end the trial before more reports of serious adverse events could be recorded?

Meeting a milestone does not necessarily mean the end of a clinical trial. Yet, having reached the efficacy endpoint, in December 2020 Pfizer unblinded the placebo group and offered the mRNA injection to original placebo recipients (p. 4, Interim Protocol). By March 2021, nearly 90% of the original placebo group had received at least one dose of the Pfizer mRNA injection (p. 3, Safety Tables). Thus, Pfizer effectively lost the ability to assess safety of the mRNA injection in comparison to a true control group.

Without a control group, data on adverse events are very difficult to interpret correctly, especially if one identifies only the *original* placebo and experimental ("vaccine") groups without considering what happens to the placebo group after they receive the mRNA injection. If someone originally receives a placebo, later receives the mRNA injection, and then has a serious side effect, how is this event counted—as occurring in the placebo group or in the experimental group? If the event is counted among the placebo recipients, then potential harms caused by the mRNA injection are masked, making it difficult to get a true picture of the cause-effect relation.

Those reporting the data must pay close attention to the date a placebo recipient received the mRNA injection (dose 3) and the date they experienced side effects. For example, Fig. 1 shows cardiac events in the original placebo (light blue) and original "vax" (light red) groups. When these groups are adjusted (dark blue and dark red) to account for cardiac events in the unblinded placebo recipients after they received the mRNA injection (dose 3), cardiac events in the original placebo group (light blue) are shifted to the adjusted "vax" group (dark red). Thus, the adjustment *increases* cardiac events in the "vax" group; conversely, cardiac events in the placebo group *decrease* when the onset date of the adverse event is determined to be after the placebo recipient was unblinded and given the mRNA injection (dose 3).

Pfizer graded these cardiac events as least serious (toxicity grade 1) to most serious (toxicity grade 4). In all grades, the effect of unblinding the placebo group was the same (Fig. 1). In all but grade 3, those who received the mRNA injection had more cardiac events than those who received the placebo (for grade 3, events in both groups were equal).

Pfizer sent a second interim report on these and other serious medical events to the FDA on April 1, 2021 (p. 77, Table 16.2.7.4.1, Interim Adverse Events). When Pfizer presented these data, did they distinguish between the original placebo group and the unblinded, mRNA-injected placebo group? When FDA personnel reviewed this report, did they consider that most of the original placebo group had already received at least one dose of the Pfizer mRNA injection?

Why did the FDA authorize emergency use of Pfizer's mRNA injection based on COVID-19 diagnosis in such a small fraction (0.4%) of the total study population? Why did the FDA approve the inaccurate PCR test for use in diagnosing COVID-19? Why did the FDA allow Pfizer to unblind the control group and abandon the gold standard in its clinical trial even as more and more mRNA recipients were seriously injured every day?

To Pfizer and the FDA, the clinical trial was a success; a "historically unprecedented achievement" according to Pfizer's press release. But when Pfizer unblinded the placebo recipients and ruined the control group, the clinical trial of Pfizer's mRNA injection did, in fact, fail. The ability to gather conclusive evidence of the long-term effects of Pfizer's mRNA injection was destroyed, and no valid conclusions can be drawn about safety and effectiveness from Pfizer's flawed clinical trial. When will Pfizer and the FDA admit this failure and stop promoting the mRNA "vaccine" as safe and effective?

Figure 1. Cardiac adverse events (AEs) adjusted for placebo group unblinding (dose 3)

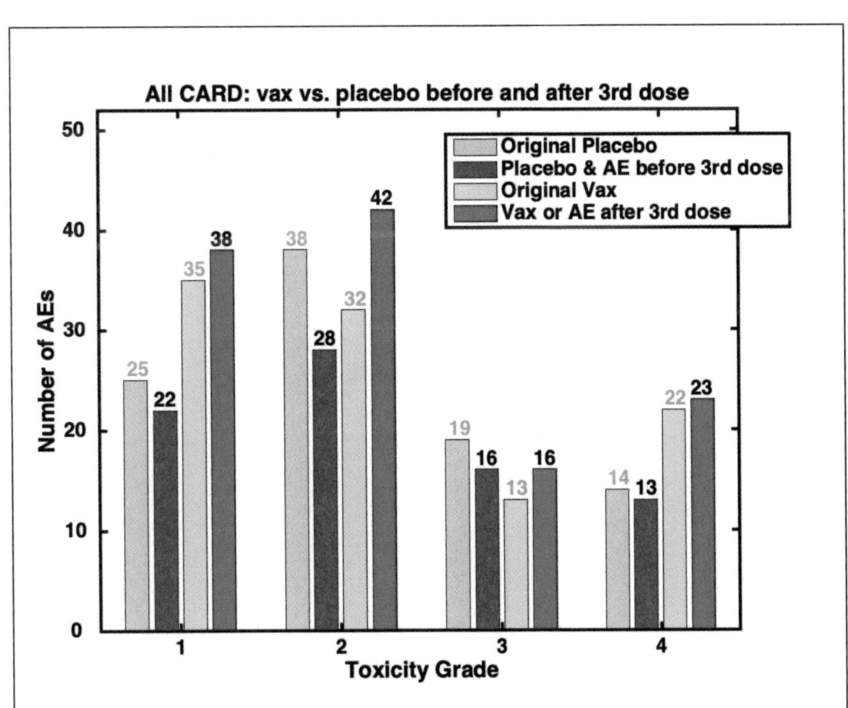

Summary

Most important finding: The clinical trial of Pfizer's mRNA "vaccine" failed, despite Pfizer's claims that the mRNA injection has been proven to be safe and effective.

Key detail leading to finding: Pfizer researchers deliberately stopped collecting data on the control group soon after the trial began.

Events of concern: Pfizer stopped the trial

- long before the planned end date
- when only a small fraction of participants had contracted COVID-19
- after COVID-19 diagnoses made based on inaccurate PCR tests
- while other participants were still reporting serious side effects

Further investigation: Why did the FDA authorize emergency use of Pfizer's mRNA injection based on COVID-19 infection in a small fraction (0.4%) of the total study population? Why did the FDA approve the flawed PCR test for use in diagnosing COVID-19? Why did the FDA allow Pfizer to unblind the control group and abandon the gold standard in its clinical trial even as more and more mRNA recipients were seriously injured every day? When Pfizer presented subsequent data, did they distinguish between the original placebo group and the unblinded, mRNA-injected placebo group? When FDA personnel reviewed subsequent data, did they consider that most of the original placebo group had already received at least one dose of the Pfizer mRNA injection?

Scale of situation: The FDA approved Pfizer's mRNA injection based on a flawed clinical trial.

Plain language explanation of key scientific term: A double-blind randomized controlled trial is an experiment in which neither patients nor researchers know who receives the drug (or intervention) being tested and who receives a placebo (double blind); the trial participants are randomly chosen to receive either the drug or placebo (randomized); and the results of those who received the drug being tested (the experimental group) are compared to those who received the placebo (the control group).

Nine: "449 Patients Suffer Bell's Palsy Following Pfizer mRNA COVID Vaccination in Initial Three Months of Rollout. A One-Year-Old Endured Bell's Palsy After Unauthorized Injection."

—Joseph Gehrett, MD; Barbara Gehrett, MD; Chris Flowers, MD; and Loree Britt

The WarRoom/DailyClout Pfizer Documents Analysis Project Post-Marketing Group (Team 1)—Joseph Gehrett, MD; Barbara Gehrett, MD; Chris Flowers, MD; and Loree Britt—produced a disturbing review of the Facial Paralysis System Organ Class (SOC) adverse events found in Pfizer document *5.3.6 Cumulative Analysis of Post-Authorization Adverse Event Reports of PF-07302048 (BNT162B2) Received Through 28-FEB-2021* (a.k.a., "*5.3.6*"). (https://www.phmpt.org/wp-content/uploads/2022/04/reissue_5.3.6-postmarketing-experience.pdf) This SOC includes facial paralysis and facial paresis, commonly known as Bell's palsy.

It is important to note that the adverse events (AEs) in the *5.3.6* document were reported to Pfizer for **only a 90-day period** starting on December 1, 2020, the date of the United Kingdom's public rollout of Pfizer's COVID-19 experimental mRNA "vaccine" product.

Key points in this report include:

- **Facial paralysis and facial paresis diagnoses made up 1.07% of the total patient post-marketing population**, or 449 total persons, reporting adverse events from December 1, 2020, to February 28, 2021.
- A **one-year-old infant** developed a Bell's palsy one day after vaccination. It was unresolved at the time of the 5.3.6 report. **The vaccine was not approved for use in children or infants at the time.**
- **399 cases (88%) were classified as serious.**
- Cases included: 295 **(66%) female**, 133 male (30%), and 21 (5%) not reported.
- Of events where time of onset was recorded, the time from vaccination to the adverse event becoming apparent ranged from **within the first 24 hours to 46 days, *with half of the facial events observed within two days***.
- Only one clinical finding in these cases: **damage to the 7th cranial nerve resulting in weakness or paralysis of the side of the face that is supplied by that nerve**.
- Consequences of that nerve damage can include eye damage from inability to close the eyelid, impaired speech, impaired mouth closure (drooling) when eating.
- Pfizer identified that ". . . noninterventional post-authorisation safety studies, C4591011 and C4591012 are expected to capture data on a sufficiently large vaccinated population to detect an increased risk of Bell's palsy in vaccinated individuals. The timeline for conducting these analyses will be established based on the size of the vaccinated population captured in the study data sources by the first interim reports (due 30 June 2021)."

Pfizer concluded: "This cumulative review does not raise new safety issues. Surveillance will continue." However, since finalizing the 5.3.6 report at the end of February 2021, **there has been no further summary data released for outside review**. Furthermore, a search on https://clinicaltrials.gov/ for the cited studies (C4591011 and C4591012) yielded no studies found (accessed February 23, 2023).

Please read this important report below.

War Room/DailyClout Pfizer Document Analysis

Post-Marketing Team Micro-Report 8:

Facial Paralysis System Organ Class (SOC) Review of 5.3.6

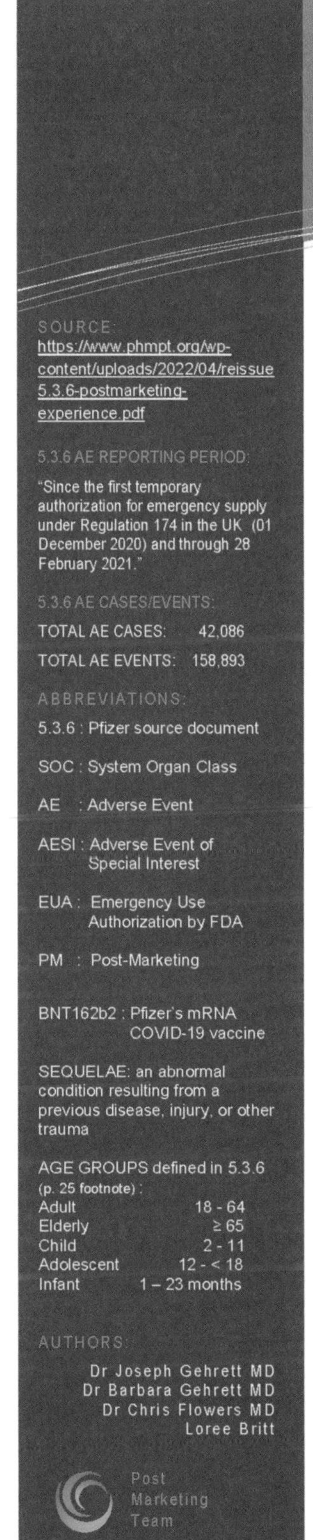

SOURCE:
https://www.phmpt.org/wp-content/uploads/2022/04/reissue 5.3.6-postmarketing-experience.pdf

5.3.6 AE REPORTING PERIOD:
"Since the first temporary authorization for emergency supply under Regulation 174 in the UK (01 December 2020) and through 28 February 2021."

5.3.6 AE CASES/EVENTS:
TOTAL AE CASES: 42,086
TOTAL AE EVENTS: 158,893

ABBREVIATIONS:
5.3.6 : Pfizer source document
SOC : System Organ Class
AE : Adverse Event
AESI : Adverse Event of Special Interest
EUA : Emergency Use Authorization by FDA
PM : Post-Marketing
BNT162b2 : Pfizer's mRNA COVID-19 vaccine
SEQUELAE: an abnormal condition resulting from a previous disease, injury, or other trauma

AGE GROUPS defined in 5.3.6 (p. 25 footnote) :

Adult	18 - 64
Elderly	≥ 65
Child	2 - 11
Adolescent	12 - < 18
Infant	1 – 23 months

AUTHORS:
Dr Joseph Gehrett MD
Dr Barbara Gehrett MD
Dr Chris Flowers MD
Loree Britt

Post Marketing Team

08Mar23

Facial Paralysis *Search criteria: PTs Facial paralysis, Facial paresis*	• Number of cases: 449[a] (1.07% of the total PM dataset), 314 medically confirmed and 135 non-medically confirmed; • Country of incidence: US (124), UK (119), Italy (40), France (27), Israel (20), Spain (18), Germany (13), Sweden (11), Ireland (9), Cyprus (8), Austria (7), Finland and Portugal (6 each), Hungary and Romania (5 each), Croatia and Mexico (4 each), Canada (3), Czech Republic, Malta, Netherlands, Norway, Poland and Puerto Rico (2 each); the remaining 8 cases originated from 8 different countries; • Subjects' gender: female (295), male (133), unknown (21); • Subjects' age group (n=411): Adult (313), Elderly (96), Infant and Child (1 each); • Number of relevant events[b]: 453, of which 399 serious, 54 non-serious; • Reported relevant PTs: Facial paralysis (401), Facial paresis (64); • Relevant event onset latency (n = 404): Range from <24 hours to 46 days, median 2 days; • Relevant event outcome: resolved/resolving (184), resolved with sequelae (3), not resolved (183) and unknown (97); Overall Conclusion: This cumulative case review does not raise new safety issues. Surveillance will continue. Causality assessment will be further evaluated following availability of additional unblinded data from the clinical study C4591001, which will be unblinded for final analysis approximately mid-April 2021. Additionally, non-interventional post-authorisation safety studies, C4591011 and C4591012 are expected to capture data on a sufficiently large vaccinated population to detect an increased risk of Bell's palsy in vaccinated individuals. The timeline for conducting these analyses will be established based on the size of the vaccinated population captured in the study data sources by the first interim reports (due 30 June

Table 7. AESIs Evaluation for BNT162b2

AESIs Category	Post-Marketing Cases Evaluation Total Number of Cases (N=42086)
	2021). Study C4591021, pending protocol endorsement by EMA, is also intended to inform this risk.

https://www.phmpt.org/wpcontent/uploads/2022/04/reissue 5.3.6-postmarketiexperience.pdf

- **Adverse Events were reported to Pfizer during a 90-day period, following the December 1, 2020, public rollout of its COVID-19 experimental "vaccine" product.**
- **In the Pfizer 5.3.6 document, these AEs were categorized by System Organ Classes (SOC) – in other words, by systems in the body.**
- **The demographics of the cases included 295 (66%) female, 133 male (30%), and 21 (5%) not reported.**
- **Age was reported for 411 persons with elderly 96 (21%), adult 313 (70%), and one each in the infant and child age groups.**

The data collected and reported here are commonly referred to both medically and in lay terminology as Bell's Palsy, though the search criteria was limited to facial paralysis and facial paresis. Under these terms, Pfizer reports **449 persons with this syndrome or 1.07% of the total patients reporting adverse events during this time period.**

Of the "relevant events," 399 (88%) were classified as serious. Of the 404 events where time of onset was recorded, the time from vaccination to the adverse event becoming apparent ranged from within the first 24 hours to 46 days, with **half of the facial events observed within two days.**

The outcomes of this SOC were as follows:
184 (39%) were reported as "resolved or resolving," though we do not know about the final degree of resolution in the "resolving" patients. Three (<1%) resolved with sequelae, **183 (39%) were not resolved, and in 97 (21%) there was no assessment as to resolution.**

In this SOC there is only one clinical finding: damage to the 7th cranial nerve resulting in weakness or paralysis of the side of the face that is supplied by that nerve. The consequences can be eye damage from inability to close the eyelid, impaired speech and impaired mouth closure (drooling) with eating, impaired social engagement, and potentially impaired employment.

A one-year-old infant developed a Bell's palsy one day after vaccination. It was unresolved at the time of the 5.3.6 report. These details were discovered in a footnote at the end of Table 7. *The vaccine was not approved for use in children or infants at the time.*

Emerging Patterns in 5.3.6 Table 7 Patients:

- Adverse Events in Women > Men at the rate of 2:1
- Inconsistent SOC categorization diluting full impact of adverse events
- Well over half the AEs with latency reported occurred within five days

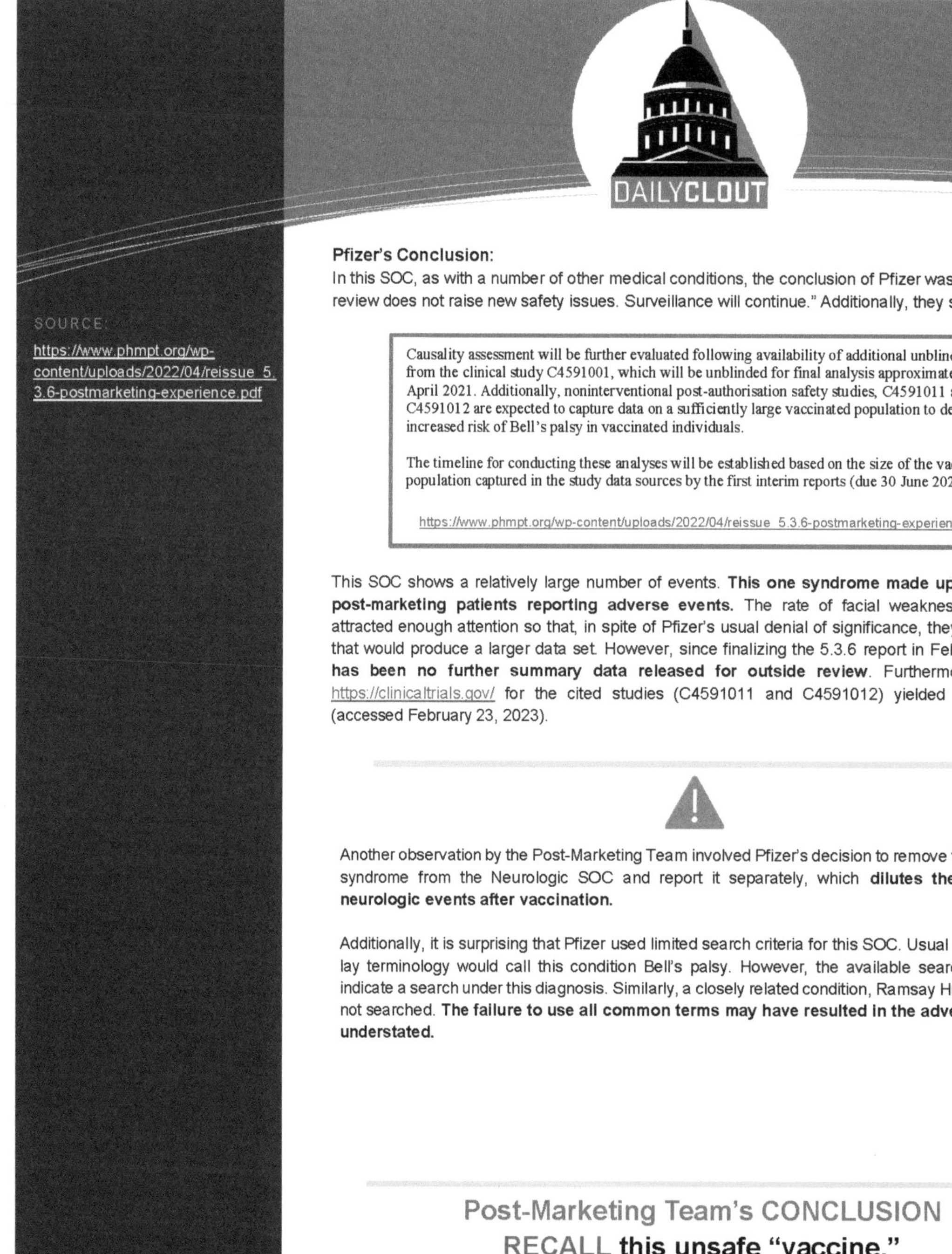

SOURCE:
https://www.phmpt.org/wp-content/uploads/2022/04/reissue_5.3.6-postmarketing-experience.pdf

Pfizer's Conclusion:
In this SOC, as with a number of other medical conditions, the conclusion of Pfizer was, "This cumulative review does not raise new safety issues. Surveillance will continue." Additionally, they state:

> Causality assessment will be further evaluated following availability of additional unblinded data from the clinical study C4591001, which will be unblinded for final analysis approximately mid-April 2021. Additionally, noninterventional post-authorisation safety studies, C4591011 and C4591012 are expected to capture data on a sufficiently large vaccinated population to detect an increased risk of Bell's palsy in vaccinated individuals.
>
> The timeline for conducting these analyses will be established based on the size of the vaccinated population captured in the study data sources by the first interim reports (due 30 June 2021)
>
> https://www.phmpt.org/wp-content/uploads/2022/04/reissue_5.3.6-postmarketing-experience.pdf

This SOC shows a relatively large number of events. **This one syndrome made up 1% of the entire post-marketing patients reporting adverse events.** The rate of facial weakness/paralysis clearly attracted enough attention so that, in spite of Pfizer's usual denial of significance, they identified studies that would produce a larger data set. However, since finalizing the 5.3.6 report in February 2021, **there has been no further summary data released for outside review**. Furthermore, a search on https://clinicaltrials.gov/ for the cited studies (C4591011 and C4591012) yielded no studies found (accessed February 23, 2023).

Another observation by the Post-Marketing Team involved Pfizer's decision to remove this facial paralysis syndrome from the Neurologic SOC and report it separately, which **dilutes the larger issue of neurologic events after vaccination.**

Additionally, it is surprising that Pfizer used limited search criteria for this SOC. Usual medical as well as lay terminology would call this condition Bell's palsy. However, the available search terms nowhere indicate a search under this diagnosis. Similarly, a closely related condition, Ramsay Hunt syndrome, was not searched. **The failure to use all common terms may have resulted in the adverse events being understated.**

Post-Marketing Team's CONCLUSION
RECALL this unsafe "vaccine."

Post Marketing Team

Ten: "Ute Krüger, MD, Breast Cancer Specialist, Reveals Increase in Cancers and Occurrences of 'Turbo Cancers' Following Genetic Therapy 'Vaccines'."

—Robert W. Chandler, MD, MBA

Ute Krüger, MD, DMedSci

Sabotage?
A pathologist reports

Ute Krüger, MD

Stockholm 22 JAN 2023

(https://lakarupproppet.se/international-conference-pandemic-strategies/)

Ute Krüger, MD, DMedSci, is a pathologist educated in Germany now living and working in Sweden. She wrote her doctoral thesis at the Humboldt University in Berlin on the analysis of 7,500 autopsy protocols. With over 25 years of experience in Clinical Pathology and a special interest in breast cancer, Ute has been researching breast cancer at Lund University for almost eight years. She was a member of the Board of the Swedish Association of Pathologists and was previously Medical Chief of the Department of Clinical Pathology in Växjö. She works as a Senior Physician in the Department of Clinical Pathology in Kalmar.

Note: A rough draft was prepared using voice recognition that was then was edited for clarity with an effort to retain Dr. Krüger's original meaning. The text was then added to the presentation graphics.

Dr. Krüger, MD, is a pathologist who specializes in diseases of the breast. Her lecture concerned changes in breast cancer cases following introduction of Spike generating genetic drug therapy.

Ute Krüger, MD

I'm very pleased to have the opportunity to speak as a member of the Physician's Appeal, Läkaruppropet. Thank you to all my colleagues who organized this important conference and to all of you for coming.

You may be wondering about the title of my lecture. But more on that later.

At the beginning of the COVID-19 outbreak, I researched on the internet and found that the mortality rate following infection with the coronavirus was around 2%. In comparison, sarcoidosis is an inflammatory disease, and the most important interstitial lung disease in Western Europe is described as having a mortality rate of just under 5%.

Have you ever heard anything in the media about sarcoidosis, which has a much higher mortality rate than COVID-19?

In addition, in medical school, I was given a different definition of the term 'pandemic' than we use today. In the period that followed, I observed the events very closely.

I've been working in pathology for 25 years. For the last 19 years, I was mainly involved in the diagnosis of breast specimens. Almost eight years ago, I got the opportunity to work in breast cancer research at Lund University. Here I have reevaluated 1,500 breast cancer cases according to currently valid guidelines.

With this experience, I know approximately the distribution of average age, tumor size, and degree of malignancy that I can expect in the daily input material of breast cancer.

Already, in Autumn 2021, I had the impression that I was suddenly receiving more material

1. From **younger patients**, often 30 to 50 years old.
2. The tumors were **growing more aggressively and faster**.
3. And that they were **larger**. More than four centimeters was not uncommon. I saw tumors up to 16 centimeters in size in the breast.
4. I also had the impression that **multifocal tumor growth** and **bilateral tumor growths** were more frequent.

My efforts at the second pathology conference in Germany, in December 2021, to find colleagues to help me prove or disprove my hypothesis of **Turbo Cancer** after vaccination against COVID-19 met with little response.

I would like to go into detail. Since I love my microscope and pictures, I would like to show you some pictures from histology. Agent E stands for Hematoxylin Eosin *(H&E)* and is a routine stain in clinical histopathology.

Tumor Aggressiveness

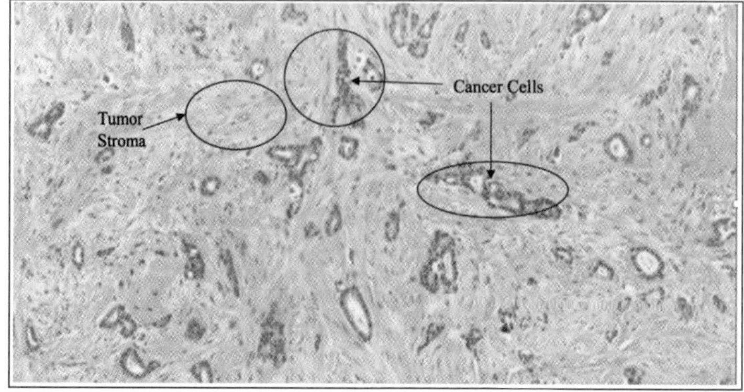

Here we see a breast carcinoma that is well differentiated (previous image). I think for most here it's difficult to interpret histologic slides. So here are a few explanations.

1. You see, small tumor glands, the dark ones here, in a relative abundance.
2. The tumor cells are relatively small and about the same size.

In the next image you see a poorly differentiated, more aggressively growing tumor. (Below) You see a solid tumor mass with a very small amount of stroma.

The tumor cells are slightly larger.

There are also cells with **huge cell nuclei**.

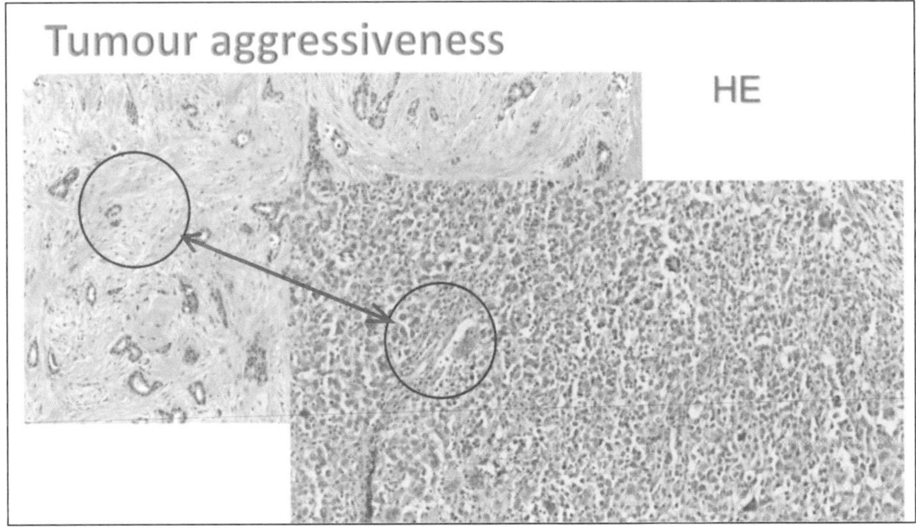

But I chose this example because of the proliferation activity. You will see the next picture is very important here to show how many tumor cells are dividing. There is an immunohistochemical stain, Ki67, which is a proliferation marker. About two-thirds of all breast cancers have a low proliferation. One-third of breast cancers are highly proliferative, meaning they grow faster and more aggressively.

I would like to show you the tumor shown in the previous picture stained with Ki67.

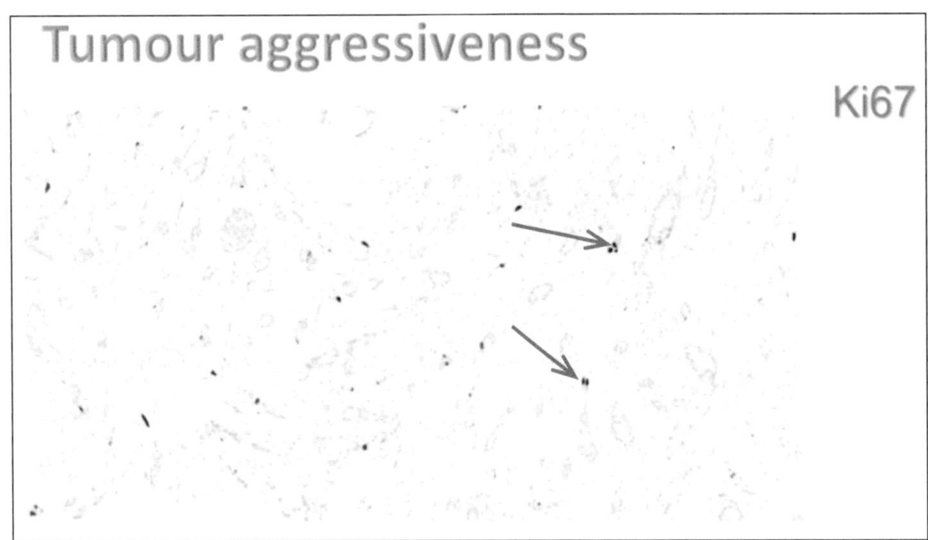

The first tumor (above) showed low proliferative activity. Less than 10% of the nuclei are stained brown, meaning they are in proliferation. You see the dots **here**.

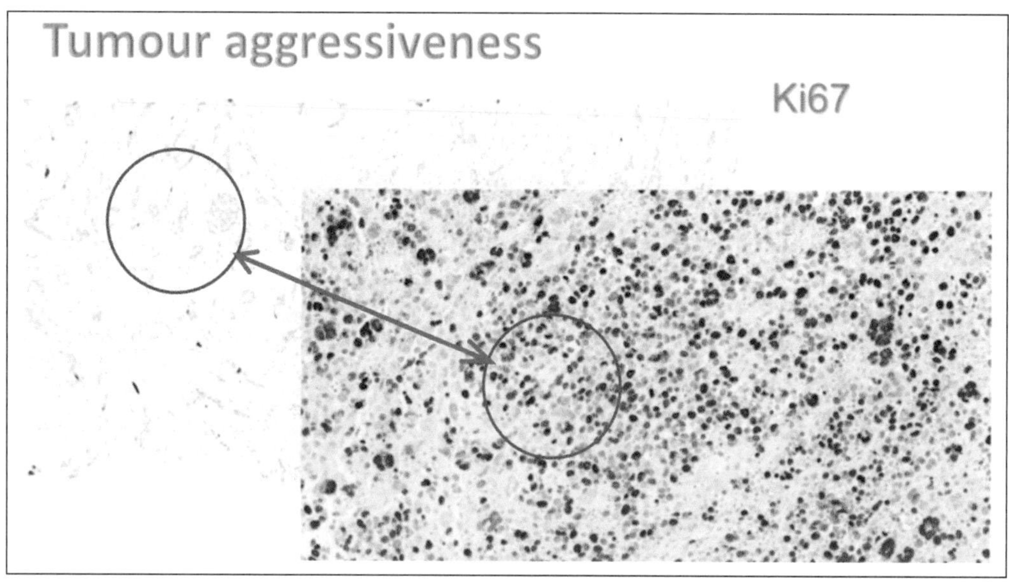

Here (above) is the very fast-growing tumor with about 70% tumor cells staining brown. That means more than two-thirds of the tumor cells are in cell division. Such aggressively growing tumors seem to be more common in my routine material now.

Of course, the aggressiveness is also reflected (below) in the increased number of mitosis (arrows).

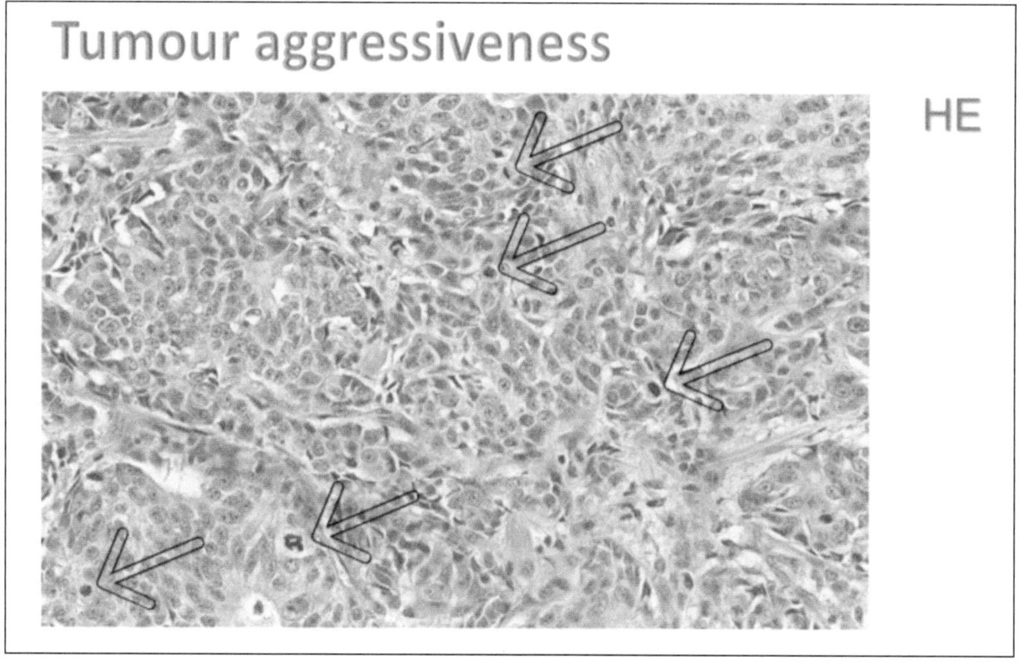

This is another breast carcinoma.

I see more and more multifocal tumors.

Multifocality

Multifocality - Case 1

- 55-year-old women
- Vaccination
- Sectoral resection:

  130 mm ductal carcinoma in situ

  20 invasive foci, max 35 mm

I found in her breast specimen a 130-millimeter ductal carcinoma in situ. That means a tumor growth in the mamillary ducts. In the same area, I found 20 invasive foci of poorly differentiated ductal breast carcinoma. And the largest invasive focus was 35 millimeters.

Here we see one of the slides. The area was ductal carcinoma in situ marked with blue. In this picture, you see seven different invasive foci marked with red. The tumor had short, pronounced growth in the lymphatic vessels, even far outside the actual invasive event in the breast, and four lymph nodes were already affected.

Recurrences

Another anomaly I think I see is recurrences. These are patients who have had breast cancer before and were more or less considered cured. It can be carcinoma from 20 years ago. **Relatively soon after the vaccination against COVID-19, the tumor growth explodes, and there is a pronounced spread of the tumor in the body; and some of the patients die within a few months.**

I will show you the case of an 80-year-old woman. Seven years ago, she had a sectoral resection, which means that part of her breast with the tumor was removed.

Three months after the vaccination against COVID-19, she was able to detect the tumor in her breast. It was a fast-growing tumor with a size of 55 millimeters in the surgical specimen.

At the same time, several skin metastases were found in the breast skin of the same site. This is also very unusual for a patient with a recurrence in the breast and skin metastasis at the same time.

Here you can see the current specimen:

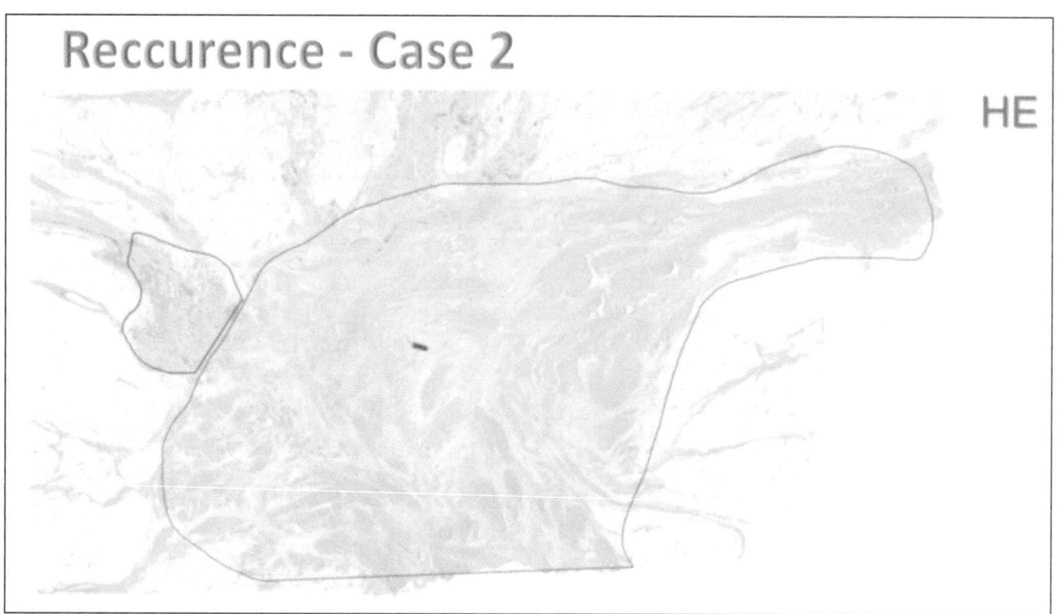

The area of the previous resection is marked in green and shows scar tissue. I hope you can see. This green is not the best color here. Right next into it, you can see the red area, and this is the recurrence. In this image, the direct proximity of the recurrence to the scar tissue was impressive.

Heterogeneity

The next aspect I would like to highlight is heterogeneity of the tumors. That means the tumor has different patterns.

I would like to show you an example of a 70-year-old woman. She had a lobular, that means relatively slow growing breast carcinoma, but she also had metastases for several years. She had been living with her metastases in the urinary bladder, the intestinal mucosa, bones, and liver for three years, and her body was in a certain balance.

Shortly after vaccination against COVID-19, the tumor growth in the liver exploded, and the patient died within a month. The doctor who sent me the liver sample wrote on the referral that it was unusual that the metastasis in the liver grew extremely fast, but the tumors in the other metastatic sites in her body did not grow at the same rate.

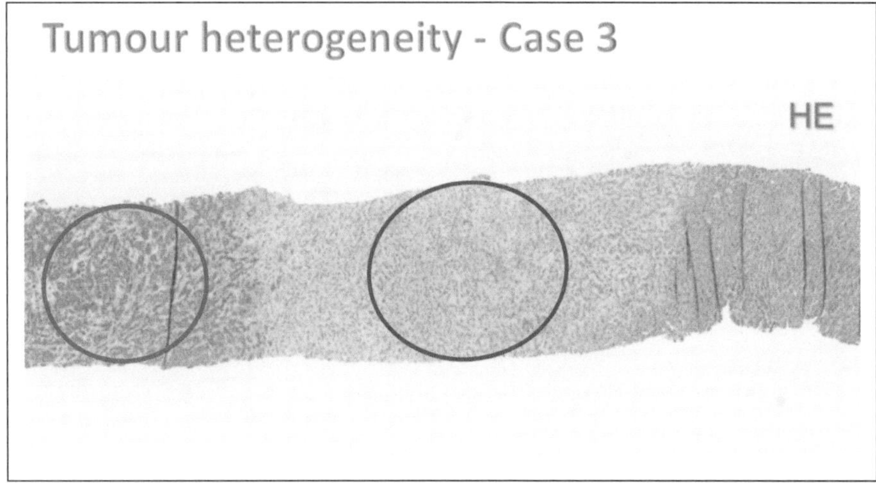

In the liver biopsy, in the microscope image on the right (referring to projection screen) here, you can see predominantly tumor-free liver tissue in the middle. (Blue circle) You see the known metastasis of lobular breast carcinoma with relatively small cell nuclei and clearly visible stroma. And on the left side you see the aggressively growing, newly added tumor with more compact tumor growth. (Red circle)

Coincident Tumors

Another anomaly is coincident tumors. Last year I had three patient cases within three weeks with three concurrent carcinomas in different organs (*inaudible*). This a few months after vaccination against COVID-19.

For example, I received a sample of breast carcinoma from a woman who had been diagnosed with primary lung carcinoma and primary pancreatic carcinoma at the same time. This is very unusual and remarkable. The tumor conferences are becoming more and more complicated. At my last breast cancer conference, I had 35 persons, some which were guests.

I have always worked in such smaller hospitals like this and never had so many patients at a tumor conference in the last 19 years before. A Swedish colleague, someone who works mainly as a neuropathologist, confirmed to me that tumor conferences are becoming more and more complicated. It's not unusual for patients to have several malignancies in different organs at the same time.

Benign Tumors

Now I would like to talk about benign tumors in the breast. Another anomaly I think I see is appearance of larger fibroepithelial tumors.

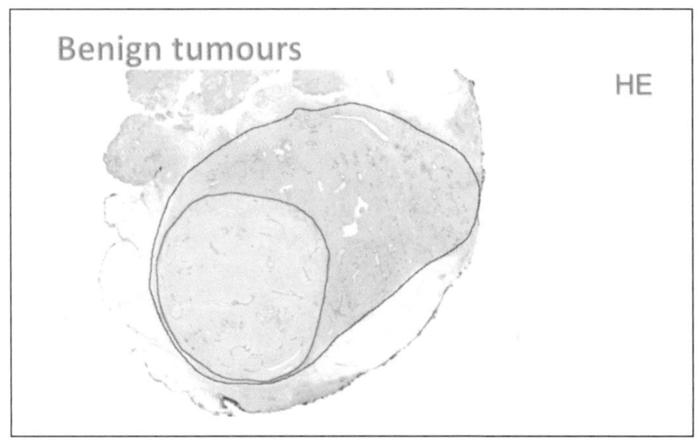

Here we see an example of fibroadenoma marked in blue (previous image). More than two-thirds of the fibroadenoma shows a little proliferative stroma in this part of the fibroadenoma. It has not grown in the last few months.

The area highlighted in red is the proliferative area in the tumor. The stroma was loosened (sic) with proliferative activity.

I see such changes with proliferative areas in already existing fibroadenomas more frequently now. I also see fibroadenomas that are proliferative active through.

Inflammation

Sometimes I get the clinical information that the patient comes to the doctor because of pain and the palpation finding, and it's possible that the cause of the pain is inflammation.

It's not uncommon for me to see inflammatory cells both in the tumor and in the surrounding tissue.

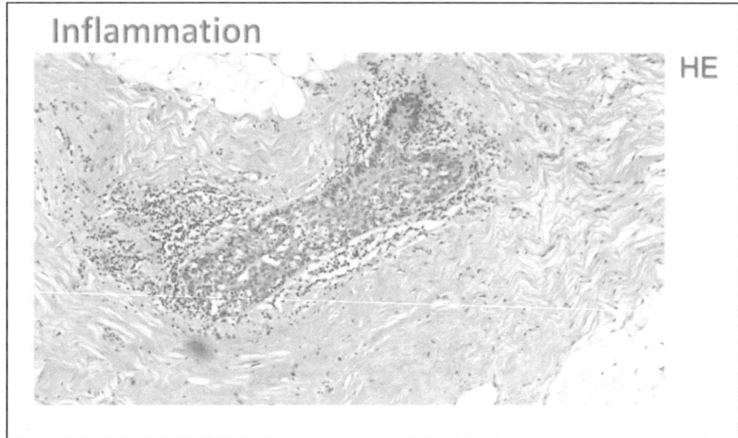

Here you see a milk duct with a periductal here in the stroma and intraepithelial, lymphocytic infiltrates. The lymphocytes are the small dark dots you see.

It's not uncommon for me to see vasculitis like changes in the mammary parenchyma. That means inflammation in the small vessels in the breast.

Now back to the tumors.

After my interview with Miriam Reichel one year ago about **turbo cancer** and vaccination against COVID-19 (previous image), there are more than 1,300 comments. Among them are **numerous reports of turbo cancer** cases from personal experience, friends, or relatives.

I have picked out one comment, and I would like you to read it for yourself. I hope you can see this in the back. Yes.

Vielleicht ist es ein Zufall vielleicht auch nicht... ich habe nach der 2. Impung Beschwerden bekommen. Nach zwei Monaten kam die Diagnose Lungengrebs im Endstadium. Aktuell geht es mir gut... obwohl die Ärzte meinten ich hätte nur noch 6-12 Monate zu leben... ich bin männlich, knapp 40 Jahre alt, Vater von Kleinkindern, Nichtraucher, trinke seit x Jahren kein Alkohol. Ich werde es nie rausfinden ob die Impfung daran Schuld ist, ...
Read more

3 REPLY

"Maybe it's a coincidence, maybe not... I developed symptoms after the second vaccination. After two months, I was diagnosed with terminal lung cancer. I am currently doing well... although the doctors said I only had 6-12 months to live...
I am male, just under 40 years old, father of small children, non-smoker, have not drunk alcohol for x years. I will never find out if the vaccination is to blame..."

Again, you can see that the patient's age and lifestyle habits do not match such an aggressive lung cancer, and it's a very unusual course.

After my lecture in Oslo to Norwegian doctors and nurses in June last year, I also received a lot of feedback. On the one hand, there are case descriptions from affected patients or their relative. And on the other hand, colleagues who confirm my theory that rapid tumor growth seem to be related to the vaccination against COVID-19.

Unfortunately, I'm also contacted by patients who urgently need help because of severe side effects or turbo cancer cases after these vaccinations. They are simply not taken seriously by the doctors treating them, because, in the eyes of most of doctors, there are no side effects of the vaccinations against COVID-19.

Unfortunately, I cannot help these patients.

Statistics

What about the statistics? I try to evaluate the number of breast carcinoma cases in our institute over the last few years. At first, I was disappointed because I could not find any trend in the number of cases, the average age of the patients and the average tumor size. On second thought, however, I realized that the evaluation of data cannot be that simple.

Fortunately, the suspected link between COVID-19 vaccination and turbo cancer does not apply to everyone. In the cases of so-called turbo cancer, the patients often receive neoadjuvant chemotherapy—that means chemotherapy before surgery because of the large tumor size. In our statistical system at the institute, only the tumor size of the surgical specimen and not the original tumor size as recorded.

In all breast cancer cases, the so-called HER-2 (https://www.mayoclinic.org/breast-cancer/expert-answers/faq-20058066) status and the hormone receptor status are examined by pathologists. HER-2 is a protein to help tumor cells grow faster, so HER-2 positive breast cancer grows quickly. There are also so-called triple-negative tumors, which means that hormone receptors and HER-2 status are negative.

Triple-negative tumors also grow quickly. These HER-2 positive and also triple-negative tumors are often treated with neoadjuvant chemotherapy. It's not uncommon to see pathologic complete response. That means the tumor has disappeared, and the tumor site on the surgical specimen is zero millimeters.

Of course, this affects the statistics.

Actually, we should investigate all cases with tumor growth after vaccination against COVID-19. Multifocal and bilateral tumor growth would need to be recorded. For reliable statistics we would need more case numbers from multiple institutes—preferably cross-national collecting and analyzing data to prove or disprove my hypothesis of post-vaccination turbo cancer.

Now, you may wonder why I choose such a strange title for my lecture. I found the painting by the Swedish artist *(unrecognizable Swedish name)*. That fits very well. But after 18 years in Sweden, I can tell you this is not the usual way of fishing here in Sweden.

In September 2022, the press in Sweden became aware of my performance in Oslo. Regional Radio interviewed the head of the Kalma Hospital, Johan Rosenquist, and Mr. Rosenquist commanded that it would take many more cases to investigate my statement and that there would be no consequences for me after this talk, as I was allowed to speak as a private person.

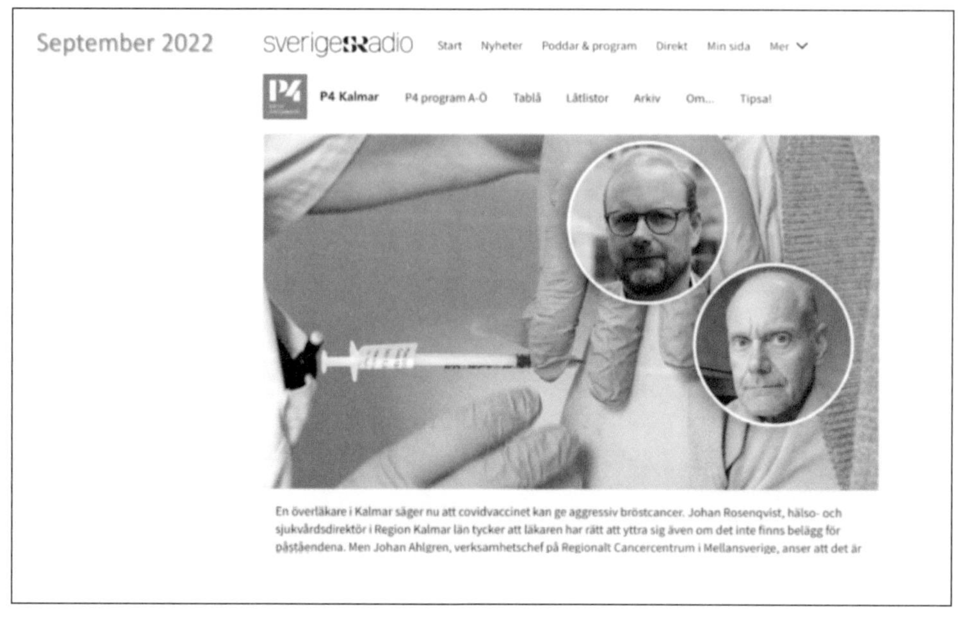

But this man here (previous image), Johan Ahlgren, the head of the mid-Sweden regional cancer center, a lecturer in oncology, calls my approach sabotaging the region's efforts to get everyone vaccinated.

I think if my theory had been wrong, no one would have cared what I said.

Now pathologists are usually associated with autopsies, but these make up a very small part of a pathologist's work these days. I have only done single autopsies recently, but there are three deaths that I strongly suspect are related to the vaccination against COVID.

I published one of these cases four months ago.

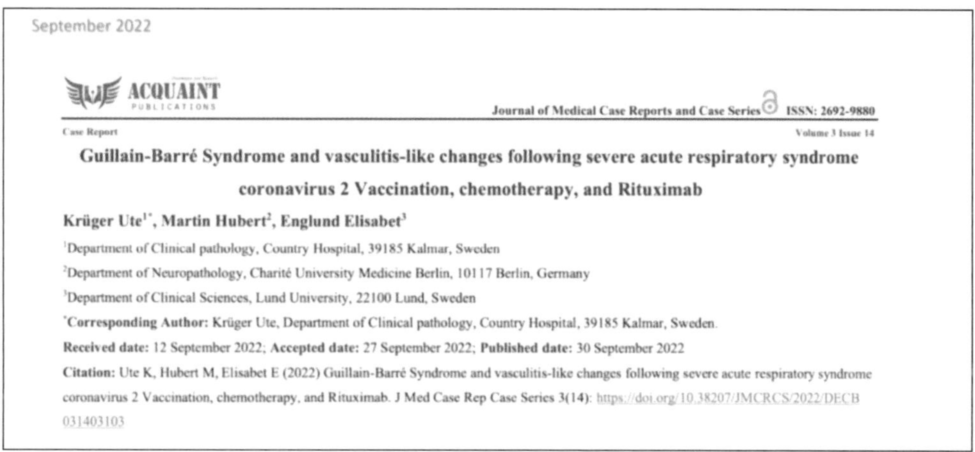

September 2022

ACQUAINT PUBLICATIONS
Journal of Medical Case Reports and Case Series ISSN: 2692-9880
Case Report
Volume 3 Issue 14

Guillain-Barré Syndrome and vasculitis-like changes following severe acute respiratory syndrome coronavirus 2 Vaccination, chemotherapy, and Rituximab

Krüger Ute[1*], Martin Hubert[2], Englund Elisabet[3]
[1]Department of Clinical pathology, Country Hospital, 39185 Kalmar, Sweden
[2]Department of Neuropathology, Charité University Medicine Berlin, 10117 Berlin, Germany
[3]Department of Clinical Sciences, Lund University, 22100 Lund, Sweden
[*]**Corresponding Author:** Krüger Ute, Department of Clinical pathology, Country Hospital, 39185 Kalmar, Sweden.

Received date: 12 September 2022; **Accepted date:** 27 September 2022; **Published date:** 30 September 2022
Citation: Ute K, Hubert M, Elisabet E (2022) Guillain-Barré Syndrome and vasculitis-like changes following severe acute respiratory syndrome coronavirus 2 Vaccination, chemotherapy, and Rituximab. J Med Case Rep Case Series 3(14): https://doi.org/10.38207/JMCRCS/2022/DECB031403103

(https://www.researchgate.net/publication/366466028_Journal_of_Medical_Case_Reports_and_Case_Series_Guillain-Barre_Syndrome_and_vasculitis-like_changes_following_severe_acute_respiratory_syndrome_coronavirus_2_Vaccination_chemotherapy_and_Rituximab)

Unfortunately, there's not enough time to go into detail. Here's the title if you want to go know more about it. The other cases, short, similar findings to those that Professor Burkhardt, I will show you in a moment.

However, I see four major problems with autopsies.

1. No postmortem examination is carried out. It's the first one, the second one.
2. Incorrect information given by the clinician when asked whether the deceased have been vaccinated. I have seen more than one case in our hospital where the clinician noted on the referral for autopsy that the deceased had not been vaccinated. However, I found in the patient's journal that the patient had indeed been vaccinated against COVID-19.
3. The third is many of the pathologist colleagues do not take samples for histological examination—that means no microscopic examination is performed. Without microscopy, you cannot see, for example, myocarditis or vasculitis.
4. And the fourth is the lack of knowledge in the assessment of microscopic findings.

You only see what you know.
Goethe

Goethe lived 200 years ago. Goethe was a German poet and naturalist. He's considered one of the most important creators of German language poetry. And, slightly modified by me,

You only see what you know.
And
What you want to see.

The changes I have described, which seem to be related to the COVID-19 vaccination are only a small part of what happens in the body.

I can say that I haven't given up. I studied medicine because I want to help people.

But now it feels like I'm watching people being killed and I cannot do anything. As a pathologist, I diagnose tumors that maybe have been caused by another colleague with a shot.

This is completely pointless and absurd.

Over the last year, I have been exploring many alternative healing methods, our nutrition, the power of our nature, and our wild herbs.

I want to bring you closer to the beauty of our nature. Here, for dinner last summer, I collected leaves and flowers from 35 different bushes, perennials, and wild herbs in my garden to eat, and I am still alive.

Among other things, these give me strength and confidence for the future.

Thank you very much for your attention.

—Ute Krüger, MD, DMedSci

Summary

Following widespread distribution and injection of Spike-producing therapeutics, Dr. Krüger noticed a number of changes on which she was asked to consult:

1. **Younger patients** are being seen, often 3 to 50 years old.
2. Tumors are **growing more aggressively and faster.**
3. Tumors are **larger**.
4. Tumors exhibit **heterogeneity**.
5. **Multifocal tumor growth** and **bilateral tumor growths** are more frequent.
6. Co-temporal onset of **more than one type of cancer.**
7. Benign tumors have **accelerated growth** possibly signifying malignant transformation.
8. The physiologic process of **inflammation** was noted as a possible source of breast pain.

Eleven: "Acute Kidney Injury and Acute Renal Failure Following Pfizer mRNA COVID Vaccination. 33% of Patients Died. Pfizer Concludes, 'No New Safety Issue.'"

—Joseph Gehrett, MD; Barbara Gehrett, MD; Chris Flowers, MD; and Loree Britt

The WarRoom/DailyClout Pfizer Documents Analysis Project Post-Marketing Group (Team 1)—Joseph Gehrett, MD; Barbara Gehrett, MD; Chris Flowers, MD; and Loree Britt—produced an alarming review of the Renal (Kidney) System Organ Class (SOC) adverse events found in Pfizer document *5.3.6 Cumulative Analysis of Post-Authorization Adverse Event Reports of PF-07302048 (BNT162B2) Received Through 28-FEB-2021* (a.k.a., "*5.3.6*"). This SOC includes acute kidney injury and acute renal failure.

It is important to note that the adverse events (AEs) in the *5.3.6* document were reported to Pfizer for **only a 90-day period** starting on December 1, 2020, the date of the United Kingdom's public rollout of Pfizer's COVID-19 experimental mRNA "vaccine" product.

Key points in this report include:

- **69 patients, including one infant**, suffered acute kidney injury or acute renal failure. *The vaccine was not authorized for infants during this time.*
- Pfizer's renal adverse event reports screen only for the most severe damage but miss important, less severe kidney damage. Thus, **Pfizer's post-marketing kidney adverse events are likely significantly underreported**.
- **Half of the severe renal adverse events were reported within four days of vaccination**.
- **67% of kidney adverse event patients were women**, and 33% were men.
- The **very short range of latency** shows the severity of the damage in this SOC.
- Pfizer reported that surveillance would continue for this SOC, yet **no information on subsequent surveillance has been publicly released** to date.

War Room/DailyClout Pfizer Document Analysis

Post-Marketing Team Micro-Report 9:
Renal (Kidney) Adverse Events of Special Interest Review of 5.3.6

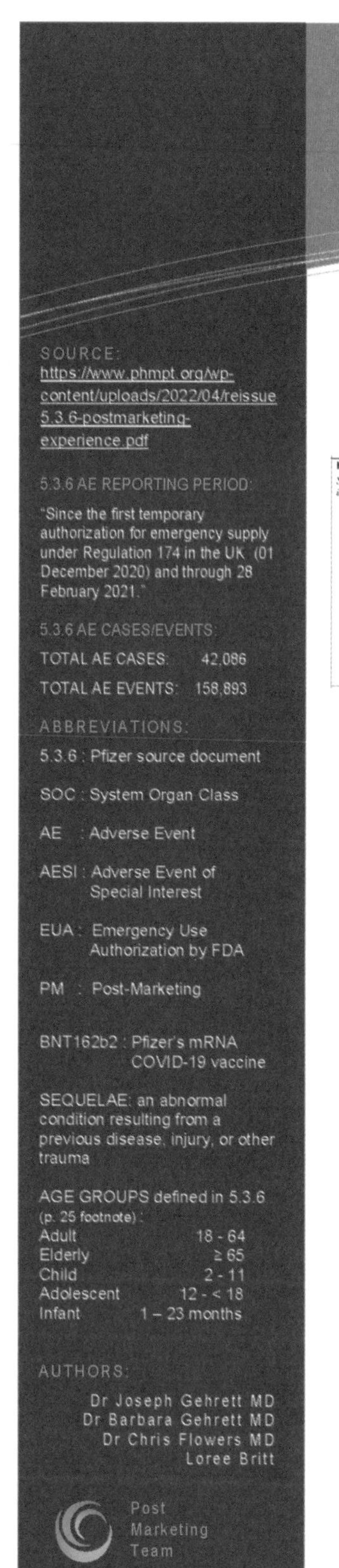

SOURCE:
https://www.phmpt.org/wp-content/uploads/2022/04/reissue 5.3.6-postmarketing-experience.pdf

5.3.6 AE REPORTING PERIOD:

"Since the first temporary authorization for emergency supply under Regulation 174 in the UK (01 December 2020) and through 28 February 2021."

5.3.6 AE CASES/EVENTS:

TOTAL AE CASES: 42,086

TOTAL AE EVENTS: 158,893

ABBREVIATIONS:

5.3.6 : Pfizer source document

SOC : System Organ Class

AE : Adverse Event

AESI : Adverse Event of Special Interest

EUA : Emergency Use Authorization by FDA

PM : Post-Marketing

BNT162b2 : Pfizer's mRNA COVID-19 vaccine

SEQUELAE: an abnormal condition resulting from a previous disease, injury, or other trauma

AGE GROUPS defined in 5.3.6 (p. 25 footnote):

Adult	18 - 64
Elderly	≥ 65
Child	2 - 11
Adolescent	12 - < 18
Infant	1 – 23 months

AUTHORS:

Dr Joseph Gehrett MD
Dr Barbara Gehrett MD
Dr Chris Flowers MD
Loree Britt

Post Marketing Team

10Mar23

Renal AESIs *Search criteria: PTs Acute kidney injury, Renal failure.*	• Number of cases: 69 cases (0.17% of the total PM dataset), of which 57 medically confirmed, 12 non-medically confirmed; • Country of incidence: Germany (17), France and UK (13 each), US (6), Belgium, Italy and Spain (4 each), Sweden (2), Austria, Canada, Denmark, Finland, Luxembourg and Norway (1 each); • Subjects' gender: female (46), male (23); • Subjects' age group (n=68): Adult (7), Elderly (60), Infant (1); • Number of relevant events: 70, all serious; • Reported relevant PTs: Acute kidney injury (40) and Renal failure (30); • Relevant event onset latency (n = 42): Range from <24 hours to 15 days, median 4 days. • Relevant event outcome: fatal (23), resolved/resolving (10), not resolved (15) and unknown (22). Conclusion: This cumulative case review does not raise new safety issues. Surveillance will continue.

https://www.phmpt.org/wpcontent/uploads/2022/04/reissue 5.3.6-postmarketiexperience.pdf

The renal AESIs were collected by searching **only events labeled as Acute Kidney Injury or Acute Renal Failure.** While these two conditions result from severe underlying disease processes, they name only the ultimate conditions. In clinical medicine, the kidneys have a number of important functions. They are involved in blood pressure control, maintenance of electrolyte balance, fluid volume stabilization, and elimination of waste products. The narrow search criteria do not detect damage to any of these basic kidney functions. Thus, **Pfizer's adverse event reports screen only for the most severe damage but miss important, less severe, kidney damage.**

There were 69 patients with one of these diagnoses. Forty-two conditions had a time interval reported from injection to onset (60%). The range varied from **within 24 hours to 15 days**, with **half of the reports within four days.** Ten (14%) were reported as resolved/resolving, though this is not further defined. Fifteen (21%) were unresolved, and 22 (31%) had no status reported. **All conditions were assessed as serious. There were 23 (33%) deaths reported.**

This SOC has a relatively small number of patients. However, as suggested above, the search method is very insensitive. As with all adverse events, the reporting system is voluntary and thus, likely very underreported. In some instances, a general chemistry panel ordered for other reasons would detect a kidney problem. In other cases, where there is serious kidney injury and rapid impairment of function, serious symptoms would develop. Permanent damage could occur and only at a later time become symptomatic or even diagnosed.

- **Adverse Events were reported to Pfizer during a 90-day period, following the December 1, 2020, public rollout of its COVID-19 experimental "vaccine" product.**
- **In the Pfizer 5.3.6 document, these AEs were categorized by System Organ Classes (SOC) – in other words, by systems in the body.**
- **Of the 69 patients, 46 were women (67%) and 23 were men (33%).**
- **There was one *infant* in this group. It should be noted the vaccine was not authorized for infants. The other age groups were 60 elderly (88%), seven adults (10%), and one unreported.**

Short of a total shutdown of the kidneys with no urine output, it generally takes several days for the standard labs of kidney function (blood urea nitrogen, creatinine, potassium) to rise to a level that causes alarm.

Kidney function tests are not done routinely after vaccination. Symptoms typically do not develop immediately but occur over a period of days as toxins accumulate. **The very short range of latency shows the severity of damage.**

Pfizer's conclusion: "This cumulative case review does not raise new safety issue. Surveillance will continue." There has been no public release of any subsequent surveillance nor further clarity as to frequency or severity of post-vaccine kidney impairment. To date there is no indication that the FDA has requested more than this assessment. **We have no information on less than catastrophic degrees of kidney injury.**

69 patients. 23 deaths.

RECALL this unsafe "vaccine."

Twelve: "In the First Three Months of Pfizer's mRNA 'Vaccine' Rollout, Nine Patients Died of Anaphylaxis. 79% of Anaphylaxis Adverse Events Were Rated as 'Serious.'"

—Joseph Gehrett, MD; Barbara Gehrett, MD; Chris Flowers, MD; and Loree Britt

The WarRoom/DailyClout Pfizer Documents Analysis Project Post-Marketing Group (Team 1)—Joseph Gehrett, MD; Barbara Gehrett, MD; Chris Flowers, MD; and Loree Britt—produced a shocking review of anaphylaxis (severe allergy reaction) adverse events found in Pfizer document *5.3.6 Cumulative Analysis of Post-Authorization Adverse Event Reports of PF-07302048 (BNT162B2) Received Through 28-FEB-2021* (a.k.a., "*5.3.6*"). Severe allergic reaction is typically triggered by latex, foods such as peanuts, bee stings, or medications (injected or taken by mouth), among other things, and is generally considered a **medical emergency**.

It is important to note that the adverse events (AEs) in the *5.3.6* document were reported to Pfizer for **only a 90-day period** starting on December 1, 2020, the date of the United Kingdom's public rollout of Pfizer's COVID-19 experimental mRNA "vaccine" product.

Key points in this report include:

- There were **nine reported deaths**.
  - Only four patients who died were reported to have serious, underlying medical conditions that "likely contributed to their deaths."
- There were 1,833 potential anaphylaxis patients reported; but, after screening using a tool called Brighton Collaboration, 831 did not meet anaphylaxis criteria, leaving **1,002 cases reported in 90 days**. Pfizer reported 2,958 "potentially relevant events" from those 1,002 individuals that included the signs and symptoms of anaphylaxis.
  - Where did the other 831 patients and their allergic adverse events go/get assigned in the post-marketing data, if anywhere?
- **Pfizer reported only the most frequent anaphylaxis signs and symptoms**: anaphylactic reaction (435), shortness of breath (356), rash (190), redness of the skin (159), hives (133), cough (115), **respiratory distress (97),** throat tightness (97), swollen tongue (93), low blood pressure (72), **low blood pressure severe enough to threaten organ function (shock) (80)**, chest discomfort (71), swelling face (70), throat swelling (68), and lip swelling (64).
- The events were rated as **serious in 2,341 (79%)** and non-serious in 617 (21%).
- The **ratio of females to males affected was over 8:1**. Of those cases with gender specified, **876 (89%) were female**, 106 (11%) were male.
- Half of patients with this adverse event were **younger than 43.5 years old**.

- Pfizer's Conclusion: Evaluation of BC *(Brighton Collaborative)* cases Level 1–4 **did not reveal any significant new safety information**. Anaphylaxis is appropriately described in the product labeling, as are non-anaphylactic hypersensitivity events. Surveillance will continue.

Read this important report below.

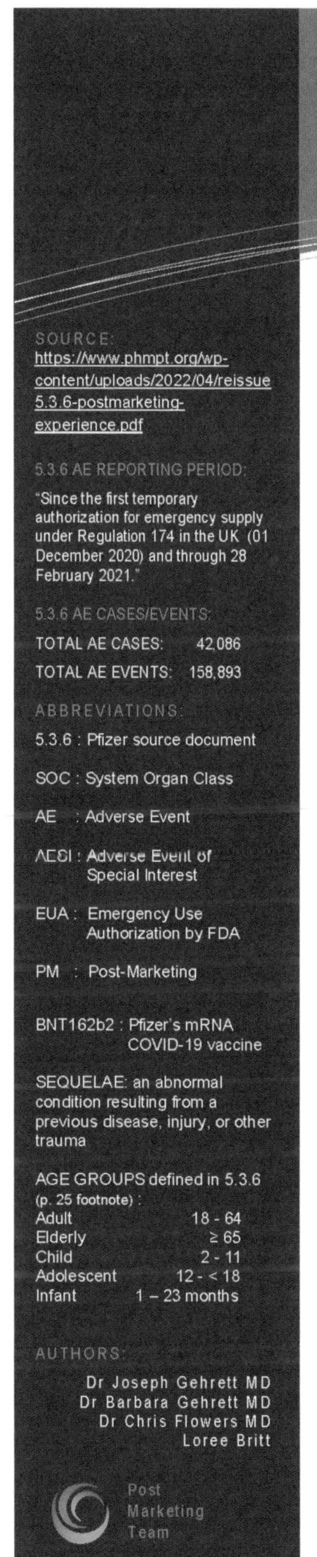

War Room/DailyClout Pfizer Document Analysis

Post-Marketing Team Micro-Report 10:

Anaphylaxis – Important Identified Risk Review of 5.3.6

Anaphylaxis | Since the first temporary authorization for emergency supply under Regulation 174 in the UK (01 December 2020) and through 28 February 2021, 1833 potentially relevant cases were retrieved from the Anaphylactic reaction SMQ (Narrow and Broad) search strategy, applying the MedDRA algorithm. These cases were individually reviewed and assessed according to Brighton Collaboration (BC) definition and level of diagnostic certainty as shown in the Table below:

Brighton Collaboration Level	Number of cases
BC 1	290
BC 2	311
BC 3	10
BC 4	391
BC 5	831
Total	1833

Level 1 indicates a case with the highest level of diagnostic certainty of anaphylaxis, whereas the diagnostic certainty is lowest for Level 3. Level 4 is defined as "reported event of anaphylaxis with insufficient evidence to meet the case definition" and Level 5 as not a case of anaphylaxis.

There were 1002 cases (54.0% of the potentially relevant cases retrieved), 2958 potentially relevant events, from the Anaphylactic reaction SMQ (Broad and Narrow) search strategy, meeting BC Level 1 to 4:

Country of incidence: UK (261), US (184), Mexico (99), Italy (82), Germany (67), Spain (38), France (36), Portugal (22), Denmark (20), Finland, Greece (19 each), Sweden (17), Czech Republic , Netherlands (16 each), Belgium, Ireland (13 each), Poland (12), Austria (11); the remaining 57 cases originated from 15 different countries.
Relevant event seriousness: Serious (2341), Non-Serious (617);
Gender: Females (876), Males (106), Unknown (20);
Age (n=961) ranged from 16 to 98 years (mean = 54.8 years, median = 42.5 years);
Relevant even outcome[a]: fatal (9)[b], resolved/resolving (1922), not resolved (229), resolved with sequela (48), unknown (754);
Most frequently reported relevant PTs (≥2%), from the Anaphylactic reaction SMQ (Broad and Narrow) search strategy: Anaphylactic reaction (435), Dyspnoea (356), Rash (190), Pruritus (175), Erythema (159), Urticaria (133), Cough (115), Respiratory distress, Throat tightness (97 each), Swollen tongue (93), Anaphylactic shock (80), Hypotension (72), Chest discomfort (71), Swelling face (70), Pharyngeal swelling (68), and Lip swelling (64).

Conclusion: Evaluation of BC cases Level 1 - 4 did not reveal any significant new safety information. Anaphylaxis is appropriately described in the product labeling as are non-anaphylactic hypersensitivity events. Surveillance will continue.

https://www.phmpt.org/wpcontent/uploads/2022/04/reissue 5.3.6-postmarketiexperience.pdf

Anaphylaxis is a condition most often referred to as a "severe allergic reaction" and is triggered by latex, foods such as peanuts, bee stings, or medications (injected or taken by mouth), among other things. Immunoglobulin E (IgE, a human antibody) is involved in this dangerous and explosive reaction. It causes mast cells to immediately release histamine and other chemicals.

The life-threatening symptoms that follow can include swelling such as hives, respiratory difficulties, drop in blood pressure and rapid heart rate, and abdominal symptoms such as pain, nausea and vomiting. These symptoms are typically rapid in both onset and progression. Treatment similarly has to be rapid (e.g., Epi-Pen) and often requires emergency room or hospital treatment.

Pfizer has used a tool, the Brighton Collaboration (BC), to assess whether the symptoms of a patient are correctly identified as anaphylaxis. In the three-month data collection, there were 1,833 potential anaphylaxis patients reported; but, after screening, 831 did not meet criteria, leaving 1,002 cases reported in this time period. Pfizer reported 2,958 "potentially relevant events" from those 1,002 individuals that included the signs and symptoms of anaphylaxis.

- **Adverse Events were reported to Pfizer during a 90-day period, following the December 1, 2020, public rollout of its COVID-19 experimental "vaccine" product.**
- **In the Pfizer 5.3.6 document, these AEs were categorized by System Organ Classes (SOC) – in other words, by systems in the body.**
- **Of those with gender specified, 876 (89%) were female, 106 (11%) were male.**
- **The reported age range was 16 to 98 years old with half being less than 43.5 years old.**

Pfizer reported only the most frequent anaphylaxis signs and symptoms: anaphylactic reaction (435), shortness of breath (356), rash (190), redness of the skin (159), hives (133), cough (115), **respiratory distress (97),** throat tightness (97), swollen tongue (93), low blood pressure (72), **low blood pressure severe enough to threaten organ function (shock) (80)**, chest discomfort (71), swelling face (70), throat swelling (68), and lip swelling (64).

The events were rated as **serious in 2,341 (79%)** and non-serious in 617 (21%).1922 events were reported as resolved or resolving (65%), 229 not resolved (8%), 48 resolved with sequelae (1.6%), and 754 had no outcome known (25%).

There were nine deaths reported.

IMPORTANT NOTE:

There are **831 patients, 45% of the total 1,833,** who were determined not to be anaphylaxis and dropped from this category. Where did those patients and their adverse events go?

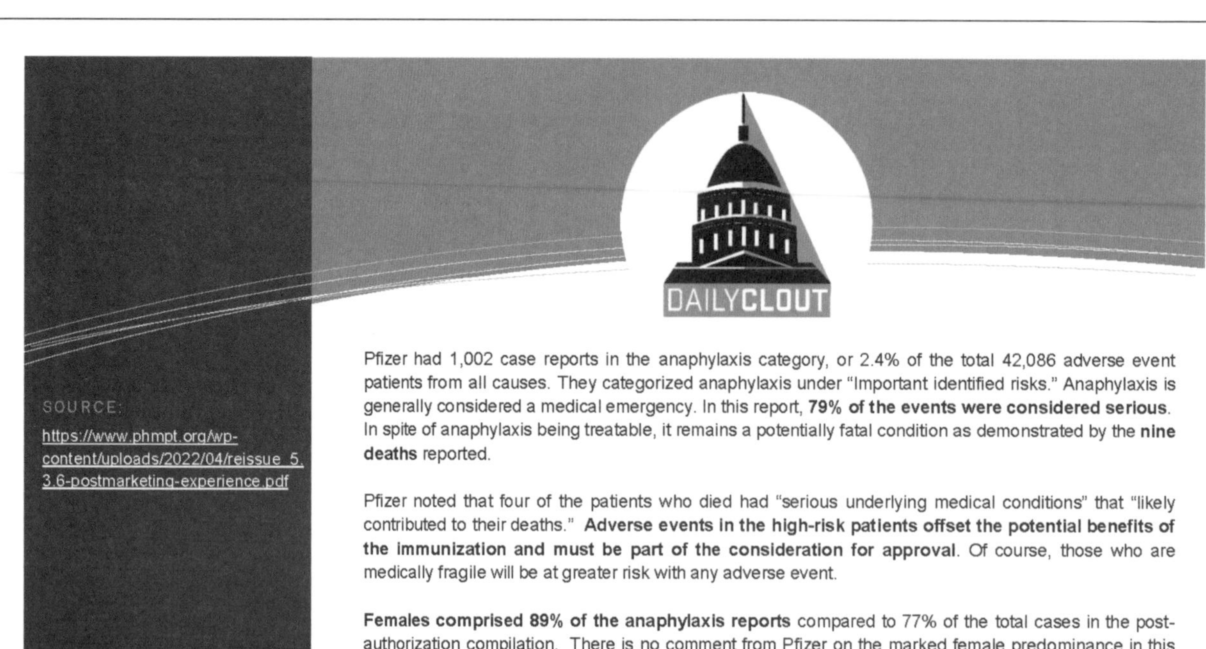

SOURCE:
https://www.phmpt.org/wp-content/uploads/2022/04/reissue_5.3.6-postmarketing-experience.pdf

Pfizer had 1,002 case reports in the anaphylaxis category, or 2.4% of the total 42,086 adverse event patients from all causes. They categorized anaphylaxis under "Important identified risks." Anaphylaxis is generally considered a medical emergency. In this report, **79% of the events were considered serious**. In spite of anaphylaxis being treatable, it remains a potentially fatal condition as demonstrated by the **nine deaths** reported.

Pfizer noted that four of the patients who died had "serious underlying medical conditions" that "likely contributed to their deaths." **Adverse events in the high-risk patients offset the potential benefits of the immunization and must be part of the consideration for approval**. Of course, those who are medically fragile will be at greater risk with any adverse event.

Females comprised 89% of the anaphylaxis reports compared to 77% of the total cases in the post-authorization compilation. There is no comment from Pfizer on the marked female predominance in this adverse event category, let alone any stated plan to address it.

Pfizer's Conclusion:

> Evaluation of BC *(Brighton Collaborative)* cases Level 1 - 4 did not reveal any significant new safety information. Anaphylaxis is appropriately described in the product labeling as are non-anaphylactic hypersensitivity events. Surveillance will continue.

Post-Marketing Team's CONCLUSION

RECALL this unsafe "vaccine."

Thirteen: "1,077 Immune-Mediated/Autoimmune Adverse Events in First 90 Days of Pfizer mRNA "Vaccine" Rollout, Including 12 Fatalities. Pfizer Undercounted This Category of Adverse Events by 270 Occurrences."

—Joseph Gehrett, MD; Barbara Gehrett, MD; Chris Flowers, MD; and Loree Britt

The WarRoom/DailyClout Pfizer Documents Analysis Project Post-Marketing Group (Team 1)—Joseph Gehrett, MD; Barbara Gehrett, MD; Chris Flowers, MD; and Loree Britt—wrote an important review of immune-mediated/autoimmune adverse events found in Pfizer document *5.3.6 Cumulative Analysis of Post-Authorization Adverse Event Reports of PF-07302048 (BNT162B2) Received Through 28-FEB-2021* (a.k.a., "*5.3.6*"). (https://www.phmpt.org/wp-content/uploads/2022/04/reissue_5.3.6-postmarketing-experience.pdf) This category is comprised of the numerous diseases resulting from disordered immune attacks against tissues of any of the body's organs. However, it excludes anaphylaxis which has its own separate report, as well as autoimmune diseases attaching nerve tissue (e.g., Guillain-Barré), which have been explored in other 5.3.6 neurological and musculoskeletal reports. (https://dailyclout.io/category/pfizer-reports/)

It is important to note that the adverse events (AEs) in the *5.3.6* document were reported to Pfizer for only a 90-day period starting on December 1, 2020, the date of the United Kingdom's public rollout of Pfizer's COVID-19 experimental mRNA "vaccine" product.

Highlights of this report include:

- Twelve immune-mediated/autoimmune adverse events were fatal.
- Of adverse events with gender specified, 77% (526) were female, and 23% (156) were male—a female to male ratio of 3.4 to 1.
- Ages of affected patients were 19% elderly, 71% adult, <1% adolescent, and 10% not reported.
- Time from vaccination to onset was given for 75% of events with a range of 24 hours to 30 days. Half started within 24 hours of injection.
- Only immune-mediated/autoimmune diseases or conditions with over 10 cases are included in Pfizer's reporting, leaving out 270 adverse events of this type. Thus, adverse events in this category were undercounted by Pfizer.
- Adverse events in this group include:
  - Hypersensitivity (which is not further defined, though it accounts for 55% of the AEs).
  - Peripheral neuropathy [https://www.mayoclinic.org/diseases-conditions/peripheral-neuropathy/symptoms-causes/syc-20352061], which often causes weakness, numbness, and pain, usually in the hands and feet, and can also affect other areas and body functions including digestion, urination, and circulation.
  - Pericarditis (inflammation of the lining of the heart).
  - Myocarditis (immune attack against the heart muscle).

    - Encephalitis (brain inflammation disorders).
    - Diabetes.
    - Psoriasis.
    - Dermatitis.
    - Blistering Skin disorder.
    - Autoimmune disorder.
    - Raynaud's phenomenon.
- Immune rejection of transplanted organs was not mentioned whether from there being zero instances of it or, perhaps, because there were 10 or fewer instances.

Pfizer concluded, "This cumulative case review does not raise new safety issues. Surveillance will continue." However, as of the date of this report, 25.5 months after the completion of Pfizer's post-marketing analysis, no additional safety data has been released to the public.

War Room/DailyClout Pfizer Document Analysis

Post-Marketing Team Micro-Report 11:

Immune-Mediated/Autoimmune AESIs Review of 5.3.6

SOURCE:
https://www.phmpt.org/wp-content/uploads/2022/04/reissue 5.3.6-postmarketing-experience.pdf

5.3.6 AE REPORTING PERIOD:

"Since the first temporary authorization for emergency supply under Regulation 174 in the UK (01 December 2020) and through 28 February 2021."

5.3.6 AE CASES/EVENTS:

TOTAL AE CASES: 42,086

TOTAL AE EVENTS: 158,893

ABBREVIATIONS:

5.3.6 : Pfizer source document

SOC : System Organ Class

AE : Adverse Event

AESI : Adverse Event of Special Interest

EUA : Emergency Use Authorization by FDA

PM : Post-Marketing

BNT162b2 : Pfizer's mRNA COVID-19 vaccine

SEQUELAE: an abnormal condition resulting from a previous disease, injury, or other trauma

AGE GROUPS defined in 5.3.6 (p. 25 footnote) :

Adult	18 - 64
Elderly	≥ 65
Child	2 - 11
Adolescent	12 - < 18
Infant	1 – 23 months

AUTHORS:

Dr Joseph Gehrett MD
Dr Barbara Gehrett MD
Dr Chris Flowers MD
Loree Britt

12Apr23

Immune-Mediated/Autoimmune AESIs *Search criteria: Immune-mediated/autoimmune disorders (SMQ) (Broad and Narrow) OR Autoimmune disorders HLGT (Primary Path) OR PTs Cytokine release syndrome; Cytokine storm; Hypersensitivity*	• Number of cases: 1050 (2.5 % of the total PM dataset), of which 760 medically confirmed and 290 non-medically confirmed; • Country of incidence (>10 cases): UK (267), US (257), Italy (70), France and Germany (69 each), Mexico (36), Sweden (35), Spain (32), Greece (31), Israel (21), Denmark (18), Portugal (17), Austria and Czech Republic (16 each), Canada (12), Finland (10). The remaining 74 cases were from 24 different countries. • Subjects' gender (n=682): female (526), male (156). • Subjects' age group (n=944): Adult (746), Elderly (196), Adolescent (2). • Number of relevant events: 1077, of which 780 serious, 297 non-serious. • Most frequently reported relevant PTs (>10 occurrences): Hypersensitivity (596), Neuropathy peripheral (49), Pericarditis (32), Myocarditis (25), Dermatitis (24), Diabetes mellitus and Encephalitis (16 each), Psoriasis (14), Dermatitis Bullous (13), Autoimmune disorder and Raynaud's phenomenon (11 each); • Relevant event onset latency (n = 807): Range from <24 hours to 30 days, median <24 hours. • Relevant event outcome: resolved/resolving (517), not resolved (215), fatal (12), resolved with sequelae (22) and unknown (312). Conclusion: This cumulative case review does not raise new safety issues. Surveillance will continue

https://www.phmpt.org/wpcontent/uploads/2022/04/reissue_5.3.6-postmarketiexperience.pdf

This category comprises the numerous diseases resulting from disordered immune attacks against tissues of any of the body's organs. Pfizer has grouped some conditions in a very general way, such as "hypersensitivity." This is defined as an exaggerated response of the normal immune system. However, anaphylaxis, another hypersensitivity reaction, has its own separate SOC report.

Autoimmune diseases attacking nerve tissue (Guillain-Barré, multiple sclerosis, polyneuropathy, and others) are not in this SOC but are found under the Neurological and Musculoskeletal SOC reports. However, dermatitis (skin hypersensitivity) is listed here.

Only diseases or conditions with over 10 cases are described with a diagnosis or symptom. Hypersensitivity had 596 (55%) of the adverse events. The conditions in this large grouping are not further defined in the report. Peripheral neuropathy had 49 events (5%) though, and, as discussed above, other neuropathies are separately listed and presumably different patients. There are **32 diagnosed pericarditis events (3%)** with inflammation of the lining around the heart. **Myocarditis (immune attack against the heart muscle itself) had 25 events (2%).**

- **Adverse Events were reported to Pfizer during a 90-day period, following the December 1, 2020, public rollout of its COVID-19 experimental "vaccine" product.**
- **In the Pfizer 5.3.6 document, these AEs were categorized by System Organ Classes (SOC) – in other words, by systems in the body.**
- **Of those whose sex was specified, 526 (77%) were female, and 156 (23%) were male (a ratio of 3.4 to 1).**
- **Age was listed as elderly 196 (19%), adult 746 (71%), adolescent 2 (<1%), and not reported 106 (10%).**

There were 1,077 events reported with **780 (72%) serious** and 297 (28%) non-serious.

12 (1.1%) AEs were fatal.

The time from vaccination to adverse event onset was defined in 807 (75%) of the 1,077 events with a range of within 24 hours to 30 days. **Half of the adverse events started within 24 hours of the injection.** 517 (48%) had outcomes reported as "resolved" or "resolving." **215 events (20%) were not resolved, 22 (2%) resolved with sequelae, and 312 (29%) had unknown outcome.**

Further observation regarding autoimmune events:

There was no mention of immune rejection of transplanted organs. It is unknown whether Pfizer found none, or whether there were 10 or fewer and thus not listed explicitly.

There is a VAERS report of a 17-year-old male who was vaccinated 1/19/2021 and was hospitalized mid-May 2021 with heart transplant rejection and heart failure.

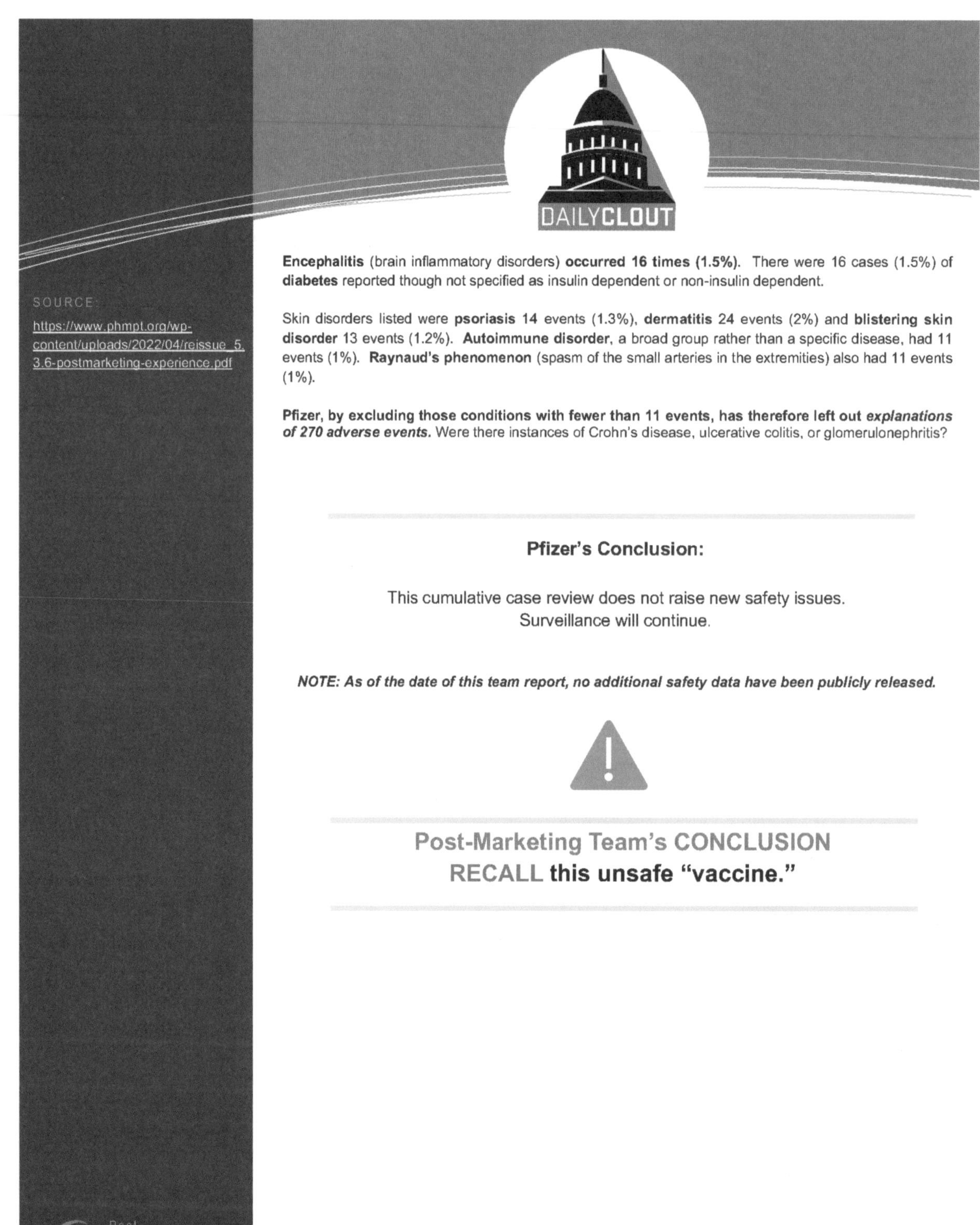

DAILYCLOUT

SOURCE:
https://www.phmpt.org/wp-content/uploads/2022/04/reissue_5.3.6-postmarketing-experience.pdf

Encephalitis (brain inflammatory disorders) **occurred 16 times (1.5%).** There were 16 cases (1.5%) of **diabetes** reported though not specified as insulin dependent or non-insulin dependent.

Skin disorders listed were **psoriasis** 14 events (1.3%), **dermatitis** 24 events (2%) and **blistering skin disorder** 13 events (1.2%). **Autoimmune disorder**, a broad group rather than a specific disease, had 11 events (1%). **Raynaud's phenomenon** (spasm of the small arteries in the extremities) also had 11 events (1%).

Pfizer, by excluding those conditions with fewer than 11 events, has therefore left out *explanations of 270 adverse events.* Were there instances of Crohn's disease, ulcerative colitis, or glomerulonephritis?

Pfizer's Conclusion:

This cumulative case review does not raise new safety issues.
Surveillance will continue.

NOTE: As of the date of this team report, no additional safety data have been publicly released.

Post-Marketing Team's CONCLUSION
RECALL this unsafe "vaccine."

Post Marketing Team

Fourteen: "34 Blood Vessel Inflammation, Vasculitis, Adverse Events Occurred in First 90 Days After Pfizer mRNA "Vaccine" Rollout, Including One Fatality. Half Had Onset Within Three Days of Injection. 81% of Sufferers Were Women."

—Barbara Gehrett, MD; Joseph Gehrett, MD; Chris Flowers, MD; and Loree Britt

The WarRoom/DailyClout Pfizer Documents Analysis Project Post-Marketing Group (Team 1)—Barbara Gehrett, MD; Joseph Gehrett, MD; Chris Flowers, MD; and Loree Britt—wrote an important review of vasculitis adverse events found in Pfizer document *5.3.6 Cumulative Analysis of Post-Authorization Adverse Event Reports of PF-07302048 (BNT162B2) Received Through 28-FEB-2021* (a.k.a., "*5.3.6*"). The search criterion for this System Organ Class was "Vasculitides." **Vasculitis, a.k.a., vasculitides, is inflammation of a blood vessel or multiple blood vessels.** Small or large vessels may be involved, and symptoms vary depending on organ involvement. **The inflammation can be related to a direct immune attack on the cells of the blood vessel or to deposits of complexes of antibody and an antigen (virus or other protein) that is not part of the blood vessel itself.**

It is important to note that the adverse events (AEs) in the *5.3.6* document were reported to Pfizer for **only a 90-day period** starting on December 1, 2020, the date of the United Kingdom's public rollout of Pfizer's COVID-19 experimental mRNA "vaccine" product.

Highlights of this report include:

- **34 vasculitis adverse events were reported among 32 cases** (i.e., patients). **One adverse event was fatal.**
- **81% of vasculitis sufferers were women**, and 19% were men.
- Onset time from injection to symptom onset was <24 hours to 19 days, with **half occurring within three days of receiving the vaccine**.
- Systemic vasculitis is difficult to treat. **It often cannot be cured and can require permanently being on medication to manage it.**
- 32% of vasculitis adverse events were related to **skin rashes**, including cutaneous vasculitis, vasculitic rash, hypersensitivity vasculitis, and palpable purpura.
- **35% of these adverse events were marked as "not resolved"** at the end of the post-marketing period.
- Pfizer received reports of **three cases of Giant cell arteritis, a serious autoimmune disease of the large blood vessels** that can lead to blindness if not quickly treated.
- **Three cases of peripheral ischemia, inflammation of blood vessels to the point of impairing blood flow**, were reported.

- **Two instances of Behçet's syndrome—a type of vasculitis with mouth, skin, and genital sores, often accompanied by eye inflammation and blood clots**—were reported.
- **One instance of Takayasu's arteritis, a very serious and rare disease where the aorta and its main branches are typically inflamed**—was recorded.

Pfizer concluded, "This case review does not raise new safety issues. Surveillance will continue."

Please read this important report below.

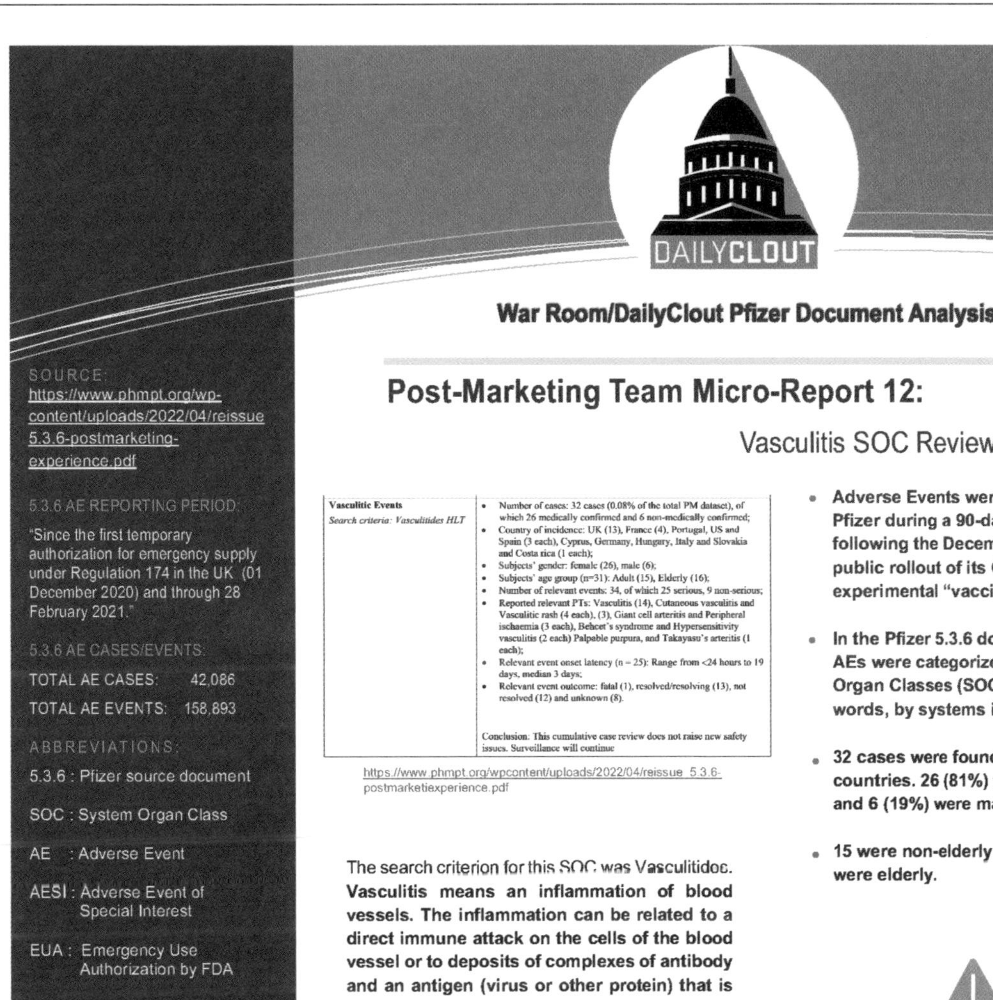

War Room/DailyClout Pfizer Document Analysis

Post-Marketing Team Micro-Report 12:

Vasculitis SOC Review of 5.3.6

SOURCE:
https://www.phmpt.org/wp-content/uploads/2022/04/reissue 5.3.6-postmarketing-experience.pdf

5.3.6 AE REPORTING PERIOD:

"Since the first temporary authorization for emergency supply under Regulation 174 in the UK (01 December 2020) and through 28 February 2021."

5.3.6 AE CASES/EVENTS:

TOTAL AE CASES: 42,086

TOTAL AE EVENTS: 158,893

ABBREVIATIONS:

5.3.6 : Pfizer source document

SOC : System Organ Class

AE : Adverse Event

AESI : Adverse Event of Special Interest

EUA : Emergency Use Authorization by FDA

PM : Post-Marketing

BNT162b2 : Pfizer's mRNA COVID-19 vaccine

SEQUELAE: an abnormal condition resulting from a previous disease, injury, or other trauma

AGE GROUPS defined in 5.3.6 (p. 25 footnote) :

Adult	18 - 64
Elderly	≥ 65
Child	2 - 11
Adolescent	12 - < 18
Infant	1 – 23 months

AUTHORS:

Dr Barbara Gehrett MD
Dr Joseph Gehrett MD
Dr Chris Flowers MD
Loree Britt

Post Marketing Team

18Apr23

Vasculitic Events *Search criteria: Vasculitides HLT*	• Number of cases: 32 cases (0.08% of the total PM dataset), of which 26 medically confirmed and 6 non-medically confirmed; • Country of incidence: UK (13), France (4), Portugal, US and Spain (3 each), Cyprus, Germany, Hungary, Italy and Slovakia and Costa rica (1 each); • Subjects' gender: female (26), male (6); • Subjects' age group (n=31): Adult (15), Elderly (16); • Number of relevant events: 34, of which 25 serious, 9 non-serious; • Reported relevant PTs: Vasculitis (14), Cutaneous vasculitis and Vasculitic rash (4 each), (3), Giant cell arteritis and Peripheral ischaemia (3 each), Behcet's syndrome and Hypersensitivity vasculitis (2 each) Palpable purpura, and Takayasu's arteritis (1 each); • Relevant event onset latency (n = 25): Range from <24 hours to 19 days, median 3 days; • Relevant event outcome: fatal (1), resolved/resolving (13), not resolved (12) and unknown (8). Conclusion: This cumulative case review does not raise new safety issues. Surveillance will continue

https://www.phmpt.org/wpcontent/uploads/2022/04/reissue 5.3.6-postmarketiexperience.pdf

The search criterion for this SOC was Vasculitidoc. **Vasculitis means an inflammation of blood vessels. The inflammation can be related to a direct immune attack on the cells of the blood vessel or to deposits of complexes of antibody and an antigen (virus or other protein) that is not part of the blood vessel itself.** Sometimes small blood vessels are involved, and other times large vessels are involved. Symptoms vary, depending on which organ the inflamed blood vessels feed.

Thirty-four adverse events were reported. Fourteen cases of unspecified vasculitis were recorded. Without further detail, it is impossible to draw conclusions about these **14** events, except to say that systemic vasculitis is not easy to treat and often cannot be cured but, instead, has to be **permanently** medicated to control. The 32 individuals had 34 relevant adverse events, with **25 (74%) classified as serious.**

Eleven AEs were related to skin rashes which were vasculitic in nature. These include: cutaneous vasculitis, vasculitic rash, hypersensitivity vasculitis, and palpable purpura. A vasculitic rash is often described as "palpable purpura," a slightly elevated bluish-red rash, often on the legs, related to disruption of small vessels.

- **Adverse Events were reported to Pfizer during a 90-day period, following the December 1, 2020, public rollout of its COVID-19 experimental "vaccine" product.**
- **In the Pfizer 5.3.6 document, these AEs were categorized by System Organ Classes (SOC) – in other words, by systems in the body.**
- **32 cases were found, from multiple countries. 26 (81%) were female and 6 (19%) were male.**
- **15 were non-elderly adults, and 16 were elderly.**

One event was fatal.

Thirteen AEs are described as "resolved/resolving" but not broken down any further. **Twelve were "not resolved,"** and eight were outcome "unknown."

No mention is given for the outcome category of "resolved with sequelae," which appears in most of the other SOC outcome lists. Does this suggest some severe sequelae for this serious set of diseases?

The time from injection to symptom onset was noted for 25 adverse events.

This time ranged from < 24 hours to 19 days, **with half of AEs occurring within three days.**

Pfizer received reports of three cases of **Giant cell arteritis, a serious syndrome that causes headache and visual symptoms. It can lead to blindness in the affected eye if it is not recognized and treated quickly.** Giant cell arteritis is often associated with **polymyalgia rheumatica**, a debilitating condition that causes weakness, aching, and stiffness in the shoulders and hips. Polymyalgia rheumatica is much commoner than Giant cell arteritis, but no cases of polymyalgia rheumatica were recorded in the musculoskeletal SOC. However, that SOC had 3,525 instances of arthralgia (joint aching); and Table 2 of 5.3.6 recorded **4**,915 instances of myalgia (muscle aching), which might easily have included developing polymyalgia rheumatica.

Three cases of peripheral ischemia were reported. This indicates that blood vessels feeding an arm, leg or digit were inflamed enough to cause **impairment of blood flow.** This threatens the loss of the affected area, with gangrene and perhaps even amputation as a result.

Two instances of Behçet's syndrome were reported. Behçet's syndrome is an unusual form of vasculitis that has mouth, skin, and genital sores, often with eye inflammation (uveitis) and blood clots. These patients may have neurologic symptoms as well.

Finally, **one case of Takayasu's arteritis** is included. This is a **very serious and very rare condition**. The NIH website Genetic and Rare Diseases states that a population estimate of Takayasu's arteritis in the U.S. is **fewer than 5,000 cases in the entire country**. The blood vessels inflamed are usually the aorta and its main branches. Takayasu's is most common among women under 40. It is also a condition that may develop slowly and manifest itself well after the start of the inflammation of the blood vessels. The fact that one case of Takayasu's arteritis was described in the first 90 days of general use of the Pfizer product is very concerning.

Takayasu arteritis - About the Disease - Genetic and Rare Diseases Information Center (nih.gov) Accessed 4.01.2023.

Pfizer's Conclusion:

This cumulative case review does not raise new safety issues.
Surveillance will continue.

NOTE: As of the date of this team report, no additional safety data have been publicly released.

Post-Marketing Team's CONCLUSION
RECALL this unsafe "vaccine."

Fifteen: "Pfizer and FDA Knew in Early 2021 That Pfizer mRNA COVID "Vaccine" Caused Dire Fetal and Infant Risks, Including Death. They Began an Aggressive Campaign to Vaccinate Pregnant Women Anyway."
—Amy Kelly, Program Director of the WarRoom/DailyClout Pfizer Documents Analysis Project

The batch of Pfizer clinical trial documents released in April 2023 by the Food and Drug Administration (FDA) under court order contains a shocking, eight-page document titled, "Pregnancy and Lactation Cumulative Review." (https://www.phmpt.org/wp-content/uploads/2023/04/125742_S2_M1_pllr-cumulative-review.pdf) The data in the Cumulative Review span ". . . from the time of drug product development to 28-FEB-2021." A Pfizer employee, Robert T. Maroko (https://www.linkedin.com/in/maroko/), approved the Review on April 20, 2021. (p. 8)

This document is among the most horrifying yet to emerge into public view. It reveals that both Pfizer and the FDA knew by early 2021 that Pfizer's mRNA COVID vaccine, BNT162b2, resulted in horrible damage to fetuses and babies. [Though I arrived at the conclusions in this article on my own from reviewing the document linked here, Sonia Elijah previously covered some of this same material on April 22nd on TrialSiteNews (https://www.trialsitenews.com/a/pfizers-pregnancy-lactation-cumulative-review-reveals-damning-data-2b15c969) and on April 26th on Substack (https://soniaelijah.substack.com/p/pfizers-pregnancy-and-lactation-cumulative) and Redacted (https://rumble.com/v2ko732-breaking-new-bombshell-pfizer-documents-reveal-damning-data-redacted-with-c.html).] Pfizer tabulated:

- **Adverse events in over 54% of cases of "maternal exposure" to vaccine** (248 out of 458). "Maternal exposure" is defined on pp. 1–2 as: "PTs Maternal exposure timing unspecified, Maternal exposure during pregnancy, Maternal exposure before pregnancy, Exposure during pregnancy." These definitions imply that Pfizer may have been looking at damage to women and babies that could result from intercourse, inhalation, and skin contact prior to pregnancy, as Pfizer defines "exposure" including all three in its protocol (Protocol Amendment 14, https://www.phmpt.org/wp-content/uploads/2022/03/125742_S1_M5_5351_c4591001-interim-mth6-protocol.pdf, pp. 213, 246, 398, 431, 575, 607, 751, 783, 918, 948, 1073, 1103, 1226, 1255, 1378, 1406, 1522, 1549, 1663, 1688, 1813, 1836, 1949, 1969, 2081, 2100, 2211, 2228, and 2337.)
- Pfizer's tally of damages to fetuses and babies includes:
  - **"53 reports [or 21%—53/248] of spontaneous abortion (51)/ abortion (1)/ abortion missed (1) following BNT162b2 vaccination."** (p. 4). A "missed abortion" is "an empty gestational sac, blighted ovum, or a fetus or fetal pole without a heartbeat prior to completion of 20 weeks 0 days gestation." (https://www.acog.org/practice-management/coding/coding-library/billing-for-interruption-of-early-pregnancy-loss)

- **Fetal tachycardia** (irregular heart rate faster than 180 beats per minute) that required **early delivery and hospitalization** of the affected newborn for five days. "The clinical outcome of fetal tachycardia was unknown." (p. 2)
- **Six premature labor and delivery cases** (p. 3) resulting in:
  - **Two newborn deaths**. Cause of death for one baby "was cited as extreme prematurity with severe respiratory distress and pneumothorax." Pfizer stated the other death was due to "premature baby less than 26 weeks and severe respiratory distress and pneumothorax." Note that **newborn pneumothorax** is a condition where air leaks out of the lung and collects between the lung and the chest wall.
  - **Newborn severe respiratory distress**. (Note: California midwife, Ellen Jasmer, has described exactly this phenomena happening in her practice in a recent DailyClout interview—https://dailyclout.io/calcified-placentas-a-nurse-midwifes-disturbing-testimony/.)

It is not just fetuses and newborn babies that Pfizer calmly noted were being damaged in the company's internal records. Pfizer also recorded multiple harms to babies through the milk of vaccinated mothers. According to Pfizer in the Cumulative Review, **19% (41/215) of babies in Pfizer's records exposed to the company's COVID mRNA vaccine via their mothers' breast milk were recorded as suffering from 48 different categories of adverse events*** (pp. 6–7). These included:

Preferred Term	Explanation	# of Events
Pyrexia	Fever	9
Off label use		8
Product use issue		7
Infant irritability		5
Headache		5
Rash		5
Diarrhoea		3
Illness		3
Insomnia		3
Suppressed lactation		3
Breast milk discolouration		2
Infantile vomiting		2
Lethargy		2
Pain		2
Peripheral coldness		2
Urticaria	Hives	2
Vomiting		2
Abdominal discomfort		1

(Continued on next page)

Preferred Term	Explanation	# of Events
Agitation		1
Allergy to vaccine		1
Angioedema	An area of swelling of the lower layer of skin and tissue just under the skin or mucous membranes. The swelling may occur in the face, tongue, larynx, abdomen, or arms and legs. Often it is associated with hives, which are swelling within the upper skin.	1
Anxiety		1
Axillary pain		1
Breast pain		1
Breast swelling		1
Chills		1
Cough		1
Crying		1
Dysgeusia	Also known as parageusia. A distortion of the sense of taste.	1
Dysphonia	Hoarseness, most frequently caused by a problem with a person's vocal cords or larynx.	1
Eructation	Belching.	1
Epistaxis	Nosebleeds.	1
Eyelid ptosis	Droopy eyelid.	1
Facial paralysis		1
Fatigue		1
Increased appetite		1
Lymphadenopathy	Swollen lymph nodes.	1
Myalgia		1
Nausea		1
Paresis		1
Poor feeding infant		1
Poor quality sleep		1
Pruritis	Itchy skin.	1
Restlessness		1
Rhinorrhoea	Runny nose.	1
Roseola	An infection that can cause a high fever followed by a rash.	1
Skin exfoliation		1
Vision blurred		1

*From Table 2—Number of Adverse Events Reported in Infants with 'Exposure via Lactation' (pp. 6–7)]

Some of the babies' suffering was serious: there were ten "Serious Adverse Events" (SAEs) from "Exposure via Lactation." The Review outlines six of them (p. 7):

1. "A 15-month old infant with medical history of vomiting experienced **skin exfoliation and infant irritability** while being breastfed (latency <7 days). **The outcome of the event 'skin exfoliation' was not recovered and outcome of event 'infant irritability' was unknown.** No causality was reported by the physician."
2. "A 9-month old infant with a medical history of meningococcal vaccine and no history of allergies, asthma, eczema or anaphylaxis experienced **rash and urticaria** a day after exposure via lactation. The outcome of the events was 'resolved' and event did not happen after the second day. No causality assessment was provided."
3. "A day after the mother received vaccination, a baby developed a **rash after breastfeeding**. At the time of the report, the event was '**not recovered**. [Sic] A causality assessment was not provided."
4. "An 8-month old infant experienced **angioedema** [an area of swelling of the lower layer of skin and tissue just under the skin or mucous membranes] one day after his mother received vaccination. The event was considered non-serious by health authority and the **outcome at the time of the report was unknown**. No causality was provided."
5. "There were **2 cases reporting 'illness' after exposure via breast milk**. In the first case, a 6-month old infant developed an **unspecified sickness 2 days post mother's vaccination**. The outcome of the event sickness was recovered, and no causality assessment was provided. The second case, **a 3-month old infant developed an unspecified illness and required hospitalization for 6 days post exposure via breast milk (>7 days latency)**. The event outcome was reported as 'recovering' and no causality assessment was provided."

Pfizer's Summary and Conclusion section of the Cumulative Review states, "The cases reviewed above are indicative of what is in the Pfizer safety database as of 28 February 2021. The sponsor (Pfizer/BioNTech) will continue to monitor and report on all pregnancy exposure and lactation cases. It is important to note that the spontaneous safety database is intended for hypothesis generation and not hypothesis testing." (p. 7)

Despite Pfizer and the FDA knowing by April 20, 2021, the extent of damage to fetuses and babies, including the fact that fetuses and newborns had died, **on April 23, 2021, inexplicably Dr. Rochelle Walensky held a White House press briefing where she recommended pregnant women get vaccinated.** (https://www.verywellhealth.com/pregnant-women-covid-vaccine-5092509)

Please read the damning "Pregnancy and Lactation Cumulative Review" available at (https://www.phmpt.org/wp-content/uploads/2023/04/125742_S2_M1_pllr-cumulative-review.pdf) below.

Document Approval Record

Document Name:	COVID-19 Vaccine - Safety Review for PLLR Label Update
Document Title:	COVID-19 Vaccine - Safety Review for PLLR Label Update

Signed By:	**Date(GMT)**	**Signing Capacity**
Maroko, Robert T	20-Apr-2021 16:11:58	Business Line Approver

090177e196d38ee9\Approved\Approved On: 20-Apr-2021 16:11 (GMT)

FDA-CBER-2021-5683-0779752

Sixteen: "Histopathology Series Part 4c—Autoimmunity: A Principal Pathological Mechanism of COVD-19 Gene Therapy Harm (CoVax Diseases) and a Central Flaw in the LNP/mRNA Platform"
—Robert W. Chandler, MD, MBA

Introduction

A new class of illness, termed within this report as **CoVax Disease**, is now warranted. CoVax Disease encompasses multiple pathological mechanisms of which **autoimmunity** occupies a central position.

In this report, the reader will learn that autoimmunity cases reported to the Vaccine Adverse Events Reporting System in the U.S. increased 24-fold from 2020 to 2021, and annual autoimmunity-related fatalities increased 37x in the same time period.

mRNA "vaccines" are now associated with changeable disease manifestations that come from the widespread biodistribution of the mRNA gene therapy products that enter host cells and then translate mRNA code into Spike proteins that the host immune system does not recognize. Those unrecognized proteins prompt an immune response from the host that targets and destroys Spike proteins in host tissues and organs, sometimes dramatically and even fatally. mRNA "vaccines" have made the "vaccinated" human body its own worst enemy.

I: Autoimmunity: Definition, Prevalence, and Clinical Variation

"Auto" is from ancient Greek word αὐτός meaning "self." (https://www.perseus.tufts.edu/hopper/morph?l=au%29to%5C&la=greek&can=au%29to%5C0&prior=ou)k&d=Perseus:text:1999.01.0167:section=363a&i=1#lexicon) When combined with immunity the term applies to medical conditions resulting from a destructive immune response to oneself. This response comes in the form of both antibody and cellular attack on one's own tissues and varies from involvement of a single organ to multiple organs or widespread systemic manifestations. (https://pathology.jhu.edu/autoimmune/prevalence/)

Johns Hopkins Medical School Department of Pathology gives the prevalence of autoimmune illnesses in the United States as 10 million, or approximately three percent of the population behind obesity 115 million, cardiovascular disease 66 million, Type 2 diabetes 50 million, and all cancers 13 million. (https://pathology.jhu.edu/autoimmune/prevalence)

Johns Hopkins lists the top ten illnesses in the autoimmune disease category (below).

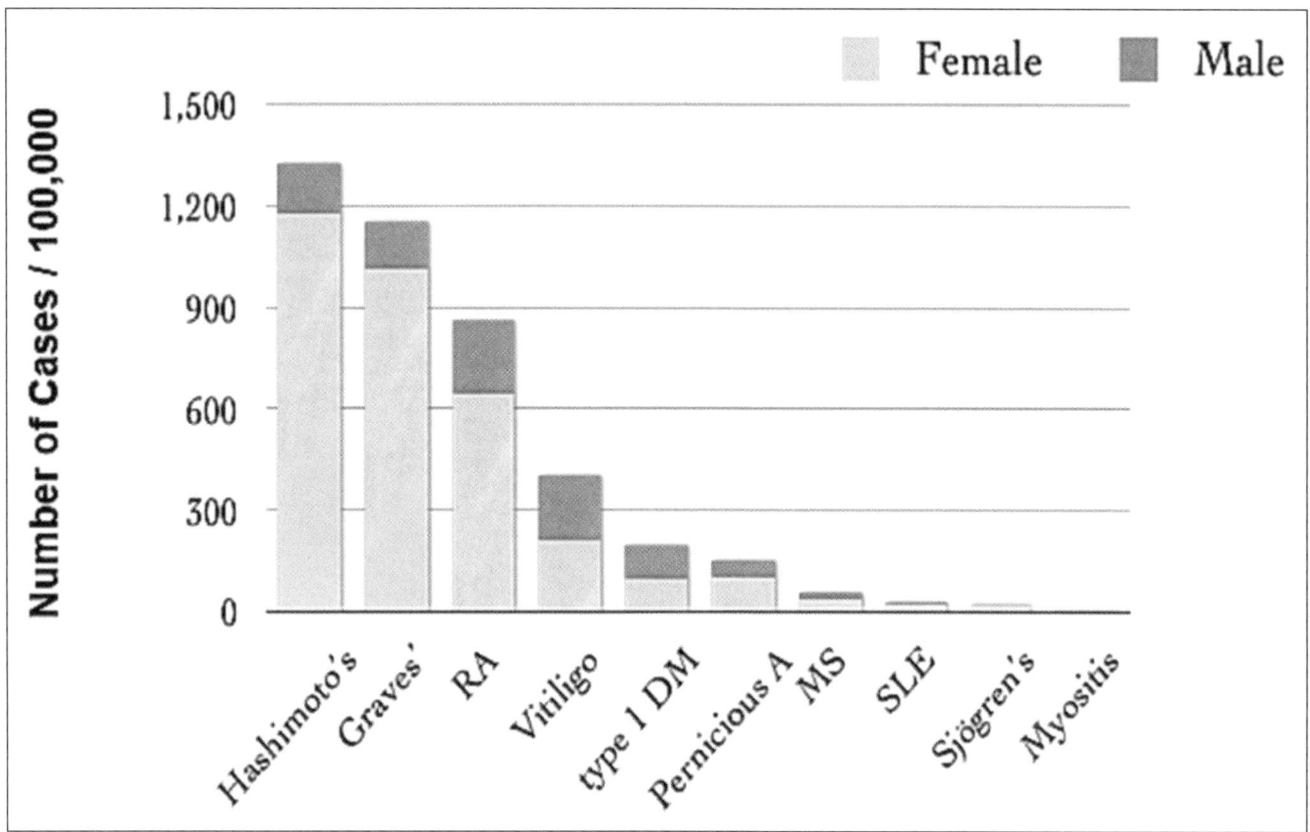

The distribution of autoimmune diseases varies from institution to institution or series to series in published literature. These illnesses may involve a single organ or multiple organs with local and/or systemic signs and symptoms.

A review of these ten conditions demonstrates the wide variation of autoimmune disease identified prior to the widespread use of mRNA products.

1 and 2. Thyroiditis: Autoimmunity can present as either thyroid underactivity, or Hypothyroid (Hashimoto's Disease). https://www.mayoclinic.org/diseases-conditions/hashimotos-disease/symptoms-causes/syc-20351855 or overactivity, Hyperthyroid (Graves' Disease) (https://www.mayoclinic.org/diseases-conditions/graves-disease/symptoms-causes/syc-20356240). Autoimmune thyroiditis is an example of single organ disease with systemic symptoms.

3. Rheumatoid arthritis (RA): RA involves joints and supportive tissues with inflammation that can affect multiple joints and regions simultaneously. It is a debilitating disease. (https://www.mayoclinic.org/diseases-conditions/rheumatoid-arthritis/symptoms-causes/syc-20353648)

Recall that the number one adverse event (AE) by system organ category (after "Other") reported in Pfizer's confidential post-marketing document 5.3.6, AEs from the first 10 weeks of emergency use inoculation of BNT162b2, was "Musculoskeletal." (https://www.phmpt.org/wp-content/uploads/2022/04/reissue_5.3.6-postmarketing-experience.pdf) The data collection for that report ended on February 28, 2021.

Note that Autoimmune illness in the Pfizer data is number five.

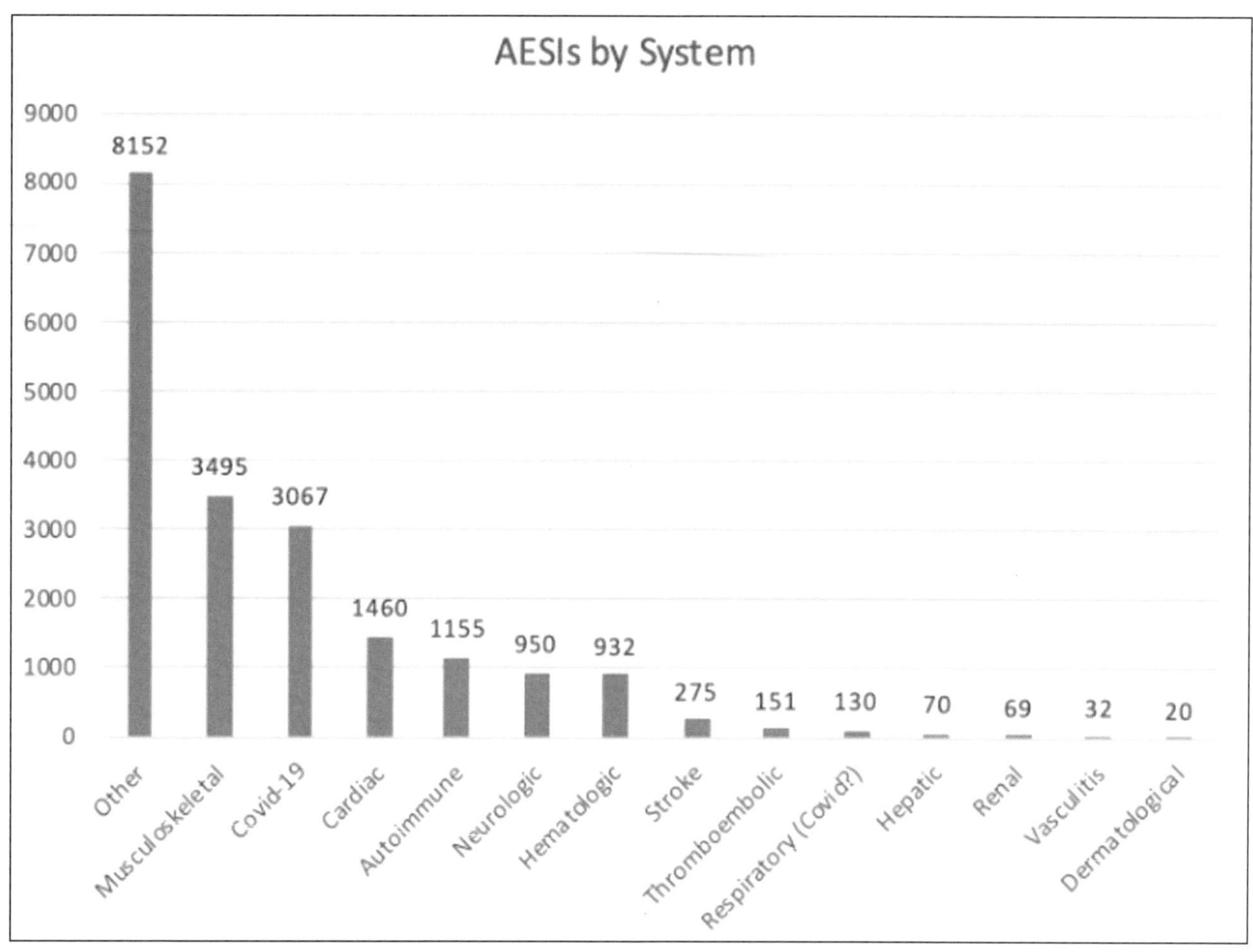

(https://dailyclout.io/pfizer-evidence-so-far-coverups-heart-damage-and-more/, p. 19 of PDF)

4. Vitiligo: A dermatologic condition. (https://www.niams.nih.gov/health-topics/vitiligo)
5. Type 1 diabetes is a result of autoimmune attack on insulin producing cells in the pancreas. (https://www.cdc.gov/diabetes/basics/what-is-type-1-diabetes.html)
6. Pernicious anemia: Chronic anemia from reduced Vitamin B12 absorption. (https://www.ncbi.nlm.nih.gov/books/NBK540989/)
7. Multiple sclerosis: Demyelinating disease of the central nervous system. (https://www.mayoclinic.org/diseases-conditions/multiple-sclerosis/symptoms-causes/syc-20350269)
8. Systemic lupus erythematosus: Discussed below.
9. Sjogren's syndrome: Involves salivary glands and musculoskeletal symptoms. (https://www.ninds.nih.gov/health-information/disorders/sjogrens-syndrome)
10. Myositis: Inflammation of muscles. (https://www.hopkinsmyositis.org/myositis/)

This topic was explored along with rhabdomyolysis in a previous report. https://dailyclout.io/report-67-part-4b-rhabdomyolysis-a-k-a-jellied-muscle-after-mrna-gene-therapy-injections/

Systemic Lupus Erythematosus

Signs and symptoms of autoimmune illness can vary over time and can manifest as multiorgan diseases such as Systemic Lupus Erythematosus, referred to as SLE or "Lupus."

Lupus can appear with a diverse array of symptoms as reported by the Centers for Disease Control and Prevention (CDC). (https://www.cdc.gov/lupus/basics/symptoms.htm)

Lupus symptoms include:

- **Muscle and joint pain.** One may experience pain and stiffness, with or without swelling. This affects most people with lupus. Common areas for muscle pain and swelling include the neck, thighs, shoulders, and upper arms.
- A fever higher than 100 degrees Fahrenheit affects many people with lupus. The fever is often caused by inflammation or infection. Lupus medicine can help manage and prevent fever.
- One may get rashes on any part of the body that is exposed to the sun, such as face, arms, and hands. One common sign of lupus is a red, butterfly-shaped rash across the nose and cheeks.
- **Chest pain.** Lupus can trigger inflammation in the lining of the lungs. This causes chest pain when breathing deeply.
- **Hair loss.** Patchy or bald spots are common. Hair loss may also be caused by some medicines or infection.
- **Sun or light sensitivity.** Most people with lupus are sensitive to light, a condition called photosensitivity. Exposure to light can cause rashes, fever, fatigue, or joint pain in some people with lupus.
- **Kidney problems.** Half of people with lupus also have kidney problems, called lupus nephritis.[3] Symptoms include weight gain, swollen ankles, high blood pressure, and decreased kidney function.
- **Mouth sores.** Also called ulcers, these sores usually appear on the roof of the mouth but can also appear on the gums, inside the cheeks, and on the lips. They may be painless, or one may have soreness or dry mouth.
- **Prolonged or extreme fatigue.** One may feel tired or exhausted even when getting enough sleep. Fatigue can also be a warning sign of a lupus flare.
- Fatigue can be a sign of anemia, a condition that happens when one's body does not have red blood cells to carry oxygen throughout the body.
- **Memory problems.** Some people with lupus report problems with forgetfulness or confusion.
- **Blood clotting.** One may have a higher risk of blood clotting. This can cause blood clots in the legs or lungs, stroke, heart attack, or repeated miscarriages.
- **Eye disease.** One may get dry eyes, eye inflammation, and eyelid rashes.

Giant Cell Arteritis

GCA or Giant Cell Arteritis (a.k.a., Temporal Arteritis) is another autoimmune disease that commonly involves the temporal artery and can impair vision,

Giant cell arteritis is an inflammation of the lining of your arteries. Most often, it affects the arteries in your head, especially those in your temples. For this reason, giant cell arteritis is sometimes called temporal arteritis. (https://www.mayoclinic.org/diseases-conditions/giant-cell-arteritis/symptoms-causes/syc-20372758)

GCA was the leading autoimmune illness following COVID-19 gene therapy products in 27 patients in a study from Limoges, France, reported by Liozon, et al. [Liozon, E., et al. (Immune-Mediated Diseases Following COVID-19 Vaccination: Report of a Teaching Hospital-Based Case-Series. J. Clin. Med. 2022, 11, 7484.) https://doi.org/10.3390/jcm11247484.]

This paper is discussed in some detail later in this article.

The point being made here is the diversity and complexity of this very broad and heterogenous group of diseases can present both diagnostic and therapeutic challenges.

The graphic below illustrates some of the manifestations of autoimmunity.

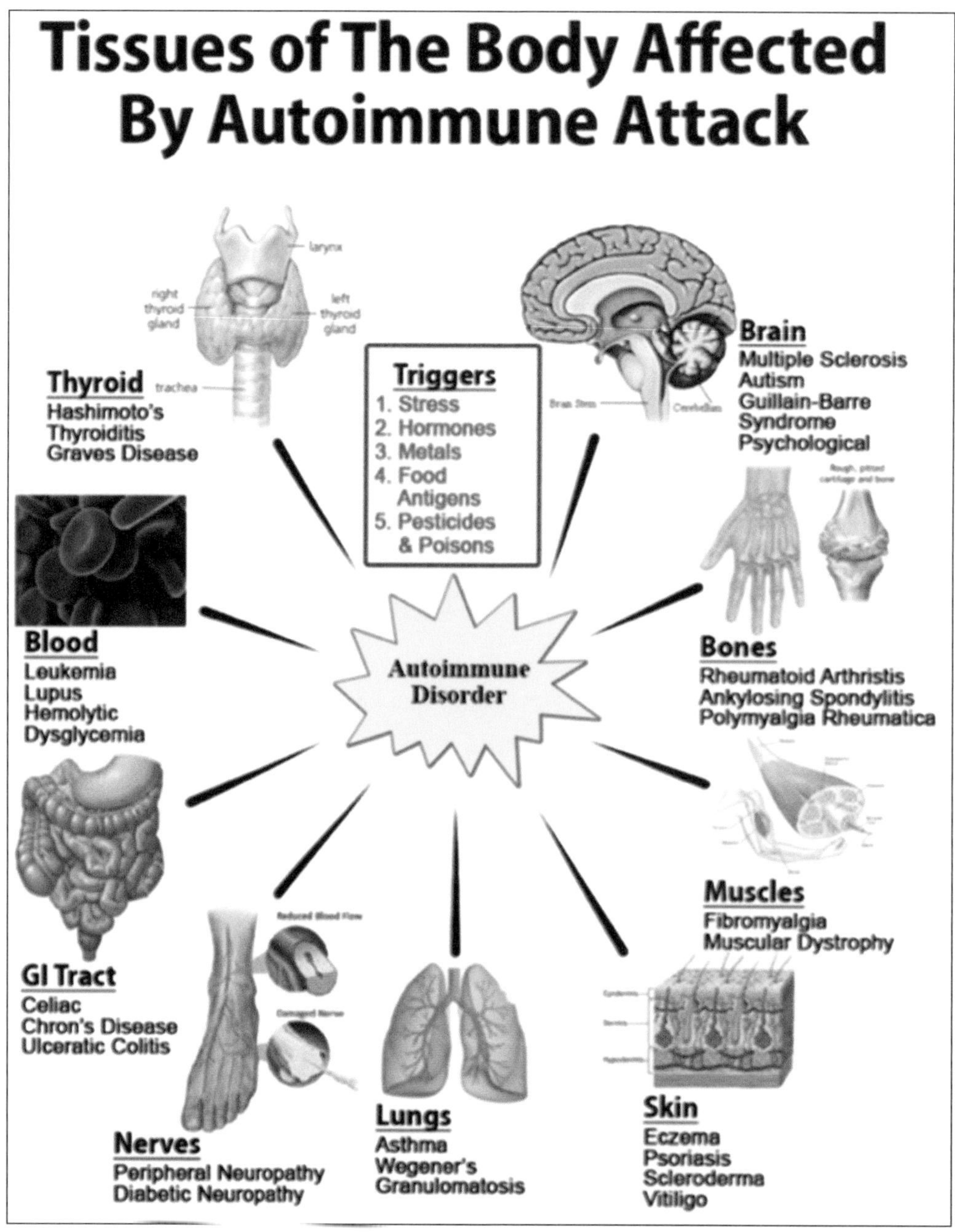

https://inimmune.com/wp-content/uploads/2016/05/Autoimmunity-768x995.png

The Autoimmune Registry provides a list of autoimmune diseases with symptoms and links to additional references: https://www.autoimmuneregistry.org/the-list.

This article will not explore the complexities of autoimmune immunology, or the wide array of illnesses thought to be caused by this attack by self on self. Rather, it will explore the link between Spike-producing Gene Therapy drugs and autoimmune-like illnesses using a series report from France; case reports from Japan, Germany and Spain; and data and case reports from the Vaccine Adverse Events Reporting System (VAERS).

There will be some crossover between the subject of Part 4B in this series, Rhabdomyolysis (https://dailyclout.io/report-67-part-4b-rhabdomyolysis-a-k-a-jellied-muscle-after-mrna-gene-therapy-injections/), which has certain features of autoimmunity as well as Multisystem Inflammatory Syndrome in Children (MIS-C) featured in Histopathology Series Part 4D (in preparation) that concerns ***CoVax Disease*** in children, as well as this article on autoimmune disease. (The term ***CoVax Diseases (CVDs)*** is introduced here to describe the disease states associated with or caused by injection of Spike-producing gene therapy products and encompass both pathologic processes based upon histopathologic analysis and clusters of illness conditions likely caused by lipid nanoparticle (LNP) components, synthetic mRNA, E.coli Kanamycin resistant DNA Plasmids, and Adenovirus-vectored DNA, and other Spike-producing drugs.)

II. Series Report

Liozon, et al. presented a series of 27 patients diagnosed with Immune-Mediated Diseases (IMDs) associated with use of Spike-generating drugs in a French referral hospital serving a population of 723,784. Data were collected during the period from January 2021 through May 31, 2022. (https://doi.org/10.3390/jcm11247484)

The authors are careful to point out that numerators and denominators cannot be measured or estimated, so incidence and prevalence statistics are not calculated.

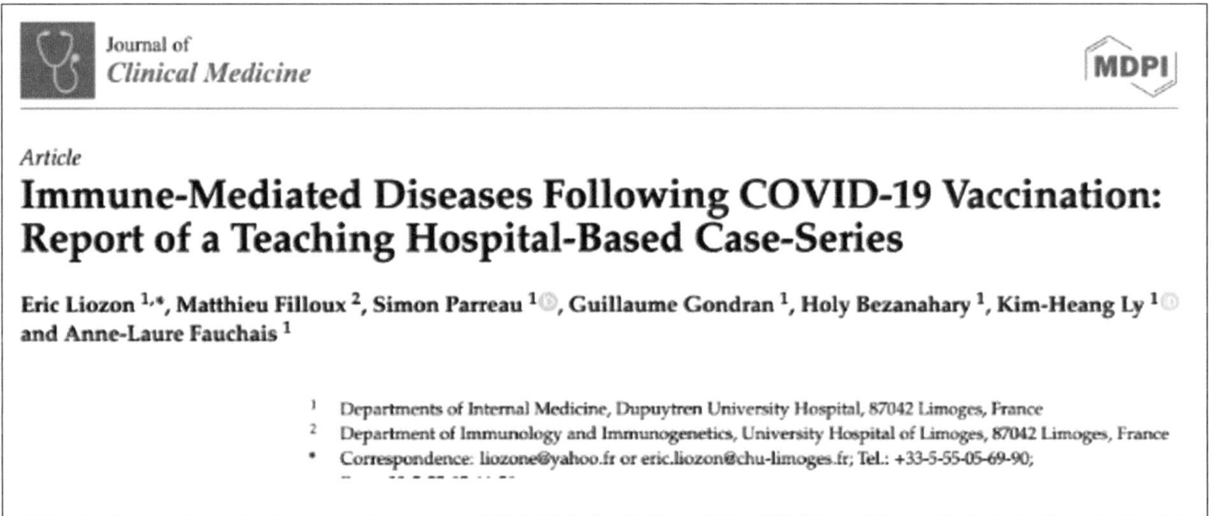

Journal of
Clinical Medicine

MDPI

Article

Immune-Mediated Diseases Following COVID-19 Vaccination: Report of a Teaching Hospital-Based Case-Series

Eric Liozon [1,*], Matthieu Filloux [2], Simon Parreau [1], Guillaume Gondran [1], Holy Bezanahary [1], Kim-Heang Ly [1] and Anne-Laure Fauchais [1]

[1] Departments of Internal Medicine, Dupuytren University Hospital, 87042 Limoges, France
[2] Department of Immunology and Immunogenetics, University Hospital of Limoges, 87042 Limoges, France
* Correspondence: liozone@yahoo.fr or eric.liozon@chu-limoges.fr; Tel.: +33-5-55-05-69-90;

Comparing the very limited crossover of this distribution with that from Johns Hopkins illustrates how varied these illnesses can be.

The 27 cases of autoimmune disease associated with Spike-producing drugs in this series are listed by category (following image). Each category is listed with criteria used to make the diagnosis.

27 Cases of Autoimmune Illness Following Spike Producing Gene Therapy

Diagnosis	N =	Diagnostic Criteria
Giant Cell Arteritis	12	https://doi.org/10.1002/art.1780330810
Polymyalgia Rheumatica	4	http://dx.doi.org/10.1136/annrheumdis-2011-200329
Myositis/Rhabdomyolysias	3	http://dx.doi.org/10.1136/annrheumdis-2017-211468
Henoch-Shönlein Purpura (SHP)	2	https://doi.org/10.1002/art.1780330809
Idiopathic Thrombocytopenic Purpura	2	Platelet count persistently <100 G/L, exclusion of other causes of thrombocytopenia.
ANCA-associated Vasculitis	1	History of allergic asthma, blood eosinophilia 3.6 × 10^9/L, multiple mononeuropathy, and ANCA positivity (anti-myeloperoxydase 49 UI/L).
Anti-Synthetase (PL7+)	1	https://www.myositis.org/about-myositis/complications/antisynthetase-syndrome/
Rheumatoid Arthritis Relapse	1	https://doi.org/10.1002/art.27584
AOSD Relapse (Still's Disease)	1	The Journal of Rheumatology, 01 Mar 1992, 19(3):424-430
Total =	27	

The discussion by the authors is interesting as the authors delicately handle the topic of safety.

> "Phase-II trials have shown reassuring safety profiles in mRNA-based vaccines and viral vector-based vaccines." (https://doi.org/10.3390/jcm11247484, p. 1.)

They go on to say caution is warranted because of "population wide surveys" which have suggested "significant side effects."

> Nevertheless, recent safety data from population-wide surveys have suggested significant side effects, notably severe thrombotic events via vaccine-induced immune thrombotic thrombocytopenia. (https://doi.org/10.3390/jcm11247484, p. 1.)

The outcome for these patients was not favorable. Only 11% recovered, 60% well to poorly controlled on medications, one died, and 26% (7/27) were lost.

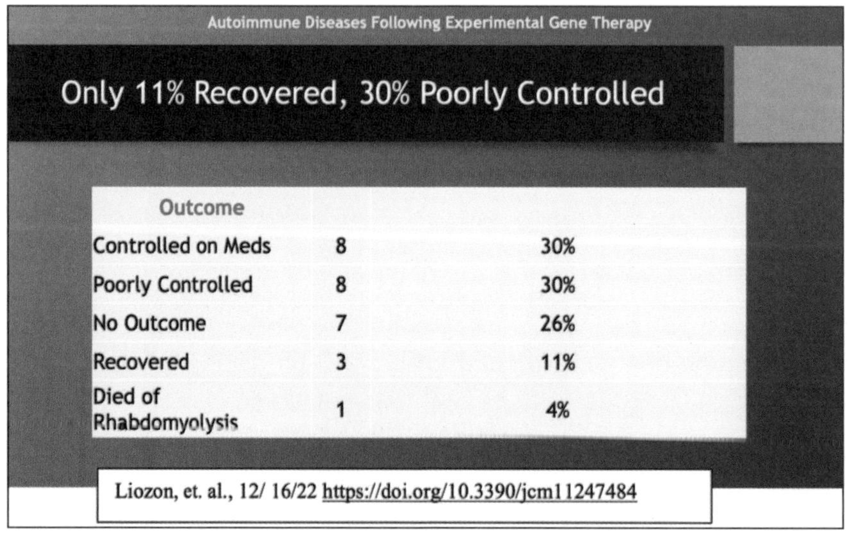

Autoimmune Diseases Following Experimental Gene Therapy

Only 11% Recovered, 30% Poorly Controlled

Outcome		
Controlled on Meds	8	30%
Poorly Controlled	8	30%
No Outcome	7	26%
Recovered	3	11%
Died of Rhabdomyolysis	1	4%

Liozon, et. al., 12/ 16/22 https://doi.org/10.3390/jcm11247484

Treatment was not effective. Steroids were often the mainstay of treatment and were used in 93% (25/27) of the cases. Only three patients were "recovered." Quotation marks are used as autoimmune illnesses can have periods of remission and reactivation. The positive ANA means that the disease is not gone.

Re-Inoculation after Autoimmune Disease Diagnosis: 11 patients

Remarkably, even when faced with the possibility of making the autoimmune condition worse, 11 patients were injected with up to three additional doses, ". . . without significant harm in 9 . . . " (82%) which means two (18%) had significant harm that was not specified.

Final outcome after additional injections is given below for this subgroup. Limited data from 10 patients in this subgroup are given with no explanation of what happened to the eleventh patient.

Final Outcome After Additional 1 or 2 Doses of EGT: 1 Lost 3 Relapsed, 4 Controlled, 2 Recovered

2	GCA	X2	Controlled
6	GCA	X2	Controlled
15	PMR	X1	Controlled
16	PMR	X1	Controlled
4	GCA	X1	Lost
1	GCA	X2	Recovered
25	ITP	X2	Recovered
13	PMR	X1	Relapsed
20	EGPA	X2	Relapsed
21	SHP	X2	Relapsed

A third of these patients relapsed, 22% recovered, and 44% were chronically ill.

Conclusions: 27 patients

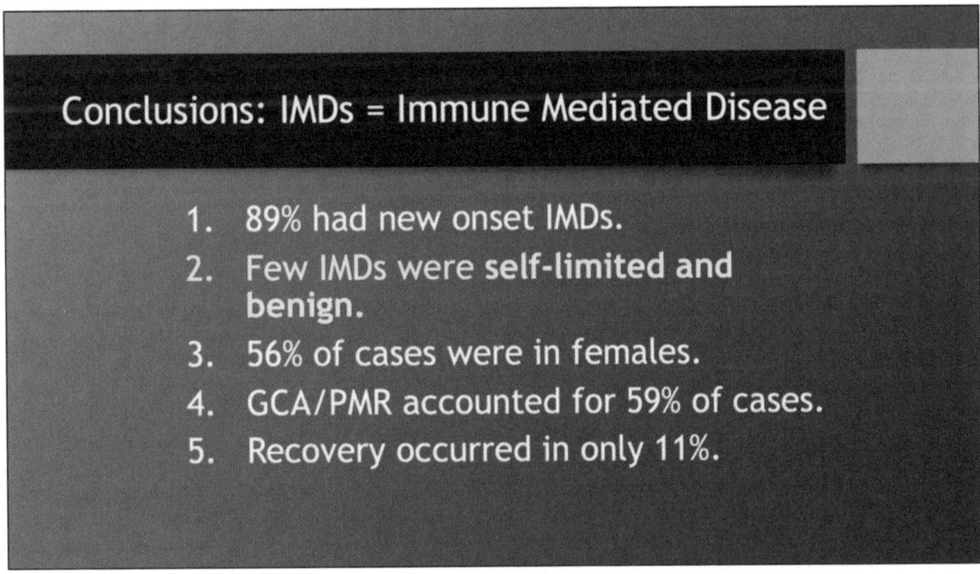

The authors concluded that 89% had new disease (previous image), and the rest were reactivation of pre-existing disease. Number two is in yellow to draw attention to the word choice used, "Few . . . were self-limited and benign." The treatment records of these patients should be reviewed to understand the course of these illnesses, a key part of assessing the severity of a disease.

As with most data associated with the use of these products, the majority of cases occurred in women, 59% were Giant Cell arteritis or Polymyalgia Rheumatica, and only 11% were considered recovered.

Causation

The authors present a confusing set of statements concerning the potential causal relationship between "vaccination" and Immune-Mediated Disease.

Here they say a causal relationship is "conjectural," and, in the same sentence, they note that a flare from re-inoculation reinforced the hypothesis of causal relationship.

> A causative relationship between vaccination and subsequent IMD is conjectural in the 27 patients, although rechallenge in two patients (Cases 1 and 20) resulted in a limited disease flare, reinforcing the hypothesis of a causal relationship between COVID-19 vaccination and GCA. (https://doi.org/10.3390/jcm11247484, p. 7.)

This is not the only example of what could be called forced ambiguity, because there are other examples of similar ambiguous phraseology.

Here they reinforce the notion that a **causal** relationship ". . . is not supported by our data" (emphasis added),

> Therefore, a causal relationship between COVID-19 vaccination and subsequent IMDs is not supported by our data. (https://doi.org/10.3390/jcm11247484, p. 7.)

This is a clear statement of **no causation** between Spike therapies and IMDs.

Then in the next sentence,

> Nevertheless, the strong temporal relationship between the two events raises legitimate safety concerns about COVID-19 vaccines in these patients. (https://doi.org/10.3390/jcm11247484, p. 7.)

The hedging continues,

> Other cases of GCA or PMR temporally associated with COVID-19 vaccination have been published, suggesting the association to be **not casual** and that post-vaccine onset of GCA/PMR is not an exceptional occurrence. (Emphasis added.) (https://doi.org/10.3390/jcm11247484, p. 8.)

Now the association between the injection and IMDs is ". . . not casual . . . " Is this the same as saying causal? A lot hinges on which side of the "s" the "u" lands.

> From our observations, unadjuvanted vaccines such as the currently marketed SARS-CoV-2 **vaccines can induce** GCA/PMR by acting as non specific triggers in genetically predisposed subjects (notably those with HLA-DRB1*04) possibly by strongly activating Toll-like receptor signaling. (Emphasis added.) (https://doi.org/10.3390/jcm11247484, p. 9.)

Merriam-Webster Dictionary gives "cause" as a synonym for induce.

https://www.merriam-webster.com/dictionary/induce

The authors even supply a mechanism whereby the gene therapy drugs activate Toll-like receptors that lead to the disease states called IMDs.

Translation: SARS-CoV-2 "vaccines" cause IMDs or "light up" an underlying pre-existing IMDs with high enough probability to raise a safety signal. Why such tortured reasoning? Are the authors being watched?

III. Literature Clinical Pathological Case (CPC) Reports
Literature Case Report 1: 14-Year-Old with Eight Organs Involved Fatality 45 hours after BNT162b2 Dose 3

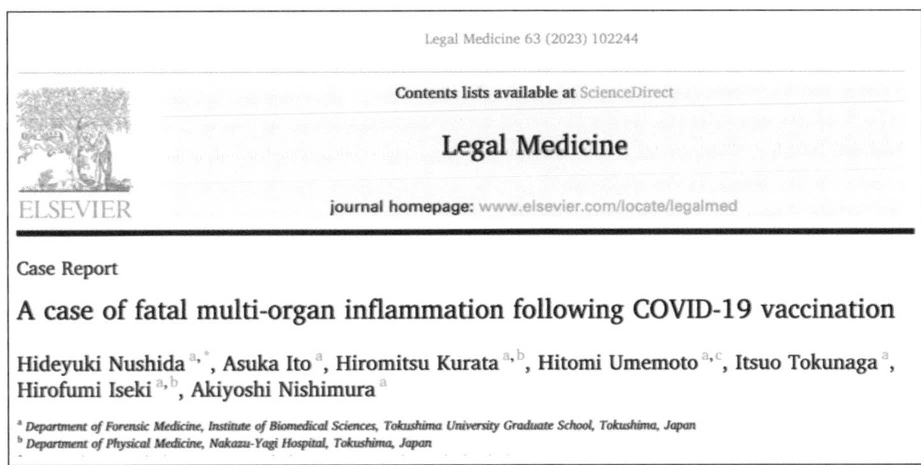

Legal Medicine 63 (2023) 102244

Contents lists available at ScienceDirect

Legal Medicine

journal homepage: www.elsevier.com/locate/legalmed

ELSEVIER

Case Report

A case of fatal multi-organ inflammation following COVID-19 vaccination

Hideyuki Nushida[a,*], Asuka Ito[a], Hiromitsu Kurata[a,b], Hitomi Umemoto[a,c], Itsuo Tokunaga[a], Hirofumi Iseki[a,b], Akiyoshi Nishimura[a]

[a] Department of Forensic Medicine, Institute of Biomedical Sciences, Tokushima University Graduate School, Tokushima, Japan
[b] Department of Physical Medicine, Nakazu-Yagi Hospital, Tokushima, Japan

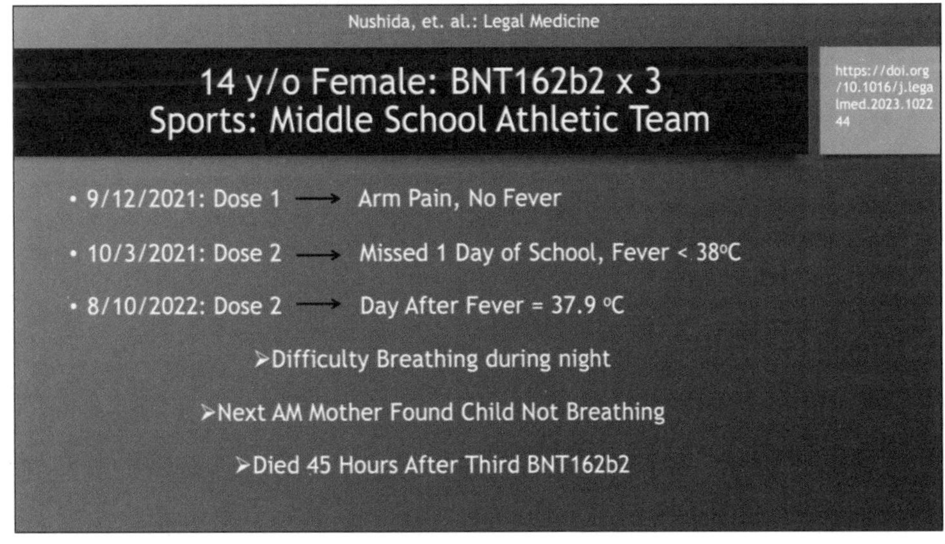

This case (previous image) authored by Nushida, et al., comes from Tokushima, Japan, and tells the medical story of a healthy, 14-year-old, athletic girl who had the following chronology beginning with "reactogenicity" after her first two doses of BNT162b2 (arm pain, fever, and malaise); then, 45 hours after her third dose, was found dead by her mother.

Her postmortem findings showed no evidence of virus.

Pertinent Negatives

- COVID-19 antigen quantification test: negative.
- Serum for adenovirus, cytomegalovirus, influenza virus (A, B), respiratory syncytial virus (RSV), Epstein-Barr virus, enterovirus (70, 71), parvovirus, and human immunodeficiency virus: negative.
- Quantitative testing for the COVID-19 antigen using nasopharyngeal swabs yielded negative results.
- The results of polymerase chain reaction (PCR) tests performed for COVID-19 using swabs from the lung, heart, liver, kidney, stomach, duodenum, diaphragm, and cerebrum after formalin fixation were also negative.
- Blood at autopsy was tested for drug toxicity using LC-MS/MS, and the results were negative.

It is interesting that the authors did some blood tests and proclaimed there was no drug toxicity ***even after they concluded that the experimental drug that she had been injected with three times was responsible for the child's demise.*** Did they not consider the "vaccine" a drug? Is this an example of classical conditioning or perhaps a form of hypnosis.

Biochemical Analysis

Presumably the SARS-CoV-2 antibody reported here was to Spike and not to nucleocapsid as the blood tests found no evidence of the virus past or present.

Elevated C Reactive Protein (CRP) and IL-6 are consistent with a systemic inflammatory reaction.

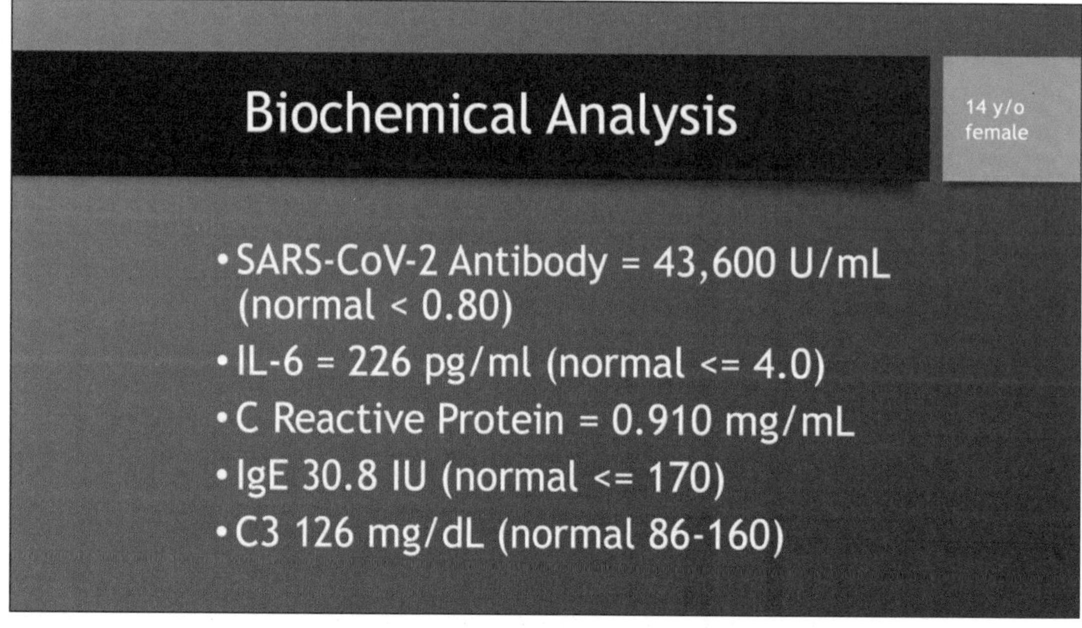

An autopsy was performed with the following gross findings. "Gross" refers to inspection with the naked eye.

At Autopsy: Gross Pathology Findings

- Heart: no degeneration or scarring on the grossly superficial surface or cross-sections.
- *Lungs: showed severe pulmonary edema.*

Histopathology with Special Stains

There are 24 histopathology sections from eight organs with three different stains indicating that the different organs attract different cell concentration and composition of activated white cells.

The viewer is dependent on the pathologist to present representative sections to indicate important findings. Sampling error should be taken into consideration.

Histopathology from eight organs shows infiltration with inflammatory lymphocytes.

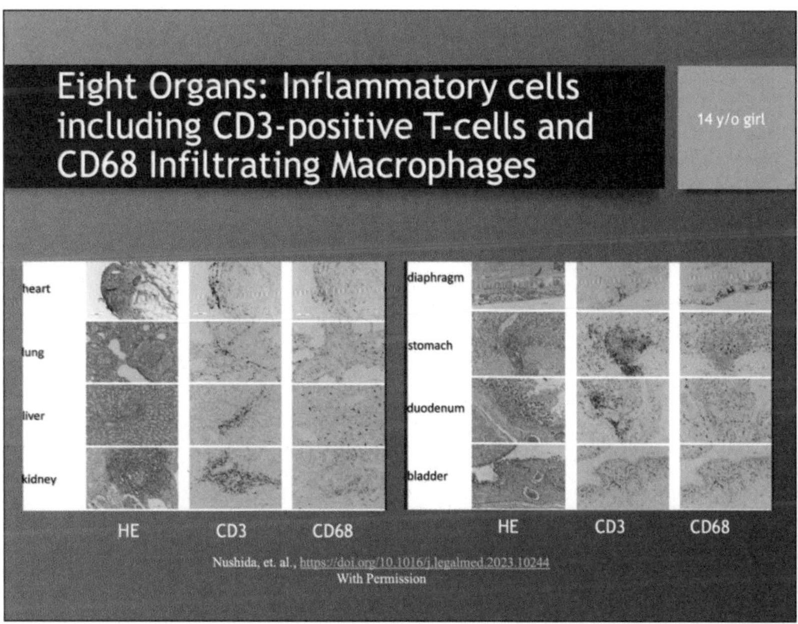

Fig. 1. Histopathology of the heart (left atrium), lung, liver, kidney, diaphragm, stomach, duodenum, and bladder. All images are × 200 magnification. HE: Hematoxylin and Eosin staining showing lymphocytic infiltration.

CD3: Immunohistochemical staining for CD3 showing inflammatory cells including CD3-positive T-cells. CD68: Immunohistochemical staining for CD68 shows the infiltrating cells include macrophages. (Used with permission).

The authors concluded (below) that the third dose of BNT162b2 produced a massive release of inflammatory chemicals from at least eight organs that was rapidly fatal. Like ***Turbocancer*** described by Dr. Burkhardt (https://dailyclout.io/report-58-part-2-autopsies-reveal-medical-atrocities-of-genetic-therapies-being-used-against-a-respiratory-virus/) and Dr. Krüger (https://dailyclout.io/report-61-ute-kruger-md-breast-cancer-specialist-reveals-increase-in-cancers-and-occurrences-of-turbo-cancers-following-genetic-therapy-vaccines/), ***Turboautoimmunity*** may be a diagnostic consideration.

Pay attention to #3 in the slide below. Common adverse reactions are characterized in the Pfizer documents, as well as publications in the medical literature, as "reactogenicity," an industry invention that inhibits detailed follow-up to define whether there is an association between early reactions to vaccines and deleterious long-term medical problems.

> Reactogenicity refers to a subset of reactions that occur soon after vaccination and are a physical manifestation of the inflammatory response to vaccination. In clinical trials, information on expected signs and symptoms after vaccination is actively sought (or 'solicited'). These symptoms may include pain, redness, swelling or induration for injected vaccines, and systemic symptoms, such as fever, myalgia, headache, or rash. The broader term 'safety' profile refers to all adverse events (AEs) that could potentially be caused/ triggered or worsened at any time after vaccination, and includes AEs, such as anaphylactic reactions, diseases diagnosed after vaccination and ***autoimmune events.*** (Emphasis added.) [https://www.nature.com /articles/s41541–019-0132–6 "The how's and what's of vaccine reactogenicity", Herve, et al.www.nature.com / npj/Vaccines, (2019) 4:39.]

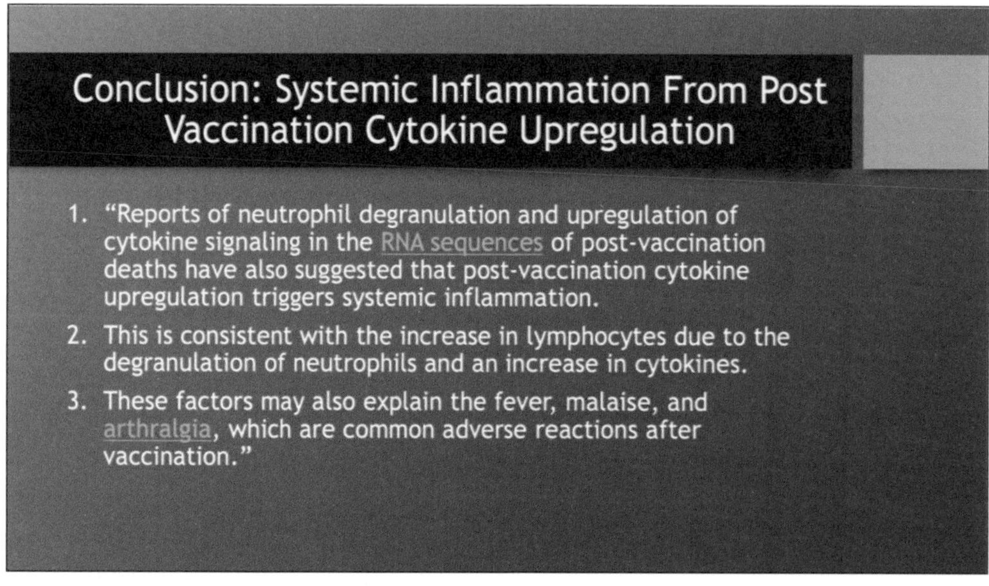

Case 1 Analysis

This case is not a typical autoimmunity case (See SLE below.) but rather is similar to the cytokine storm witnessed in patients with severe COVID-19. Catastrophic, multiorgan system failure with little advance warning is common to both conditions.

This presentation is also similar to Multisystem Inflammatory Syndrome in Children, MIS-C, or MIS-A in Adults. The rare condition known as Catastrophic Antiphospholipid Syndrome involves multiple organs but tends to occur in older females. (Cervera, Rodríguez-Pinto, Espinosa, Catastrophic Antiphospholipid Syndrome, Chapter 45 in *Mosaic of Autoimmunity* by Pericone and Schoenfeld, Academic Press, London, San Diego, Cambridge, Oxford, 2019. P484.)

The illness in this 14-year-old deceased may have a similar mass release of toxic substances from an autoimmune mediated process resulting from successive doses of BNT162b2 and a final, mass event after a third presentation of Spike antigen following Dose 3.

The explanation for the mechanism producing the fatal outcome in this case offered by the authors in the conclusions above describes the terminal phase of this process, massive degranulation of lymphocytes.

The initial phase occurs when mRNA from BNT162b2 stimulates production of foreign proteins that draw a host reaction manifest as extensive infiltration of killer lymphocytes, C3 inflammatory cells, and C68 macrophages, as was seen in the histopathology from all eight organs of this young teenager. [https://www.abcam.com/primary-antibodies/human-cd-antigen-guide is a good source to identify the different antigens on leucocyte cell surfaces (CD or Cluster of Differentiation).]

BNT162b2 has known dose-related effects, and it is possible that each successive dose increases the magnitude of the cellular response. That cellular response consists of multiorgan infiltration of activated lymphocytes that suddenly and massively dump cytokines and other agents resulting in sudden death upon presentation of a third dose of BNT162b2. The terminal event may have been cardiac arrest.

Case 2: Literature Histopathology Series Lymphocytic Infiltration and Muscle

Schwab, et al. presented histopathology from five cases of fatal myocarditis which they concluded were likely or possibly causally related to mRNA gene therapy products. (Schwab C, Domke LM, Hartmann L, Stenzinger A, Longerich T, Schirmacher P. Autopsy-based histopathological characterization of myocarditis after anti-SARS-CoV-2-vaccination. Clin Res Cardiol. 2023 Mar;112(3):431–440. doi: 10.1007/s00392–022-02129–5. Epub 2022 Nov 27. PMID: 36436002; PMCID: PMC9702955)

The authors provide the following schematic linking injected mRNA with CD4+ T cells and cardiac disease.

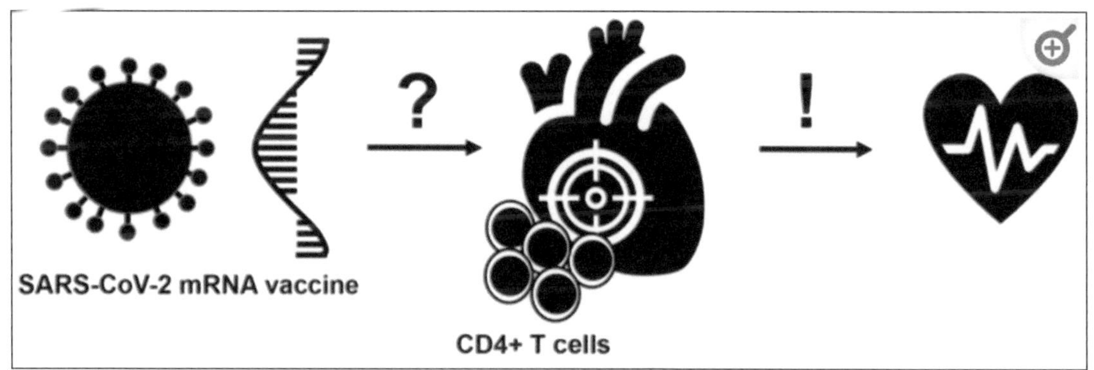

(https://www.ncbi.nlm.nih.gov/pmc/articles/PMC9702955/pdf/392_2022_Article_2129.pdf, p. 1)

Their findings were:

> All cases showed a consistent phenotype: (A) focal interstitial lymphocytic myocardial infiltration, in three cases accompanied by demonstrable microfocal myocyte destruction. (B) T-cell dominant infiltrate with CD4 positive T-cells outnumbering CD8 positive T-cells by far; (C) frequently associated with T-cell infiltration of epicardium and subepicardial fat tissue revealing a similar immune phenotype(CD4 > > CD8).(https://www.ncbi.nlm.nih.gov/pmc/articles/PMC9702955/pdf/392_2022_Article_2129.pdf, p. 7.)

The inflammation observed in heart (smooth) muscle bears similarities with that seen in other muscle tissues following use of the gene therapy products.

The slide below is a compilation of histopathology from published or presented histopathology studies that share findings of lymphocytic infiltration in heart, deltoid, intestine, uterus, and aorta.

Lymphocytic Infiltration and Tissue Necrosis In Striated and Smooth muscle Following C-19 Gene Therapy Products

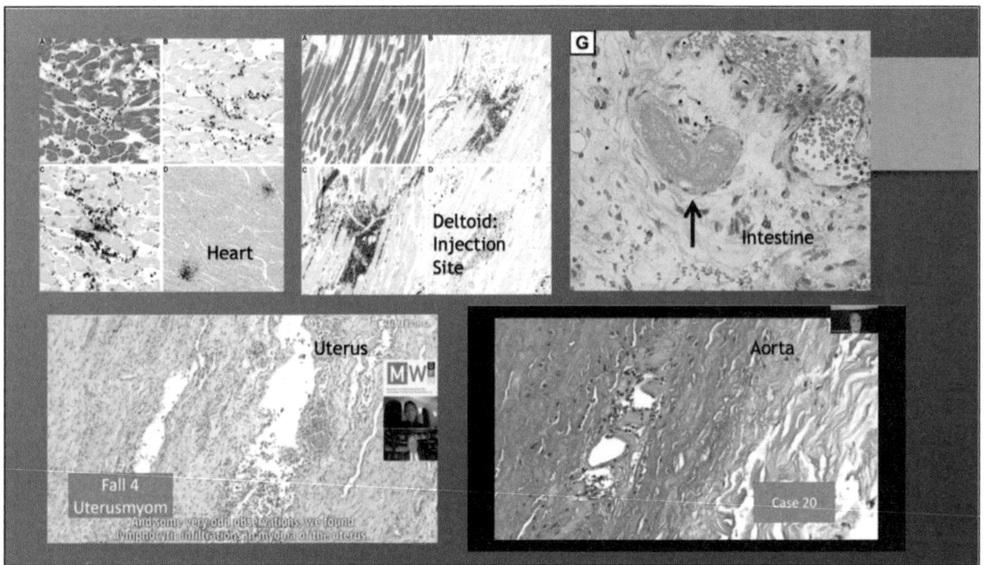

Top Left: Myocarditis Top Center: Myositis Deltoid (Schwab, et al.); Top Right: Vasculitis Intestine, (Kamura, et al. Case 1)

Bottom Left: Uterus (Burkhardt Collection, https://dailyclout.io/report-58-part-2-autopsies-reveal-medical-atrocities-of-genetic-therapies-being-used-against-a-respiratory-virus/)

Bottom Right: Muscular Layer of Aorta (Burkhardt Collection, https://dailyclout.io/report-56-autopsies-reveal-the-medical-atrocities-of-genetic-therapies-being-used-against-a-respiratory-virus/)]

Further analysis is necessary to understand more about lymphocyte activation and targeting in these cases.

The authors conclude by pointing to the uniqueness of the findings in autopsies on persons dying shortly after receiving Spike-Producing Gene Therapy Products compared with their past experience.

> During the last 20 years of autopsy service at Heidelberg University Hospital ***we did not observe comparable myocardial inflammatory infiltration.*** This was validated by histological re-evaluation of age- and sex-matched cohorts from three independent periods, which did not reveal a single case showing a comparable cardiac pathology. (Emphasis added.) (https://www.ncbi.nlm.nih.gov/pmc/articles/PMC9702955/pdf/392_2022_Article_2129.pdf, p. 5.)

Case 2 Analysis

CoVax Diseases resemble their non-Spike implicated counterparts except in the severity and distribution of the former compared with the latter. Sudden death is not a common finding in non-Spike autoimmune disease.

Time will tell if there is some common pathomechanism in muscle tissues, both smooth and striated. The histopathology is similar in muscle tissue in the human body from aorta, to heart, to uterus, to intestine, to skeletal muscles. Rich vascularity is characteristic of smooth and striated muscle. Is the commonality vascularity or muscle? Or both?

Is the uterus responsible for the consistent predominance of women in having Adverse Events that has been identified in multiple LNP/mRNA data sets? In nine months, women can birth a seven- to eight-pound baby. Does this organ act like an LNP/mRNA magnet and Spike generator? Possibly.

Literature Case Report 3: Systemic Lupus Erythematosus (SLE)

Lupus was introduced earlier in this article. Sogbe, et al. diagnosed a case of Systemic Lupus Erythematosus (SLE) in a 72-year-old woman with chronic renal failure from membranoproliferative glomerulonephritis (MPGN). [Sogbe, Miguel, et al. "Systemic Lupus Erythematosus Myocarditis after COVID-19 Vaccination." *Reumatología Clínica*, Elsevier, 1 Feb. 2023, https://www.reumatologiaclinica.org/es-systemic-lupus-erythematosus-myocarditis-after-articulo-S1699258X22001553.]

Autoimmune disease had been ruled out before she was injected with BNT162b2.

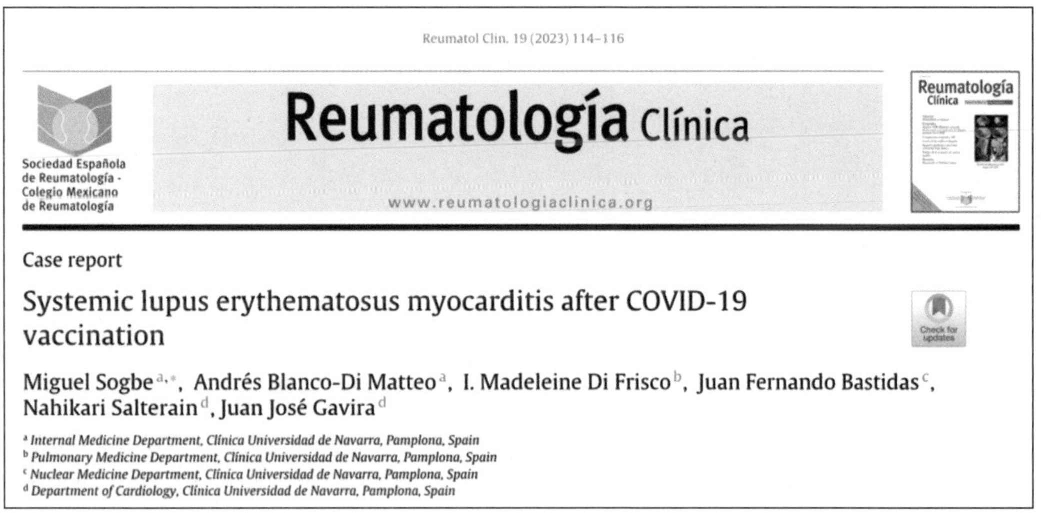

Reumatol Clin. 19 (2023) 114–116

Sociedad Española de Reumatología - Colegio Mexicano de Reumatología

Reumatología Clínica

www.reumatologiaclinica.org

Case report

Systemic lupus erythematosus myocarditis after COVID-19 vaccination

Miguel Sogbe[a,*], Andrés Blanco-Di Matteo[a], I. Madeleine Di Frisco[b], Juan Fernando Bastidas[c], Nahikari Salterain[d], Juan José Gavira[d]

[a] Internal Medicine Department, Clínica Universidad de Navarra, Pamplona, Spain
[b] Pulmonary Medicine Department, Clínica Universidad de Navarra, Pamplona, Spain
[c] Nuclear Medicine Department, Clínica Universidad de Navarra, Pamplona, Spain
[d] Department of Cardiology, Clínica Universidad de Navarra, Pamplona, Spain

The patient has been in chronic hemodialysis since 2017 after renal graft dysfunction due to chronic rejection. She presented to the emergency room with pleuritic chest pain one week after vaccination with the third dose of BNT162b2 mRNA. (https://www.reumatologiaclinica.org/es-systemic-lupus-erythematosus-myocarditis-after-articulo-S1699258X22001553)

Blood Studies (*Elevated):

- Haemoglobin 12.4 g/dl (12–16 g/dl),
- Total leucocyte count 4.86 × 10E9/L (4.8–10.8 × 10E9/L),
- Total lymphocyte count 0.58 × 10E9/L (1.2–4.5 × 10E9/L)*,
- Urea 91 mg/dl (16.6–48.5 mg/dl)*,
- Troponin I 231.5 ng/L (0–14 ng/L)* -peak value-,

- Erythrocyte sedimentation rate 17 mm/h (0–10 mm/h)*,
- C-reactive protein 0.58 mg/dl (≤0.5 mg/dl)*.
- SARS-CoV-2 PCR test was negative.
- Serological tests for cardiotropic pathogens were negative.

Autoimmune Serology:

- ANA (anti-nuclear antibody) IFA (indirect immunofluorescence assay) positive in 1:160 dilution*,
- Anti-dsDNA, and anti-histone antibodies were positive*,
- Low serum C3 level 75.50 mg/dl, (NV: 79–152 mg/dl)*,
- normal C4 level 27.50 mg/dl, (NV: 16–38 mg/dl) and
- normal CH50 level 537 U/ml, (NV: 392–1019 U/ml)

Her positive ANA IFA (indirect immunofluorescence assay), Anti-dsDNA, and anti-histone antibodies were essential to support the diagnosis of Lupus Myocarditis.

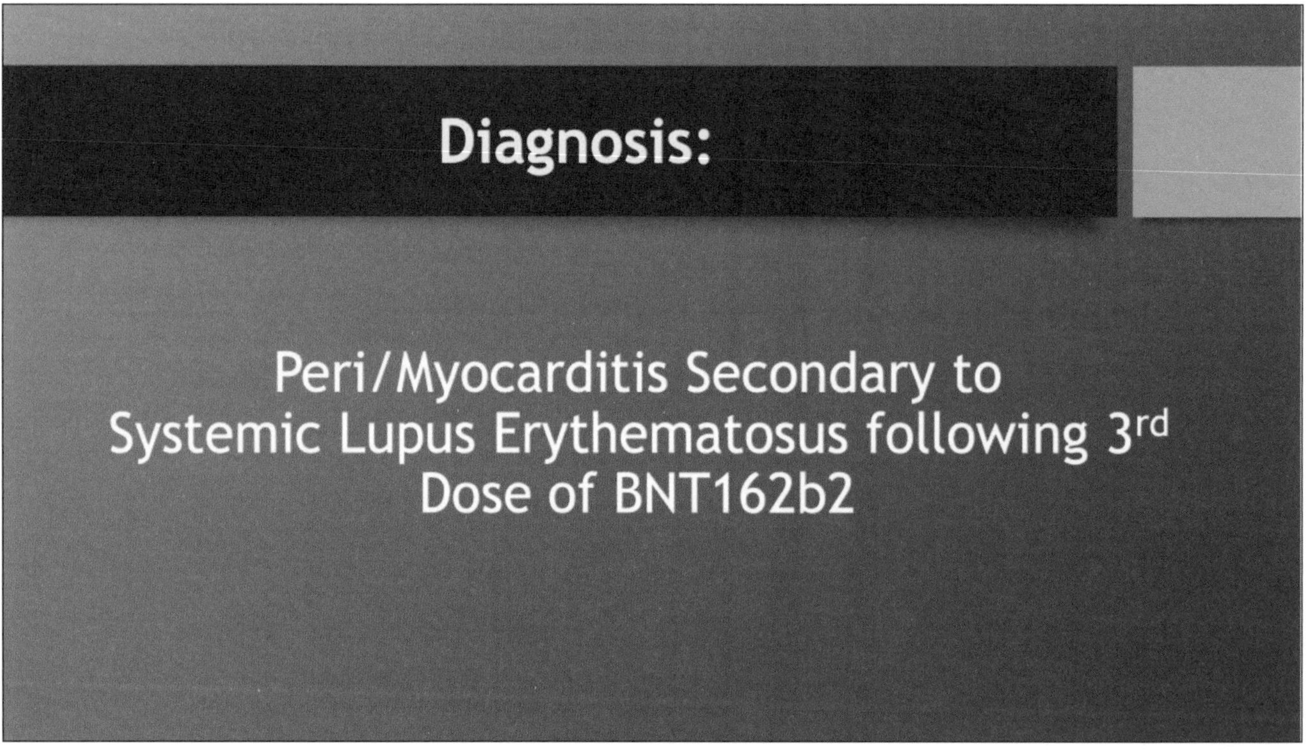

Diagnostic criteria for Lupus are given in the Appendix. Aringer, et al. provide the criteria for diagnosis used in the paper. [Aringer, Martin, et al. "2019 European League Against Rheumatism/American College of Rheumatology Classification Criteria for Systemic Lupus Erythematosus." *Wiley Online Library*, Arthritis and Rheumatology, 6 Aug. 2019, https://onlinelibrary.wiley.com/doi/10.1002/art.40930.]

She was successfully treated with prednisone and beta-blockers. At follow-up three months after discharge, she appeared to be in remission but with positive serology indicating a potential for recurrent illness.

The authors implicate injected mRNA as the cause of her cardiac disease (following image).

Causation

1. "The immune system might detect the mRNA in the vaccine as an antigen, resulting in the activation of proinflammatory cascades and immunological pathways in the heart.

2. Molecular mimicry between the spike protein of SARS-CoV-2 and cardiac self-antigens is another possible mechanism."

Case 3 Analysis

What sets this case apart is the clear serologic evidence of systemic lupus erythematosus (SLE). In spite of many morbidities, she responded to steroids and a medication to control her heart rate.

This individual might have had a subclinical case of systemic lupus erythematosus that was brought out by BNT162b2. Alternatively, the drug might have been the sole cause of her SLE.

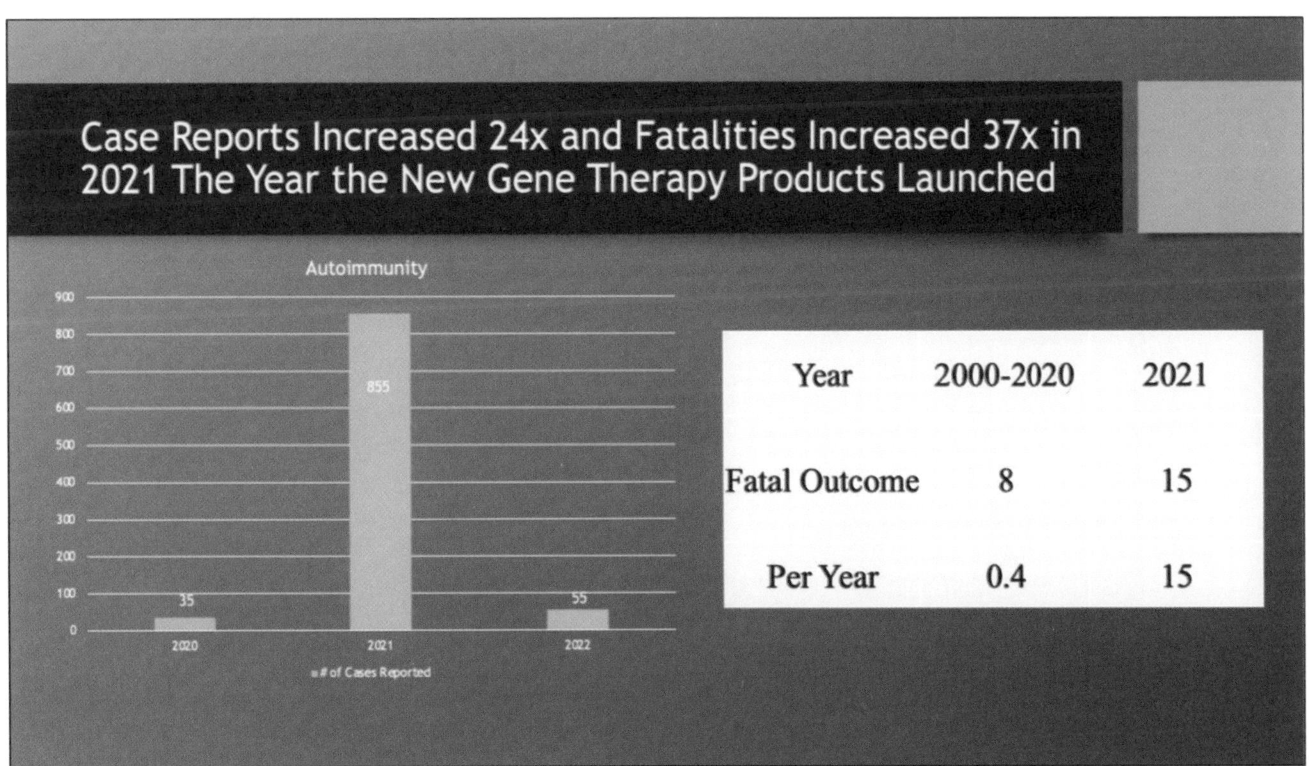

IV. VAERS Data 2000–2022: Reports of Autoimmunity

There was a 24-fold increase in reported cases of autoimmunity in VAERS going from only COVID-19 disease in 2020 to COVID-19 gene therapy injections and COVID-19 in 2021. The reporting then dropped off precipitously in 2022.

Fatalities associated with these reported cases of autoimmunity averaged 0.4 per year for the two decades prior to the rollout of the COVID-19 gene therapy products before they jumped 37-fold, to 15, in 2021.

VAERS Case 1: Autoimmune MIS-C
Previously Healthy 12-Year-Old Male

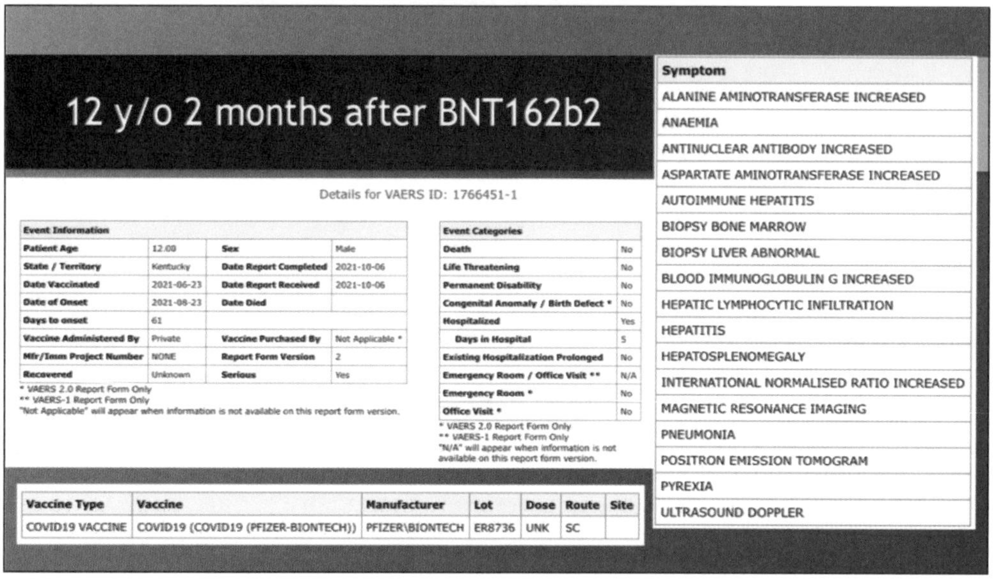

12 y/o 2 months after BNT162b2
Details for VAERS ID: 1766451-1
Event Information
Patient Age 12.00
Sex Male
State / Territory Kentucky
Date Report Completed 2021-10-06
Date Vaccinated 2021-06-23
Date Report Received 2021-10-06
Date of Onset 2021-08-23
Date Died
Days to onset 61
Vaccine Administered By Private
Vaccine Purchased By Not Applicable *
Mfr/Imm Project Number NONE
Report Form Version 2
Recovered Unknown
Serious Yes
* VAERS 2.0 Report Form Only
** VAERS-1 Report Form Only
"Not Applicable" will appear when information is not available on this report form version.
Event Categories
Death No
Life Threatening No
Permanent Disability No
Congenital Anomaly / Birth Defect * No
Hospitalized Yes
Days in Hospital 5
Existing Hospitalization Prolonged No
Emergency Room / Office Visit ** N/A
Emergency Room * No
Office Visit * No
* VAERS 2.0 Report Form Only
** VAERS-1 Report Form Only
"N/A" will appear when information is not available on this report form version.
Vaccine Type | Vaccine | Manufacturer | Lot | Dose | Route | Site
COVID19 VACCINE | COVID19 (COVID19 (PFIZER-BIONTECH)) | PFIZER\BIONTECH | ER8736 | UNK | SC
Symptom
ALANINE AMINOTRANSFERASE INCREASED
ANAEMIA
ANTINUCLEAR ANTIBODY INCREASED
ASPARTATE AMINOTRANSFERASE INCREASED
AUTOIMMUNE HEPATITIS
BIOPSY BONE MARROW
BIOPSY LIVER ABNORMAL
BLOOD IMMUNOGLOBULIN G INCREASED
HEPATIC LYMPHOCYTIC INFILTRATION
HEPATITIS
HEPATOSPLENOMEGALY
INTERNATIONAL NORMALISED RATIO INCREASED
MAGNETIC RESONANCE IMAGING
PNEUMONIA
POSITRON EMISSION TOMOGRAM
PYREXIA
ULTRASOUND DOPPLER

This 12-year-old boy spent five days in the hospital with multiple organ involvement including anemia, hepatitis, pneumonia, lymphoblastic hepatic infiltrate, and an enlarged spleen following BNT162b2 injection. The brief entry in VAERS included the comment that this presentation of autoimmune hepatitis was unusual.

#1766451 Fevers, Multifocal Pneumonia, Severe Anemia, Autoimmune Hepatitis, Massive Lymphoblastic Hepatic Infiltrate, Splenomegaly
12 y/o with Autoimmune MIS-C
Adverse Event Description
Patient received Covid 19 vaccinations on 6/23/2021 and on 7/14/2021 without immediate complications. He developed fevers on 8/23 and was diagnosed with multifocal pneumonia, severe anemia, and hepatitis with elevated AST/ALT and INR on 8/30. Subsequent work-op including imaging with PET-CT , bone marrow biopsy, and liver biopsy revealed severe autoimmune hepatitis based on massive lymphoplasmocytic hepatic infiltration by histology, elevated serum IgG and ANA. He was treated with prednisone and rituximab. His condition is currently improving, The degree of liver infiltration with lymphocytes and severity of hepatosplenomegaly is atypical what I usually see in new diagnosis of autoimmune hepatitis.
Lab Data
Hopsital admissions, imaging with ultrasound, MRI, PET CT, liver and bone marrow biopsies.
Current Illness
none
Adverse Events After Prior Vaccinations
Medications At Time Of Vaccination
none
History/Allergies
none,none
Dx: Multiorgan Autoimmune vs. MIS-C or Both

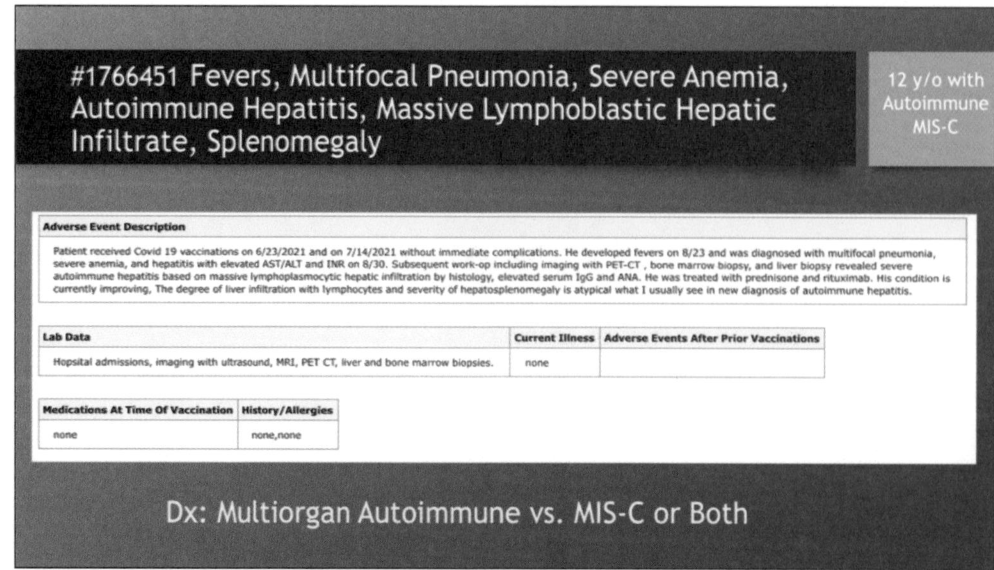

"The degree of liver infiltration with lymphocytes and severity of hepatosplenomegaly is atypical for what I usually see in new diagnosis of autoimmune hepatitis."

Unfortunately, the information contained in this entry was limited, and the histopathology findings were not reported (previous image).

He was hospitalized for five days and had an extensive medical workup including liver and bone biopsies, MRI, and PET scans. His ANA was positive like the last case. More information would be helpful, but VAERS is a Registry and not a detailed medical record.

VAERS Case 2: Positive ANA, Vestibular Dysfunction, Myocarditis, Chronic Fatigue, Menstrual Dysfunction . . . Permanent Disability

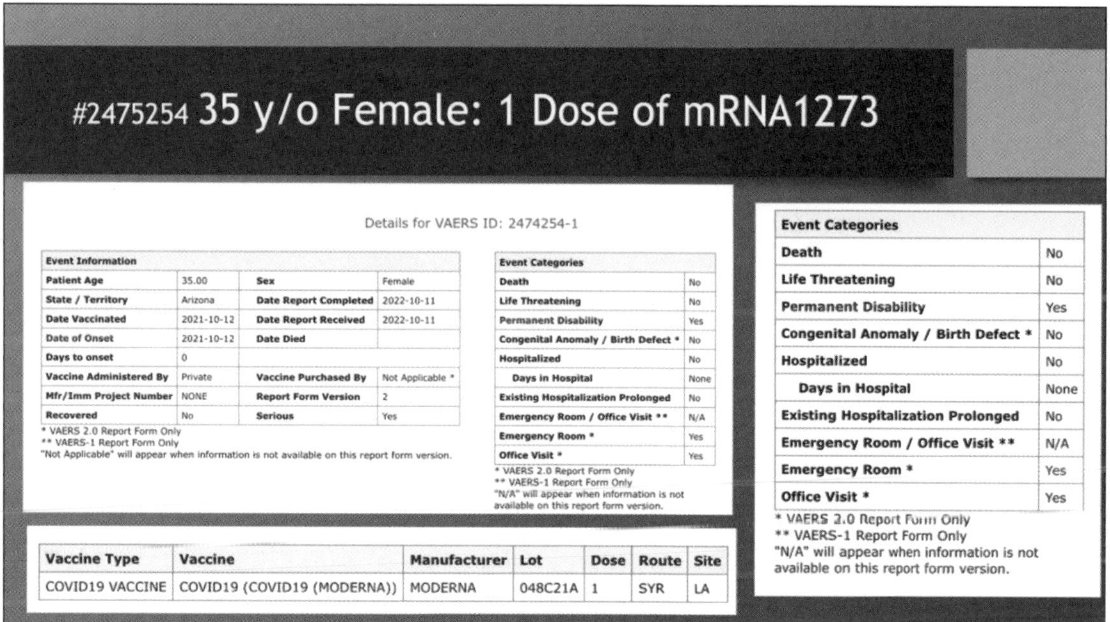

#2475254 35 y/o Female: 1 Dose of mRNA1273

Details for VAERS ID: 2474254-1

Event Information			
Patient Age	35.00	Sex	Female
State / Territory	Arizona	Date Report Completed	2022-10-11
Date Vaccinated	2021-10-12	Date Report Received	2022-10-11
Date of Onset	2021-10-12	Date Died	
Days to onset	0		
Vaccine Administered By	Private	Vaccine Purchased By	Not Applicable *
Mfr/Imm Project Number	NONE	Report Form Version	2
Recovered	No	Serious	Yes

* VAERS 2.0 Report Form Only
** VAERS-1 Report Form Only
"Not Applicable" will appear when information is not available on this report form version.

Event Categories	
Death	No
Life Threatening	No
Permanent Disability	Yes
Congenital Anomaly / Birth Defect *	No
Hospitalized	No
Days in Hospital	None
Existing Hospitalization Prolonged	No
Emergency Room / Office Visit **	N/A
Emergency Room *	Yes
Office Visit *	Yes

* VAERS 2.0 Report Form Only
** VAERS-1 Report Form Only
"N/A" will appear when information is not available on this report form version.

Vaccine Type	Vaccine	Manufacturer	Lot	Dose	Route	Site
COVID19 VACCINE	COVID19 (COVID19 (MODERNA))	MODERNA	048C21A	1	SYR	LA

Event Categories	
Death	No
Life Threatening	No
Permanent Disability	Yes
Congenital Anomaly / Birth Defect *	No
Hospitalized	No
Days in Hospital	None
Existing Hospitalization Prolonged	No
Emergency Room / Office Visit **	N/A
Emergency Room *	Yes
Office Visit *	Yes

* VAERS 2.0 Report Form Only
** VAERS-1 Report Form Only
"N/A" will appear when information is not available on this report form version.

The 35-year-old woman had loss of her period for nine months, and then she reported having had a period lasting three months. Her symptoms began the day she was injected.

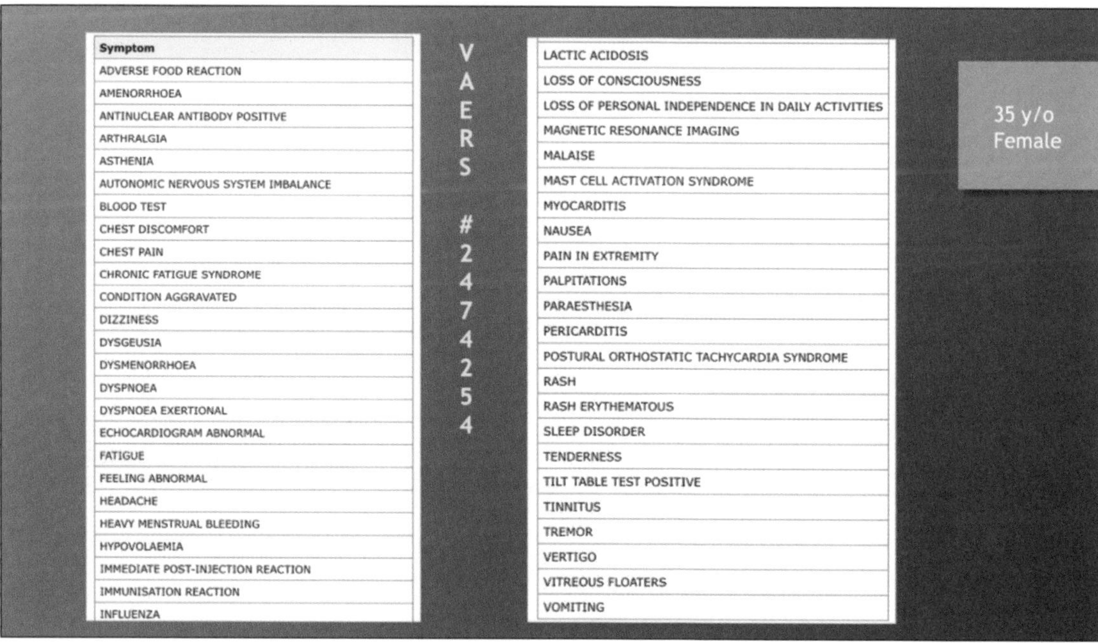

VAERS #2474254

35 y/o Female

Symptom
ADVERSE FOOD REACTION
AMENORRHOEA
ANTINUCLEAR ANTIBODY POSITIVE
ARTHRALGIA
ASTHENIA
AUTONOMIC NERVOUS SYSTEM IMBALANCE
BLOOD TEST
CHEST DISCOMFORT
CHEST PAIN
CHRONIC FATIGUE SYNDROME
CONDITION AGGRAVATED
DIZZINESS
DYSGEUSIA
DYSMENORRHOEA
DYSPNOEA
DYSPNOEA EXERTIONAL
ECHOCARDIOGRAM ABNORMAL
FATIGUE
FEELING ABNORMAL
HEADACHE
HEAVY MENSTRUAL BLEEDING
HYPOVOLAEMIA
IMMEDIATE POST-INJECTION REACTION
IMMUNISATION REACTION
INFLUENZA
LACTIC ACIDOSIS
LOSS OF CONSCIOUSNESS
LOSS OF PERSONAL INDEPENDENCE IN DAILY ACTIVITIES
MAGNETIC RESONANCE IMAGING
MALAISE
MAST CELL ACTIVATION SYNDROME
MYOCARDITIS
NAUSEA
PAIN IN EXTREMITY
PALPITATIONS
PARAESTHESIA
PERICARDITIS
POSTURAL ORTHOSTATIC TACHYCARDIA SYNDROME
RASH
RASH ERYTHEMATOUS
SLEEP DISORDER
TENDERNESS
TILT TABLE TEST POSITIVE
TINNITUS
TREMOR
VERTIGO
VITREOUS FLOATERS
VOMITING

She had an extensive problem list with multiple organ system representation (previous image). Her ANA titers were elevated.

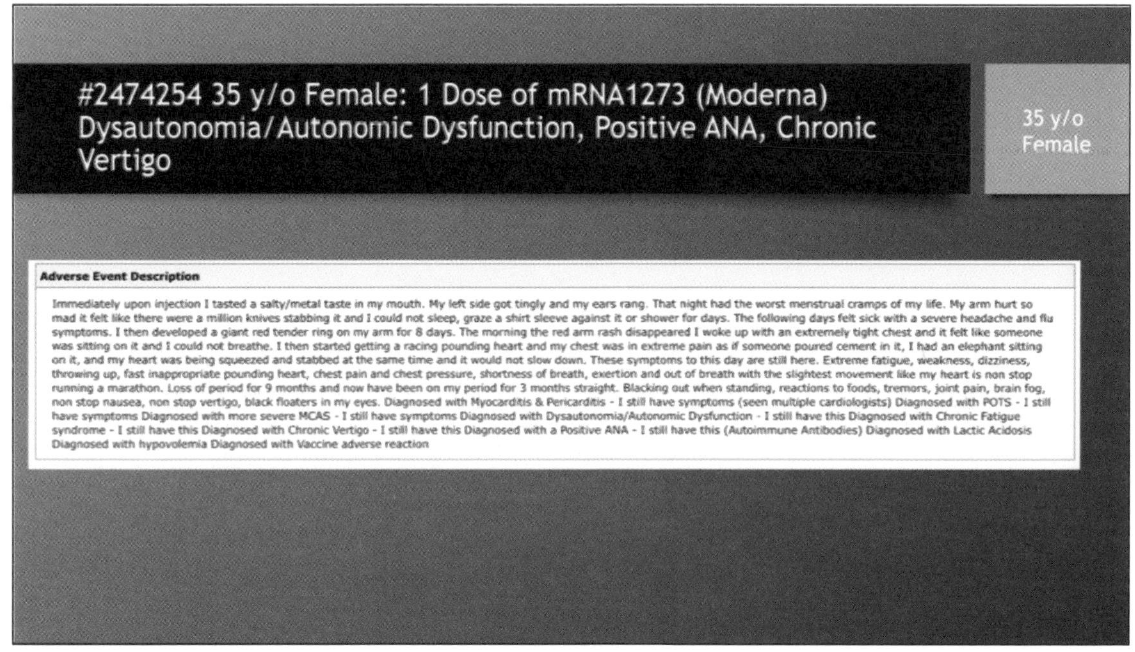

Adverse Event Description

Immediately upon injection I tasted a salty/metal taste in my mouth. My left side got tingly and my ears rang. That night had the worst menstrual cramps of my life. My arm hurt so mad it felt like there were a million knives stabbing it and I could not sleep, graze a shirt sleeve against it or shower for days. The following days felt sick with a severe headache and flu symptoms. I then developed a giant red tender ring on my arm for 8 days. The morning the red arm rash disappeared I woke up with an extremely tight chest and it felt like someone was sitting on it and I could not breathe. I then started getting a racing pounding heart and my chest was in extreme pain as if someone poured cement in it, I had an elephant sitting on it, and my heart was being squeezed and stabbed at the same time and it would not slow down. These symptoms to this day are still here. Extreme fatigue, weakness, dizziness, throwing up, fast inappropriate pounding heart, chest pain and chest pressure, shortness of breath, exertion and out of breath with the slightest movement like my heart is non stop running a marathon. Loss of period for 9 months and now have been on my period for 3 months straight. Blacking out when standing, reactions to foods, tremors, joint pain, brain fog, non stop nausea, non stop vertigo, black floaters in my eyes. Diagnosed with Myocarditis & Pericarditis - I still have symptoms (seen multiple cardiologists) Diagnosed with POTS - I still have symptoms Diagnosed with more severe MCAS - I still have symptoms Diagnosed with Dysautonomia/Autonomic Dysfunction - I still have this Diagnosed with Chronic Fatigue syndrome - I still have this Diagnosed with Chronic Vertigo - I still have this Diagnosed with a Positive ANA - I still have this (Autoimmune Antibodies) Diagnosed with Lactic Acidosis Diagnosed with hypovolemia Diagnosed with Vaccine adverse reaction

She reported she was left with permanent disability one year after the onset of her illness.

VAERS Case 3: 29-Year-Old Female with Central and Peripheral Nervous System Dysfunction and Peri-Myocarditis . . . Permanent Disability

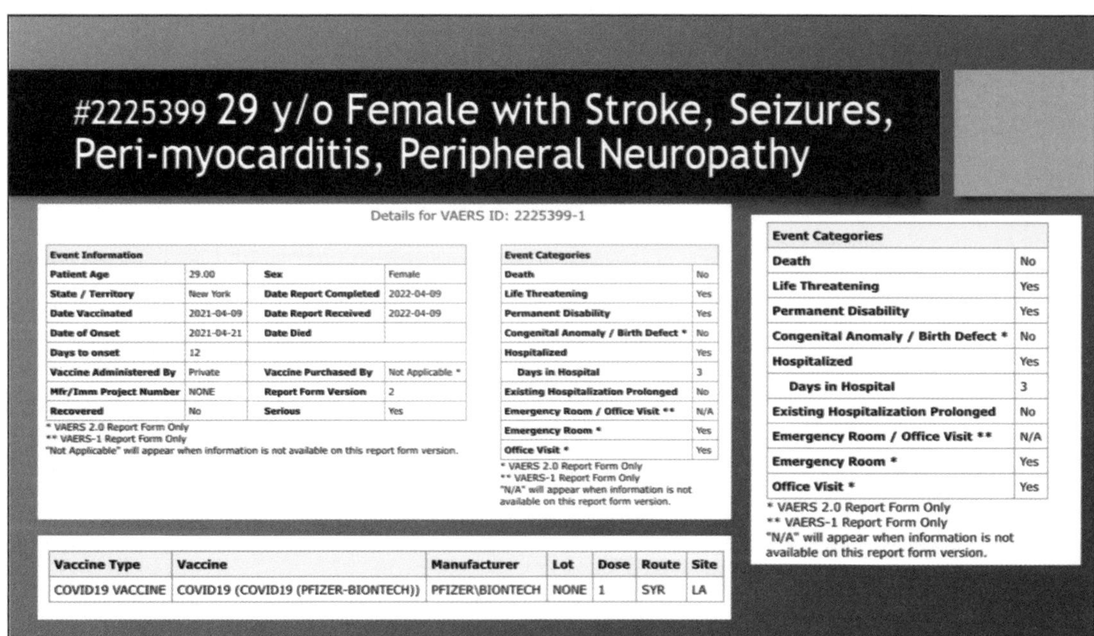

Details for VAERS ID: 2225399-1

Event Information			
Patient Age	29.00	Sex	Female
State / Territory	New York	Date Report Completed	2022-04-09
Date Vaccinated	2021-04-09	Date Report Received	2022-04-09
Date of Onset	2021-04-21	Date Died	
Days to onset	12		
Vaccine Administered By	Private	Vaccine Purchased By	Not Applicable *
Mfr/Imm Project Number	NONE	Report Form Version	2
Recovered	No	Serious	Yes

* VAERS 2.0 Report Form Only
** VAERS-1 Report Form Only
"Not Applicable" will appear when information is not available on this report form version.

Event Categories	
Death	No
Life Threatening	Yes
Permanent Disability	Yes
Congenital Anomaly / Birth Defect *	No
Hospitalized	Yes
Days in Hospital	3
Existing Hospitalization Prolonged	No
Emergency Room / Office Visit **	N/A
Emergency Room *	Yes
Office Visit *	Yes

* VAERS 2.0 Report Form Only
** VAERS-1 Report Form Only
"N/A" will appear when information is not available on this report form version.

Vaccine Type	Vaccine	Manufacturer	Lot	Dose	Route	Site
COVID19 VACCINE	COVID19 (COVID19 (PFIZER-BIONTECH))	PFIZER\BIONTECH	NONE	1	SYR	LA

Event Categories	
Death	No
Life Threatening	Yes
Permanent Disability	Yes
Congenital Anomaly / Birth Defect *	No
Hospitalized	Yes
Days in Hospital	3
Existing Hospitalization Prolonged	No
Emergency Room / Office Visit **	N/A
Emergency Room *	Yes
Office Visit *	Yes

* VAERS 2.0 Report Form Only
** VAERS-1 Report Form Only
"N/A" will appear when information is not available on this report form version.

This 29-year-old woman received one dose of BNT162b2 followed 12 days later by numerous adverse events, including cardiac and neurological problems that required three days of hospitalization.

She reported central and peripheral nervous system involvement as well as peri-myocarditis and arrhythmia.

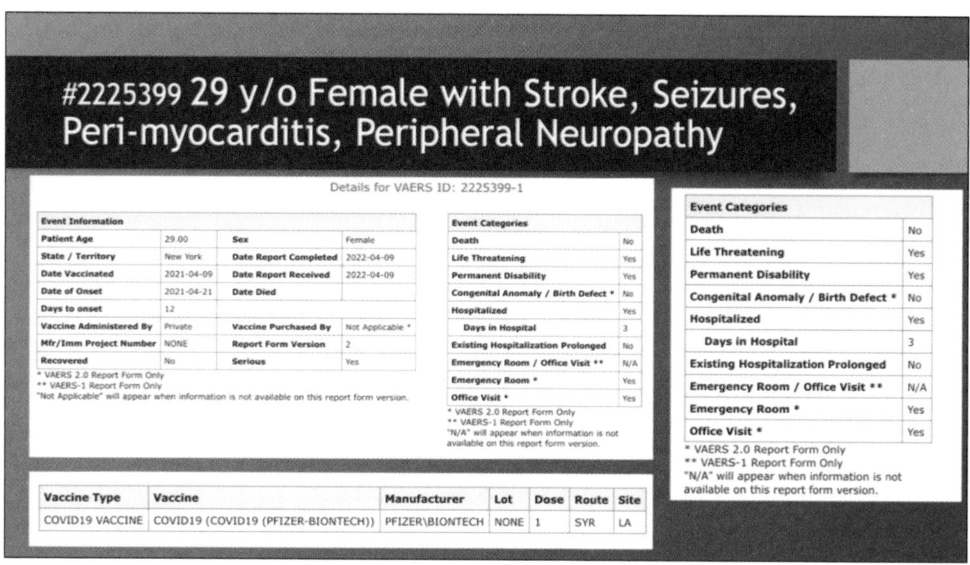

Event Information			
Patient Age	29.00	Sex	Female
State / Territory	New York	Date Report Completed	2022-04-09
Date Vaccinated	2021-04-09	Date Report Received	2022-04-09
Date of Onset	2021-04-21	Date Died	
Days to onset	12		
Vaccine Administered By	Private	Vaccine Purchased By	Not Applicable *
Mfr/Imm Project Number	NONE	Report Form Version	2
Recovered	No	Serious	Yes

* VAERS 2.0 Report Form Only
** VAERS-1 Report Form Only
"Not Applicable" will appear when information is not available on this report form version.

Event Categories	
Death	No
Life Threatening	Yes
Permanent Disability	Yes
Congenital Anomaly / Birth Defect *	No
Hospitalized	Yes
Days in Hospital	3
Existing Hospitalization Prolonged	No
Emergency Room / Office Visit **	N/A
Emergency Room *	Yes
Office Visit *	Yes

* VAERS 2.0 Report Form Only
** VAERS-1 Report Form Only
"N/A" will appear when information is not available on this report form version.

Vaccine Type	Vaccine	Manufacturer	Lot	Dose	Route	Site
COVID19 VACCINE	COVID19 (COVID19 (PFIZER-BIONTECH))	PFIZER\BIONTECH	NONE	1	SYR	LA

A year later, she was receiving ongoing treatment. She reported that she was permanently disabled. **A 29-year-old with rapid onset of hormonal, cardiac, neurologic, and vascular disease is stunning.**

V. Mechanism of Injury

BNT162b2 and mRNA1273, Moderna's mRNA COVID "vaccine," are complicated products designed to be delivered throughout the human body, including crossing the blood-brain barrier and placental barrier.

Adverse events can potentially be related to five components of these drugs: the whole preparation upon delivery into the deltoid, the lipid coating, the mRNA, Spike and related proteins produced by the mRNA, and E. coli plasmid contamination. Autoimmunity most likely comes from foreign proteins translated from the injected synthetic mRNA (#3 and #4).

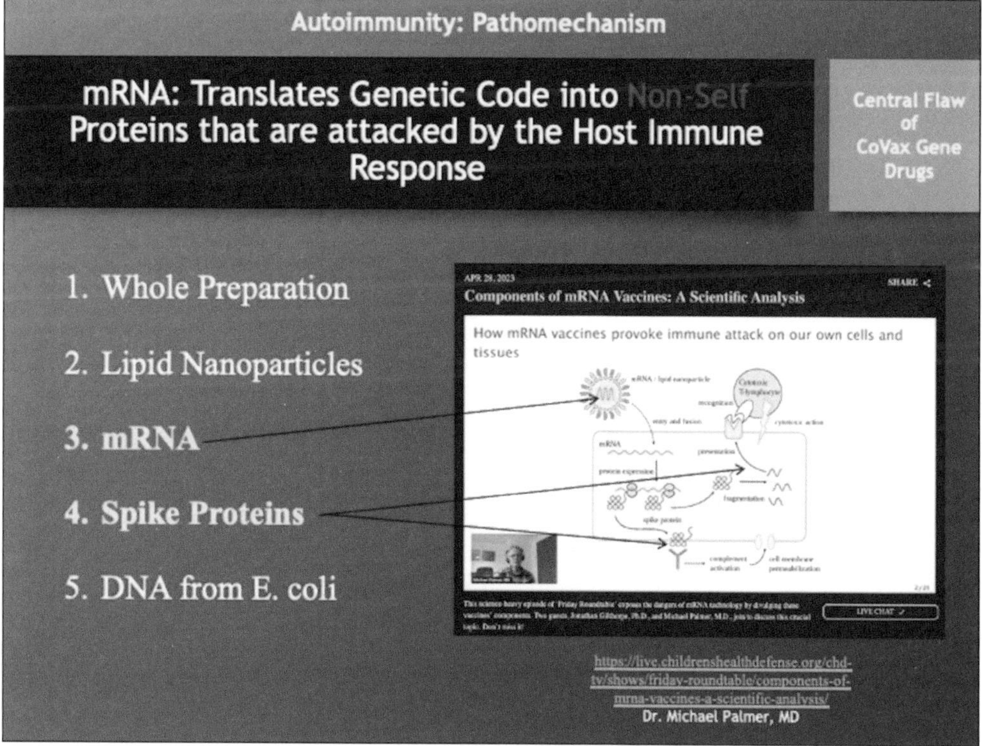

The inset diagram **above** illustrates how the mRNA component may cause autoimmune disease. [This schematic was presented by Dr. Michael Palmer, MD on Children's Health Defense TV on April 28, 2023, and is used with permission. (https://live.childrenshealthdefense.org/chd-tv/shows/friday-roundtable/components-of-mrna-vaccines-a-scientific-analysis/)]

The lipid nanoparticle (LNP) vehicle delivers mRNA to cells. mRNA is then released and commandeers the host's cellular machinery to produce Spike and related proteins. These proteins are then recognized by the host immune system as foreign or not self. The host then launches an immune attack to destroy cells and extracellular material containing non-self-antigens causing an inflammatory reaction as well as associated tissue damage and clinical signs and symptoms.

VI. Discussion:

Conspicuously absent from the medical literature are large series reports of inflammatory conditions following injection of novel gene therapies for COVID-19. This includes the final report of the two-year findings of the original Phase 2/3 Clinical Trial of over 40,000 subjects which, unfortunately, was unblinded, thus ending the control group.

No prospective studies are in evidence, and there are few retrospective studies in spite of the fact that Pfizer confidential document 5.3.6 identified autoimmunity associated with the use of its LNP/mRNA product following widespread use of this product under the Emergency Use Authorization (EUA) granted by the Food and Drug Administration (FDA) in December 2020 until data collection for 5.3.6 ended on February 28, 2021.

Case reports, small autopsy series, and a small clinical series have documented a wide array of autoimmune illnesses ranging from limited organ involvement, such as Giant Cell Arteritis, to severe, sometimes catastrophic, cascading progressive multiple organ failure involving skin, brain, peripheral nervous system, liver, spleen, kidney, smooth and striated muscle, heart, intestines, uterus, testes, blood vessels, and skeletal muscle.

Polykretis, et al. published a paper March 8, 2023, in which they reviewed current evidence linking autoimmunity and COVID-19 genetic vaccines. (doi:10.20944/preprints202303.0140.v1)

They concluded:

1. Autoimmunity is a central flaw in the mRNA platform. Genetic vaccines instruct human cells to manufacture viral protein in order to obtain an immune response, thereby producing an autoimmune response in organs.
2. Prolonged persistence of mRNA in circulation and tissues maintains exposure to mRNA longer than the advertised few days.
3. Adequate biodistribution studies were not performed before widespread inoculation of billions of people. Polykretis, et al. cite studies showing persistence of synthetic mRNA in blood for two weeks and persistence in lymph nodes up to eight weeks.
4. Exosomes (https://www.annualreviews.org/doi/pdf/10.1146/annurev-biochem-013118–111902) are small transport vehicles that can distribute intact mRNA throughout the body by the vascular and lymphatic systems and illustrate possible distribution from origin in the spleen to heart, liver, and brain.

5. Strong histological evidence supports COVID-19 vaccine-induced inflammation in tissues from T-lymphocytic infiltration.

VII. Conclusions

This review presents clinical evidence in support of the hypothesis that autoimmunity is one of the principal pathological mechanisms of harm from COVID-19 gene therapy drugs and, as such, represents a central flaw in the LNP/mRNA platform.

VAERS reporting of autoimmunity cases jumped from 35 in COVID-19 year one, 2020, to 840 in year one of the COVID-19 gene therapy products—a 24x increase in reports and a 37x increase in annual fatalities.

Incompletely tested, novel gene therapy is now associated with protean disease manifestations that derive from widespread biodistribution of products that enter host cells and translate mRNA code into Spike proteins that the host immune system has never seen.

Foreign proteins call forth an immune response from the host that target and destroy Spike proteins in host tissues and organs, sometimes dramatically and fatally.

The widespread organ involvement and severity of illness warrant consideration of a new class of illness, **CoVax Disease**, encompassing multiple pathological mechanisms of which autoimmunity occupies a central position.

Clotting and vascular disorders, neurological disease, direct toxicity, and neoplasia should also be considered part of this collection of harms following LNP/mRNA therapy at this early stage in the study of these disorders.

Appendix: SLE

The difficulty in recognizing many of the autoimmune diseases is illustrated well by the disease called systemic lupus erythematosus. The complexity criteria developed by the combined European League Against Rheumatism and the American College of Rheumatology points to the diagnostic challenge these illnesses present, not to mention the task of developing effective treatments.

Special Article | Free Access

2019 European League Against Rheumatism/American College of Rheumatology Classification Criteria for Systemic Lupus Erythematosus

Martin Aringer MD, Karen Costenbader MD, MPH, David Daikh MD, PhD, Ralph Brinks PhD, Marta Mosca MD, PhD, Rosalind Ramsey-Goldman MD, DrPH, Josef S. Smolen MD ... See all authors

First published: 06 August 2019 | https://doi.org/10.1002/art.40930 | Citations: 734

Entry criterion
Antinuclear antibodies (ANA) at a titer of ≥1:80 on HEp-2 cells or an equivalent positive test (ever)

If absent, do not classify as SLE
If present, apply additive criteria

↓

Additive criteria
Do not count a criterion if there is a more likely explanation than SLE.
Occurrence of a criterion on at least one occasion is sufficient.
SLE classification requires at least one clinical criterion and ≥10 points.
Criteria need not occur simultaneously.
Within each domain, only the highest weighted criterion is counted toward the total score§.

Clinical domains and criteria	Weight	Immunology domains and criteria	Weight
Constitutional		***Antiphospholipid antibodies***	
Fever	2	Anti-cardiolipin antibodies OR	
Hematologic		Anti-β2GP1 antibodies OR	
Leukopenia	3	Lupus anticoagulant	2
Thrombocytopenia	4	***Complement proteins***	
Autoimmune hemolysis	4	Low C3 OR low C4	3
Neuropsychiatric		Low C3 AND low C4	4
Delirium	2	***SLE-specific antibodies***	
Psychosis	3	Anti-dsDNA antibody* OR	
Seizure	5	Anti-Smith antibody	6
Mucocutaneous			
Non-scarring alopecia	2		
Oral ulcers	2		
Subacute cutaneous OR discoid lupus	4		
Acute cutaneous lupus	6		
Serosal			
Pleural or pericardial effusion	5		
Acute pericarditis	6		
Musculoskeletal			
Joint involvement	6		
Renal			
Proteinuria >0.5g/24h	4		
Renal biopsy Class II or V lupus nephritis	8		
Renal biopsy Class III or IV lupus nephritis	10		

Total score:

Classify as Systemic Lupus Erythematosus with a score of 10 or more if entry criterion fulfilled.

(https://onlinelibrary.wiley.com/doi/10.1002/art.40930)

Seventeen: "Musculoskeletal Adverse Events of Special Interest Afflicted 8.5% of Patients in Pfizer's Post-Marketing Data Set, Including Four Children and One Infant. Women Affected at a Ratio of Almost 4:1 Over Men."

—Barbara Gehrett, MD; Joseph Gehrett, MD; Chris Flowers, MD; and Loree Britt

The WarRoom/DailyClout Pfizer Documents Analysis Project Post-Marketing Group (Team 1)—Barbara Gehrett, MD; Joseph Gehrett, MD; Chris Flowers, MD; and Loree Britt—wrote a review of musculoskeletal adverse events of special interest (AESIs) found in Pfizer document *5.3.6 Cumulative Analysis of Post-Authorization Adverse Event Reports of PF-07302048 (BNT162B2) Received Through 28-FEB-2021* (a.k.a., "*5.3.6*"). This group of AESIs includes diagnoses of arthralgia (joint pain), arthritis (joint inflammation), arthritis/bacterial, chronic fatigue syndrome, polyarthritis (inflammation of multiple joints), post-viral fatigue syndrome, and rheumatoid arthritis (an autoimmune and inflammatory disease).

It is important to note that the AESIs in the *5.3.6* document were reported to Pfizer for **only a 90-day period** starting on December 1, 2020, the date of the United Kingdom's public rollout of Pfizer's COVID-19 experimental mRNA "vaccine" product.

Highlights of this report include:

- 3,600 cases of musculoskeletal AESIs were reported, which equates to **8.5% of the post-marketing data set** of 42,086 cases/patients.
- The 3,600 patients reported 3,640 adverse events. **1,614 (44%) were classified as serious.**
- The time from administration to adverse event ranged from less than 24 hours to 32 days. **50% of the events started within the first 24 hours after injection.**
- Of the cases where gender was reported, 2,760 individuals were female, and 711 were male—an almost **4:1 ratio of female to male**.
- Though mostly adults were affected with these AESIs, **two adolescents, four children,** and **one infant** also reported musculoskeletal AESIs during a time frame when **Pfizer's BNT162b2 mRNA COVID "vaccine" was not approved for use in individuals under 16 years of age.**
- The most common adverse event was **arthralgia/joint pain (3,525 or 97%)**, followed by **70 arthritis AESIs (2%), 26 rheumatoid arthritis AESIs (<1%) and 5 AESIs (<1%) polyarthritis**.
- Outcome for 3,662 of the adverse events were: 1,801 (49%) resolved or resolving, **959 (26%) not resolved**, **49 (1%) resolved with sequelae**, and **853 (23%) were unknown**.

Pfizer concluded, "This cumulative case review does not raise new safety issues. Surveillance will continue."

Please read the full report below.

War Room/DailyClout Pfizer Document Analysis

SOURCE:
https://www.phmpt.org/wp-content/uploads/2022/04/reissue 5.3.6-postmarketing-experience.pdf

5.3.6 AE REPORTING PERIOD:

"Since the first temporary authorization for emergency supply under Regulation 174 in the UK (01 December 2020) and through 28 February 2021."

5.3.6 AE CASES/EVENTS:

TOTAL AE CASES: 42,086

TOTAL AE EVENTS: 158,893

ABBREVIATIONS:

5.3.6 : Pfizer source document

SOC : System Organ Class

AE : Adverse Event

AESI : Adverse Event of Special Interest

EUA : Emergency Use Authorization by FDA

PM : Post-Marketing

BNT162b2 : Pfizer's mRNA COVID-19 vaccine

SEQUELAE: an abnormal condition resulting from a previous disease, injury, or other trauma

AGE GROUPS defined in 5.3.6 (p. 25 footnote) :

Adult	18 - 64
Elderly	≥ 65
Child	2 - 11
Adolescent	12 - < 18
Infant	1 – 23 months

AUTHORS:

Dr Barbara Gehrett MD
Dr Joseph Gehrett MD
Dr Chris Flowers MD
Loree Britt

10May23

Post-Marketing Team Micro-Report 14:
Musculoskeletal Adverse Events of Special Interest Review of 5.3.6

Musculoskeletal AESIs *Search criteria: PTs Arthralgia; Arthritis; Arthritis bacterial[a]; Chronic fatigue syndrome; Polyarthritis; Polyneuropathy; Post viral fatigue syndrome; Rheumatoid arthritis*	• Number of cases: 3600 (8.5% of the total PM dataset), of which 2045 medically confirmed and 1555 non-medically confirmed; • Country of incidence: UK (1406), US (1004), Italy (285), Mexico (236), Germany (72), Portugal (70), France (48), Greece and Poland (46), Latvia (33), Czech Republic (32), Israel and Spain (26), Sweden (25), Romania (24), Denmark (23), Finland and Ireland (19 each), Austria and Belgium (18 each), Canada (16), Netherlands (14), Bulgaria (12), Croatia and Serbia (9 each), Cyprus and Hungary (8 each), Norway (7), Estonia and Puerto Rico (6 each), Iceland and Lithuania (4 each); the remaining 21 cases originated from 11 different countries; • Subjects' gender (n=3471): female (2760), male (711); • Subjects' age group (n=3372): Adult (2850), Elderly (515), Child (4), Adolescent (2), Infant (1); • Number of relevant events: 3640, of which 1614 serious, 2026 non-serious; • Reported relevant PTs: Arthralgia (3525), Arthritis (70), Rheumatoid arthritis (26), Polyarthritis (5), Polyneuropathy, Post viral fatigue syndrome, Chronic fatigue syndrome (4 each), Arthritis bacterial (1); • Relevant event onset latency (n = 2968): Range from <24 hours to 32 days, median 1 day;

https://www.phmpt.org/wpcontent/uploads/2022/04/reissue_5.3.6-postmarketiexperience.pdf

- **Adverse Events were reported to Pfizer during a 90-day period, following the December 1, 2020, public rollout of its COVID-19 experimental "vaccine" product.**
- **In the Pfizer 5.3.6 document, these AEs were categorized by System Organ Classes (SOC) – in other words, by systems in the body.**

The category of musculoskeletal adverse events of special interest (AESI) is made up of these different diagnoses: arthralgia (joint pain), arthritis (joint inflammation), arthritis/bacterial, chronic fatigue syndrome, polyarthritis (inflammation of multiple joints), post viral fatigue syndrome, and rheumatoid arthritis.

Fibromyalgia is not included in this report but was listed under the category of "Neurological AESIs." Polyneuropathy (symmetrical damage to peripheral nerves) *is* included in this report rather than under "Neurological AESIs."

3,600 cases were reported, **8.5% of the post-marketing data set** of 42,086 cases. These 3,600 individuals reported 3,640 events. **1,614 (44%) were classified as serious.** The most common adverse event was **arthralgia (3,525 or 97%)**, followed by 70 arthritis (2%), 26 rheumatoid arthritis (<1%) and 5 (<1%) polyarthritis.

The time from administration to adverse event ranged from < 24 hours to 32 days. **50% of the events started within the first 24 hours.**

Of the cases where gender was reported, 2,760 individuals were female, and 711 were male. Once again, the pattern from 5.3.6 SOCs continues, with an almost **4:1 ration of female to male.**

Age distribution was **predominantly adult (2,850, 85%)**, with 515 (15%) elderly, **two adolescents, four children, and one infant. During these dates, BNT162b2 was not approved for use in children or infants. Approval was only for adolescents 16 and above (not clarified by Pfizer).**

Outcomes were reported for 3,662 events: 1,801 (49%) resolved or resolving, **959 (26%) not resolved, 49 (1%) resolved with sequelae**, and **853 (23%) were unknown.**

Pfizer's Conclusion:

"This cumulative case review does not raise new safety issues. Surveillance will continue."

Note: As of the date of this team report, no follow-up safety surveillance data have been publicly released.

RECALL

this unsafe "vaccine."

Eighteen: "'Other AESIs' Included MERS, Multiple Organ Dysfunction Syndrome (MODS), Herpes Infections, and 96 DEATHS. 15 Patients Were Under Age 12, Including Six Infants."

—Joseph Gehrett, MD; Barbara Gehrett, MD; Chris Flowers, MD; and Loree Britt

The WarRoom/DailyClout Pfizer Documents Analysis Project Post-Marketing Group (Team 1)—Joseph Gehrett, MD; Barbara Gehrett, MD; Chris Flowers, MD; and Loree Britt—wrote a report about Other Adverse Events of Special Interest (AESIs) found in Pfizer document *5.3.6 Cumulative Analysis of Post-Authorization Adverse Event Reports of PF-07302048 (BNT162B2) Received Through 28-FEB-2021* (a.k.a., "*5.3.6*"). This category of AESIs is not related to a specific set of medical conditions or a specific organ. Rather, it contains medical conditions such as herpes virus infections, MERS (Middle East Respiratory Syndrome), MODS (Multiple Organ Dysfunction Syndrome); symptoms such as fever and inflammation; and non-medical-related issues like manufacturing issues.

It is important to note that the AESIs in the *5.3.6* document were reported to Pfizer for **only a 90-day period** starting on December 1, 2020, the date of the United Kingdom's public rollout of Pfizer's COVID-19 experimental mRNA "vaccine" product.

Key points of this report include:

- **Death was listed as the "relevant [adverse] event outcome" for 96 individuals** in this category.
- **Fifteen patients were under 12 years of age, including six infants**; and Pfizer's mRNA "vaccine" was not approved for use in people under age 16 at the time of *5.3.6*'s data collection.
- Of those reports with gender known, **76% were female and 24% were male, a greater than 3:1 female to male ratio.**
- There were **391 herpes infections** reported, including shingles, herpetic eye infections, and non-shingles herpes infections.
- There were **18 Multiple Organ Dysfunction (MODS) adverse events.**
- Onset of adverse events, a.k.a. latency, was from within 24 hours to 61 days with **half occurring within one day.**
- This category included 8,152 patients/cases, which is 19.4% of the total cases reported to Pfizer during its 90-day post-marketing safety surveillance.
- Non-elderly adults had almost six times the number of adverse events seen in elderly adults.
- Fever was the most common adverse event.
- Pfizer concluded: "This cumulative case review does not raise new safety issues. Surveillance will continue." To date, no follow-up, updated, and comprehensive safety report has been publicly released.

Please read this important two-page report below.

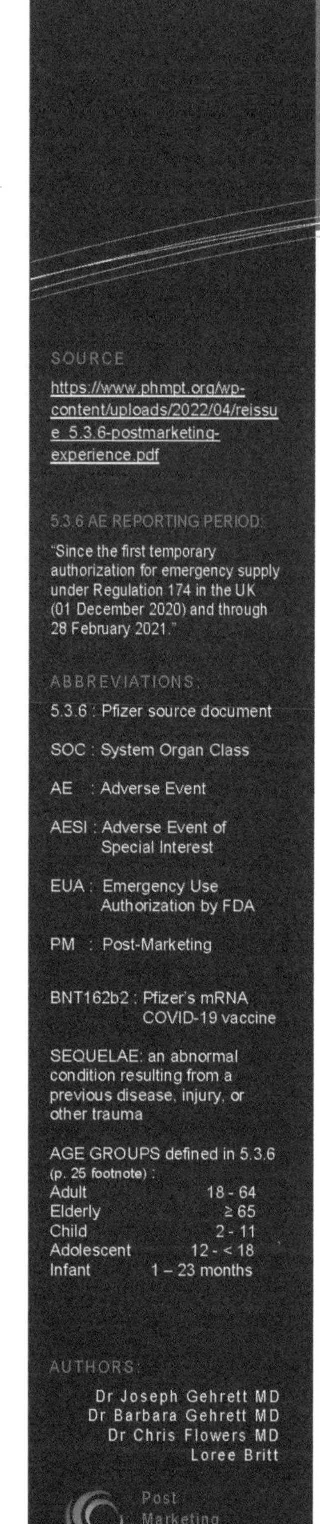

SOURCE:
https://www.phmpt.org/wp-content/uploads/2022/04/reissue_5.3.6-postmarketing-experience.pdf

5.3.6 AE REPORTING PERIOD:
"Since the first temporary authorization for emergency supply under Regulation 174 in the UK (01 December 2020) and through 28 February 2021."

ABBREVIATIONS:
- 5.3.6 : Pfizer source document
- SOC : System Organ Class
- AE : Adverse Event
- AESI : Adverse Event of Special Interest
- EUA : Emergency Use Authorization by FDA
- PM : Post-Marketing
- BNT162b2 : Pfizer's mRNA COVID-19 vaccine
- SEQUELAE: an abnormal condition resulting from a previous disease, injury, or other trauma

AGE GROUPS defined in 5.3.6 (p. 25 footnote) :

Adult	18 - 64
Elderly	≥ 65
Child	2 - 11
Adolescent	12 - < 18
Infant	1 – 23 months

AUTHORS:
Dr Joseph Gehrett MD
Dr Barbara Gehrett MD
Dr Chris Flowers MD
Loree Britt

Post Marketing Team
22May23

War Room/DailyClout Pfizer Document Analysis

Post-Marketing Team Micro-Report 13:
Other Adverse Events of Special Interest Review of 5.3.6

Death was the "relevant event outcome" for 96 individuals, which is 7.8% of 1,223 deaths from all causes reported in this 90-day report.

Other AESIs	
Search criteria: Herpes viral infections (HLT) (Primary Path) OR PTs Adverse event following immunisation; Inflammation; Manufacturing laboratory analytical testing issue; Manufacturing materials issue; Manufacturing production issue; MERS-CoV test; MERS-CoV test negative; MERS-CoV test positive; Middle East respiratory syndrome; Multiple organ dysfunction syndrome; Occupational exposure to communicable disease; Patient	• Number of cases: 8152 (19.4% of the total PM dataset), of which 4977 were medically confirmed and 3175 non-medically confirmed; • Country of incidence (> 20 occurrences): UK (2715), US (2421), Italy (710), Mexico (223), Portugal (210), Germany (207), France (186), Spain (183), Sweden (133), Denmark (127), Poland (120), Greece (95), Israel (79), Czech Republic (76), Romania (57), Hungary (53), Finland (52), Norway (51), Latvia (49), Austria (47), Croatia (42), Belgium (41), Canada (39), Ireland (34), Serbia (28), Iceland (25), Netherlands (22). The remaining 127 cases were from 21 different countries; • Subjects' gender (n=7829): female (5969), male (1860); • Subjects' age group (n=7479): Adult (6330), Elderly (1125), Adolescent, Child (9 each), Infant (6);

AESIs[a] Category	**Post-Marketing Cases Evaluation[b] Total Number of Cases (N=42086)**
isolation; Product availability issue; Product distribution issue; Product supply issue; Pyrexia; Quarantine; SARS-CoV-1 test; SARS-CoV-1 test negative; SARS-CoV-1 test positive	• Number of relevant events: 8241, of which 3674 serious, 4568 non-serious; • Most frequently reported relevant PTs (≥6 occurrences) included: Pyrexia (7666), Herpes zoster (259), Inflammation (132), Oral herpes (80), Multiple organ dysfunction syndrome (18), Herpes virus infection (17), Herpes simplex (13), Ophthalmic herpes zoster (10), Herpes ophthalmic and Herpes zoster reactivation (6 each); • Relevant event onset latency (n =6836): Range from <24 hours to 61 days, median 1 day; • Relevant events outcome: fatal (96), resolved/resolving (5008), resolved with sequelae (84), not resolved (1429) and unknown (1685). Conclusion: This cumulative case review does not raise new safety issues. Surveillance will continue

https://www.phmpt.org/wp-content/uploads/2022/04/reissue_5.3.6-postmarketing-experience.pdf

- **Adverse Events were reported to Pfizer during a 90-day period, following the December 1, 2020, public rollout of its COVID-19 experimental "vaccine" product.**
- **In the Pfizer 5.3.6 document, these AEs were categorized by System Organ Classes (SOC) – in other words, by systems in the body.**
- **Of those with known age, 1,125 were elderly; 6,330 were non-elderly adult; nine were adolescents; nine were children; and six were infants.**
- **Of those whose gender was known (7,829), 5,969 (76%) were female and 1,860 (24%) were male with a ratio of 3.2:1 female to male.**

IMPORTANT NOTE:
This category of AESIs is not related to a particular set of medical conditions or a specific organ of the body. The search criteria range from medical conditions (Herpes virus infections, Middle East Respiratory Syndrome, Multiple Organ Dysfunction Syndrome) to symptoms (fever, inflammation) to non-medical-related issues (manufacturing laboratory analytical testing issues, manufacturing production issues). The Pfizer report only specifically identified those conditions that had **six or more occurrences.**

This SOC included 8,152 patients, 19.4% of the total cases/patients in the 90-day period of this report. The cases were reported from 48 countries. In this total group of patients, there were 8,241 adverse events of which **3,674 (45%) were serious** and 4,568 (55%) were non-serious.

Fever was the most common event (7,666). **Herpes virus infection, including shingles, herpetic eye infections, and non-shingles herpes infections, comprised 391 events.** Inflammation was listed with 132 events, but these were not defined further as to type or bodily location. **Multiple organ dysfunction syndrome had 18 events.**

Death was the "relevant event outcome" for 96 individuals, which is 7.8% of 1,223 deaths from all causes reported in this 90-day report and for 1% of the patients in this SOC. Of the 8,241 total events, outcomes of "resolved/resolving" were listed for 5,008 events (61%), "resolved with sequelae" for 84 (1%), "not resolved" for 1,429 (17%), and "unknown" for 1,685 (20%). **The time from injection to adverse event onset was from within 24 hours to 61 days with half occurring within one day.**

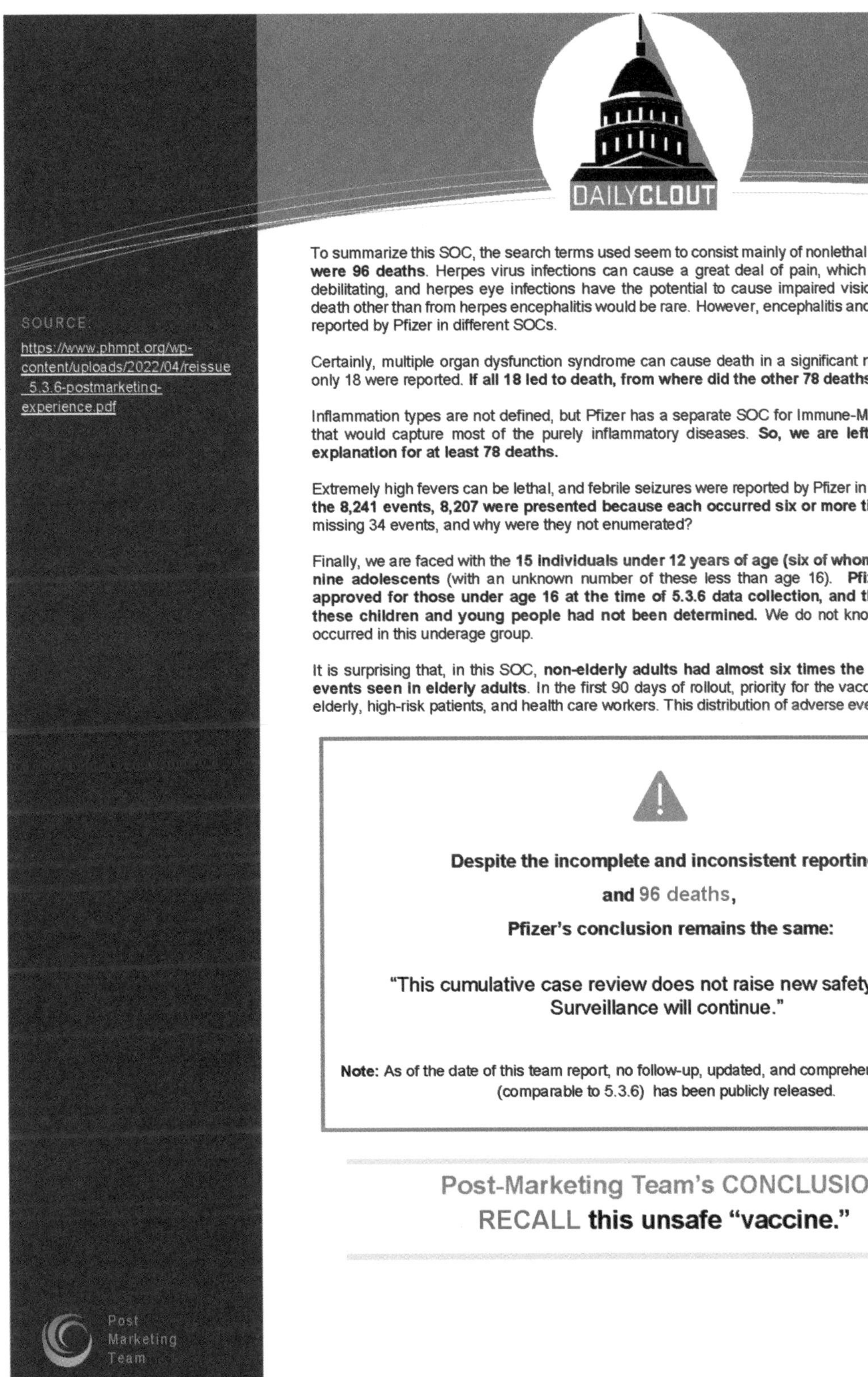

To summarize this SOC, the search terms used seem to consist mainly of nonlethal conditions, yet **there were 96 deaths**. Herpes virus infections can cause a great deal of pain, which can be chronic and debilitating, and herpes eye infections have the potential to cause impaired vision or blindness. Yet, death other than from herpes encephalitis would be rare. However, encephalitis and encephalopathy are reported by Pfizer in different SOCs.

Certainly, multiple organ dysfunction syndrome can cause death in a significant number of cases, but only 18 were reported. **If all 18 led to death, from where did the other 78 deaths come?**

Inflammation types are not defined, but Pfizer has a separate SOC for Immune-Mediated/Autoimmune that would capture most of the purely inflammatory diseases. **So, we are left with no clarity or explanation for at least 78 deaths.**

Extremely high fevers can be lethal, and febrile seizures were reported by Pfizer in a separate SOC. **Of the 8,241 events, 8,207 were presented because each occurred six or more times.** What were the missing 34 events, and why were they not enumerated?

Finally, we are faced with the **15 individuals under 12 years of age (six of whom were infants), and nine adolescents** (with an unknown number of these less than age 16). **Pfizer's drug was not approved for those under age 16 at the time of 5.3.6 data collection, and the dosage given to these children and young people had not been determined.** We do not know how many deaths occurred in this underage group.

It is surprising that, in this SOC, **non-elderly adults had almost six times the number of adverse events seen in elderly adults**. In the first 90 days of rollout, priority for the vaccine was given to the elderly, high-risk patients, and health care workers. This distribution of adverse events is unexplained.

Despite the incomplete and inconsistent reporting

and 96 deaths,

Pfizer's conclusion remains the same:

"This cumulative case review does not raise new safety issues. Surveillance will continue."

Note: As of the date of this team report, no follow-up, updated, and comprehensive safety report (comparable to 5.3.6) has been publicly released.

Post-Marketing Team's CONCLUSION
RECALL **this unsafe "vaccine."**

Nineteen: "Pfizer Knew by November 2020 That Its mRNA COVID Vaccine Was Neither Safe Nor Effective. Here Is What Pfizer's Employees and Contractors Knew and When They Knew It."

—Lead Author: L.D. LaLonde, MS Contributors: Loree Britt; Michelle Cibelli, RN, BSN; Barbara Gehrett, MD; Joseph Gehrett, MD.; and Chris Flowers, MD Editors: Amy Kelly, Chris Flowers, MD, and David Shaw

Introduction

Through the review of two documents—Pharmacovigilance Plan for Biologic License Application #125742 Of Covid-19 mRNA vaccine (nucleoside modified) (BNT162b2, PF-07302048) and 5.3.6 Cumulative Analysis of Post-Authorization Adverse Event Reports of PF-07302048 (BNT162b2) Received Through 28-Feb-2021—referred to below as "PV" and "5.3.6," the contributors to this report came to understand Pfizer knows its product does not work and that it poses a danger to the public. In this report, they have demonstrated these admissions using Pfizer's own words. When those documents are overlaid with the Emergency Use Authorization (EUA) from 2020 and the EUA from late 2021, it becomes apparent that the Company ignored safety signals and used weak statistics to justify product use. When these documents are viewed together, there is sufficient evidence to say Pfizer understood that there were problems with its mRNA COVID product before the original EUA was submitted in November 2020.

Abbreviations

PV = Pharmacovigilance Plan for Biologic License Application #125742 Of Covid-19 mRNA vaccine (nucleoside modified) (BNT162b2, PF-07302048). Date of Report: 28 July 2021, Version 1.1

EUA 2020 = Emergency Use Authorization (EUA) for an Unapproved Product Review Memorandum. Date of Document: 20 November 2020, Author: Marion F. Gruber, Ph.D., Director, CBER/OVRR

5.3.6 = Reissue of 5.3.6 Cumulative Analysis of Post-Authorization Adverse Event Reports of PF-07302048 (BNT162b2) Received Through 28-Feb-2021. Approval Date: 30 April 2021.

EUA 5–11 = Emergency Use Authorization (EUA) for an Unapproved Product Review Memorandum. Date of Document: 06 October 2021, Author: Peter Marks, MD, PhD, Director, CBER, and Acting Director, CBER/OVRR.

SOC = System Organ Class

AE = Adverse Event

Executive Summary in chronological order:

- In November 2020 (EUA 2020), Pfizer dismissed safety signals in its clinical trial C4591001 (ages 16+). Moreover, although Pfizer considered any adverse event (AE) within six weeks of product use to be reasonably associated with the product (EUA 2020, p. 10), it dismissed the observed safety signals in EUA 2020, 5.3.6, PV, and EUA 5–11.

- In November 2020 (EUA 2020), Pfizer had a weak demonstration of efficacy based on very few occurrences (eight cases in the vaccinated cohort versus 162 cases in the unvaccinated cohort). C4591001 may be invalid because investigators are unclear about 3,410 suspected COVID cases (1,594 vaccinated and 1,816 placebo). If COVID cases occurred in the thousands and investigators used only 170 cases for efficacy, their statistics did not reflect reality. Investigators then destroyed their clinical trial by unblinding and vaccinating all placebo cohort participants (PV, p. 13, pp. 18–19). In effect, this act terminated the trial. Pfizer acknowledged unblinding and vaccinating the placebo cohort would adversely affect the data (EUA 2020, p. 53). The company cut off data collection the day after placebo participants were vaccinated (EUA 5–11, p. 12).
- Through December 2020 to February 2021 (5.3.6) field reports, Pfizer observed AEs including deaths and permanent harms. Per Pfizer's own standard of AEs within six weeks of product use being considered product-related (EUA 2020, p. 10), Pfizer de facto recognized its product caused AEs, because many of the AEs in 5.3.6 occurred within hours or days of product use.
- In its report dated July 28, 2021 (PV), Pfizer still planned to use C4591001 (a portion of which was due April 2023) to reach final conclusions on its mRNA COVID product's efficacy and safety. The cut off of data collection on March 12, 2021, should be understood as Pfizer's acknowledgement of the termination of its clinical trial. Pfizer attempted to substitute titer-based lab tests for efficacy, but later admitted lab titers do not represent disease protection (i.e., efficacy) (EUA 5–11, p. 13).
- In Pfizer's July 2021 report (PV), Pfizer acknowledged pericarditis and myocarditis as risks of product use. Pfizer did not call it a dose-response, but it reported pericarditis and myocarditis risks as higher after dose #2 (PV, p. 50). Pfizer reported a similar dose-dependent pattern elsewhere (EUA 2020, p. 6, p. 42, p. 56; EUA 5–11, p. 46). All other AEs noted in the EUA 2020, from study C4591001, and AEs reported from the field in 5.3.6 were ignored. Additional studies listed by Pfizer in PV seem to not exist online.
- In October 2021 (EUA 5–11), efficacy was weakly demonstrated. Investigators did not draw upon C4591001 for support. Rather, they substituted titers for efficacy.
- In Pfizer's October 2021 EUA 5–11 submission, Pfizer described a dose-response relationship between its product and AEs in both dosage and dose number. Investigators speculated that subclinical damages would manifest in the long-term. The implication is that continued doses with subclinical damages would eventually manifest as clinical damages. Pfizer admitted a young male subject's AE, previously dismissed, was actually related to product use months after initial signal detection. This event represented a pattern of behavior: no matter what AE occurred, investigators concluded it was unrelated to Pfizer's product.
- EUA 5–11 introduced unsupported points to push product use in children. Pfizer introduced claims on transmission prevention and attacked the unvaccinated. Investigators did not provide clinical trial evidence for support. The product did not have well-demonstrated benefits, so any risks (and there are many) immediately rendered a poor risk-benefit ratio.

Emergency Use Authorization (EUA) for an Unapproved Product Review Memorandum. Date of Document: 20 November 2020, Author: Marion F. Gruber, PhD, Director, CBER/OVRR. EUA 2020 Regarding Efficacy

(https://www.fda.gov/media/144416/download)

Pfizer's original efficacy claim was based upon ratios between very small numbers over a short period of time (six weeks), representing extremely weak evidence. The vaccinated group had eight COVID-19 cases, and the placebo group had 162 cases (EUA 2020, p. 20). Investigators used this simple ratio to determine high efficacy, as 162 is around 20 times greater than eight. Compare these occurrences against the 17,411 in the vaccine cohort and the 17,511 in the placebo cohort used for the statistical evaluation (EUA 2020, p. 23). Eight and 162 were infinitesimal. If an individual took the vaccine, it dropped their risk of a positive PCR test from 0.92% to 0.045% in a six-week period. To put it another way, one should consider the result as doses needed to treat the population. Investigators vaccinated about 17,500 individuals (35,000 doses) to prevent approximately 150 COVID cases. For the other 17,350, the benefit was effectively zero during the six weeks. For them, vaccination was only risk.

This analysis described the purest meaning of the investigators' results. They arrived at a statistic derived under a narrow set of parameters, the most important of which was the very short-term nature of six weeks. In this context, the fraction of a percentage drop in COVID risk was inconsequential to the population. Pfizer failed to discuss the alternative conclusions based on few occurrences in a short time span. Pfizer would have understood that 35,000 doses to save about 150 cases was not practical for a public health intervention. This approximation of doses-needed-to-treat is just as valid as the efficacy claim in the context of a six-week period. It is the same result at which Pfizer arrived, drawn from the same evidence; however, it is rephrased in more practical language. A reasonable person would not take an experimental drug if the benefit was a 0.88% drop in COVID risk.

To create strength in statistical evaluation, the trial needed to run for two years to allow occurrences to build up in the placebo and experimental cohorts. Only then could valid conclusions be made. The result would either hold up and become stronger with time as vaccinated participants resisted disease over the long term, or investigators would find that COVID cases also accumulated in the vaccinated cohort just as they did in the placebo cohort. The practical reality was that this short-term cultivation of data was enough to perform a statistical math exercise only. Investigators did not demonstrate 95% efficacy over a year or longer period of time. If efficacy waned in the short-, middle-, or long-terms, it would not be captured by this preliminary analysis. For a short, preliminary, investigative trial with further follow-up planned, Pfizer's conclusion was technically acceptable, despite issues, as long as the clinical trial continued, unaltered, to the planned 24-month completion date.

On page 41 (EUA 2020), the investigators reported there was a testing issue in their clinical trial, which could have affected even their preliminary efficacy assessment. *There were suspected COVID cases numbering in the thousands that were not PCR-confirmed.* The authors discussed this finding in the context of safety, discussing both reactogenicity and adverse events, but they did not provide commentary on efficacy.

<u>Suspected COVID-19 Cases</u>

As specified in the protocol, suspected cases of symptomatic COVID-19 that were not PCR-confirmed were not recorded as adverse events unless they met regulatory criteria for seriousness. Two serious cases of suspected but unconfirmed COVID-19 were reported, both in the vaccine group, and narratives were reviewed. In one case, a 36-year-old male with no medical comorbidities experienced fever, malaise, nausea, headache and myalgias beginning on the day of Dose 2 and was hospitalized 3 days later for further evaluation of apparent infiltrates on chest radiograph and treatment of dehydration. A nasopharyngeal PCR test for SARS-CoV-2 was negative on the day of admission, and a chest CT was reported as normal. The participant was discharged from the hospital 2 days after admission. With chest imaging findings that are difficult to reconcile, it is possible that this event represented reactogenicity following the second vaccination, a COVID-19 case with false negative test that occurred less than 7 days after completion of the vaccination series, or an unrelated infectious process. In the other case, a 66-year-old male with no medical comorbidities experienced fever, myalgias, and shortness of breath beginning 28 days post-Dose 2 and was hospitalized one day later with abnormal chest CT showing a small left-sided consolidation. He was discharged from the hospital 2 days later, and multiple nasopharyngeal PCR tests collected over a 10-day period beginning 2 days after symptom onset were negative. It is possible, though highly unlikely, that this event represents a COVID-19 case with multiple false negative tests that occurred more than 7 days after completion of the vaccination regimen, and more likely that it represents an unrelated infectious process.

Among 3,410 total cases of suspected but unconfirmed COVID-19 in the overall study population, 1,594 occurred in the vaccine group vs. 1816 in the placebo group. Suspected COVID-19 cases that occurred within 7 days after any vaccination were 409 in the vaccine group vs. 287 in the placebo group. It is possible that the imbalance in suspected COVID-19 cases occurring in the 7 days postvaccination represents vaccine reactogenicity with symptoms that overlap with those of COVID-19. Overall though, these data do not raise a concern that protocol-specified reporting of suspected, but unconfirmed COVID-19 cases could have masked clinically significant adverse events that would not have otherwise been detected.

(https://www.fda.gov/media/144416/download, p. 41.)

They unwittingly admitted in this section that they did not obtain clear results on large numbers of participants with suspected cases of COVID. Since testing was a critical procedure to determine efficacy, it brings serious questions to the legitimacy of the clinical trial. Based on this information, the EUA clinical trial C4591001 results may not be valid. Personnel operating these trials should provide important context and relevant information stating otherwise.

The EUA 2020: Implications of Failure to Test Suspected COVID-19 Cases

Investigators reported 1,595 suspected COVID cases in the vaccine group and another 1,816 suspected COVID cases in the placebo group (EUA 2020, p. 41). Remember, investigators determined efficacy on 170 total COVID cases between the cohorts. If they thought they had thousands of other COVID cases and never confirmed them through testing, they would not have reached the correct determination of efficacy. If what the investigators reported was true, the C4591001 study would have been invalid by November 2020. The section to follow will highlight the implications of this testing problem regarding efficacy.

If the investigators were correct about missing COVID cases and these 3,410 cases were not included in their analysis, the real comparison could have been 1,602 vaccinated against 1,978 placebos. The risk to placebo participants could have been 11.3% compared to 9.2% in the vaccinated cohort for a 2.1% drop in risk of COVID. Practically speaking, it would not be a great difference in scale of occurrences between the cohorts. Most importantly, their efficacy would be closer to 19% with these numbers. Consider how this incidence rate would affect the clinical trial. If investigators witnessed thousands of cases of COVID in both cohorts in this short period, then they were on track to run out of trial participants in about a year if that rate of infection continued. Efficacy in that scenario would approach zero, and investigators would have been able to see that inevitability if thousands of participants were contracting COVID in both cohorts.

The true efficacy could be 95%, 19%, 0%, or some other figure. Hypothetically, there could have been more COVID cases in the vaccinated group, which would have represented negative efficacy. We cannot know because the investigators are unsure what some symptomatic cases meant. The arrival at only eight cases of COVID in the vaccinated versus 162 cases of COVID in the unvaccinated among thousands of symptomatic patients is concerning. If there is an explanation for what it means, the public deserves to hear it from the investigators.

EUA 2020 Regarding Safety

The standard for considering AEs to be potentially related to the product are as follows: "From a safety perspective, a 2-month median follow-up following completion of the full vaccination regimen will allow identification of potential adverse events that were not apparent in the immediate post-vaccination period. Adverse events considered plausibly linked to the vaccination generally start within 6 weeks of vaccine receipt" (EUA 2020, p. 10). For reference, the EUA findings from C4591001 represented six weeks of follow-up on average per patient.

In the vaccine group, investigators reported occurrences of myocardial infarction (MI) as 0.02% (four to five patients, p. 40), cerebrovascular accident (CV) as 0.02% (four to five patients, p. 40), appendicitis as 12 patients (0.04%) (p. 40), and Bell's palsy as four patients (~0.02%) (p. 37). The standard of using few occurrences to make conclusions, as used for efficacy, applied here, too. During the short, six-week study, the risk of MI or CV quadrupled or quintupled in the vaccine group as compared to the one placebo death from MI and the one placebo death from hemorrhagic stroke (EUA 2020, p. 40). Risk of appendicitis increased 50% with vaccination (12 versus eight). Bell's palsy did not occur in any placebo participants. These observations were safety signals.

Investigators reported six deaths during the trial (two vaccine versus four placebo). One vaccine subject was over 55 and experienced cardiac arrest 62 days after dose #2. The other subject was over 55 and died of unlisted causes three days after dose #1, but investigators noted he was obese with atherosclerotic disease. The placebo deaths were one MI, one hemorrhagic stroke, and two unknown causes. Of these six, one was under 55 years old, and the specific age is not disclosed. Investigators assured the public that "all deaths represent events that occur in the general population of the age groups where they occurred, at a similar rate" (EUA 2020, p. 40).

The investigators took time in the EUA to declare the AEs as chance events consistent with the general population at large. This acknowledgement is extended to deaths (p. 43), appendicitis (p. 43), and Bell's palsy

(p. 52), yet no commentary accompanies MI and CV. These assertions are **not valid** per their own standard from page 10—i.e., "From a safety perspective, a 2-month median follow-up following completion of the full vaccination regimen will allow identification of potential adverse events that were not apparent in the immediate post-vaccination period. Adverse events considered plausibly linked to the vaccination generally start within 6 weeks of vaccine receipt"—where they noted any occurrences within their six-week trial period would be plausibly linked to product use. It was also **not valid** because the investigators were charged with running a clinical trial where findings from the vaccine group were compared specifically to the placebo group. It was the entire purpose of the clinical trial. Rather than doing this analysis in an open and honest way, the investigators, who realized there could be significant safety issues, blamed chance. **Nonetheless, investigators used very small numbers to determine that efficacy was high. *They then ignored the same small numbers to determine safety,* which included dismissal of adverse events that occurred within a short time after doses. *The methods that were good enough for efficacy were suddenly not good enough for safety.***

The Fate of the Placebo Cohort

In light of the problems highlighted above with statistics based on small numbers, the investigators had one course of action to pursue truth in their clinical trial. They needed to run the 24-month clinical trial to completion. The missed COVID cases were an issue, but they could potentially make up for it with due diligence by tracking down these cases and by following both cohorts to the two-year completion date. In the event the product worked very well with an excellent safety profile, the evidence over a longer span would tell that truth despite imperfections in the process. It was in the best interests of Pfizer and the world's patients to witness this truth. If it turned out the product did not work or that it was not safe or both, the integrity of the clinical trial C4591001 was critically important to stop product use.

On page 53 of the EUA 2020, the investigators discussed the consideration to unblind and to vaccinate the placebo cohort. The Vaccines and Related Biological Products Advisory Committee (VRBPAC) provided discussion.

> The committee discussed potential implications of loss of blinded, placebo-controlled follow-up in ongoing trials including how this may impact availability of safety data to support a Biologics License Application (BLA). Some pointed out the importance of long-term safety data for the Pfizer-BioNTech COVID-19 vaccine as it is made using technology not used in previously licensed vaccines. In response to the question whether the ongoing Phase 3 study would still be sufficiently powered if eligible placebo recipients were vaccinated, **Pfizer asserted** that, even with an anticipated **loss of placebo-controlled follow-up of 20%,** the study would maintain adequate statistical power and would be positioned to accrue additional data on vaccine efficacy, including efficacy against severe disease, as well as safety, **although unblinding of the study would reduce interpretability of results**. (Bold added, EUA 2020, p. 53)

Pfizer already had statistical issues documented above and acknowledged within the EUA 2020 that they were open to reducing their study's power further by unblinding and vaccinating the placebo cohort participants. There was no rubric for how they would choose which participants would be among the unblinded 20%, but they had a solution in mind. Nonetheless, with this 20% standard established by Pfizer in this November 2020 EUA, **Pfizer vaccinated their entire placebo cohort**. Pfizer documented it outside the view and knowledge of the world's patients (Table 5, PV, pp. 18–19). **Pfizer reported the vaccination of 19,696 placebo participants, representing the entirety of their placebo cohort. Pfizer completed this process rapidly, finishing on *12 March 2021*.**

Investigators moved to unblind and vaccinate placebo participants immediately after the EUA 2020 was approved. Per Pfizer's own 20% standard established in the EUA 2020 (p.53), the power of this study was effectively destroyed on March 12, 2021 (PV, ps.18–19). Thus, Pfizer essentially ended its clinical trial, C4591001, in March 2021. Whatever continued on was something else approximating an observational study. If the product was highly efficacious and safe, it was not in Pfizer's interest to manipulate the placebo cohort. A complete clinical trial with clean data, free of manipulation, was in the best interest of patients and society, because it was much more likely to conclude the truth. Pfizer committed this act before it had valid efficacy and safety data. As a result, the trial cannot produce an accurate efficacy analysis.

EUA 2020—Conclusion Summary Statement

By the completion of the EUA 2020, the investigators knew they had significant shortcomings in their efficacy assessment. They had safety signals that they refused to acknowledge as product related. Yet, Pfizer pushed an efficacy statement it could not support and declared a high level of safety that was refuted by its own reported observations. If the limited data were sufficient for efficacy, the same limited data were sufficient to acknowledge significant safety signals. Furthermore, Pfizer's failure to capture COVID cases in its study cohorts rendered any efficacy outputs invalid. The investigators were subject matter experts in these areas. The construction of statistics in the EUA, combined with selective observations, indicated they very likely knew or at least suspected the product had limited or zero efficacy and significant safety concerns by November 2020. Their termination of the clinical trial before valid data became available did not serve the interest of society; it seemingly served to hide data from the public.

5.3.6 Cumulative Analysis of Post-Authorization Adverse Event Reports of PF-07302048 (BNT162b2) Received Through 28-Feb-2021.

(https://www.phmpt.org/wp-content/uploads/2022/04/reissue_5.3.6-postmarketing-experience.pdf)

FDA Approval Date: 30 April 2021

Obtained by Court Order

(https://phmpt.org/pfizer-court-documents/)

5.3.6 Regarding Safety

The 5.3.6 document (38 pages) was a safety-monitoring report authored by Worldwide Safety at Pfizer (WSP). The findings represented adverse events submitted voluntarily to Pfizer's safety database from various sources, including medical providers and clinical studies, between 01 December 2020 and 28 February 2021. The AEs consisted of 42,086 cases reporting 158,895 total adverse events. The AEs were broken into System Organ Classes (SOCs) with each SOC further divided into individual conditions observed in the field. The report described AEs with percentages representing proportions of reports received. Any percentages should not be taken as incidence rates of occurrence, as this observational data was not a clinical trial. Nonetheless, it should have been evident to Pfizer that its product harmed patients, which included permanent harms and 1,223 deaths.

Within the first three months after rollout of product, providers in the field reported damages across all organ systems to Pfizer. Reference the table below. This table includes special concern areas being tracked by Pfizer through 2020 and 2021. The first special concern, anaphylaxis, is considered an "Identified Risk" (IR), Vaccine-Associated Enhanced Disease (VAED) is considered a "Potential Risk" (PR). The third category of "Missing Information" (MI) concerns "Pregnancy and Lactation," "Use in Pediatric Individuals," and "Vaccine Effectiveness." These IR, PR, and MI categories were predetermined categories of interest from the EUA 2020 that garnered more information in 5.3.6. All other SOCs charted below fell outside those original categories.

SOC	Page	Number, %	Serious (N, [%])	Non-Serious (N, [%])	Report Author's Notations
Anaphylaxis (IR)	10	2,958 7.0%	2,341 5.6%	617 1.5%	
VAED (PR)	11	-	-	-	75 potential cases
Pregnancy and Lactation (MI)	12–13	413 0.98%	84 0.2%	329 0.78%	Spontaneous abortions and neonatal deaths reported; alterations to breastfeeding
Pediatric (MI)	13	34 0.08%	24 0.05%	10 0.02%	One Facial Paralysis
Vaccine Effectiveness (MI)	13–15	1,665* 4.0%	1625 3.9%	21 0.05%	"Serious" is considered a case of COVID; no immunity conferred
Cardiovascular	16	1,403* 3.3%	946 2.2%	495 1.2%	130 myocardial infarctions, 91 cardiac failures
COVID-19	17	3,067* 7.4%	2,585 6.1%	774 1.8%	Unremarkable; deals with positive cases
Dermatological	17	20 0.05%	16 0.04%	4 0.01%	Unremarkable; Reactions
Haematological	18	932* 2.2%	681 1.6%	399 0.95%	Numerous examples of spontaneous bleeding from mucous membranes
Hepatic	18–19	70* 0.2%	53 0.13%	41 0.1%	Metabolic alterations within the liver
Facial Paralysis	19–20	449* 1.07%	399 0.95%	54 0.12%	Authors refer to studies C4591001, C4591011, C4591012, C4591021
Immune-Mediated and Autoimmune	20	1,050* 2.5%	780 1.9%	297 0.70%	32 Pericarditis, 25 Myocarditis
Musculoskeletal	20–21	3,600* 8.5%	1,614 3.8%	2,026 4.8%	3,525 Arthralgia
Neurological	21	501* 1.2%	515 1.2%	27 0.06%	204 Seizure, 83 Epilepsy
Other	21–22	8,152* 19.4%	3,674 8.7%	4,568 10.8%	7,666 Pyrexia Herpetic conditions
Pregnancy Related	22	-	-	-	Refers to pages 12–13
Renal	22	69* 0.17%	70 0.17%	0 0%	All serious: 40 acute kidney injury, 30 renal failure

SOC	Page	Number, %	Serious (N, [%])	Non-Serious (N, [%])	Report Author's Notations
Respiratory	22–23	130* 0.3%	126 0.3%	11 0.03%	44 respiratory failures
Thromboembolic Events	23	151* 0.3%	165 0.4%	3 0.007%	60 Pulmonary Embolism, 39 Thrombosis, 35 Deep Vein Thrombosis (DVT)
Stroke	23–24	275* 0.6%	300 0.7%	0 0%	All serious; Ischaemic and Haemorrhagic conditions reported
Vasculitis	24	32* 0.08%	25 0.06%	9 0.02%	Specific condition leading to one fatality not noted
Medication Error	26	2056* 4.9%	124 0.29%	1932 4.6%	Seven fatalities not categorized as "Serious." Authors lack information leading to fatalities, considered noncontributory

(N, [%]): Annotation refers to number of cases (N) and the proportion of AE reports [%]

*: denotes counting discrepancies within the 5.3.6 report

Report Author's Annotations: Any commentary in this column represents sample highlights from each SOC. All readers are encouraged to read the 5.3.6 document to understand the scale, depth, and width of Pfizer's aggregated safety reports from the field.

Accounting was not well-done in this Pfizer report and was best illustrated by Table 1 (5.3.6, p.7). The authors reported the adverse events by age brackets that were not standardized in age range, which led to potential issues in understanding age-related effects. The age groupings were <17, 18–30, 31–50, 51–64, and >75. This non-standardized approach obscured any age-related effects among AEs. Most AEs occurred in the 31–50 range, but this age range was also the widest age range. When this document first became available for review, it was difficult to make sense of how data was gathered and grouped. More information on this topic emerged later in the PV document. Table 1 did relay important findings. **There were 1,223 deaths in the field that providers thought were product related.** There were also 520 reports of AEs with sequelae, 11,361 reports of "not recovered at the time of report," and another 9,400 events without known resolution criteria.

There was one concept Pfizer confirmed in their reporting system regarding latency. When aggregated, **it was apparent that reported AEs developed immediately after product use. The median latency for each category is less than a week**. See the table below. By Pfizer's own standard from the EUA 2020 ("From a safety perspective, a 2-month median follow-up following completion of the full vaccination regimen will allow identification of potential adverse events that were not apparent in the immediate post-vaccination period. Adverse events considered plausibly linked to the vaccination generally start within 6 weeks of vaccine receipt"), this realization alone should have been enough to suggest AEs were product related. Yet very consistently and predictably throughout the 5.3.6 report, Pfizer stated, "Conclusion: This cumulative case review does not raise new safety issues. Surveillance will continue." It begs the question when Pfizer would admit there were significant safety issues with its product and when they would notify the public.

SOC	AE Development Range	AE Development Median
Cardiovascular	<24 hours—21 days	<24 hours
Covid-19	<24 hours—374 days	5 days
Dermatological	<24 hours—17 days	3 days
Haematological	<24 hours—33 days	1 day
Hepatic	<24 hours—20 days	3 days
Facial Paralysis	<24 hours—46 days	2 days
Immune-Mediated and Autoimmune	<24 hours—30 days	<24 hours
Musculoskeletal	<24 hours—32 days	1 day
Neurological	<24 hours—48 days	1 day
Other	<24 hours—61 days	1 day
Renal	<24 hours—15 days	4 days
Respiratory	<24 hours—18 days	1 day
Thromboembolic	<24 hours—28 days	4 days
Stroke	<24 hours—41 days	2 days
Vasculitic	<24 hours—19 days	3 days

5.3.6—Conclusion Summary Statement

The 5.3.6 document was reviewed elsewhere in the WarRoom/DailyClout Pfizer Documents Analysis Project, because it was dense and required further exploration as a result. **In the context of what Pfizer knew about safety and efficacy in March 2021 and remembering 5.3.6 was not available to the public without a court order, Pfizer confirmed its product caused significant, severe AEs across all organ systems.** What could have been chance AEs in the EUA 2020 C4591001 study were substantiated by field reporting. There were many more AEs than MI, CV, appendicitis, and Bell's palsy. Death was confirmed as an adverse event based on field reports. Per Pfizer's EUA 2020, any findings within six weeks would reasonably have been linked to the product. These AEs were often reported within days of product administration. By March 2021, Pfizer knew its product had safety issues, and it knew from the EUA that its efficacy was questionable at best, and invalid or null at worst.

Pharmacovigilance Plan for Biologic License Application (PV)

(https://www.phmpt.org/wp-content/uploads/2023/01/125742_S21_M1_pharmacovigilance-plan.pdf)

Report Date: 28 July 2021

Obtained by Court Order

(https://phmpt.org/pfizer-court-documents/)

The PV document updated and tracked Pfizer's plans to detect and to address safety signals. The 99-page document summarized studies and findings up to the date it was published. It added myocarditis and pericarditis as concerning adverse events (AEs) related to the product. Other System Organ Classes' (SOCs) AEs were on the same scale as pericarditis and myocarditis, yet they were ignored as important risks. After the EUA 2020,

Pfizer should have been curious about C4591001 AEs, specifically MI, CV, and facial paralysis (Bell's Palsy). In 5.3.6 reporting, it identified 130 MI, 275 strokes, and 449 paralyses among many other AEs compared to just 32 cases of pericarditis and 25 cases of myocarditis. There were 165 serious thrombolytic events reported as a separate category in 5.3.6 as well. No AEs were addressed from 5.3.6 other than the predetermined list from the EUA 2020 (IR, PR, MI), and the newly added cardiac AEs (listed under "Immune-Mediated/ Autoimmune" on p. 20 in 5.3.6). PV does not provide updated data on MI, CV, paralyses, or thrombolytic events. For reference, appendicitis does not even appear in 5.3.6. What was once witnessed and discussed in the EUA 2020 C4591001 clinical trial and witnessed in field reporting received no further mention in PV. No warnings reached the public on potential harms. Claims of efficacy remained high, and no additional safety signals were addressed from other SOCs.

PV identified ongoing studies that may develop knowledge on efficacy and safety. When a search for those studies was completed on clinicaltrials.gov, many studies did not appear (last checked May 22, 2023). C4591001 was listed as completed on February 10, 2023. No results are available. C4591015, a clinical trial focused on pregnant women, was completed on July 15, 2022. It listed "Primary Endpoints" as 4-30-2023. No results are available. BNT-162–01 showed the results were submitted for review on April 11, 2023. No results are available. C4591007 was listed as pending completion on October 3, 2023. The following clinical trials were listed in PV and were not found on clinicaltrials.gov: C4591008, C4591009, C4591011, C4591012, C4591022, W1235284, and W1255886. PV listed pending report dates for many of these studies. No interim results appear online, as many studies likewise do not appear. Notes on these studies appear in Appendix 1 of this report.

The most important pages of the PV report dealt with vaccinations to the placebo cohort in the EUA study, C4591001. In the EUA 2020, Pfizer outlined the statistical evaluation problems if it vaccinated more than 20% of the placebo cohort (EUA 2020, p. 53). Table 5, "Exposure to BNT162b2 by Age Group and Dose (C4591001)—Open Label Follow-up Period—Subjects Who Originally Received Placebo and Then Received BNT162b2 After Unblinding," showed Pfizer vaccinated 19,696 placebo participants, representing the entirety of their placebo cohort, by March 12, 2021 (PV, p. 18–19). Pfizer continued to cite the C4591001 study throughout PV as an ongoing clinical trial although Pfizer knew the study was no longer valid per its own standards as laid out in the EUA 2020 (p. 53).

Pharmacovigilance Regarding Safety

Pfizer's acknowledgement of myocarditis and pericarditis set a precedent for what AEs Pfizer took seriously as safety signals. Yet, Pfizer ignored other AEs. Reference the chart below to compare other SOCs from 5.3.6 against myocarditis and pericarditis as reported in PV. Hundreds of serious AE reports occurred across all SOCs including fatalities and unresolved conditions. There were just 32 cases of pericarditis and 25 cases of myocarditis in 5.3.6. All other SOCs exceeded myocarditis and pericarditis in 5.3.6 and are not mentioned in PV. Other AEs were on scale with myocarditis and pericarditis yet were not added as publicly acknowledged AEs for informed consent. Pfizer seemingly broke from its own standard by ignoring other significant product harms that it observed at the degree as accepted harms.

Pfizer does acknowledge a serious risk pattern from its product through additional product doses. "Evaluation by the US CDC has found reports [of myocarditis and pericarditis] to be most frequent in adolescent and young adult male patients following the **second** dose of vaccine" (Bold added. PV, p. 50). The appendix of the EUA 5–11 noted the emergence of AEs after additional doses as acknowledgement of a dose-response effect (EUA 5–11, p. 46). The EUA 2020 acknowledged higher rates of AEs after dose two and also noted more AEs in younger participants (EUA 2020, p. 6, p. 42, and p. 56). Pfizer understood there was a relationship between AEs and continued product exposures, and it was observed across the documents. This example with myocarditis and pericarditis was the only place Pfizer admitted the connection between additional doses and the risks of significant AEs. Within the context of the serious AEs across all organ systems, it is reasonable to assume additional doses increase the risks of other types of AEs. This assumption would require a mechanism to explain how the product damaged all organ systems as opposed to narrower, specific types of damage.

System Organ Class	Document	Serious	Fatal	Unresolved
Myocarditis (added)	PV	459	14	106
Pericarditis (added)	PV	370	3	63
Cardiovascular	5.3.6	946	136	140
Haematological	5.3.6	681	34	267
Hepatic	5.3.6	53	5	14
Facial Paralysis	5.3.6	399	0	183
Immune-mediated or Autoimmune ***	5.3.6	780	12	215
Musculoskeletal	5.3.6	1,614	0	959
Neurological	5.3.6	515	16	89
Other	5.3.6	3,674	96	1,429
Renal	5.3.6	70	23	15
Respiratory	5.3.6	126	41	18
Thromboembolic	5.3.6	165	18	49
Stroke	5.3.6	300	61	85
Vasculitic	5.3.6	25	12	1
(IIR) Anaphylaxis	PV	2341	9	229
(IPR) VAED	PV	138	38	65
(MI) Pregnancy	PV	75	38	-
(MI) Lactation	PV	5	-	-
(MI) Pediatric	PV	24	0	16

This table demonstrates that AEs from all SOCs are on the same risk scale as the added AEs of myocarditis and pericarditis. Other SOCs from 5.3.6, in fact, exceed them.

(added) AEs now included as safety signals. The occurrences are not from 5.3.6.

*** This category from 5.3.6 contained results for myocarditis and pericarditis.

(IIR) Important Identified Risk—considered an important safety signal.

(IPR) Important Potential Risk—considered a potential safety signal.

(MI) Missing Information Category

Pfizer delivers on a possible mechanism through its discussion on lab-derived efficacy measures, where the company acknowledged it knew about systemic spread of the product. Pfizer knew from rat studies (pp. 9–10) that the product ingredients did travel away from the injection site and aggregated elsewhere (liver, spleen, adrenal glands, ovaries). Pfizer reassured the public that fertility was not affected, and the company touted immunity in offspring, too (PV, p. 11). Nonetheless, this important piece served as a mechanism for breadth of AEs witnessed in its documents. Pfizer may not have had a singular type of AE in large excess, but it witnessed and documented a variety of AEs across SOCs. Pfizer's documentation of systemic spread should have allowed them to connect its product to harms. Harms occurred in any organ system exposed to Pfizer's product, and harms occurred with additional exposures to the product.

For reference before the EUA 5–11, Pfizer did review animal studies and introduced lab values in animal models to determine efficacy. Investigators claimed 100% efficacy in immune response in Rhesus Macaques based on chemical immune reaction (PV, p. 9). Although provocative, this reaction would not necessarily indicate human immunity to COVID. Although not evident in this time frame, Pfizer's celebration of 100% efficacy based on lab titers in animals served as the preamble to using lab-based titers as a substitute for clinical trial data. The upcoming EUA 5–11 expanded this concept of replacing clinical trial data Pfizer presumably knew were not valid.

A discrepancy noted in 5.3.6 received some clarifying information in PV. The age brackets for AE reporting were unusual in 5.3.6 with non-standardized intervals. There was a large age bracket of ages 31 to 50, while other brackets covered about 10 years or less. When authors shared statistics from their safety database, notable coincidences emerged. Myocarditis in ages over 16 occurred most often in young men with a mean age of 37.2 years old and a median age of 32.0 years old (PV, p. 48). For pericarditis in ages over 16, there was no gender difference, and the mean age was 51.5 years old, while the median age was 51.0 years old. The way ages were assembled in 5.3.6, split and diluted myocarditis AEs. In the upcoming EUA 5–11, it was shown again that myocarditis occurred most often in males under age 40 with no incidence rate provided by the investigators (EUA 5–11, pp. 14–15). Investigators did provide incidence rates for these AEs for patients between the ages of 12 and 17. It was striking how Pfizer reported these demographics across documents and how it grouped these cardiac conditions under a different category in 5.3.6. **It hinted at something specific with myocarditis in men ages 18 to 39, but there was never an explanation about it.** Elaboration by Pfizer investigators would be helpful for understanding how they chose to report these findings and if there were important findings in this age group. With investigators speculating about subclinical, long-term damages in EUA 5–11 (p. 15), and through documentation of various severe AEs leading to death, Pfizer should share what it knows about this avoided age group.

Pharmacovigilance Plans

Section III (PV, pp. 71–92) dealt with the actual Pharmacovigilance plan. This section outlined the courses for current and future studies. Pfizer reviewed the categories of focus. There were Important Risks (Anaphylaxis, Myocarditis, and Pericarditis), Important Potential Risks (VAED/VAERD), and Missing Information (Pregnancy/Lactation, Vaccine Effectiveness, Use in Pediatrics <12). Pfizer outlined its sources for signal detection on PV pages 71–72, which included references to literature and to Web-based reporting systems. Pfizer documented that it knew what was happening with its product in scientific literature, in the field, and within its own reports. Pfizer planned to perform future studies for each category above. Studies of other SOCs were not planned. Perhaps safety signal detection would take place coincidentally, but Pfizer had already ignored safety signals to date.

Pages 73–84 outlined Pfizer's intent to complete further studies to evaluate efficacy and safety. Studies were outlined by category with due dates specified. Many interim report dates had passed, without reporting, by May 22, 2023. Clinical trial C4591001 was the first study listed on the list of ongoing studies (PV, p. 92). Pfizer intended to make use of this study despite tampering with the placebo cohort months prior to this Pharmacovigilance plan.

Consider what it meant when the C4591001 clinical trial was not completed to term. The claims of efficacy and safety have never been supported. There were only sparse, preliminary results of efficacy based on statistical misrepresentation. Adverse events indicated the product perhaps was not safe in the EUA 2020 and definitely not safe in 5.3.6 by March 2021. The clinical trial was meant to run to 24 months to allow for a proper and robust evaluation of two large cohorts. Pfizer destroyed this trial before relevant results were ever realized. Whatever remained of the trial was completed on February 10, 2023, but even those results are still not available.

The problems with C4591001 made it even more imperative to complete the other studies listed within the PV document. With that in mind, our team set out to verify the status of these studies nearly two years after they were planned and promised by Pfizer. It turns out **many of these studies do not exist.** Pfizer seems to have had no intention of pursuing the relevant clinical trial data needed to determine a valid efficacy statement. Its dismissal of safety signals both in its own C4591001 trial and in field reporting suggested the company had no strong interest in product safety signals. The absence of promised studies to determine efficacy and to monitor safety completed its failure of honest evaluations.

PV—Conclusion Summary Statement

By July 2021, Pfizer observed its product traveled throughout the body and caused AEs across all organ systems in immediate timeframes after administration with additional doses increasing the likelihood of harm. It also became apparent Pfizer had no intention to report those observations to the public in those terms. Clinical trials planned and listed within PV were also abandoned. If C4591001 was going well, it would have been reported ad nauseum. Since C4591001 was altered well ahead of this report, Pfizer hoped the introduction of titers would give an alternative measure to claim efficacy regardless of disease protection. Investigators in the EUA 5–11 (p. 17) documented this lab-based evaluation was not valid for proving protection from COVID.

Consider the political environment and mandates at the time of this published report in 2021. Pfizer knew it had these problems, and yet the company allowed public statements on efficacy and safety to continue unopposed. The decision not to halt product use represented a top-to-bottom failure at Pfizer. The people compiling these reports were subject matter experts. They knew what the findings meant even as they reported a lack of safety concerns and as they reported high efficacy. They understood every problem posed so far. Even with what Pfizer learned by the time it published PV, the company continued onward to the children.

Where does this lead in the next EUA for five- to 11-year-old children in October 2021? Read this section understanding that the interim results for the young 12- to 15-year-old cohort are due within

weeks. There appears to be a rush to complete the EUA 5–11 before relevant trial information becomes available.

Emergency Use Authorization (EUA) for an Unapproved Product Review Memorandum

(https://www.fda.gov/media/153947/download)

Submission and Receipt Date: October 6, 2021
Review Completion Date: October 29, 2021

After nearly a year of product use and with investigators knowing the issues with safety and efficacy, one would hope for the EUA for five- to 11-year-olds, the EUA meant to authorize use for the youngest Americans, to lay out a very logical case for product use. This document should have been Pfizer's best effort, but it was not. The document itself appeared hastily constructed suggesting several authors assembled it quickly with disjointed opinions. It contained typos, incoherent commentary, and contradictory narratives. These narratives included claims that vaccinating children would stop spread, although investigators provided no evidence to support the claim and subsequently listed the claim itself as a gap in their knowledge. Investigators also attempted to suggest titers could represent efficacy and later suggested it was not a valid measure. Another narrative included the conclusion of a favorable risk-benefit ratio and yet showed an unfavorable risk-benefit ratio while admitting the COVID risk to children was always minimal.

The primary conclusions made by investigators in the EUA 5–11 were, again, based on weak evidence. Authors concluded efficacy using small numbers and lab values. They did not draw substantial support from C4591001. Authors concluded safety in the face of mounting evidence that the product was not safe. They consistently concluded beneficial risk-benefit ratios while demonstrating with computer modeling that they had an unfavorable risk-benefit ratio. **Tucked into the appendix is an admission that investigators understood a dose-response problem with the product (EUA 5–11, p. 46). They learned in C4591007 that AEs were related to both dosage and dose number. Investigators speculated about what these findings could mean for long-term safety (EUA 5–11, p. 15).**

EUA 5–11 Regarding Efficacy

In the clinical trial C4591001, investigators used weak evidence for efficacy. In EUA 5–11 (using study C4591007), they relied on a similar format. After two months of follow-up, they noted three COVID cases (out of 1,518 participants) in the vaccine group compared to sixteen cases (out of 750 participants) in their placebo group (p. 26). The incidence rate was 0.02% in the vaccine group and 2.13% in the control group. These percentages are statistically significant but, again, take place over a very short time span. Efficacy is not well-supported by this evidence.

Curiously, in the eleven months since the original EUA 2020, investigators did not report great increases in follow-up in C4591001. They reported around 60% of test and placebo cohorts at four or more months of follow-up, leaving around 40% of the cohorts at much less follow-up (EUA 5–11, p. 12). Pfizer cut off data collection on March 12, 2021, leaving a six-month gap before the EUA 5–11. The data cutoff is consistent with Pfizer's understanding that the clinical trial effectively ended after vaccination of the entire placebo cohort.

Efficacy claims in the October 2021 EUA for five- to 11-year-olds lack support from the original trial as a result. With the added context from PV (pp. 18–19) which was not made available to the public until after the court order, the public can now see that Pfizer abandoned its efficacy monitoring in C4591001. Pfizer, per their own standard (EUA 2020, p. 53), knew its efficacy analysis was no longer valid without a placebo cohort and terminated its data collection on March 12, 2021. If Pfizer had continued the clinical trial with blinded placebos as planned, it would have had up to six more months of data for EUA 5–11. Instead, Pfizer's investigators turned to vaccinating children knowing they destroyed what could have been the most important data to parents. The public was denied whatever truth C4591001 could have provided. The public once again was forced to accept another document lacking evidence.

The investigators understood the problems with short-term follow-up of only two months. They introduced immunobridging as a metric for efficacy. In brief, investigators used bloodwork to look for production of antibodies as a response to product use. They assumed an antibody titer implied protection. On page 17 (EUA 5–11), investigators made it clear that "the immune marker(s) used for immunobridging do not need to be scientifically established to predict protection," yet they used immunobridging to determine efficacy. **Investigators claim 100% efficacy (EUA 5–11, p. 13) based on these titers despite a subsequent admission on page 17 that they do not know what titer concentration would confer protection. Investigators used a test for efficacy that they knew was not valid.**

EUA 5–11 Regarding Safety

Pfizer identified a dose-response relationship and connected it to the potential for long-term damages. The EUA Appendix (p. 46) discussed the dosage reduction in children. Investigators found, during C4591007, two factors that led to more adverse reactions: 1) the dose number, and 2) the dosage. Investigators found a **dose-response relationship** between the product and AEs in their own trial. Furthermore, the number of doses being related to adverse events was significant because it suggested cumulative risks with continued dosages. Investigators did not report severe adverse events in the appendix like myocarditis. The solicited AEs for which they were checking became more severe. Nonetheless, these dose-dependent concepts dovetailed with potential long-term concerns that investigators had about the product (EUA 5–11, p.15). The investigators suggested that subclinical damages would aggregate over time through repeated doses and AEs would eventually manifest clinically in children. With negligible risk to children from COVID, AEs from product use posed more risk than the disease itself.

There was an explanation for the addition of pericarditis and myocarditis in this EUA that was not present in PV (EUA 5–11, p. 13). There were two cases of pericarditis in the C4591001 study by the June 2021 cutoff date. One case was a 55-year-old male 28 days ("within 6 weeks," a standard from EUA 2020) after dose #2 (risk factor "dose number" in EUA 2020, p. 6; PV, p. 50; EUA 5–11, p. 46). Investigators deemed this adverse event unrelated to product in both PV and EUA 5–11 despite the factors identified by investigators that would suggest a relationship. The second case took place in an unblinded placebo, a male 16 years of age (risk factor "young male" in PV, p.50) that developed myopericarditis two days after dose #2. After two months of symptoms, his cardiologist still recommended "limited activity" (EUA 5–11, p.13). PV, in July 2021, denied product involvement even when faced with a known AE related to product use: "Two (2) serious adverse events [PT Pericarditis] were reported, both deemed not related to study treatment by the Investigator" (PV, p.47).

An admission that the AE was related to Pfizer's product finally emerged within the October 2021 EUA 5–11. The product resulted in an unresolved condition at the last follow-up. In this case, "The investigator concluded that the there [sic] was a reasonable possibility that the myopericarditis was related to vaccine administration due to the plausible temporal relationship. FDA agrees with this assessment" (EUA 5–11, p. 13).

Investigators attempted to explain away a known AE risk in PV, got caught, and were forced to amend the report for the EUA 5–11. There was a discrepancy here. The structure of these documents suggested this 16-year-old patient's side effect was important to product risk labeling. Yet, when he was identified in the PV document as unrelated, myocarditis and pericarditis were already identified as important risk factors. It begets the question whether critical evaluation was taking place. Further information from the investigators is needed as this issue is not explained clearly in the provided documents.

Investigators should have been suspicious of product involvement with AEs per their own standard from EUA 2020, yet they continued the denial of product involvement with AEs through 5.3.6 and PV despite relevant factors learned along the way. Only in EUA 5–11 did they finally admit the product could have been related to the 16-year-old's AE. They never admitted the potential for product involvement in the 55-year-old male's AE despite relevant factors involved that they identified.

Pericarditis and myocarditis were added as label warnings based on this one case above from C4591001 and based upon VAERS reports (EUA 5–11, pp. 13–14). (PV notes "Important Identified Risk 'Myocarditis and Pericarditis'" on page 8 sourced from Pfizer Safety Database). Investigators finally acknowledged the risk of myocarditis and pericarditis from product use by the October 2021 in EUA 5–11. What finally forced this acknowledgement? Was it because the side effects took place in young males and would be more difficult to explain away than other side effects? A thorough explanation from investigators is required to eliminate this suspicion, especially after the age bracket issues identified in 5.3.6 with young patients ages 18–39.

Myocarditis and pericarditis adverse events were on scale with other AEs reported by the field in 5.3.6, yet Pfizer ignored or dismissed those additional AEs. "Review of passive surveillance AE reports and the Sponsor's periodic safety reports did not indicate any new safety concerns." They continue digging, "No unusual frequency, clusters, or other trends for AEs were identified that would suggest a new safety concern, including among the reports described as involving children 5–11 years of age" (EUA 5–11, p. 14).

The EUA investigators posed a serious set of facts revolving around pericarditis and myocarditis. The Food and Drug Administration (FDA) analysis from Optum healthcare claims database estimated incident rates in ages 16 to 17 of 200 cases per million (0.02%) and in 12- to 15-year-old of 180 cases per million (0.018%) (EUA 5–11, p.15). These rates of adverse events occurred at a similar rate as the AEs of MI, CV, appendicitis, and Bell's palsy in EUA 2020 (pp. 37, 40). Investigators suspected that the damage was more significant than the rates above**: "Information is not yet available about potential long-term sequelae and outcomes in affected individuals, or whether the vaccine might be associated initially with subclinical myocarditis (and if so, what are the long-term sequelae)."** (Bold Added, EUA 5–11, p. 15). This statement was the first time among documents reviewed that the authors turned to **long-term questions of adverse events**. Investigators went further: "A mechanism of action by which the vaccine could cause myocarditis and pericarditis has not been established." This unknown mechanism should have been a serious concern overall in light of the variety of AEs observed and in light of animal studies showing the spread of product throughout the body. Pfizer may not have known the exact cellular mechanism linking its product to AEs. However, the company should have been able to piece together that systemic spread of product caused damage across all organ systems in a dose-response relationship in at least the short term and potentially also in the long term. It suspected subclinical damages would affect patients on a significant delay. What is yet to be learned about males ages 18–39? The compilation of this set of safety concerns should have been a full-stop event for Pfizer. The constellation of evidence indicated Pfizer knew it did not have a favorable risk-benefit ratio as investigators identified significant product issues that would cause more damage than the disease itself.

EUA 5–11—Risk-Benefit Analysis

Investigators are honest regarding the minimal risks of COVID to the 5–11 age group. Authors note on page 7 (EUA 5–11) the reality that 15% to 50% of patients are asymptomatic even when they have COVID. They recover within one to two weeks and have milder symptoms than adults. By the time EUA 5–11 was published, there were 44 million identified cases of COVID in the United States with 722,000 deaths (EUA 5–11, p.7). About 8.7% (3.8 million) of cases were in the 5 to 11 age group. A rational assumption was that many more asymptomatic cases were never diagnosed and did not factor in the rates of AEs from COVID. Among the millions of known COVID cases, there were 4,300 hospitalizations and 146 deaths total included in the EUA 5–11 data. The risk of hospitalization and/or death was negligible for the 5-to-11 age group.

These statistics did not support vaccination in this cohort outright because the risk was nearly zero. The benefits would have been imperceptible as so few young children were affected by significant disease. **Even a vaccine with rare risks posed as much risk or more risk than the disease itself.** Here was what the authors wrote on page 37 (EUA 5–11): "While no cases of severe COVID-19 were accrued during study follow-up to date, it is highly likely that vaccine effectiveness against severe COVID-19 among children 5–11 years of age will be **even higher** than vaccine effectiveness against non-severe COVID-19, as is the case in adults." (Bold Added.) This conclusion was incoherent. **The data set for C4591007 cannot support this claim since there were no severe disease occurrences** (EUA 5–11, p. 26). It was a hopeful speculation. Investigators doubled down on page 38 (EUA 5–11), noting that "widespread deployment" will "have substantial effect on COVID-19 associated morbidity and mortality in this age group [5–11 years]." Their lab values did not support this claim per their own words (p. 17, EUA 5–11). Their own statistics on epidemiology refuted this statement, too. "Widespread" cannot be applied to events that rarely occur. They shared no data from C4591007 in this EUA related to transmission. Their conclusion was wrong because it was unsubstantiated in every respect.

Investigators clearly understood that COVID-19 was tolerated well in the young, and they would have understood that reality was a barrier to product deployment. Their solution was to discuss disease transmission as a new concept in EUA 5–11 (p. 8). Transmission was not discussed in the original EUA in 2020, 5.3.6, or the PV document, yet it emerged in this document. By page 9 (EUA 5–11), they argued dangers posed to adults by transmission from children. Ironically, adults were already approved and could have this allegedly highly efficacious product. Transmission from children should be of no concern to vaccinated adults if Pfizer showed the product works. Investigators went a step further to blame transmission of virus on individuals who are not vaccinated. Again, if the product works, there is limited risk to the vaccinated from the unvaccinated. Pfizer did not provide evidence from C4591007 that the product halted transmission or that unvaccinated individuals caused more transmission. Nonetheless, investigators created an argument that tried to have it both ways. The product supposedly worked well enough to have high levels of protection for adults yet did not work well enough to offer protection around children.

On page 38 (EUA 5–11), investigators documented important "Data Gaps." Investigators listed "Vaccine effectiveness against asymptomatic infection" and "Vaccine effectiveness against transmission of SARS-CoV-2" as gaps in their knowledge. The investigators, after a section where they argued the need for widespread vaccination in children and declared their product could greatly reduce symptoms and greatly reduce transmission, **listed their own conclusions as gaps in their knowledge** (EUA 5–11, p. 38). This section highlighted Pfizer's use of hopeful speculation over data to push for product approval. There can be room to speculate about potential benefits in scholarly work, but the investigators had no data to support their speculations. They had a very limited efficacy statistic from C4591007 and lab titers that they knew did not equate to disease protection. The investigators attempted to jump from two weak data points into a full-throated claim that the product would substantially reduce morbidity, mortality, and transmission. Even under the assumption the product did those things, the investigators never showed that it achieved any of those goals.

The above gaps in benefits were then overlaid with the risks posed to children. On page 38 (EUA 5–11), ". . . the risk of vaccine associated myocarditis/pericarditis among children 5–11 years of age is unknown at this time." The investigators' statement was technically true, but they could have estimated a risk of 0.02% based on myocarditis risks in ages 12–17 (EUA 5–11, p. 15). Based on this statement and the gaps in benefits, the investigators could not have made objective claims that there was a favorable risk-benefit ratio. They admitted openly that they did not know the benefits or the risks. Investigators wanted the public to believe a disease with limited risk to children (4,300 total hospitalizations and 146 total deaths reported in EUA 5–11) justified the widespread use of a product with unsubstantiated efficacy and with safety concerns that they would have known rivaled or exceeded the damage of the disease itself.

After investigators argued a case that should have denied the product approval, investigators turned to computer modeling and showed it definitely should not have been approved. Per the investigators (EUA 5–11, p. 46), for one million vaccinated children during a six-month period, the product would prevent an estimated 45,000 (4.5%) cases, reduce 200 hospitalizations (0.02%), reduce 60 to 80 ICU stays (0.0006%), and prevent zero or one death (0–0.0001%). After vaccinating one million children, a vast majority would have received no benefit. The model factored in risk of myocarditis. Investigators expected about 100 cases of myocarditis (0.01%), about 100 hospitalizations (0.01%), about 30 ICU admissions (0.003%), and zero deaths. The investigators demonstrated in their model extremely limited benefit, in the vicinity of zero percent, and they demonstrated risks on the scale of benefits. Their model did not predict a favorable risk-benefit ratio. It showed it would require tremendous numbers of vaccinations to deter a few COVID hospitalizations. If investigators factored in the numerous other AE risks from 5.3.6, this risk-benefit assessment would have rapidly degraded into the inevitable conclusion that the product risks outweighed any negligible benefits.

EUA 5-11—Conclusion Summary Statement

By the completion of the EUA 5–11, investigators still had efficacy shortcomings. Nearly a year into product use, the public should have heard about the successes of the C4591001 study in motion, yet that was not the case. Unbeknownst to the public, C4591001 was effectively destroyed by Pfizer, negating the ability to derive long-term data. The statistics from C4591007 were likewise weak. Investigators began discussions on boosters, another sign they had weak or absent efficacy. Investigators showed higher doses and cumulative doses contributed to adverse events yet refused to acknowledge risks accumulated in EUA 2020, 5.3.6, and PV. **They concluded a favorable risk-benefit ratio, yet demonstrated it was unfavorable.** Investigators introduced transmission as a reason to vaccinate and blamed unvaccinated individuals for transmission. They had no evidence from C4591007 to support either conclusion.

Questions That Need Answers

- What are the results from C4591001 and other ongoing trials?
- What process determined which adverse events were considered legitimate and which adverse events were not? The standard was not clear in Pfizer's documents. There were inconsistencies in the standards that require explanation by the investigators. The investigators are confident the product is safe. Do ongoing clinical trials support safety? Are field reports in conflict with the clinical trials? If so, reconciliation by investigators is needed.
- The criticisms in this report could be dispelled by strong efficacy and safety measures in the clinical trials. What caused Pfizer to destroy its own clinical trial, C4591001?
- Why did transmission enter in the EUA 5–11 when it was not discussed previously? Was it meant to make the case to vaccinate a population that did not have a practical benefit?
- What did Pfizer know about the profile of adverse events in males ages 18 to 39?

Commentary on the Advisory Committees and the EUAs

In pharmacovigilance, an important step before approval of a new drug is the advisory committee review process. According to the Centers for Disease Control and Prevention (CDC), "Safety is a Priority During Vaccine Development and Approval. Before vaccines are licensed by the FDA, they are tested extensively in the laboratory and with human subjects to ensure their safety." (https://www.cdc.gov/vaccinesafety/ensuring-safety/history/index.html) The Advisory Committee on Immunization Practices (ACIP) is the CDC's advisory committee recommending vaccines. VRBPAC is the FDA's vaccine/biologic products advisory board and is part of the Center for Biologics Evaluation and Research (CBER). VRBPAC ". . . reviews and evaluates data concerning the safety, effectiveness, and appropriate use of vaccines and related biological products which are intended for use in the prevention, treatment, or diagnosis of human diseases . . . " (https://www.fda.gov/advisory-committees/blood-vaccines-and-other-biologics/vaccines-and-related-biological-products-advisory-committee) These committees effectively had two chances to address product issues before the FDA EUA-approved and the CDC publicly recommended Pfizer's mRNA COVID vaccine. It does not appear that the committees did their due diligence. A report on the failures of pharmacovigilance within this these committees is planned as upcoming work in the larger WarRoom/DailyClout Pfizer Documents Analysis project.

Conclusion

Efficacy of the BNT162b2 mRNA COVID "vaccine" was not demonstrated by Pfizer during 2020 and 2021. If investigators were pleased with results after six weeks, they could have continued every six weeks with interim reports which could have rolled into 5.3.6, PV, and EUA 5–11.

Pfizer's declination to continue its own clinical trial by vaccinating placebo participants is a significant problem. There is not an intact clinical trial to show high drug efficacy over time. Maybe there is a good explanation? If so, Pfizer needs to share it, especially with C4591001 ruined and many other studies terminated. Is it possible Pfizer recognized its trial was going to produce unfavorable results and ended it before those results became more obvious? Pfizer would be unable to defend itself using C4591001, especially because it negated the clinical trial by vaccinating the entire placebo cohort in March 2021. The lack of interim trial results, the destruction of C4591001, the shift to antibody titers to try to prove effectiveness, and the addition of hopeful speculation in clinical trial documents indicate problems with BNT162b2's efficacy.

Safety was not demonstrated by Pfizer. The Company understood its product spread throughout the body, witnessed AEs across all organ systems, witnessed immediate latency, and witnessed dose-response relationships which also caused investigators to speculate about long-term AEs. None of that indicated safety. Taken together, Pfizer, based on its own written standards and its own reports, should have understood its product had significant risks and limited, if any, benefits.

Appendix 1: Study Due Dates from PV Document

Study Number	Population	PV Due Dates	Results Posted? Clinical Trials Notables
C4591001 (C)	EUA study	Final: 8–31-2023	No results available Completed: 2–10-2023
C4591001 (A)	Ages 12–15	First Report one-month of two dose: 4–30-2021 Six-Month: 10–31-2021 (report due immediately after EUA 5–11). Two Year: 4–30-2023	No results available
C4591007 (A)	Ages under 12	First report with up to one-month post-dose: 9–30-2021 Interims: Six Month: 3–31-2022 Two Year: 9–30-2023	No results available Pending Completion: 10–3-2023
C4591008**	US Healthcare Workers	Interims: 6–30-2021 12–31-2021 6–30-2022 12–31-2022 Final: 12–31-2023	Not found on clinicaltrials.gov
C4591009**	US population	Interim: 10–31-2023 Final: 10–31-2025	Not found on clinicaltrials.gov
C4591009**	Ages 5–12	First Report: 9–30-2021 Six Month: 3–31-2022 Two Year: 9–30-2023	Not found on clinicaltrials.gov
C4591011**	US Military	Interims: 10–31-2022 6–30-2022 12–31-2022 Final: 12–31-2023	Not found on clinicaltrials.gov

(Continued on next page)

Study Number	Population	PV Due Dates	Results Posted? Clinical Trials Notables
C4591012**	VA System	Interims: 6–30-2021 12–31-2021 6–30-2022 12–31-2022 Final: 12–31-2023	Not found on clinicaltrials.gov
C4591014 (R)	Efficacy by Kaiser Permanente	Final submission: 6–30-2023	No results available Pending completion: 3–25-2024
C4591015 (C)	Pregnant Women	Primary Endpoints: 4–30-2023	No results available Completed 7–15-2022
C4591022 **	Pregnancy, Infant Outcomes	Interims: 1–31-2022 1–31-2023 1–31-2024 1–31-2025 Final: 12–1-2025	Not found on clinicaltrials.gov
W1235284	Comparison to other respiratory diseases	Final submission: 6–30-2023	Not found on clinicaltrials.gov
W1255886	Lower Respiratory Study	Final submission: 6–30-2023	Not found on clinicaltrials.gov
BNT-162- 01 (A)	Cohort 13	First submission: 9–30-2021	Results not available. Results submitted 4–11-2023 for review.

Legend:
**: When searched on http://clinicaltrails.gov, these studies are "Not Found" and the site redirects to C4591001
(A): Active (R): Recruiting (C): Completed
C4591001 is a composite of 7 different studies listed as (A) or (C)
CSR: Clinical Study Report
Endpoints: The principal outcomes that are measured in a clinical trial.

Twenty: "mRNA COVID "Vaccines" Have Created a New Class of Multi-Organ/System Disease: "CoVax Disease." Children from Conception on Suffer Its Devastating Effects.
—Histopathology Series—Part 4d"—Robert W. Chandler, MD, MBA

Summary

Pfizer Report 70 (https://dailyclout.io/report-70-part-4c-clinical-evidence-supporting-the-hypothesis-that-autoimmunity-is-one-of-the-principal-pathological-mechanisms-of-harm-from-covid-19-gene-therapy-drugs/) introduced **CoVax Disease, a new class of illness that has arisen from the use of mRNA COVID vaccines**. CoVax Disease encompasses multiple pathological mechanisms, presents with multi-organ/system involvement, and affects all age groups. This article will focus on the very negative medical impacts on children's health after receiving mRNA vaccines. Additionally, sex-related effects will be presented.

COVID-19 was a disease primarily of older people and those with co-morbidities. Unfortunately, the experimental gene therapy products have devastating effects in young, healthy people of both sexes and all ages from conception upward through the age brackets.

Young men have more myo/pericardial disease and more fatalities. Females have more disability and more adverse events overall. As the Cho et al. and Barmada et al. studies document, long-term cardiac disease is one outcome from these drugs.

Disease categories associated with gene therapy products are diverse, are unusual in many cases, and can be unusually severe as in MIS-C, sudden death/cardiac arrest, stroke, and turbo cancer, as illustrated by the intracardiac epithelioid sarcoma case reviewed earlier in this report. Children are not exempt from these illnesses.

Causation is a complicated topic. However, there is ample support presented herein for concluding that some or many of the medical conditions that arise following an injection of C19 gene therapy products were caused by them.

I. Introduction

Since the early medical papers emerging from China in January 2020, it was apparent that the SARS-CoV-2 (SC2) virus preferentially produced disease, COVID-19 (C19), in the elderly with comorbidities. (doi:10.1001/jama.2020.1585) Some younger people with comorbidities also were at risk, but the risk for severe disease and death in the general population was low. (https://doi.org/10.1016/S0140–6736(20)30211–7)

Lipid-nanoparticle coated mRNA drugs (LNP/mRNA) were developed to prevent C19 disease. These drugs along with the adenovirus-vectored products will be referred to collectively as gene therapy products (GTPs). Pfizer Confidential 5.3.6 (https://www.phmpt.org/wp-content/uploads/2022/04/reissue_5.3.6-post-marketing-experience.pdf) documented COVID-19 as an Adverse Event following inoculation with Pfizer's

BNT162b2 gene therapy product as well as over 140,000 additional adverse events reported in the first ten to twelve weeks. (https://dailyclout.io/pfizer-evidence-so-far-coverups-heart-damage-and-more/)

The object of this paper is to review Adverse Event reports following LNP/mRNA gene therapy treatments pertinent to differential age and sex reporting with specific coverage of the zero- to 18-year-old demographic.

Specific clinical features of medical conditions afflicting America's youth following injection of Spike-mediated genetic treatment will be discussed. The article will conclude by reviewing the time course of Vaccine Adverse Event Reporting System (VAERS) event reporting relative to United States (U.S.) dosing data. VAERS reports are for the U.S. and its Territories unless otherwise specified.

The Centers for Disease Control and Prevention (CDC) and the Food and Drug Administration (FDA) maintain VAERS, which is a public resource. (https://wonder.cdc.gov/controller/datarequest/D8) Open VAERS harvests data from VAERS in specific categories and is useful for summary data. (https://openvaers.com/covid-data)

In addition to VAERS and Open VAERS, this article will make use of case and small series reports from the medical literature, the FDA-released Pfizer Confidential Documents archive (https://phmpt.org/pfizer-16-plus-documents/), and reporting from individuals including William Makis, MD, Professor Mark Crispin Miller, Ed Dowd and his partners, Jessica Rose, and others.

As a caution, the limitations of VAERS are spelled out on the CDC website:

> VAERS accepts reports of adverse events and reactions that occur following vaccination. Healthcare providers, vaccine manufacturers, and the public can submit reports to VAERS. While very important in monitoring vaccine safety, VAERS reports alone cannot be used to determine if a vaccine caused or contributed to an adverse event or illness. The reports may contain information that is incomplete, inaccurate, coincidental, or unverifiable. Most reports to VAERS are voluntary, which means they are subject to biases. This creates specific limitations on how the data can be used scientifically. Data from VAERS reports should always be interpreted with these limitations in mind. 1
>
> The strengths of VAERS are that it is national in scope and can quickly provide an early warning of a safety problem with a vaccine. As part of CDC and FDA's multi-system approach to post-licensure vaccine safety monitoring, VAERS is designed to rapidly detect unusual or unexpected patterns of adverse events, also known as "safety signals." If a safety signal is found in VAERS, further studies can be done in safety systems such as the CDC's Vaccine Safety Datalink (VSD) or the Clinical Immunization Safety Assessment (CISA) project. These systems do not have the same limitations as VAERS, and can better assess health risks and possible connections between adverse events and a vaccine. 2

Reports are not detailed enough for many determinations (1), but the utility of the system is as an **early warning system** (2). Much of the VAERS system activity is invisible to the user, such as what activity occurs from the time a report is filed with the CDC to when it gets posted and after its initial posting. The numbers that are given in response to queries change over time as the system is updated at least every week. Follow-up data are obtained but not posted.

Events vary in detail from diagnosis only to detailed reporting of medical data. Some categories return cryptic data like "No Adverse Event," of which there were 32,072 reports for C19 products from 2020 through mid-June 2023 for all ages with 1,003 associated events including 13 deaths.

Messages:
- VAERS data in CDC WONDER are updated every Friday. Hence, results for the same query can change from week to week.
- These results are for 905 total events.
- Rows with zero Events Reported are hidden. Use Quick Options above to show zero rows.

Symptoms	Event Category	Events Reported	Percent (of 905)
NO ADVERSE EVENT	Death	13	1.44%
	Life Threatening	1	0.11%
	Permanent Disability	1	0.11%
	Hospitalized	9	0.99%
	Existing Hospitalization Prolonged	2	0.22%
	Emergency Room / Office Visit **	1	0.11%
	Emergency Room *	127	14.03%
	Office Visit *	849	93.81%
	Total	**1,003**	**110.83%**
Total		**1,003**	**110.83%**

Here is the information of one fatal case so designated. VAERS ID 1205421 was a 61-year-old male who died less than two days after receiving his second dose of mRNA1273 (Moderna). Someone made the effort to file the report but with no details about this man's death.

Adverse Event Description
On April 8, 2021 patient received his second dose of Moderna COVID-19 vaccine at pharmacy at 1:08pm. Patient waited the appropriate 15 minutes, and then left pharmacy. He reported no adverse reactions to our staff during that time, and did not call afterward to report any adverse reactions. At approximately 4:30pm on April 10, 2021, I received notification that patient was found DOA at his residence. No other information is available at this time.

VAERS numbers are generally considered to represent only a fraction of the actual prevalence of the medical conditions reported and for this reason attempts have been made to estimate the true number by estimating what is called the Under Reporting Factor (URF). True prevalence data are hard to find.

Steve Kirsch, Jessica Rose, and Mathew Crawford performed detailed calculations of URF concluding that an **URF of 41** was justified. (https://www.skirsch.com/covid/Deaths.pdf, https://jessicar.substack.com/p/the-under-reporting-factor-in-vaers)

In summary, others estimated the URF of:

Reference	URF
Aaron Siri using anaphylaxis	50
Rose using serious adverse events	30
Vaersanalysis using CMS data	44.6

So our hypothesis is that 41X is a safe, conservative factor useful for all types of events.

II. C19 Gene Therapy: Age and Sex Effects

A. BNT162b2 and mRNA1273 (LNP/mRNA) Approval Schedule

For reference, the following dates are noted:

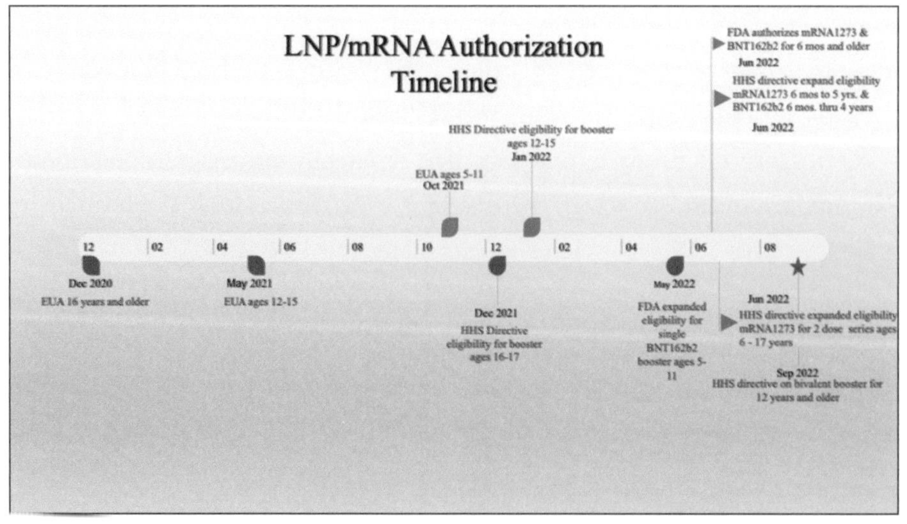

(https://www.cdc.gov/vaccines/covid-19/clinical-considerations/interim-considerations-us.html#table-01)

The staggered inoculation of age groups should be kept in mind when looking at data sets that combine age groups. This stagger adds an element of complexity when time series analyses are conducted across age groups.

Example: The age brackets in VAERS do not match the age brackets used for the "vaccine" dosing schedule. However, the six-month-old to 5-year-old group has bracketing close to the corresponding dosing brackets.

Five-year-olds became eligible to receive BNT162b2 on October 29, 2021, and both mRNA drugs were released for six-month-olds to four-year-olds for BNT162b2 and five-year-olds for mRNA1273 on June 18, 2022. The histogram below is a plot of adverse events according to the inoculation schedule showing a spike in event reports following the two authorization dates.

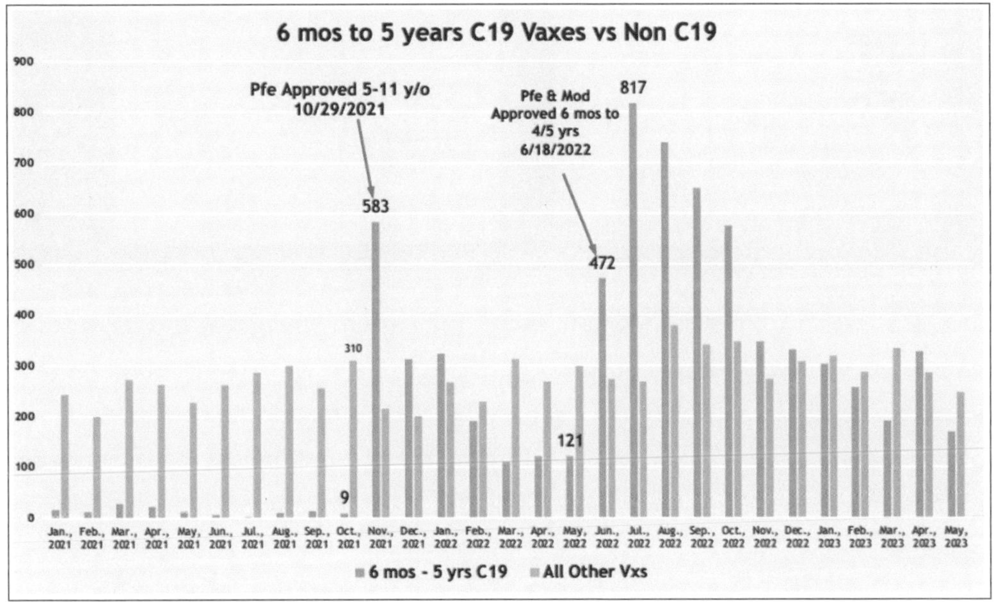

VAERS 6/11/2023

Almost immediately, adverse event reporting in VAERS jumped from nine events to 583 events following the October 21, 2021, authorization and from 121 events (full month before) to 817 events (full month afterward). June was a transition month.

B. VAERS Reports: Age and Sex Differences

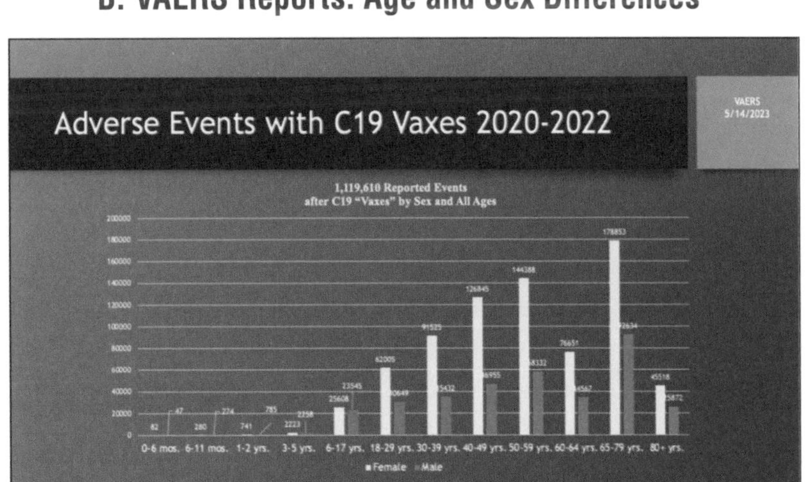

VAERS adverse event reporting was both age and sex dependent with frequency increasing with age (previous image). Dosing by age bracket or preferably by year age would be helpful if it were available.

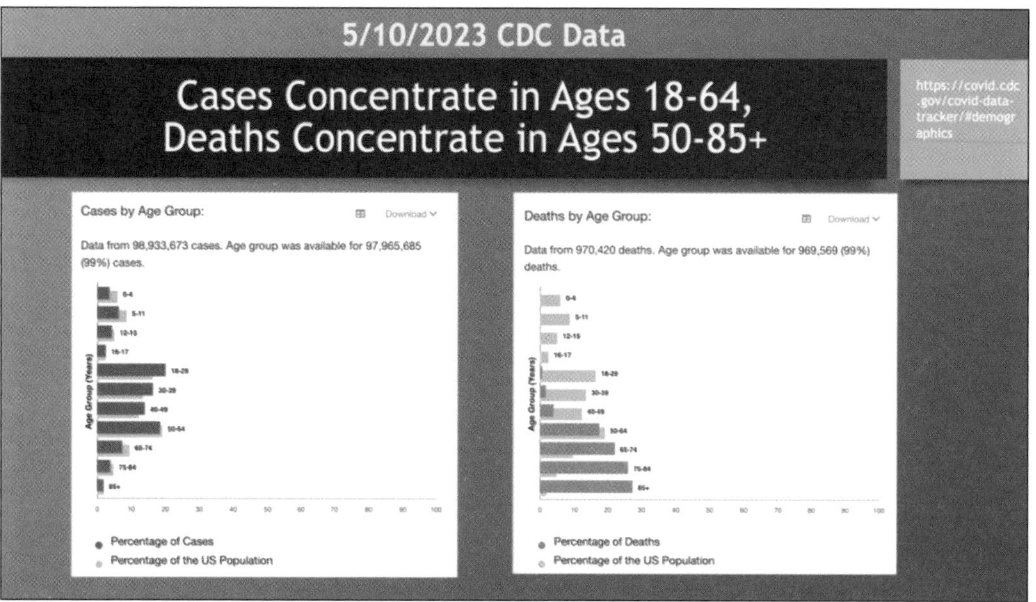

Looking at C19 statistics from the CDC, an interesting pattern emerges with C19 cases concentrated in ages 18 to 64 (46 years), while C19 deaths were more prevalent in the age 50 to 85+ bracket (~35 years), above. Like the early reports, these data identify the primary risk factor for mortality with C19 is age along with co-morbidity. The morbidity and mortality from C19 gene therapy products (GTPs) has more representation in the lower age brackets than C19.

C. VAERS Events: 29 Years of Age and Younger, Menarche

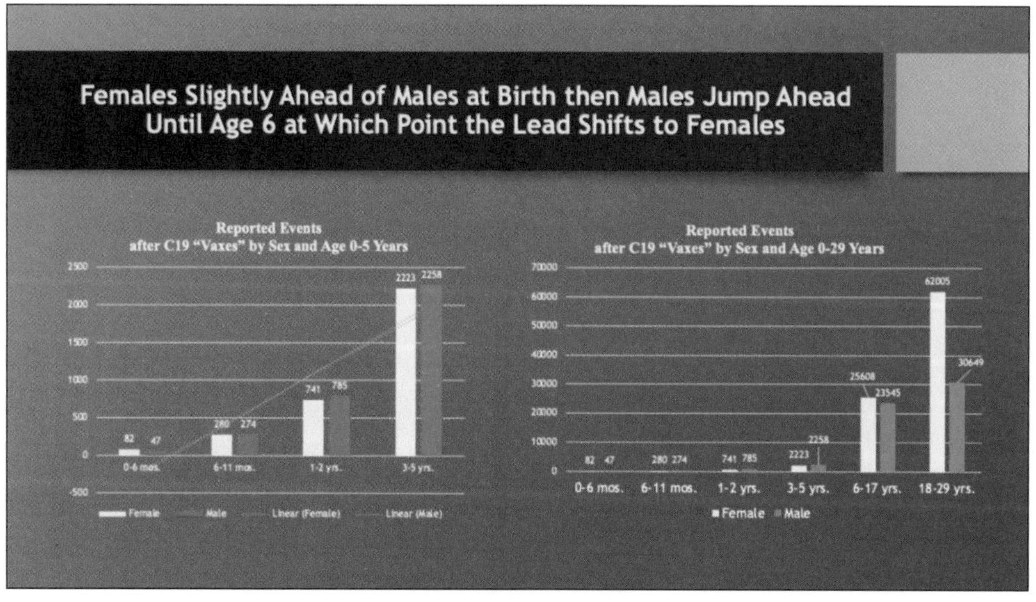

VAERS through 5/12/2023

Males and females have almost equal adverse event reporting from one to five years, at which age females take over and the lead never changes thereafter (previous image). There is a big increase in female predominance going from the six- to 17-year bracket to the 18- to 29-year bracket as the number of reports in females more than doubles. The age bracket from six to 17 years is too broad given the hormonal and physical changes occurring during this period. Age is reported in VAERS by age bracket although specific age by year data is recorded but is not directly retrievable using defined search terms.

A possible explanation for the female shift to a dominant position during adolescence and early adulthood may be related to female sexual development, specifically the onset of menarche. A CDC National Health Statistics Report from 2020 shows the age of onset of menarche follows the approximate time course of the emerging dominance of females in having reported adverse events from LNP/mRNA products. (https://www.cdc.gov/nchs/data/nhsr/nhsr146–508.pdf)

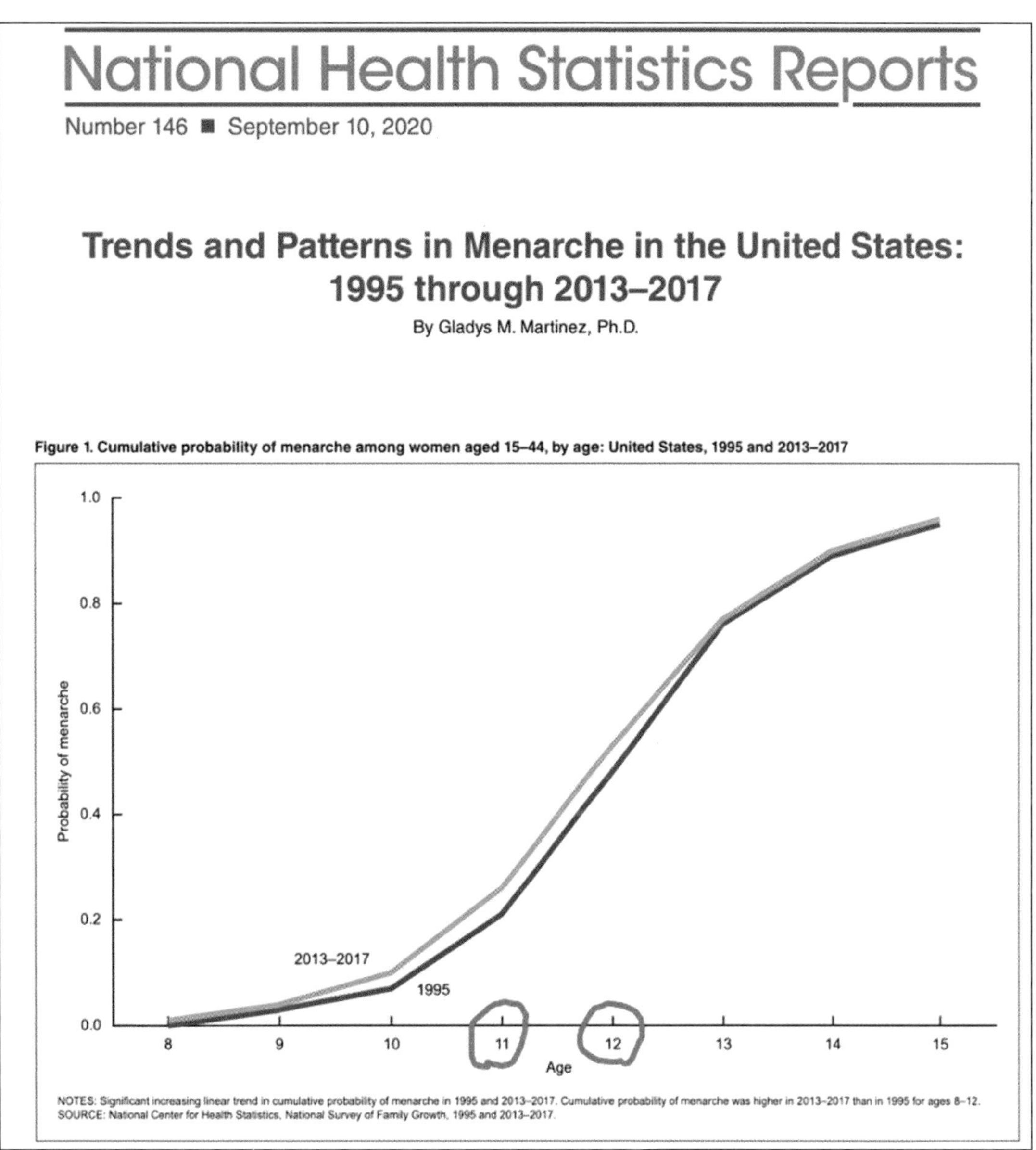

National Health Statistics Reports

Number 146 ■ September 10, 2020

Trends and Patterns in Menarche in the United States: 1995 through 2013–2017

By Gladys M. Martinez, Ph.D.

Figure 1. Cumulative probability of menarche among women aged 15–44, by age: United States, 1995 and 2013–2017

NOTES: Significant increasing linear trend in cumulative probability of menarche in 1995 and 2013–2017. Cumulative probability of menarche was higher in 2013–2017 than in 1995 for ages 8–12.
SOURCE: National Center for Health Statistics, National Survey of Family Growth, 1995 and 2013–2017.

The median age at menarche decreased from 1995 (12.1) to 2013–2017 (11.9). The cumulative probability of menarche at young ages was higher in 2013–2017 compared with 1995.

Ed Clark from the WarRoom/DailyClout Pfizer Document Analysis Project was able to retrieve age-by-year data from VAERS for December 2020 through June 2023 with no filtering by vaccine type or manufacturer. Hormonal changes for females, below left, are ramping up during the pre-teen years as females begin to dominate adverse event reporting as shown below on the right.

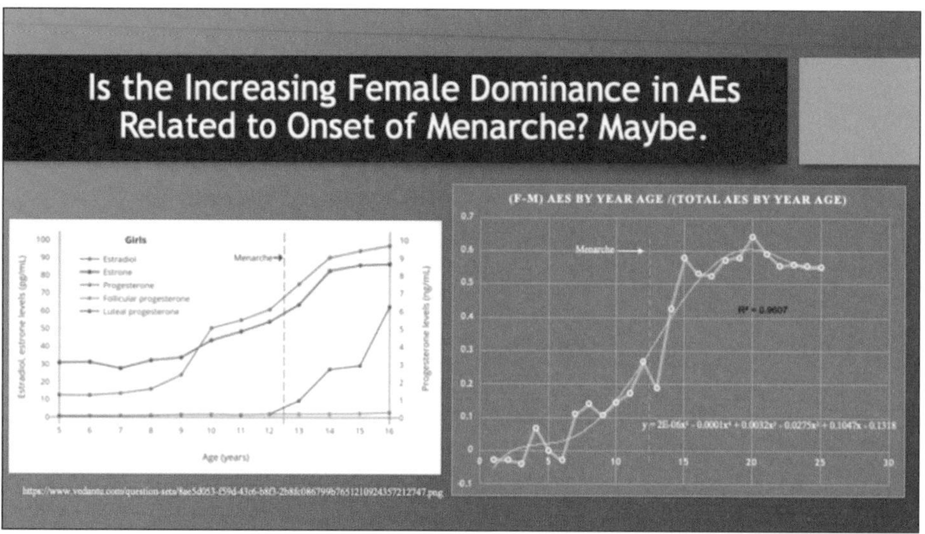

The polynomial function was used strictly to obtain the general trending of the data over the observed data range and should not be construed to predict outcomes beyond that range.

More granular data, age data by year rather than coarse age brackets, for the full data set in VAERS compared with quantitative hormonal changes might help further support or reject this hypothesis.

D. VAERS Events: Ages 40 and Above, Menopause

VAERS 5/12/2023

Women peak at 72% of adverse event reporting from ages 40 to 49 years, as menopause begins for many, and progressively drops down to 63% in the post-menopause age brackets.

Spontaneous or natural menopause is recognized retrospectively after 12 months of amenorrhea (previous image). It occurs at an average age of 52 years, but the age of natural menopause can vary widely from 40 to 58 years. (https://www.menopause.org/docs/default-source/2014/nams-recomm-for-clinical-care.pdf)

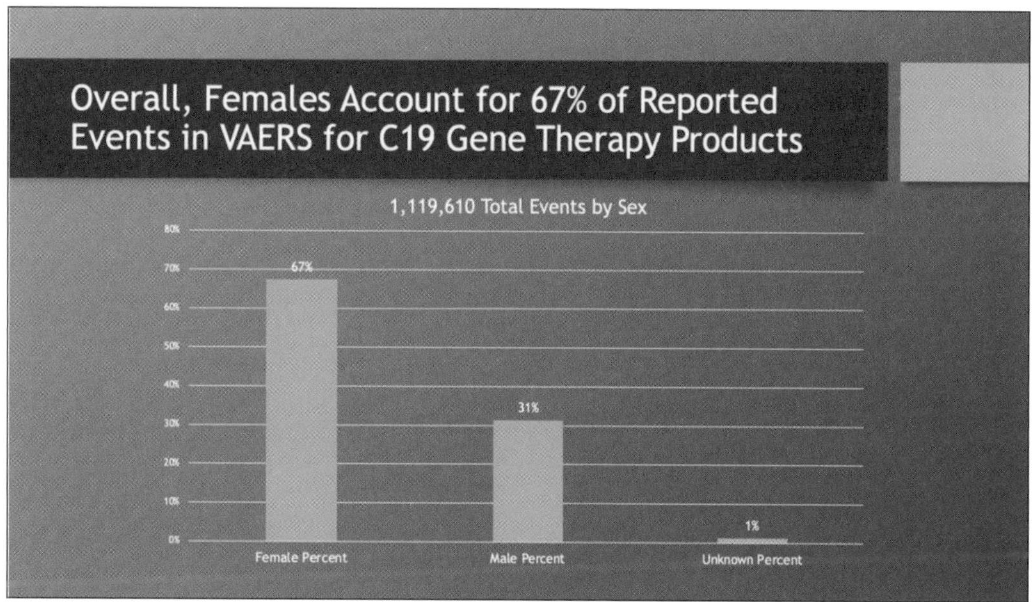

VAERS May 12, 2023

Overall, the women substantially dominate adverse event reporting in VAERS by more than a two-to-one margin.

E. VAERS Events: Disability after C19 Gene Therapy

Similar to adverse event reporting, women also have approximately a two-to-one lead in disability after adverse events after receiving C19 gene therapy products.

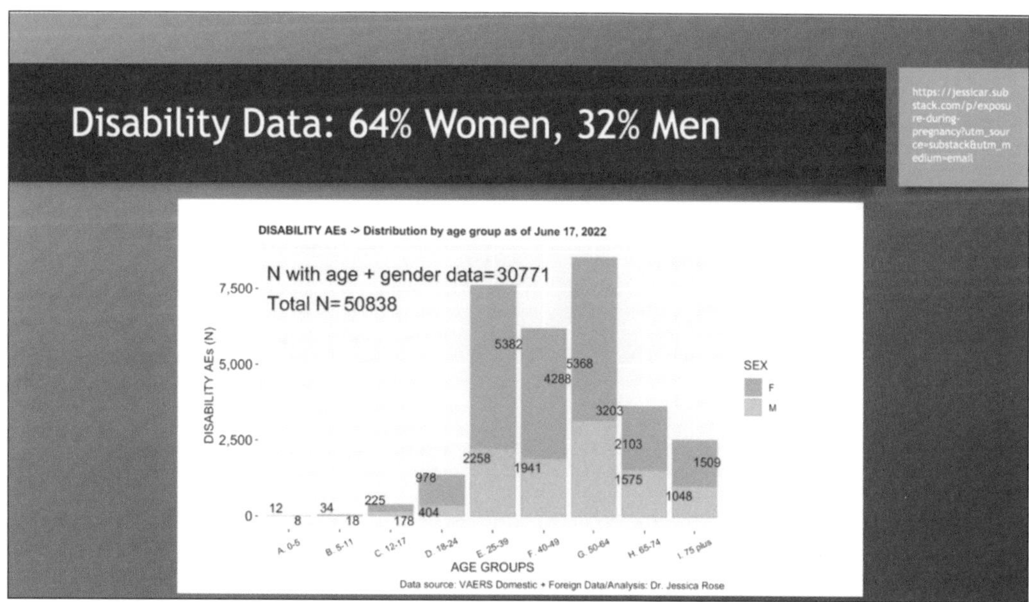

Jessica Rose prepared the chart above illustrating female preponderance in having disability from adverse events following C19 “vaccine” treatments across all age groups.

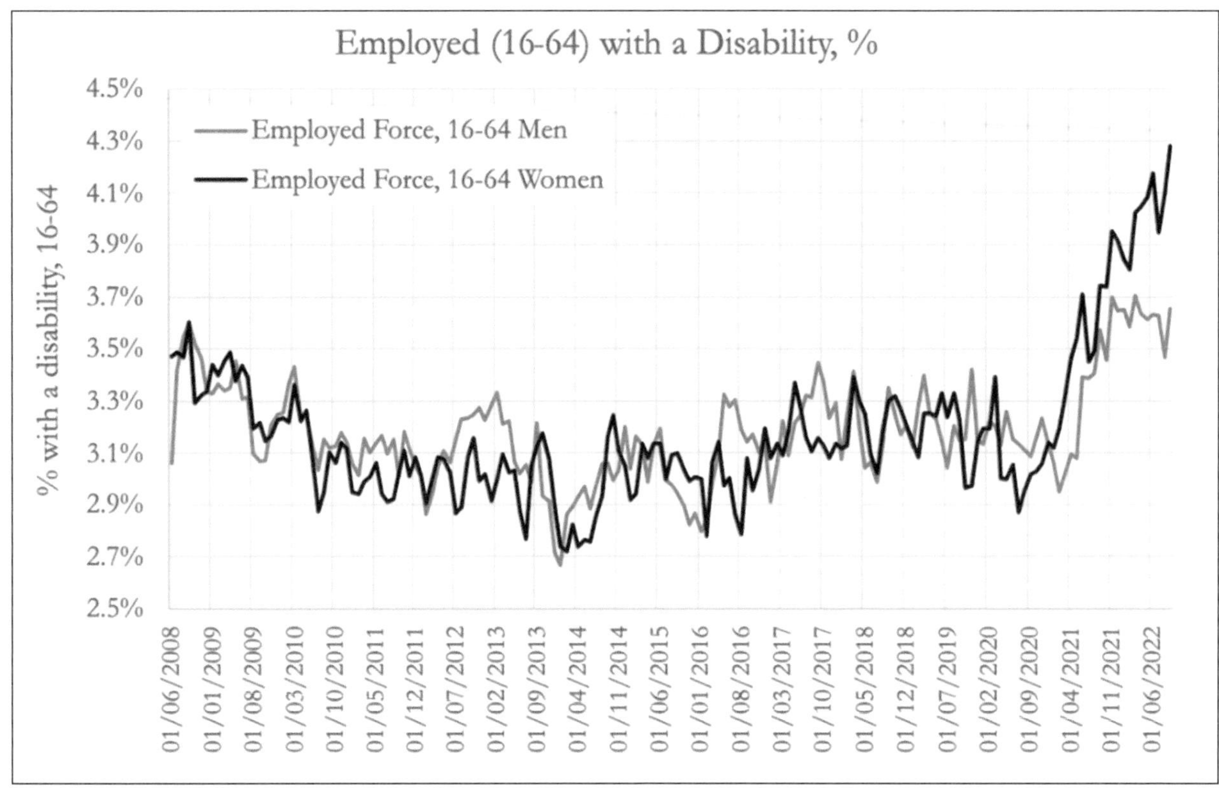

Ed Dowd’s group found similar pattern of women diverging sharply upward from men in disability event reporting in VAERS as the GTP program ramped up. (https://phinancetechnologies.com/HumanityProjects/US%20Disabilities%20-%20Part1.htm)

Below is a plot of Dr. Rose’s data showing men narrowing the gap in the 12- to 17-age bracket with respect to disability followed by female dominance of the statistic until the gap begins to narrow after age 50.

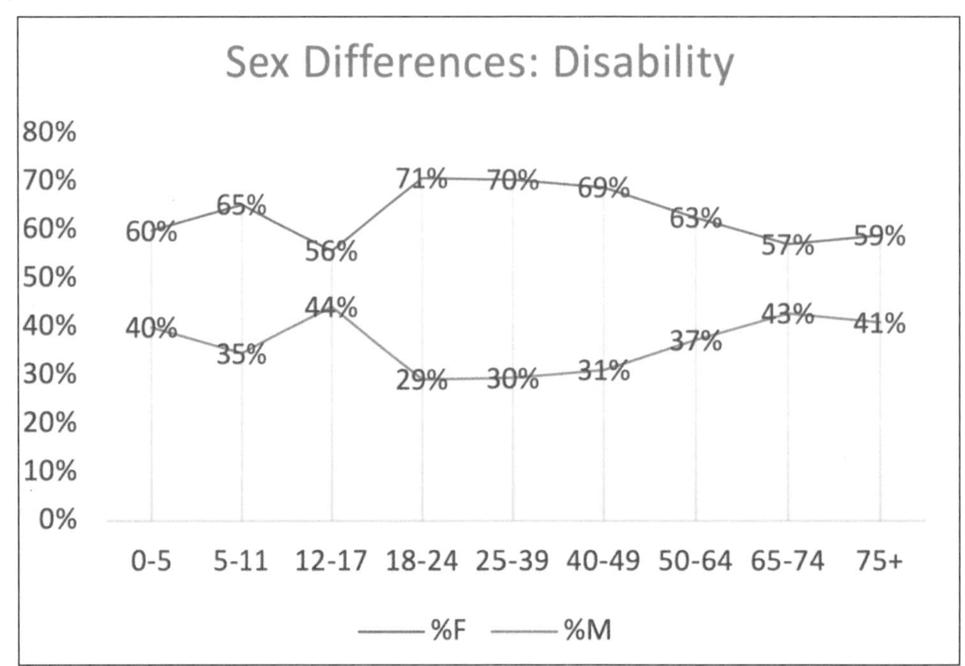

Females account for 64% of the disabled from adverse events in the entire data set of 30,532 disability event reports.

III. CoVax Disease and Youth: Conception Through Early Adulthood

This section will look at a number of significant medical problems following GTPs affecting youth beginning in utero.

A. In Utero, Miscarriage/Stillbirth

The recent release of Pfizer document, Appendix 2.2, reporting on the cumulative data collection through June 2022 and the interval of December 19, 2021, through June 18, 2022. (https://www.globalresearch.ca/wp-content/uploads/2023/05/pfizer-report.pdf) There were a total of 1,485,027 cases with 4,964,106 total adverse events.

Dr. Rose has pulled out the following data from this document. (https://open.substack.com/pub/jessicar/p/pfizer-appendix-22-document-compared)

	Total	Spontaneous Serious		
Reproductive system and breast disorders	**178353**			% Serious
		Interval	Cumulative	Cumulative
Amenorrhoea	12404	454	1143	9.2
Dysmenorrhoea	15319	920	3706	24.2
Heavy menstrual bleeding	30500	1707	6364	20.9
Menstrual disorder	24427	868	2360	9.7
Menstruation delayed	15101	600	2412	16.0
Menstruation irregular	16535	858	2813	17.0
Polymenorrhoea	10668	336	896	8.4
Vaginal hemorrhage	5034	324	1333	26.5

Female menstrual dysfunction was reported in 129,988 adverse events as of a year ago (06/18/2022 with 16% considered serious (below)). There were 35,534 reports of excessive bleeding/hemorrhage with 22% of the events considered serious.

Pfizer Appendix 2.2: After J. Rose

Uterine/Ovarian Dysfunction after BNT162b2
As of 06/18/2022

Dx	N =	% Serious Cum for DX	# Serious	% of Serious Cases
Vaginal hemorrhage	5,034	26.5%	1,334	6%
Dysmenorrhea	15,319	24.6%	3,768	18%
Heavy menstrual bleeding	30,500	20.9%	6,375	30%
Menstruation irregular	16,535	17.0%	2,811	13%
Menstruation delayed	15,101	16.0%	2,416	11%
Menstrual disorder	24,427	9.7%	2,369	11%
Amenorrhoea	12,404	9.2%	1,141	5%
Polymenorrhea	10,668	8.4%	896	4%
Total	129,988	16.2%	21,111	

Dx	N =	Serious	% Serious
Excessive Bleeding	35,534	7,709	22%

Interfering with normal ovarian/menstrual functions has consequences. Below is the data from OpenVAERS showing a spike in miscarriages and stillbirths as the LNP/mRNA dosing program ramped up.

The following table from OpenVAERS lists almost 5,000 miscarriages or stillbirths; 36,765 menstrual disorders; vaginal/uterine hemorrhage in 12,871; and over 1,000 fetal defects as of May 12, 2023.

58,022 Female Reproductive Events after C19 "Vaxes" (Open VAERS)

SYMPTOMS	CASES
Menstrual Disorders	36,765
Vaginal/Uterine Hemorrhage	12,871
Miscarriage/Stillbirth	4,995
Caesarian/Preterm Labour/Birth Difficulties/PreTerm	1,445
Fetal Defects/Fetal Cardiac Issues/Fetal Disorders	1,046
Pregnancy Difficulties	900

Applying the URF of 41 from Kirsch et al. gives the following estimates of the true numbers.

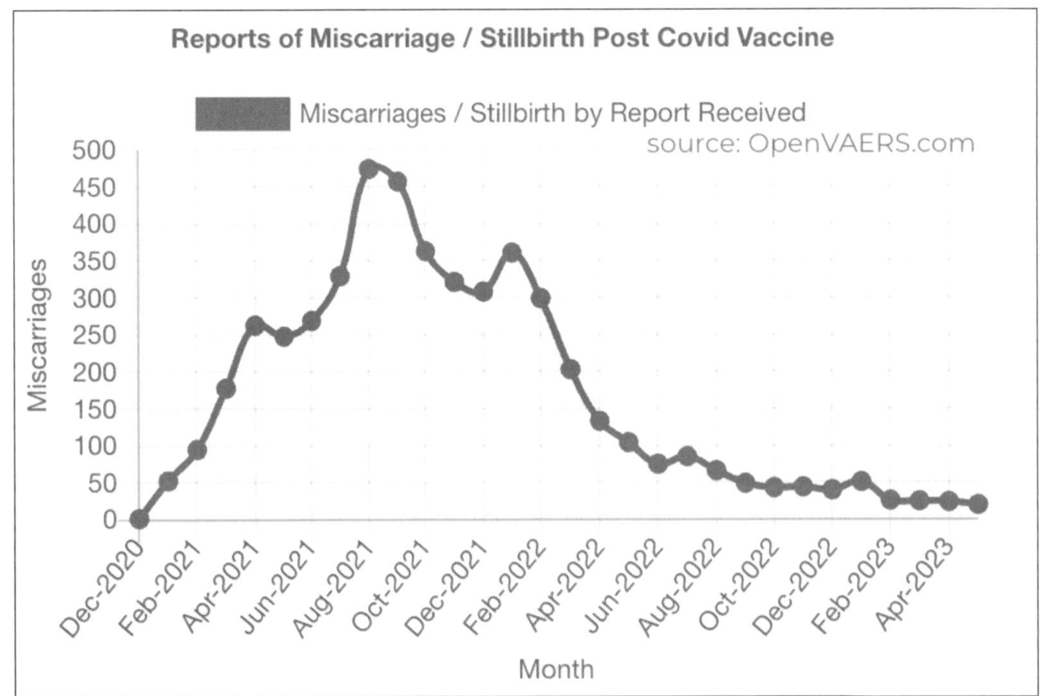

These are sobering numbers.

By comparison, it has been estimated that thalidomide caused 10,000 cases of birth defects in Europe from 1957 to 1961 before it was pulled from the market. Like spike-producing drugs, thalidomide was never tested in pregnant women and yet was aggressively marketed to them for morning sickness. (https://www.drugs.com/monograph/thalidomide.html)

Francis Kelsey of the FDA is credited with preventing use of thalidomide in the United States and, in doing so, ushered in the modern era at the FDA beginning in 1960 and ending in 2020, when the FDA/CDC shifted from protector to perpetrator. (https://www.fda.gov/about-fda/fda-history-exhibits/frances-oldham-kelsey-medical-reviewer-famous-averting-public-health-tragedy)

The miscarriage/stillbirth event reporting in VAERS shows a crescendo pattern as GTP population dosing ramped up in 2021 followed by a drop-off in 2022.

Symptom	N =	x URF = 41
Miscarriage/Stillbirths	4,995	204,795
Menstrual Disorders	36,765	1,507,365
Vaginal/Uterine Hemorrhage	12,871	527,711
Caesarean/Preterm/Premature	1,445	59,245
Fetal Defects/Fetal Cardiac/Fetal Disorder	1,046	42,886
Pregnancy Difficulties	900	36,900

What is the physiological connection between perinatal adverse events for mother and child and the C19 gene therapy drugs? Dr. Arne Burkhardt, MD, a recently deceased pathologist in Reutlingen, Germany, organized a 10-member team of pathologists, coroners, and scientists to study the histopathology of C19 vaccine organ and tissue damage in autopsy and biopsy specimens from about 130 subjects.

The Burkhardt Group's seminal work establishing the pathological basis of CoVax Disease has been presented in Parts 1 and 2 of this series. (https://dailyclout.io/report-56-autopsies-reveal-the-medical-atrocities-of-genetic-therapies-being-used-against-a-respiratory-virus/, https://dailyclout.io/report-58-part-2-autopsies-reveal-medical-atrocities-of-genetic-therapies-being-used-against-a-respiratory-virus/)

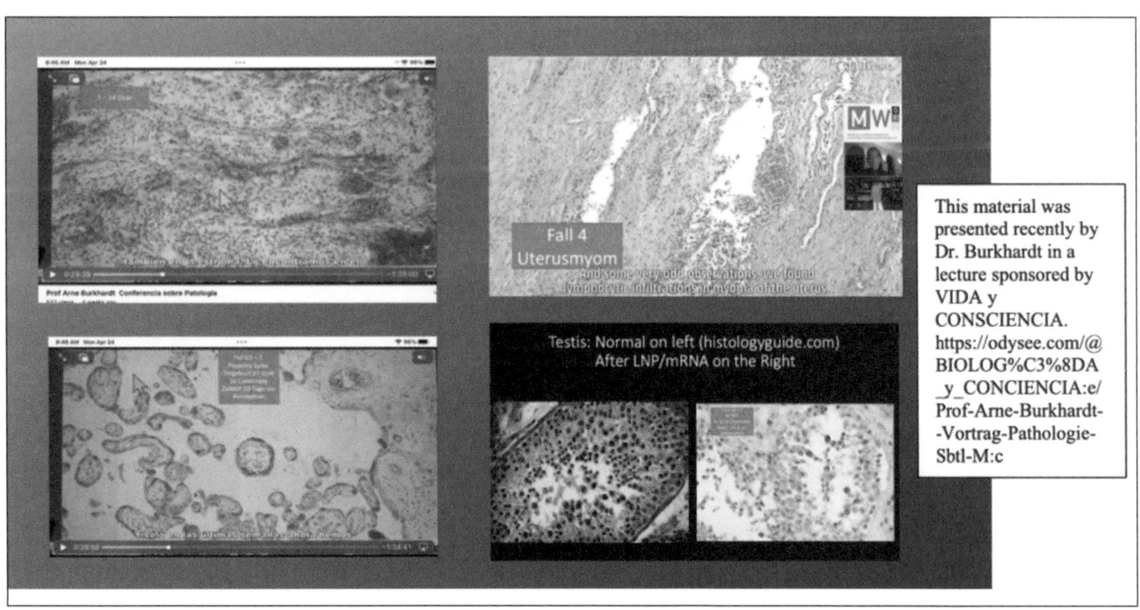

This material was presented recently by Dr. Burkhardt in a lecture sponsored by VIDA y CONSCIENCIA. https://odysee.com/@BIOLOG%C3%8DA_y_CONCIENCIA:e/Prof-Arne-Burkhardt--Vortrag-Pathologie-Sbtl-M:c

Dr. Burkhardt's group identified the histopathology of harms associated with C19 gene therapy products in **ovaries** positive staining for **spike protein** *Top Left*, **lymphocytic infiltration** in the **uterus** (endometrium) *Top Right*, **spike** protein in **placenta** *Bottom Left*, and **spermatozoa depletion** in **testes** *Bottom Right* (previous image). The mechanisms of these harms were diverse and varied from vasculitis, protein deposition, inflammation, necrosis, and neoplasia.

Not surprisingly, GTPs have been linked to a decline in live births around the globe.

"Nine Months Post-COVID mRNA 'Vaccine' Rollout, Substantial Birth Rate Drops in 13 European Countries, England/Wales, Australia, and Taiwan." (https://dailyclout.io/report-52-nine-months-post-covid-mrna-vaccine-rollout-substantial-birth-rate-drops/)

B. Neonatal/Breast Milk

Pfizer Confidential Document 5.3.6 reporting on the first 10 weeks of widespread use of BNT162b2—up to February 28, 2021, in the U.S. and 12 weeks in the United Kingdom—identified a problem with nursing mothers who received BNT162b2 while breastfeeding:

Serious Foetus/Baby Cases	4	
Fetal growth restriction/premature	2 each	
Neonatal death	1	
Breast Feeding Infants	133	
No adverse events	***116***	***87%***
Breast feeding infant child reactions	17	13%

Keep in mind that follow-up for the "No Adverse Events" is not provided by the CDC, thus calling into question the accuracy of these numbers.

There were four serious fetus baby cases and one neonatal death. Thirteen percent of the 133 breastfeeding infants had 19 different reactions. (https://dailyclout.io/pfizer-evidence-so-far-coverups-heart-damage-and-more/)

Meanwhile, breastfeeding mothers experienced the following:

Breast feeding mother cases	6
Chills, malaise, and pyrexia	1
Suppressed lactation	4
Unknown AE	1
Breast milk discoloration	1

Given these reactions were observed early in the release of BNT162b2, it was not too surprising to see the study by Hanna et al. who reported detecting mRNA from C19 "vaccines" in breast milk (following image).

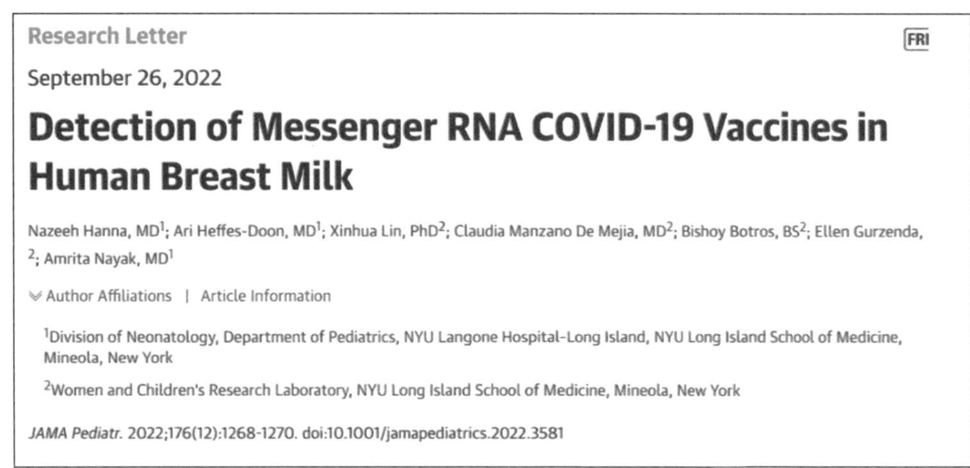

Research Letter FRI

September 26, 2022

Detection of Messenger RNA COVID-19 Vaccines in Human Breast Milk

Nazeeh Hanna, MD[1]; Ari Heffes-Doon, MD[1]; Xinhua Lin, PhD[2]; Claudia Manzano De Mejia, MD[2]; Bishoy Botros, BS[2]; Ellen Gurzenda, [2]; Amrita Nayak, MD[1]

Author Affiliations | Article Information

[1]Division of Neonatology, Department of Pediatrics, NYU Langone Hospital-Long Island, NYU Long Island School of Medicine, Mineola, New York

[2]Women and Children's Research Laboratory, NYU Long Island School of Medicine, Mineola, New York

JAMA Pediatr. 2022;176(12):1268-1270. doi:10.1001/jamapediatrics.2022.3581

(https://jamanetwork.com/journals/jamapediatrics/fullarticle/2796427)

Of 11 lactating individuals enrolled, trace amounts of BNT162b2 and mRNA-1273 COVID-19 mRNA vaccines were detected in 7 samples from 5 different participants at various times up to 45 hours postvaccination (Table 2).

Table 2. Detection of Vaccine RNA in Whole Expressed Breast Milk and Extracellular Vesicles in 5 Patients at Various Time Points Postvaccination

Participant No.	Vaccine type	Time points of vaccine mRNA detection in EBM	Concentration of vaccine mRNA detected in whole milk[a]	Concentration of vaccine mRNA detected in EBM EVs[a]
4	BNT162b2	27-h[b] Sample	Not detected	14.01 pg/mL
6	mRNA-1273	27-h and 42-h[b] Samples	11.7 pg/mL	16.78 pg/mL
7	BNT162b2	37-h[b] Sample	Not detected	4.69 pg/mL
8	BNT162b2	1-h and 3-h[b] Samples	1.3 pg/mL	6.77 pg/mL
10	mRNA-1273	45-h[b] Sample	2.5 pg/mL	2.13 pg/mL

Abbreviation: EBM, expressed breast milk; EVs, extracellular vesicles; mRNA, messenger RNA.

[a] Units for concentration are picogram of mRNA per milliliter of whole milk equivalent

[b] Sample used for vaccine mRNA concentration detection.

The sporadic presence and ***trace quantities of COVID-19 vaccine mRNA detected in EBM suggest that breastfeeding after COVID-19 mRNA vaccination is safe,*** *particularly beyond 48 hours after vaccination* [italics and bold added]. These data demonstrate for the first time to our knowledge the biodistribution of COVID-19 vaccine mRNA to mammary cells and the potential ability of tissue EVs to package the vaccine mRNA that can be transported to distant cells.

Remarkably, the authors concluded that mRNA in breast milk was safe.

CHILDREN HEALTH | EU ISSUES | PUBLIC HEALTH | VACCINATION | VACCINE SAFETY

EMA's latest bombshell instalment of damning data confirms their failure: PSUR #3, the pregnancy and lactation cases

By Sonia Elijah • May 15, 2023

In May of this year a report was issued by Sonia Elijah in Children's Health Defense Europe (above) identifying in infants **two cases of stroke** after being exposed to mRNA-containing breast milk, **three cases of severe neurologic disease**, and **four cases of respiratory Adverse Events of Special Interest (AESIs).** (https://soniaelijah.substack.com/p/emas-latest-bombshell-instalment)

Meanwhile, a Freedom of Information Act (FOIA) request for records from the CDC turned up the following email in which John Su reported seeing ". . . a fair bit of 'exposure by breast milk'- does that indicate an increase in reporting of this particular PT?" (https://jackanapes.substack.com/p/wake-up-and-smell-the-glitch-in-the)

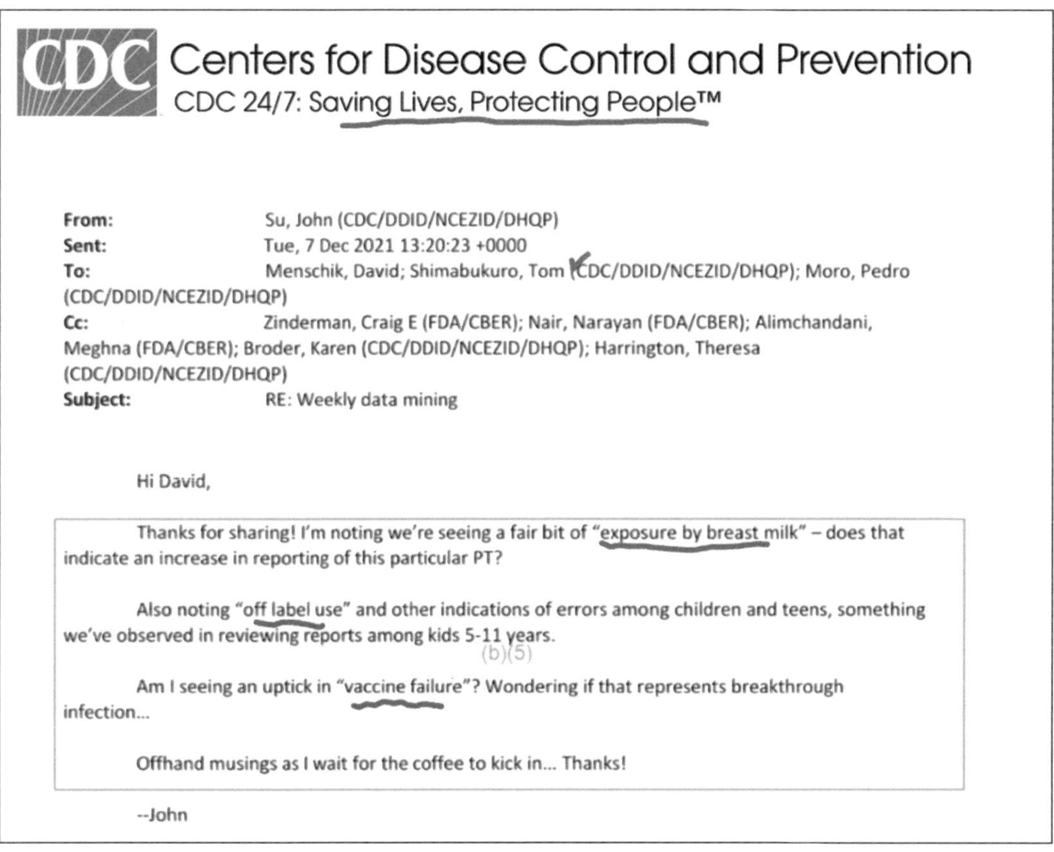

CDC Centers for Disease Control and Prevention
CDC 24/7: Saving Lives, Protecting People™

From: Su, John (CDC/DDID/NCEZID/DHQP)
Sent: Tue, 7 Dec 2021 13:20:23 +0000
To: Menschik, David; Shimabukuro, Tom (CDC/DDID/NCEZID/DHQP); Moro, Pedro (CDC/DDID/NCEZID/DHQP)
Cc: Zinderman, Craig E (FDA/CBER); Nair, Narayan (FDA/CBER); Alimchandani, Meghna (FDA/CBER); Broder, Karen (CDC/DDID/NCEZID/DHQP); Harrington, Theresa (CDC/DDID/NCEZID/DHQP)
Subject: RE: Weekly data mining

Hi David,

Thanks for sharing! I'm noting we're seeing a fair bit of "exposure by breast milk" – does that indicate an increase in reporting of this particular PT?

Also noting "off label use" and other indications of errors among children and teens, something we've observed in reviewing reports among kids 5-11 years.

(b)(5)

Am I seeing an uptick in "vaccine failure"? Wondering if that represents breakthrough infection...

Offhand musings as I wait for the coffee to kick in... Thanks!

--John

The email goes on to note errors in administering the C19 drug products to children aged five- to 11-years-old. This subject will be addressed more fully in Section IIIC, on administrative errors, to follow.

John Su goes on to note "vaccine failure" was on the increase, as was known from **Pfizer Confidential Document 5.3.6, February 28, 2021.**

VAERS contains documentation of 38 cases of symptoms reported after breastfeeding infants' exposure to mRNA from their mother's milk:

Messages:
- **VAERS data in CDC WONDER are updated every Friday. Hence, results for the same query can change from week to week.**
- **These results are for 35 total events.**
- **Rows with zero Events Reported are hidden. Use Quick Options above to show zero rows.**

Vaccine Type ⇩	Symptoms	➡ Events Reported ⇧⇩
COVID19 VACCINE (COVID19)	**EXPOSURE VIA BREAST MILK**	38
	Total	**38**
Total		**38**

VAERS ID 1415059 (below) documents a three-month-old female who received mRNA1273 in her mother's breast milk and had a seizure lasting seven minutes the same day as her mother's inoculation. She was hospitalized for two days.

The report was completed three days after the seizure with no outcome information other than the child was considered to have permanent disability. The CDC does collect outcome information but does not share it.

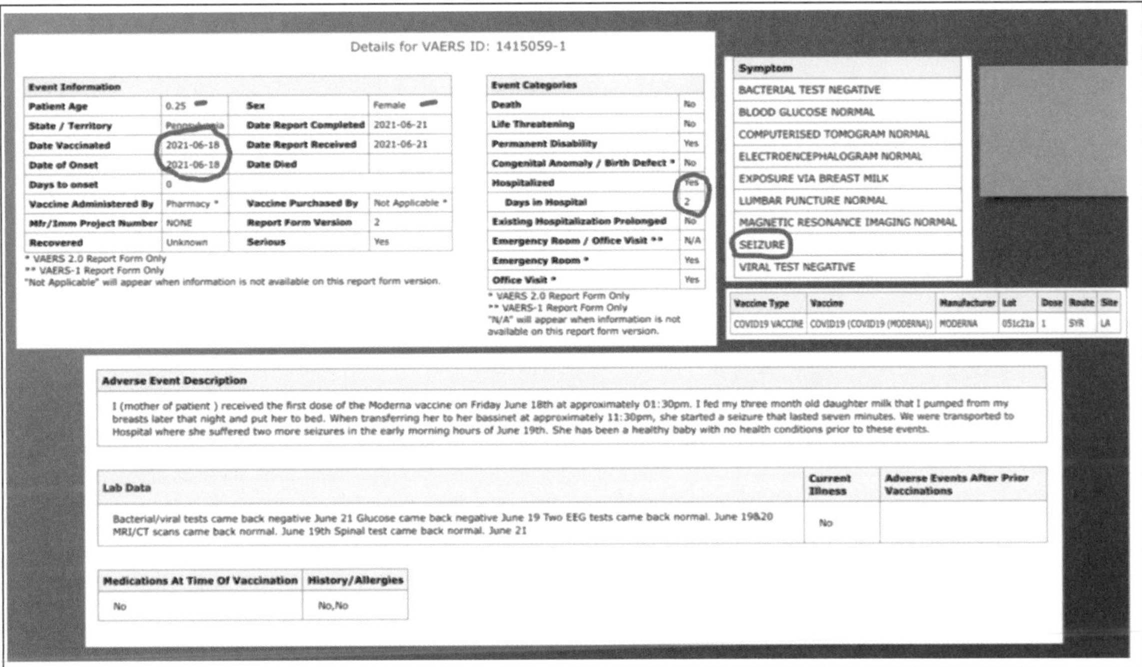

Details for VAERS ID: 1415059-1

Event Information			
Patient Age	0.25	Sex	Female
State / Territory	Pennsylvania	Date Report Completed	2021-06-21
Date Vaccinated	2021-06-18	Date Report Received	2021-06-21
Date of Onset	2021-06-18	Date Died	
Days to onset	0		
Vaccine Administered By	Pharmacy *	Vaccine Purchased By	Not Applicable *
Mfr/Imm Project Number	NONE	Report Form Version	2
Recovered	Unknown	Serious	Yes

* VAERS 2.0 Report Form Only
** VAERS-1 Report Form Only
"Not Applicable" will appear when information is not available on this report form version.

Event Categories	
Death	No
Life Threatening	No
Permanent Disability	Yes
Congenital Anomaly / Birth Defect *	No
Hospitalized	Yes
Days in Hospital	2
Existing Hospitalization Prolonged	No
Emergency Room / Office Visit **	N/A
Emergency Room *	Yes
Office Visit *	Yes

* VAERS 2.0 Report Form Only
** VAERS-1 Report Form Only
"N/A" will appear when information is not available on this report form version.

Symptom
BACTERIAL TEST NEGATIVE
BLOOD GLUCOSE NORMAL
COMPUTERISED TOMOGRAM NORMAL
ELECTROENCEPHALOGRAM NORMAL
EXPOSURE VIA BREAST MILK
LUMBAR PUNCTURE NORMAL
MAGNETIC RESONANCE IMAGING NORMAL
SEIZURE
VIRAL TEST NEGATIVE

Vaccine Type	Vaccine	Manufacturer	Lot	Dose	Route	Site
COVID19 VACCINE	COVID19 (COVID19 (MODERNA))	MODERNA	051c21a	1	SYR	LA

Adverse Event Description
I (mother of patient) received the first dose of the Moderna vaccine on Friday June 18th at approximately 01:30pm. I fed my three month old daughter milk that I pumped from my breasts later that night and put her to bed. When transferring her to her bassinet at approximately 11:30pm, she started a seizure that lasted seven minutes. We were transported to Hospital where she suffered two more seizures in the early morning hours of June 19th. She has been a healthy baby with no health conditions prior to these events.

Lab Data	Current Illness	Adverse Events After Prior Vaccinations
Bacterial/viral tests came back negative June 21 Glucose came back negative June 19 Two EEG tests came back normal. June 19&20 MRI/CT scans came back normal. June 19th Spinal test came back normal. June 21	No	

Medications At Time Of Vaccination	History/Allergies
No	No,No

VAERS ID 1166062 (below) was a four-month-old male who died three days after his mother's second dose of BNT162b2. The cause of death was failure to thrive, fever, and a hematologic condition known as thrombotic thrombocytopenic purpura.

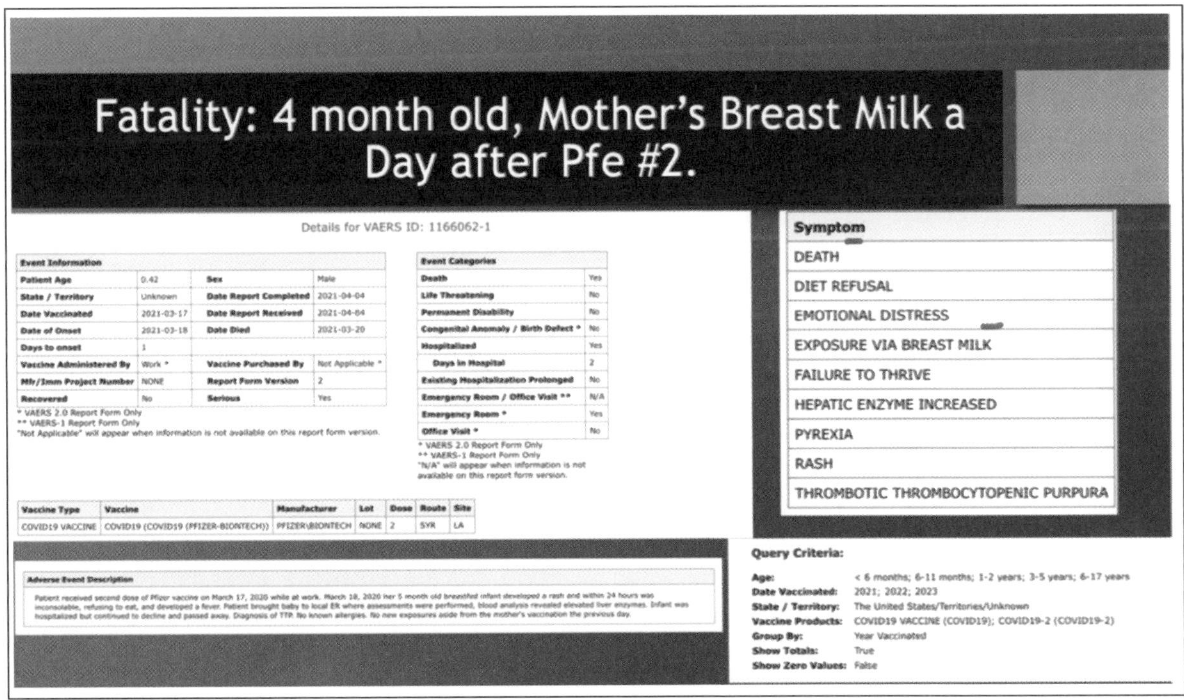

Fatality: 4 month old, Mother's Breast Milk a Day after Pfe #2.

Details for VAERS ID: 1166062-1

Event Information			
Patient Age	0.42	Sex	Male
State / Territory	Unknown	Date Report Completed	2021-04-04
Date Vaccinated	2021-03-17	Date Report Received	2021-04-04
Date of Onset	2021-03-18	Date Died	2021-03-20
Days to onset	1		
Vaccine Administered By	Work *	Vaccine Purchased By	Not Applicable *
Mfr/Imm Project Number	NONE	Report Form Version	2
Recovered	No	Serious	Yes

* VAERS 2.0 Report Form Only
** VAERS-1 Report Form Only
"Not Applicable" will appear when information is not available on this report form version.

Event Categories	
Death	Yes
Life Threatening	No
Permanent Disability	No
Congenital Anomaly / Birth Defect *	No
Hospitalized	Yes
Days in Hospital	2
Existing Hospitalization Prolonged	No
Emergency Room / Office Visit **	N/A
Emergency Room *	Yes
Office Visit *	No

* VAERS 2.0 Report Form Only
** VAERS-1 Report Form Only
"N/A" will appear when information is not available on this report form version.

Symptom
DEATH
DIET REFUSAL
EMOTIONAL DISTRESS
EXPOSURE VIA BREAST MILK
FAILURE TO THRIVE
HEPATIC ENZYME INCREASED
PYREXIA
RASH
THROMBOTIC THROMBOCYTOPENIC PURPURA

Vaccine Type	Vaccine	Manufacturer	Lot	Dose	Route	Site
COVID19 VACCINE	COVID19 (COVID19 (PFIZER-BIONTECH))	PFIZER\BIONTECH	NONE	2	SYR	LA

Adverse Event Description
Patient received second dose of Pfizer vaccine on March 17, 2020 while at work. March 18, 2020 her 5 month old breastfed infant developed a rash and within 24 hours was inconsolable, refusing to eat, and developed a fever. Patient brought baby to local ER where assessments were performed, blood analysis revealed elevated liver enzymes. Infant was hospitalized but continued to decline and passed away. Diagnosis of TTP. No known allergies. No new exposures aside from the mother's vaccination the previous day.

Query Criteria:

Age: < 6 months; 6-11 months; 1-2 years; 3-5 years; 6-17 years
Date Vaccinated: 2021; 2022; 2023
State / Territory: The United States/Territories/Unknown
Vaccine Products: COVID19 VACCINE (COVID19); COVID19-2 (COVID19-2)
Group By: Year Vaccinated
Show Totals: True
Show Zero Values: False

These are a sample of similar cases in VAERS. Dr. Makis has published a recent article on LNP/mRNA-related neonatal fatalities. (https://open.substack.com/pub/makismd/p/mrna-and-pregnancy-infants-who-died)

C. Administrative Issues Have Consequences: Ages Zero to 17 Years

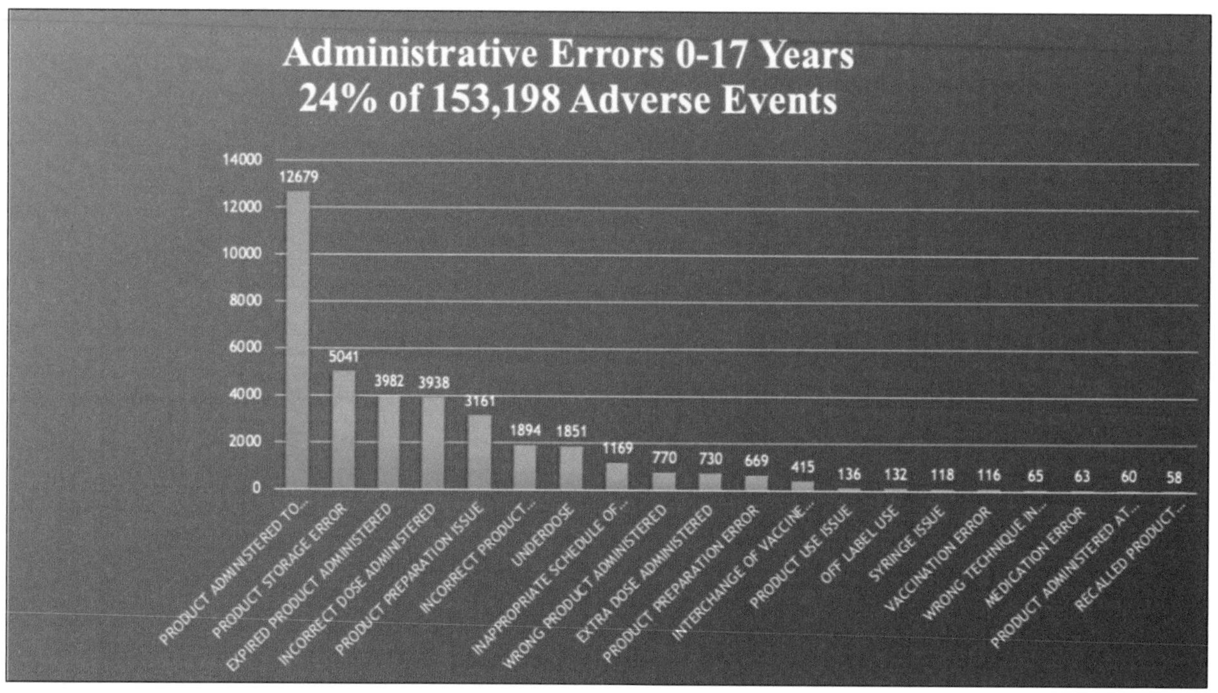

VAERS through 5/12/2023 C19 Gene Therapy Drugs U.S. and Territories

Administrative errors occurred in 24% (37,235/153,198) of all reported events in children ages zero- to 17-years-old. (VAERS) **The most common by far is administration of the product to children and adolescents too young to receive the drug.**

Was this a byproduct of the $1 billion campaign to scare and cajole parents into having their children injected to keep them safe?

Follow-up on these administrative errors, like the one below, has not been posted if it exists.

Messages:
- VAERS data in CDC WONDER are updated every Friday. Hence, results for the same query can change from week to week.
- These results are for 12,216 total events.

Symptoms	Age	Events Reported	Percent (of 12,216)
PRODUCT ADMINISTERED TO PATIENT OF INAPPROPRIATE AGE	< 6 months	19	0.16%
	6-11 months	13	0.11%
	1-2 years	81	0.66%
	3-5 years	486	3.98%
	6-17 years	11,617	95.10%
	Total	12,216	100.00%
Total		12,216	100.00%

Note: Submitting a report to VAERS does not mean that healthcare personnel or the vaccine caused or contributed to the adverse event (possible side effect).

Some of these errors have had disastrous complications.

This search result gives more detail on the "No Adverse Effects," Death, Life Threatening, and Permanent Disability (following image).

Event Category	Sex	Events Reported
Death	Female	2
	Male	11
	Total	**13**
Life Threatening	Female	1
	Male	1
	Total	**2**
Permanent Disability	Male	1
	Total	**1**
Total		**16**

VAERS ID 2457513: This 15-year-old girl had an injection that was coded "Product Administered to Patient of Inappropriate Age." After her second dose of mRNA-1273 (Moderna), she had a cardiac arrest and died. Why was a 15-year-old receiving an injection to "prevent" C19 classified as a "PATIENT"?

Details for VAERS ID: 2457513-1

Event Information			
Patient Age	15.00	Sex	Female
State / Territory		Date Report Completed	2022-09-23
Date Vaccinated		Date Report Received	2022-09-24
Date of Onset		Date Died	
Days to onset			
Vaccine Administered By	Unknown	Vaccine Purchased By	Not Applicable *
Mfr/Imm Project Number	USMODERNATX, INC.MOD20216	Report Form Version	2
Recovered	No	Serious	Yes

* VAERS 2.0 Report Form Only
** VAERS-1 Report Form Only
"Not Applicable" will appear when information is not available on this report form version.

Event Categories	
Death	Yes
Life Threatening	No
Permanent Disability	No
Congenital Anomaly / Birth Defect *	No
Hospitalized	No
Days in Hospital	None
Existing Hospitalization Prolonged	No
Emergency Room / Office Visit **	N/A
Emergency Room *	No
Office Visit *	No

* VAERS 2.0 Report Form Only
** VAERS-1 Report Form Only
"N/A" will appear when information is not available on this report form version.

Symptom
CARDIAC ARREST
PRODUCT ADMINISTERED TO PATIENT OF INAPPROPRIATE AGE

Vaccine Type	Vaccine	Manufacturer	Lot	Dose	Route	Site
COVID19 VACCINE	COVID19 (COVID19 (MODERNA))	MODERNA	NONE	2	OT	

Adverse Event Description

This spontaneous case was reported by a consumer and describes the occurrence of CARDIAC ARREST (Died of Cardiac arrest) in a 15-year-old female patient who received mRNA-1273 (Moderna COVID-19 Vaccine) for COVID-19 prophylaxis. The occurrence of additional non-serious events is detailed below. No Medical History information was reported. On an unknown date, the patient received second dose of mRNA-1273 (Moderna COVID-19 Vaccine) (unknown route) 1 dosage form. On an unknown date, the patient experienced CARDIAC ARREST (Died of Cardiac arrest) (seriousness criteria death and medically significant) and PRODUCT ADMINISTERED TO PATIENT OF INAPPROPRIATE AGE (Inappropriate age of vaccine administered). The patient died on an unknown date. The reported cause of death was Cardiac arrest. It is unknown if an autopsy was performed. At the time of death, PRODUCT ADMINISTERED TO PATIENT OF INAPPROPRIATE AGE (Inappropriate age of vaccine administered) outcome was unknown. No Concomitant and Treatment Medication was provided by reporter. This case was identified as part of a retrospective clean-up activity where certain emails received in the Moderna mailbox were missed to be booked-in. Company comment: This fatal spontaneous case concerns a 15ûyearûold female patient, with no medical history reported, who received the second dose of her COVIDû19 immunization schedule with mRNA-1273 vaccine (unknown schedule) and experienced the serious unexpected event of cardiac arrest. Vaccination date and onset date of the event were not reported; hence latency and temporal association cannot be properly assessed. No further information regarding the event, death date and cause of death have been provided for medical review. It is unknown whether an autopsy was performed. Product administered to patient of inappropriate age was also considered in the case. The benefit risk relationship of mRNA-1273 is not affected by this report. Most recent FOLLOW-UP information incorporated above includes: On 14-May-2021: Follow-up received and does not contain any new information. On 14-May-2021: Follow-up received and does not contain any new information. On 17-May-2021: Follow-up received and does not contain any new information. On 22-Jul-2021: Follow-up received and does not contain any new information. Sender's Comments: This fatal spontaneous case concerns a 15ûyearûold female patient, with no medical history reported, who received the second dose of her COVIDû19 immunization schedule with mRNA-1273 vaccine (unknown schedule) and experienced the serious unexpected event of cardiac arrest. Vaccination date and onset date of the event were not reported; hence latency and temporal association cannot be properly assessed. No further information regarding the event, death date and cause of death have been provided for medical review. It is unknown whether an autopsy was performed. Product administered to patient of inappropriate age was also considered in the case. The benefit risk relationship of mRNA-1273 is not affected by this report. Reported Cause(s) of Death: Cardiac arrest.

"The benefit risk relationship of m/RNA-1273 is not affected by this report." Why would anyone say something like that after a 15-year-old, otherwise healthy girl died after receiving a failed "vaccine"?

VAERS ID 1772015, "Inappropriate Schedule of Product Administration," also concerns a 15-year-old—this time a boy.

Four days following dose two of Pfizer's BNT162b2, the teen developed multifocal hemorrhagic lesions in his brain: cerebrum, brainstem, and cerebellum (following chart).

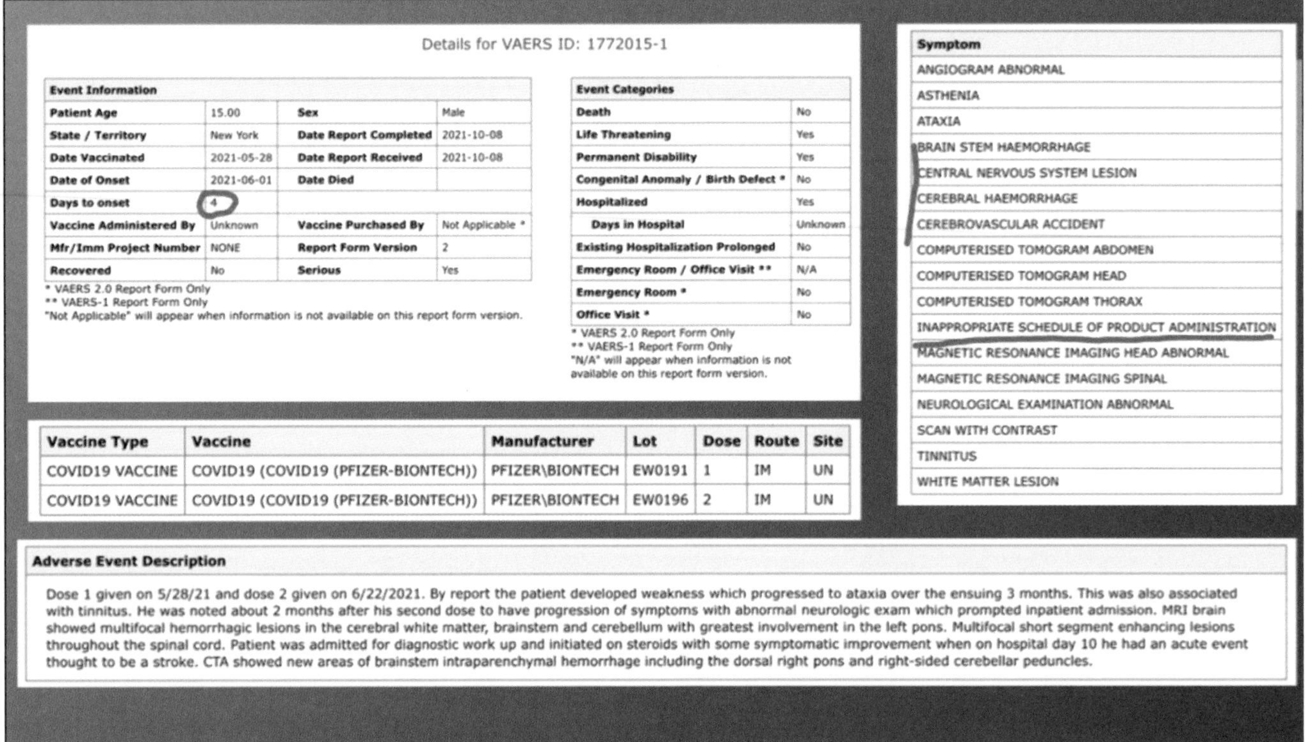

Details for VAERS ID: 1772015-1

Event Information			
Patient Age	15.00	Sex	Male
State / Territory	New York	Date Report Completed	2021-10-08
Date Vaccinated	2021-05-28	Date Report Received	2021-10-08
Date of Onset	2021-06-01	Date Died	
Days to onset	4		
Vaccine Administered By	Unknown	Vaccine Purchased By	Not Applicable *
Mfr/Imm Project Number	NONE	Report Form Version	2
Recovered	No	Serious	Yes

* VAERS 2.0 Report Form Only
** VAERS-1 Report Form Only
"Not Applicable" will appear when information is not available on this report form version.

Event Categories	
Death	No
Life Threatening	Yes
Permanent Disability	Yes
Congenital Anomaly / Birth Defect *	No
Hospitalized	Yes
Days in Hospital	Unknown
Existing Hospitalization Prolonged	No
Emergency Room / Office Visit **	N/A
Emergency Room *	No
Office Visit *	No

* VAERS 2.0 Report Form Only
** VAERS-1 Report Form Only
"N/A" will appear when information is not available on this report form version.

Vaccine Type	Vaccine	Manufacturer	Lot	Dose	Route	Site
COVID19 VACCINE	COVID19 (COVID19 (PFIZER-BIONTECH))	PFIZER\BIONTECH	EW0191	1	IM	UN
COVID19 VACCINE	COVID19 (COVID19 (PFIZER-BIONTECH))	PFIZER\BIONTECH	EW0196	2	IM	UN

Symptom
ANGIOGRAM ABNORMAL
ASTHENIA
ATAXIA
BRAIN STEM HAEMORRHAGE
CENTRAL NERVOUS SYSTEM LESION
CEREBRAL HAEMORRHAGE
CEREBROVASCULAR ACCIDENT
COMPUTERISED TOMOGRAM ABDOMEN
COMPUTERISED TOMOGRAM HEAD
COMPUTERISED TOMOGRAM THORAX
INAPPROPRIATE SCHEDULE OF PRODUCT ADMINISTRATION
MAGNETIC RESONANCE IMAGING HEAD ABNORMAL
MAGNETIC RESONANCE IMAGING SPINAL
NEUROLOGICAL EXAMINATION ABNORMAL
SCAN WITH CONTRAST
TINNITUS
WHITE MATTER LESION

Adverse Event Description

Dose 1 given on 5/28/21 and dose 2 given on 6/22/2021. By report the patient developed weakness which progressed to ataxia over the ensuing 3 months. This was also associated with tinnitus. He was noted about 2 months after his second dose to have progression of symptoms with abnormal neurologic exam which prompted inpatient admission. MRI brain showed multifocal hemorrhagic lesions in the cerebral white matter, brainstem and cerebellum with greatest involvement in the left pons. Multifocal short segment enhancing lesions throughout the spinal cord. Patient was admitted for diagnostic work up and initiated on steroids with some symptomatic improvement when on hospital day 10 he had an acute event thought to be a stroke. CTA showed new areas of brainstem intraparenchymal hemorrhage including the dorsal right pons and right-sided cerebellar peduncles.

He was considered **permanently disabled at age 15 years**.

The topic of administrative errors and their consequences is worthy of a separate report given that there were over 37,000 of them. Applying the URF of 41, that works out to 1,526,635 medication errors of various types.

D. Cardiac Arrest

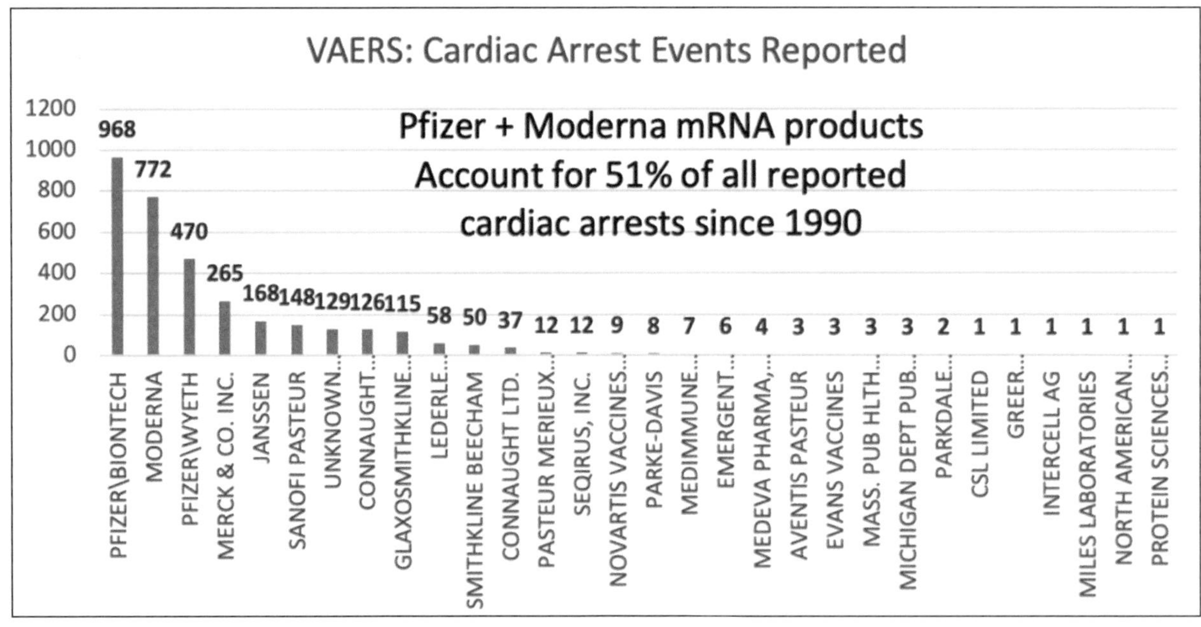

VAERS through 5/12/2023

Across all ages and since 1990, the mRNA products account for more than half of all vaccine-related cardiac arrests in VAERS (previous image).

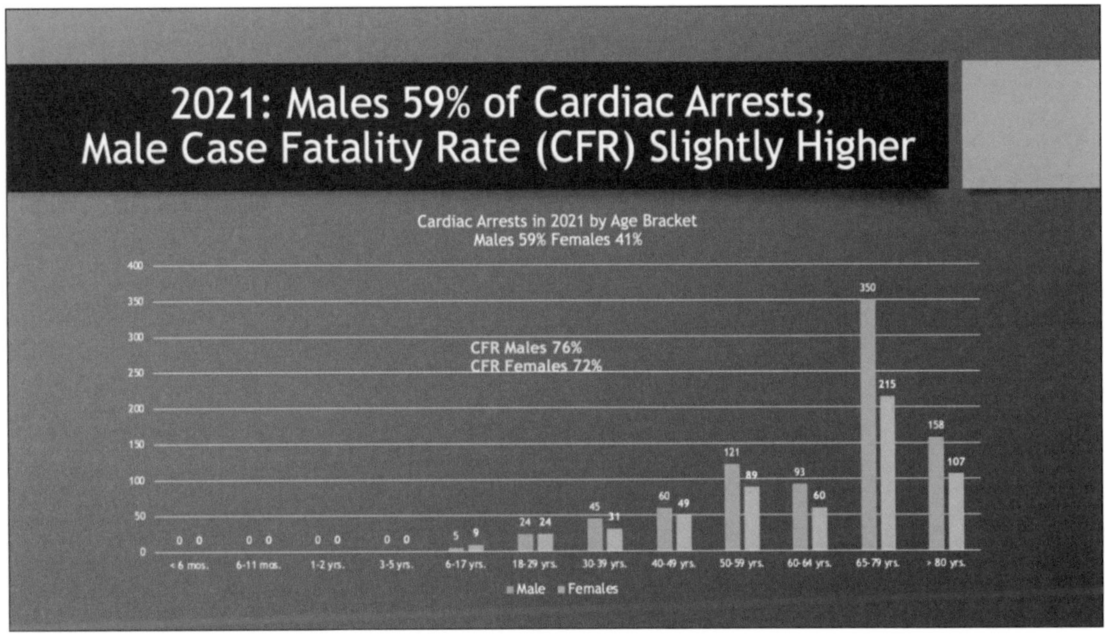

Males had 59% of the cardiac arrest events reported to VAERS in 2021, the first year of the GTPs.

The fatality rate was 72% for females and 76% for males. The age bracket from 65 to 79 years of age had the greatest number of reports. There were 62 event reports for < 30 years-of-age. With an URF of 41 that results in 2,501 cardiac arrests in this age bracket.

The following cases from VAERS are typical cardiac arrest cases, except these are children (ages six to 17), not septuagenarians.

6-17	CARDIAC ARREST	F	1199455-1	Patient reported difficulty breathing and chest pain; suffered cardiac arrest and death	
6-17	CARDIAC ARREST	F	1225942-1	Patient was a 16yr female who received Pfizer vaccine 3/19/21 at vaccine clinic and presented with ongoing CPR to the ED 3/28/21 after cardiac arrest at home. Patient placed on ECMO and imaging revealed bilateral large pulmonary embolism as likely etiology of arrest. Risk factors included oral contraceptive use. Labs have since confirmed absence of Factor V leiden or prothrombin gene mutation. Patient declared dead by neurologic criteria 3/30/21.	ECMO: Extracorpore al Membrane Oxygenator. Substitutes for the lungs.
6-17	CARDIAC ARREST	F	1420762-1	Cardiac arrest without resuscitation. Unknown cause of cardiac arrest. Awaiting autopsy report.	
6-17	CARDIAC ARREST	F	1828901-1	Patient reported symptomatic (non-severe) case of COVID-19 August 2021 and recovered fully. She reported receiving Pfizer COVID vaccine 9/3/21 and second dose 9/15/21. She present to the emergency department of my hospital 10/23/21 with chest pain and dyspnea for 48h. Was feeling completely well prior to onset of chest discomfort. Symptoms were mild. No sick contacts or family members. ED evaluation remarkable for normal exam, no hypoxia, normal blood pressure. EKG with diffuse ST elevation. Troponin elevated at 20. CTA chest negative for PE or pneumonia. SARS-CoV-PCR positive but thought to be persistent positive rather than reinfection because of lack of clinical symptoms, recent COVID-19 and recent vaccination. Cardiologist consulted, thought acute coronary syndrome unlikely based on age and lack of risk factors. STAT Echo resulted depressed EF 40-45%. Simultaneously she had become increasingly tachycardic and EKG appeared more ischemic. Cardiac cath lab was activated and she was about to be transported when she suffered cardiac arrest. Initial rhythm was VT. Received ACLS protocol CPR x 65 minutes including multiple cardioversion, amiodarone, lidocaine, magnesium and other antiarrhythmics. Unfortunately she was not able to be resuscitated and died. Cause of death possible acute myocarditis.	Administration Error 13 days, not 21
6-17	CARDIAC ARREST	F	1865389-1	Patient with progressive hypoxemia throughout the day despite multiple changes in ventilator settings/modes. HFOV discussed with family, but functional oscillator not available and was awaiting arrival of donor oscillator. She is not a candidate for ECMO due to pulmonary hemorrhage and thrombocytopenia with recent chemotherapy as well as BMI (morbidly obese). Trial on nitric oxide performed with minimal improvement (sats increased from 60% to 65-68%). She was noted to have increasing peaked T waves as well as development of Q waves concerning for hyperkalemia and worsening cardiac function consistent with multiorgan failure; perfusion was quite poor with mottled extremities and difficult to palpate central pulse	
6-17	CARDIAC ARREST	F	1912785-1	Dose 1 given 4/21/2021 Pfizer Lot # EW0172 Patient had a cardiac arrest at home and was pronounced dead at Emergency Room. Covid test was negative.	
6-17	CARDIAC ARREST	F	2327226-1	She developed inflamed lymph nodes (lymphadenitis), all over the body rash, ongoing fever for more than 3 weeks. She was diagnosed with MIS-C, her heart, intestines, lungs, skin and liver were inflamed. She was hospitalized and treated with immunoglobulin, steroids, anticoagulant, fever reducing medications, etc. By the second treatment, her belly started getting distended, her lungs were filled with liquids. She was transferred to ICU and her heart stopped beating right there.	

E. Central Nervous System and Neurological Disease: Co-VAN Cluster

Samim et al. performed an extensive literature review of neurologic diseases associated with the C19 gene therapy drugs and organized them under the heading of Co-VAN (COVID-19 vaccine-associated neurological diseases) as illustrated below.

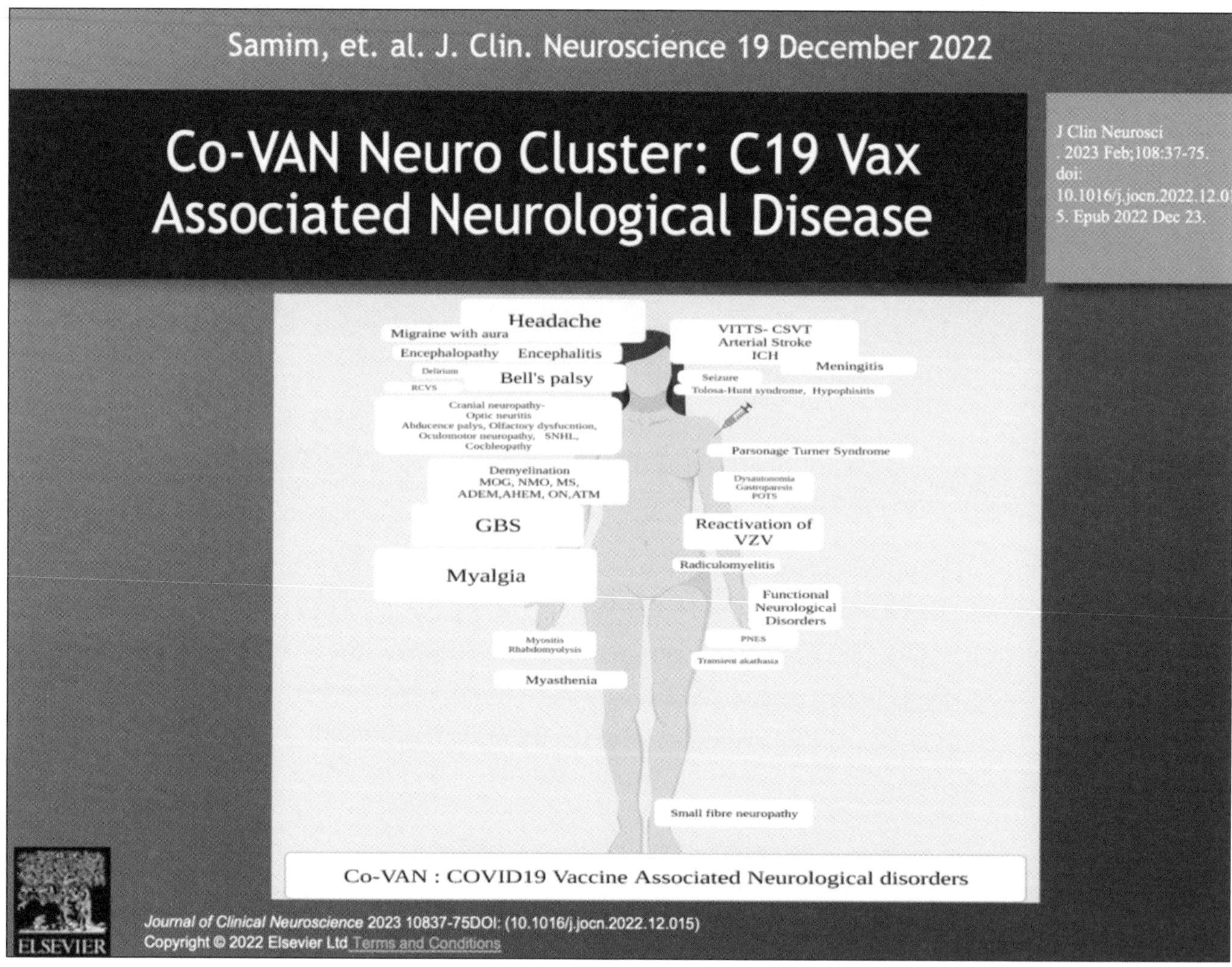

The Co-VAN Disease group fits well in a taxonomy of CoVax Disease based on organ systems and pathological processes associated with spike-generating drugs. This topic will be further developed in Part 5 of this series.

There is wide variation in the manifestations of neurological disease following spike-generating drugs from peripheral neuropathy to demyelinating disorders to stroke, seizures, encephalitis, protein deposition disease, and cranial neuropathies, such as Bell's Palsy or Ramsay Hunt Syndrome (cranial nerve VII).

Pfizer Confidential Document 5.3.6 summarized adverse events and Adverse Events of Special Interest (AESIs) from December 2020 through February 2021. The following histogram summarizes the neurological diseases identified from a pool of over 40,000 GTP subjects reporting complications after receiving BNT162b2.

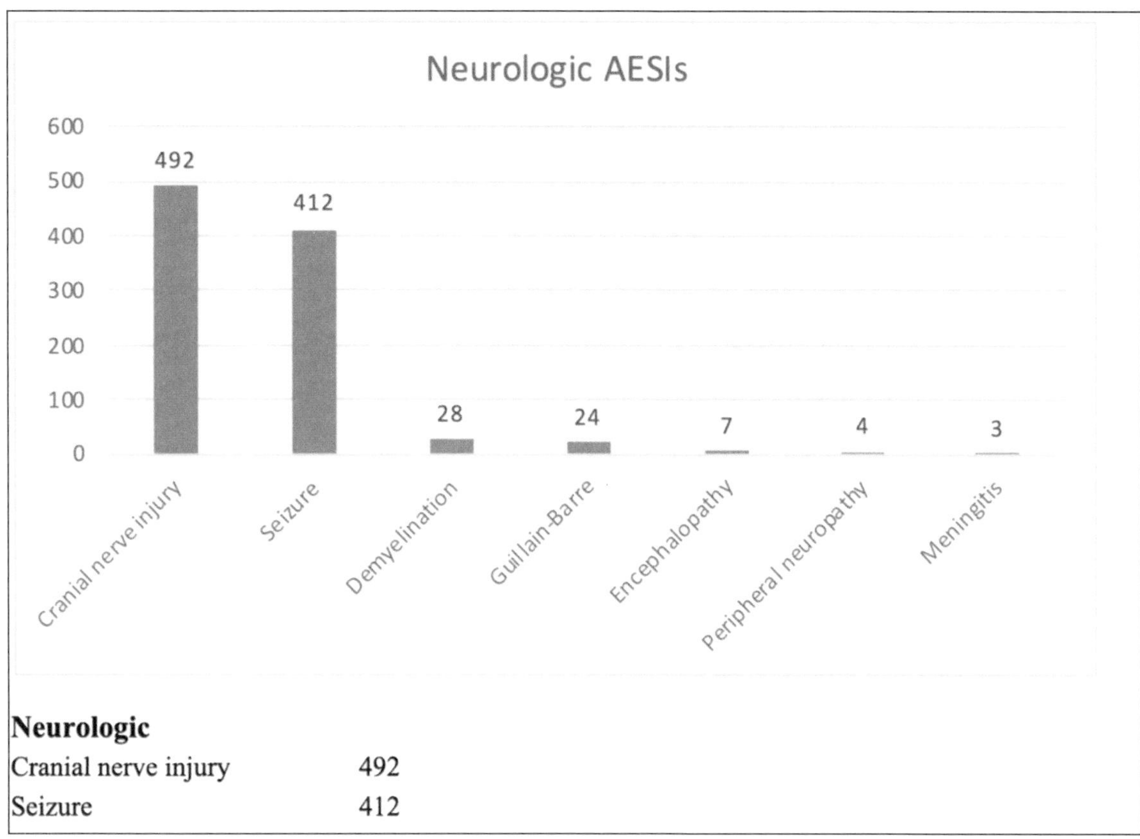

Neurologic	
Cranial nerve injury	492
Seizure	412

In spite of this significant "signal," C19 gene therapy products were mandated by governments throughout the world. Not surprisingly, neurologic complications appeared in children as OpenVAERS summarizes below.

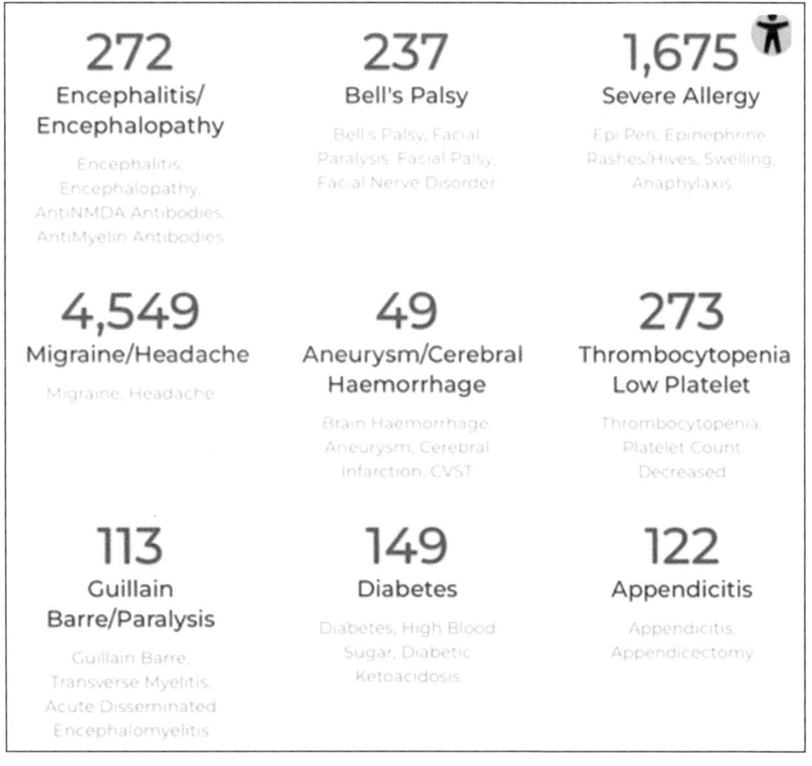

(https://openvaers.com/covid-data/child-summaries)

Overall, there were 5,220 neurological events ranging from 4,549 with migraine to 49 with cerebral hemorrhage/aneurysm in ages six months to 17 years. The estimates with an URF of 41 are given below.

These are significant medical problems that often leave permanent impairment and disability.

	N =	URF = x41
Migraine	4,549	186,509
Encephalitis	272	11,152
Bell's Palsy	237	9,717
GBP	113	4,633
Cerebral Hemorrhage/aneurysm	49	2,009
Totals	5,220	214,020

i. Acute Disseminated Encephalomyelitis (ADEM)

ADEM is a rare autoimmune demyelinating central nervous system disease, typically associated with patients younger than 15 years of age. Encephalitis is an inflammatory condition of the brain tissue known as myelin (the tissue coating nerve fibers and maintaining normal function of the nerves). Dr. Burkhardt's collection contains histopathological evidence of inflammation of both brain and the membranes around it.

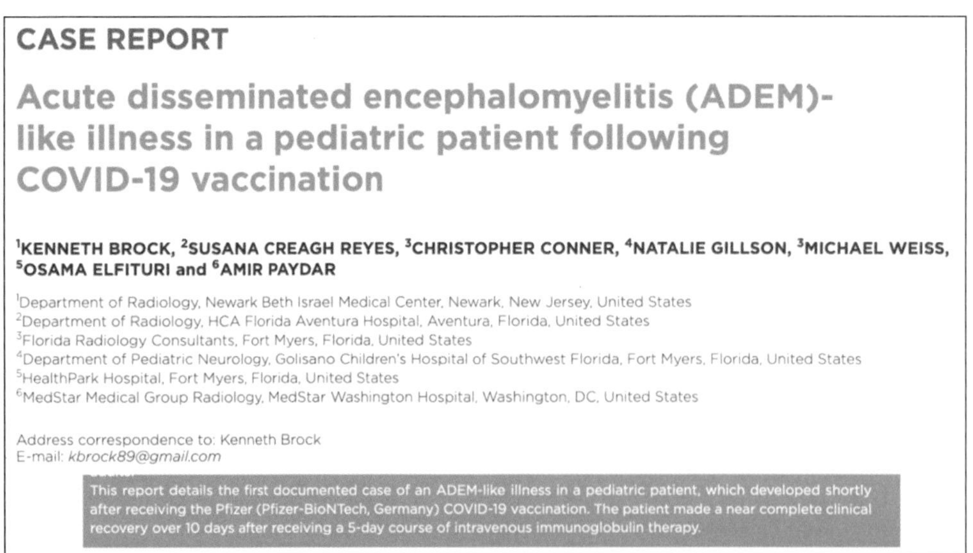

CASE REPORT

Acute disseminated encephalomyelitis (ADEM)-like illness in a pediatric patient following COVID-19 vaccination

[1]KENNETH BROCK, [2]SUSANA CREAGH REYES, [3]CHRISTOPHER CONNER, [4]NATALIE GILLSON, [3]MICHAEL WEISS, [5]OSAMA ELFITURI and [6]AMIR PAYDAR

[1]Department of Radiology, Newark Beth Israel Medical Center, Newark, New Jersey, United States
[2]Department of Radiology, HCA Florida Aventura Hospital, Aventura, Florida, United States
[3]Florida Radiology Consultants, Fort Myers, Florida, United States
[4]Department of Pediatric Neurology, Golisano Children's Hospital of Southwest Florida, Fort Myers, Florida, United States
[5]HealthPark Hospital, Fort Myers, Florida, United States
[6]MedStar Medical Group Radiology, MedStar Washington Hospital, Washington, DC, United States

Address correspondence to: Kenneth Brock
E-mail: *kbrock89@gmail.com*

This report details the first documented case of an ADEM-like illness in a pediatric patient, which developed shortly after receiving the Pfizer (Pfizer-BioNTech, Germany) COVID-19 vaccination. The patient made a near complete clinical recovery over 10 days after receiving a 5-day course of intravenous immunoglobulin therapy.

(https://www.ncbi.nlm.nih.gov/pmc/articles/PMC10043599/)

Brock et al. present the following case of a 10-year-old, previously healthy female who presented with progressive lower extremity weakness, paresthesia, and urinary retention.

Case Report

A 10-year-old female with no past medical history presented to Golisano Children's Hospital, Fort Myers, Florida, United States on December 21, 2021 [sic] with 7 days of progressive lower extremity weakness, paresthesia, and urinary retention. No recent symptoms of infection were reported.

Neurological examination showed mild lower extremity hyperreflexia, right lower extremity weakness with inability to ambulate, a mild pronator drift, and a right visual field defect. The patient received her second dose of mRNA-based COVID-19 vaccine 16 days prior to the onset of symptoms.

Upon admission, a comprehensive laboratory assay including CBC, CMP, ESR, and CRP, was negative. A respiratory viral panel which included SARS-COV2 PCR testing was negative. Contrast-enhanced MRI of the brain demonstrated multiple prominent T2/FLAIR hyperintense subcortical and deep white matter lesions with incomplete rim-enhancement, ***compatible with active demyelination,*** and avid peripheral diffusion restriction (Figure 1) *Emphasis added.*

Figure 1. Initial brain MRI findings. Left: Axial FLAIR image demonstrating multifocal supratentorial subcortical and deep white matter lesions (red arrows). Middle: Axial T1 post-contrast image demonstrating peripheral lesion enhancement. Right: Axial DWI image demonstrating avid peripheral diffusion restriction. DWI, diffusion-weighted imaging; FLAIR, fluid attenuation inversion recovery.

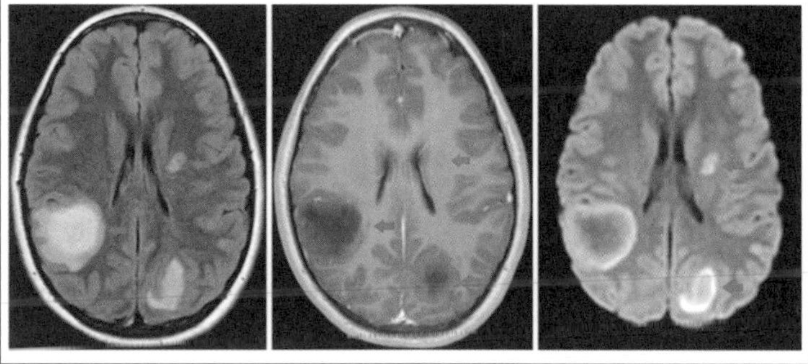

Figure 2. Initial thoracic spine MRI findings. Left: Sagittal STIR image demonstrating multiple longitudinally extensive spinal cord lesions (red arrows). Right: Sagittal T1 post-contrast image demonstrating corresponding lesion enhancement. STIR, short tau inversion recovery.

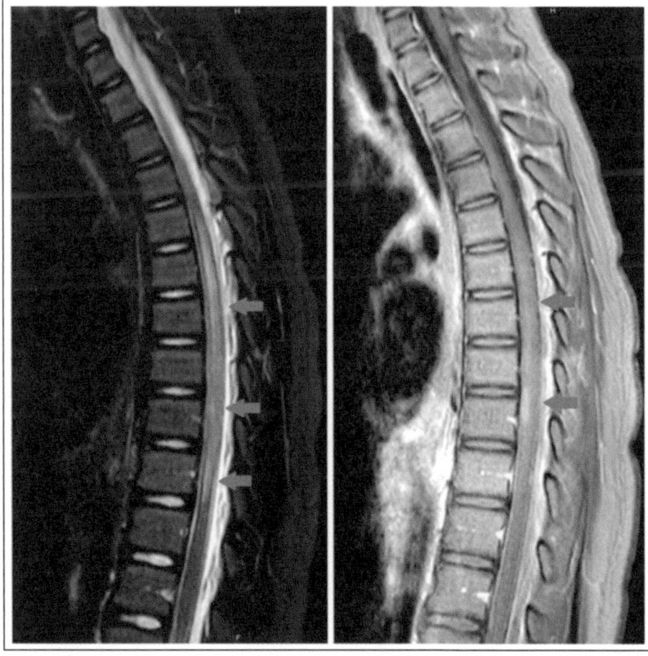

Contrast-enhanced MRI of the cervical, thoracic, and lumbar spine demonstrated a long-segment, non-expansile, partially enhancing intramedullary lesion within the thoracic spinal cord, also most likely compatible with active demyelinating process (Figure 2 above). There were no findings suggestive of Guillain-Barré syndrome. Secondary differential considerations included transverse myelitis, CNS lymphoma, and atypical multiple sclerosis.

Three months after initial evaluation, the patient returned for outpatient neurologic follow-up. She reported mild fatigue. Examination showed a mild, persistent gait instability and hip muscle weakness. All other symptoms were resolved. A few days later, final histopathology showed white matter neurons with an extensive macrophage-rich inflammatory infiltrate with small lymphocytic component forming perivascular aggregates (Figure 3).

Figure 3. Right occipital lobe brain biopsy demonstrating gliotic cerebral cortex with extensive macrophage-rich inflammatory infiltrate and small lymphocytic component forming perivascular aggregates (hematoxylin and eosin stains, magnification 20x and 60x).

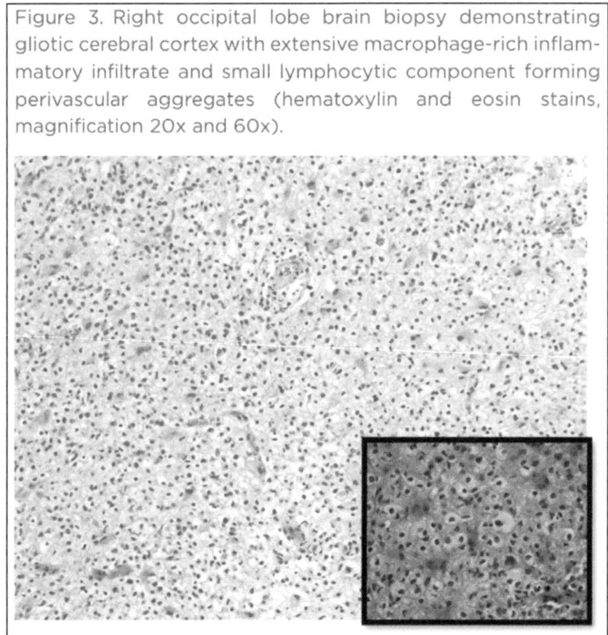

Luxol/H&E stain showed severe loss of myelin, with macrophages demonstrating phagocytosed myelin debris. Immunohistochemistry showed **depletion, but relative preservation of axons,** with scattered **perivascular accumulations of CD3+ T-cells and CD4/CD8 co-expressing T-cells**. (Bold added.)

Figure 4. Follow up brain MRI findings. Left: Axial FLAIR image demonstrating decreased FLAIR signal hyperintensity of the white matter lesions (red arrows). Middle: Axial non-contrast T1 image demonstrating lesion signal characteristics relatively isointense to grey matter. Right: Axial DWI image demonstrating decreased diffusion restriction of the lesions. DWI, diffusion-weighted imaging; FLAIR, fluid attenuation inversion recovery.

Widespread neurologic damage is well-documented in this case. Residual impairment and disability are probable.

VAERS ID 1948381 concerns an 11-year-old male with onset of ADEM three weeks after injection of flu vaccine along with BNT162b2 on November 17, 2021. MRI evaluation revealed abnormalities throughout the cerebral cortex with extension into adjacent tissues. The child was admitted to the hospital for five days during which he received high-dose steroids.

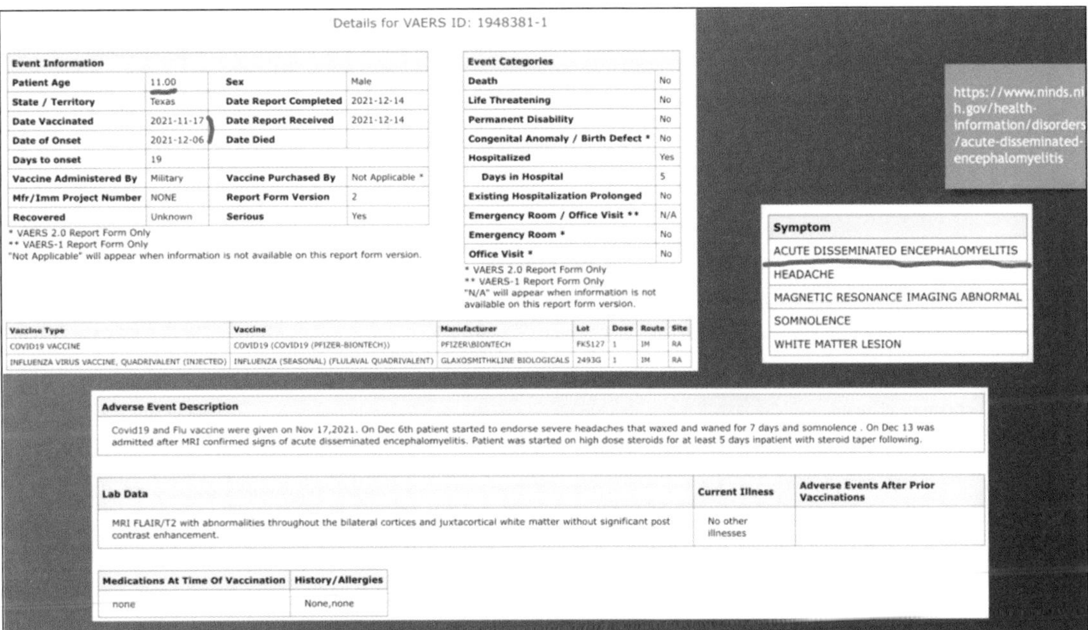

Details for VAERS ID: 1948381-1

Event Information			
Patient Age	11.00	Sex	Male
State / Territory	Texas	Date Report Completed	2021-12-14
Date Vaccinated	2021-11-17	Date Report Received	2021-12-14
Date of Onset	2021-12-06	Date Died	
Days to onset	19		
Vaccine Administered By	Military	Vaccine Purchased By	Not Applicable *
Mfr/Imm Project Number	NONE	Report Form Version	2
Recovered	Unknown	Serious	Yes

* VAERS 2.0 Report Form Only
** VAERS-1 Report Form Only
"Not Applicable" will appear when information is not available on this report form version.

Event Categories	
Death	No
Life Threatening	No
Permanent Disability	No
Congenital Anomaly / Birth Defect *	No
Hospitalized	Yes
Days in Hospital	5
Existing Hospitalization Prolonged	No
Emergency Room / Office Visit **	N/A
Emergency Room *	No
Office Visit *	No

* VAERS 2.0 Report Form Only
** VAERS-1 Report Form Only
"N/A" will appear when information is not available on this report form version.

Vaccine Type	Vaccine	Manufacturer	Lot	Dose	Route	Site
COVID19 VACCINE	COVID19 (COVID19 (PFIZER-BIONTECH))	PFIZER\BIONTECH	FK5127	1	IM	RA
INFLUENZA VIRUS VACCINE, QUADRIVALENT (INJECTED)	INFLUENZA (SEASONAL) (FLULAVAL QUADRIVALENT)	GLAXOSMITHKLINE BIOLOGICALS	2493G	1	IM	RA

https://www.ninds.nih.gov/health-information/disorders/acute-disseminated-encephalomyelitis

Symptom
ACUTE DISSEMINATED ENCEPHALOMYELITIS
HEADACHE
MAGNETIC RESONANCE IMAGING ABNORMAL
SOMNOLENCE
WHITE MATTER LESION

Adverse Event Description

Covid19 and Flu vaccine were given on Nov 17,2021. On Dec 6th patient started to endorse severe headaches that waxed and waned for 7 days and somnolence . On Dec 13 was admitted after MRI confirmed signs of acute disseminated encephalomyelitis. Patient was started on high dose steroids for at least 5 days inpatient with steroid taper following.

Lab Data	Current Illness	Adverse Events After Prior Vaccinations
MRI FLAIR/T2 with abnormalities throughout the bilateral cortices and juxtacortical white matter without significant post contrast enhancement.	No other illnesses	

Medications At Time Of Vaccination	History/Allergies
none	None,none

The report was filled out on December 14, 2021, while the child was still in the acute phase of his illness, which began on December 6, 2021, only three days following hospital discharge; so, no outcome determination was possible even though the report indicated no death or disability. Residual impairment examination after one year is required.

ii. Aneurysm/Cerebral Hemorrhage

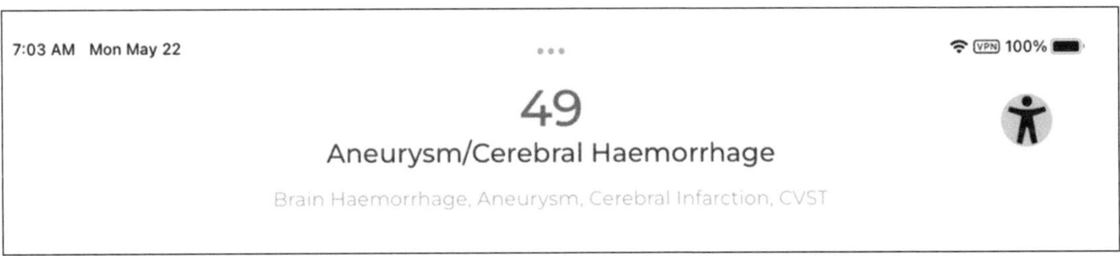

The pathomechanism of spike-producing drug-associated aneurysm was well described by Dr. Burkhardt's Group in Parts 1 and 2 of this series. Briefly, vascular endothelium (inner lining) is attacked by either spike and related proteins or by activated leukocytes that disrupt the inner lining of an artery which leads to a rupture of the vessel wall, dissection of blood by arterial pressure through the muscular layer with dilatation of the vessel wall or rupture, or both.

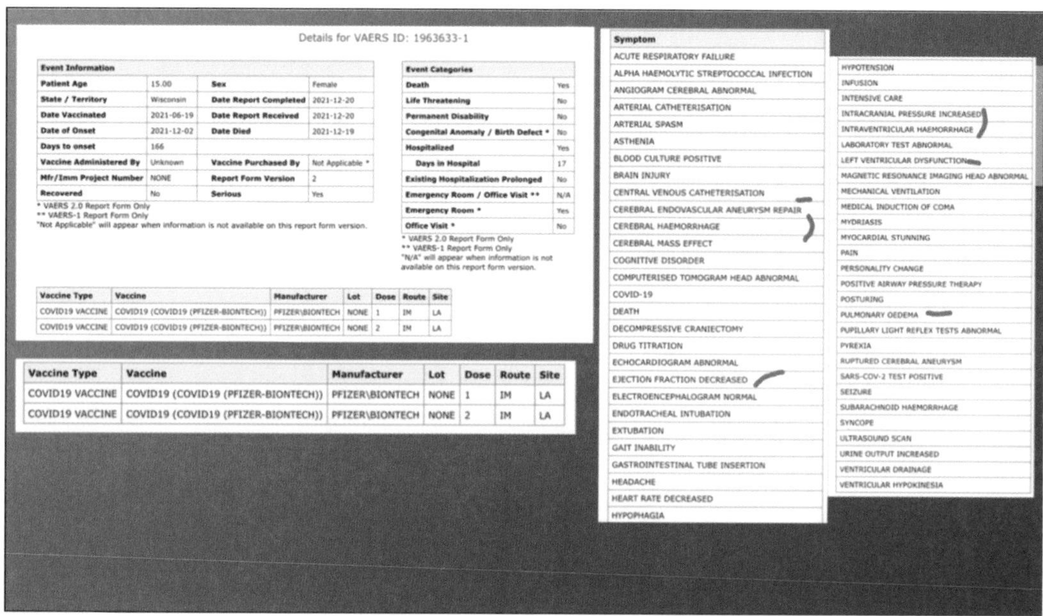

Details for VAERS ID: 1963633-1

Event Information			
Patient Age	15.00	Sex	Female
State / Territory	Wisconsin	Date Report Completed	2021-12-20
Date Vaccinated	2021-06-19	Date Report Received	2021-12-20
Date of Onset	2021-12-02	Date Died	2021-12-19
Days to onset	166		
Vaccine Administered By	Unknown	Vaccine Purchased By	Not Applicable *
Mfr/Imm Project Number	NONE	Report Form Version	2
Recovered	No	Serious	Yes

* VAERS 2.0 Report Form Only
** VAERS-1 Report Form Only
"Not Applicable" will appear when information is not available on this report form version.

Event Categories	
Death	Yes
Life Threatening	No
Permanent Disability	No
Congenital Anomaly / Birth Defect *	No
Hospitalized	Yes
Days in Hospital	17
Existing Hospitalization Prolonged	No
Emergency Room / Office Visit **	N/A
Emergency Room *	Yes
Office Visit *	No

* VAERS 2.0 Report Form Only
** VAERS-1 Report Form Only
"N/A" will appear when information is not available on this report form version.

Vaccine Type	Vaccine	Manufacturer	Lot	Dose	Route	Site
COVID19 VACCINE	COVID19 (COVID19 (PFIZER-BIONTECH))	PFIZER\BIONTECH	NONE	1	IM	LA
COVID19 VACCINE	COVID19 (COVID19 (PFIZER-BIONTECH))	PFIZER\BIONTECH	NONE	2	IM	LA

Vaccine Type	Vaccine	Manufacturer	Lot	Dose	Route	Site
COVID19 VACCINE	COVID19 (COVID19 (PFIZER-BIONTECH))	PFIZER\BIONTECH	NONE	1	IM	LA
COVID19 VACCINE	COVID19 (COVID19 (PFIZER-BIONTECH))	PFIZER\BIONTECH	NONE	2	IM	LA

Symptom
ACUTE RESPIRATORY FAILURE
ALPHA HAEMOLYTIC STREPTOCOCCAL INFECTION
ANGIOGRAM CEREBRAL ABNORMAL
ARTERIAL CATHETERISATION
ARTERIAL SPASM
ASTHENIA
BLOOD CULTURE POSITIVE
BRAIN INJURY
CENTRAL VENOUS CATHETERISATION
CEREBRAL ENDOVASCULAR ANEURYSM REPAIR
CEREBRAL HAEMORRHAGE
CEREBRAL MASS EFFECT
COGNITIVE DISORDER
COMPUTERISED TOMOGRAM HEAD ABNORMAL
COVID-19
DEATH
DECOMPRESSIVE CRANIECTOMY
DRUG TITRATION
ECHOCARDIOGRAM ABNORMAL
EJECTION FRACTION DECREASED
ELECTROENCEPHALOGRAM NORMAL
ENDOTRACHEAL INTUBATION
EXTUBATION
GAIT INABILITY
GASTROINTESTINAL TUBE INSERTION
HEADACHE
HEART RATE DECREASED
HYPOPHAGIA
HYPOTENSION
INFUSION
INTENSIVE CARE
INTRACRANIAL PRESSURE INCREASED
INTRAVENTRICULAR HAEMORRHAGE
LABORATORY TEST ABNORMAL
LEFT VENTRICULAR DYSFUNCTION
MAGNETIC RESONANCE IMAGING HEAD ABNORMAL
MECHANICAL VENTILATION
MEDICAL INDUCTION OF COMA
MYDRIASIS
MYOCARDIAL STUNNING
PAIN
PERSONALITY CHANGE
POSITIVE AIRWAY PRESSURE THERAPY
POSTURING
PULMONARY OEDEMA
PUPILLARY LIGHT REFLEX TESTS ABNORMAL
PYREXIA
RUPTURED CEREBRAL ANEURYSM
SARS-COV-2 TEST POSITIVE
SEIZURE
SUBARACHNOID HAEMORRHAGE
SYNCOPE
ULTRASOUND SCAN
URINE OUTPUT INCREASED
VENTRICULAR DRAINAGE
VENTRICULAR HYPOKINESIA

VAERS ID 1963633: This 15-year-old girl received her second dose of BNT162b2 on June 19, 2021, and passed away on December 17, 2021, after a very complicated 17 days in the hospital.

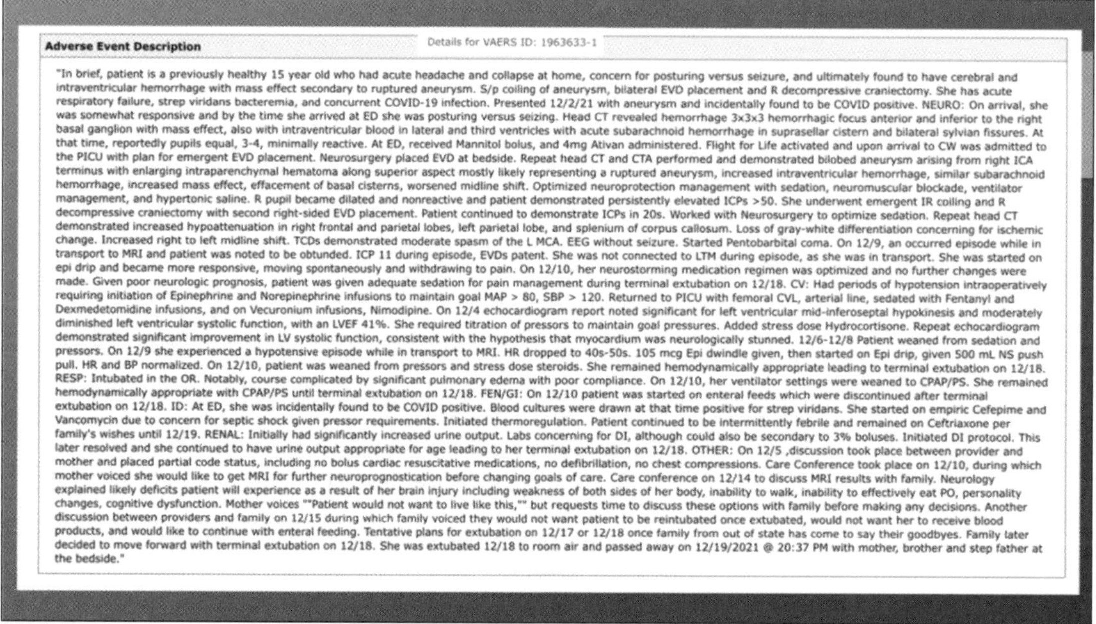

Adverse Event Description

Details for VAERS ID: 1963633-1

"In brief, patient is a previously healthy 15 year old who had acute headache and collapse at home, concern for posturing versus seizure, and ultimately found to have cerebral and intraventricular hemorrhage with mass effect secondary to ruptured aneurysm. S/p coiling of aneurysm, bilateral EVD placement and R decompressive craniectomy. She has acute respiratory failure, strep viridans bacteremia, and concurrent COVID-19 infection. Presented 12/2/21 with aneurysm and incidentally found to be COVID positive. NEURO: On arrival, she was somewhat responsive and by the time she arrived at ED she was posturing versus seizing. Head CT revealed hemorrhage 3x3x3 hemorrhagic focus anterior and inferior to the right basal ganglion with mass effect, also with intraventricular blood in lateral and third ventricles with acute subarachnoid hemorrhage in suprasellar cistern and bilateral sylvian fissures. At that time, reportedly pupils equal, 3-4, minimally reactive. At ED, received Mannitol bolus, and 4mg Ativan administered. Flight for Life activated and upon arrival to CW was admitted to the PICU with plan for emergent EVD placement. Neurosurgery placed EVD at bedside. Repeat head CT and CTA performed and demonstrated bilobed aneurysm arising from right ICA terminus with enlarging intraparenchymal hematoma along superior aspect mostly likely representing a ruptured aneurysm, increased intraventricular hemorrhage, similar subarachnoid hemorrhage, increased mass effect, effacement of basal cisterns, worsened midline shift. Optimized neuroprotection management with sedation, neuromuscular blockade, ventilator management, and hypertonic saline. R pupil became dilated and nonreactive and patient demonstrated persistently elevated ICPs >50. She underwent emergent IR coiling and R decompressive craniectomy with second right-sided EVD placement. Patient continued to demonstrate ICPs in 20s. Worked with Neurosurgery to optimize sedation. Repeat head CT demonstrated increased hypoattenuation in right frontal and parietal lobes, left parietal lobe, and splenium of corpus callosum. Loss of gray-white differentiation concerning for ischemic change. Increased right to left midline shift. TCDs demonstrated moderate spasm of the L MCA. EEG without seizure. Started Pentobarbital coma. On 12/9, an occurred episode while in transport to MRI and patient was noted to be obtunded. ICP 11 during episode, EVDs patent. She was not connected to LTM during episode, as she was in transport. She was started on epi drip and became more responsive, moving spontaneously and withdrawing to pain. On 12/10, her neurostorming medication regimen was optimized and no further changes were made. Given poor neurologic prognosis, patient was given adequate sedation for pain management during terminal extubation on 12/18. CV: Had periods of hypotension intraoperatively requiring initiation of Epinephrine and Norepinephrine infusions to maintain goal MAP > 80, SBP > 120. Returned to PICU with femoral CVL, arterial line, sedated with Fentanyl and Dexmedetomidine infusions, and on Vecuronium infusions, Nimodipine. On 12/4 echocardiogram report noted significant for left ventricular mid-inferoseptal hypokinesis and moderately diminished left ventricular systolic function, with an LVEF 41%. She required titration of pressors to maintain goal pressures. Added stress dose Hydrocortisone. Repeat echocardiogram demonstrated significant improvement in LV systolic function, consistent with the hypothesis that myocardium was neurologically stunned. 12/6-12/8 Patient weaned from sedation and pressors. On 12/9 she experienced a hypotensive episode while in transport to MRI. HR dropped to 40s-50s. 105 mcg Epi dwindle given, then started on Epi drip, given 500 mL NS push pull. HR and BP normalized. On 12/10, patient was weaned from pressors and stress dose steroids. She remained hemodynamically appropriate leading to terminal extubation on 12/18. RESP: Intubated in the OR. Notably, course complicated by significant pulmonary edema with poor compliance. On 12/10, her ventilator settings were weaned to CPAP/PS. She remained hemodynamically appropriate with CPAP/PS until terminal extubation on 12/18. FEN/GI: On 12/10 patient was started on enteral feeds which were discontinued after terminal extubation on 12/18. ID: At ED, she was incidentally found to be COVID positive. Blood cultures were drawn at that time positive for strep viridans. She started on empiric Cefepime and Vancomycin due to concern for septic shock given pressor requirements. Initiated thermoregulation. Patient continued to be intermittently febrile and remained on Ceftriaxone per family's wishes until 12/19. RENAL: Initially had significantly increased urine output. Labs concerning for DI, although could also be secondary to 3% boluses. Initiated DI protocol. This later resolved and she continued to have urine output appropriate for age leading to her terminal extubation on 12/18. OTHER: On 12/5 ,discussion took place between provider and mother and placed partial code status, including no bolus cardiac resuscitative medications, no defibrillation, no chest compressions. Care Conference took place on 12/10, during which mother voiced she would like to get MRI for further neuroprognostication before changing goals of care. Care conference on 12/14 to discuss MRI results with family. Neurology explained likely deficits patient will experience as a result of her brain injury including weakness of both sides of her body, inability to walk, inability to effectively eat PO, personality changes, cognitive dysfunction. Mother voices ""Patient would not want to live like this,"" but requests time to discuss these options with family before making any decisions. Another discussion between providers and family on 12/15 during which family voiced they would not want patient to be reintubated once extubated, would not want her to receive blood products, and would like to continue with enteral feeding. Tentative plans for extubation on 12/17 or 12/18 once family from out of state has come to say their goodbyes. Family later decided to move forward with terminal extubation on 12/18. She was extubated 12/18 to room air and passed away on 12/19/2021 @ 20:37 PM with mother, brother and step father at the bedside."

The consequences of an aneurysm can be dire not only for the injured but the loved ones who suffer while they helplessly watch as their child dies.

iii. Stroke

The father's words in the image below tell the story pretty well.

Childhood stroke has been reported to occur in 1 to 13 per 100,000 children. (Pediatric Stroke: A Review, Tsze and Valente, *Emerg Med Int.* 2011; 2011: 734506. Published online 2011 Dec 27. doi: 10.1155/2011/734506 PMCID: PMC3255104 PMID: 22254140, https://www.ncbi.nlm.nih.gov/pmc/articles/PMC3255104/.) Unfortunately, denominators to properly calculate prevalence rates are lacking with GTPs.

Pfizer Confidential Document 5.3.6 reports a seven-year-old child who received BNT162b2 long before pediatric dosing was developed and long before the emergency use authorization (EUA) was extended to children. This is another example of the sloppy administration of the "vaccine" program.

Dr. Makis and Professor Miller, independently of one another, maintain archives of CoVax Disease cases culled from the media (links below).

(https://makismd.substack.com) and (https://markcrispinmiller.substack.com)

Edward Dowd's book, Dowd, Ed. *"Cause Unknown": The Epidemic of Sudden Deaths in 2021 & 2022.* Skyhorse Publishing, 2023 is a similar archive of sudden deaths along with statistical data.

F. Autoimmunity and Immunologic Effects

Autoimmunity was featured in Part 4C of this series. (https://dailyclout.io/report-70-part-4c-clinical-evidence-supporting-the-hypothesis-that-autoimmunity-is-one-of-the-principal-pathological-mechanisms-of-harm-from-covid-19-gene-therapy-drugs/)

Briefly, autoimmunity is a process in which one's immune system attacks "self" thus producing an illness through the process of inflammation, often with system-wide effects although more localized varieties exist such as thyroiditis and diabetes type I.

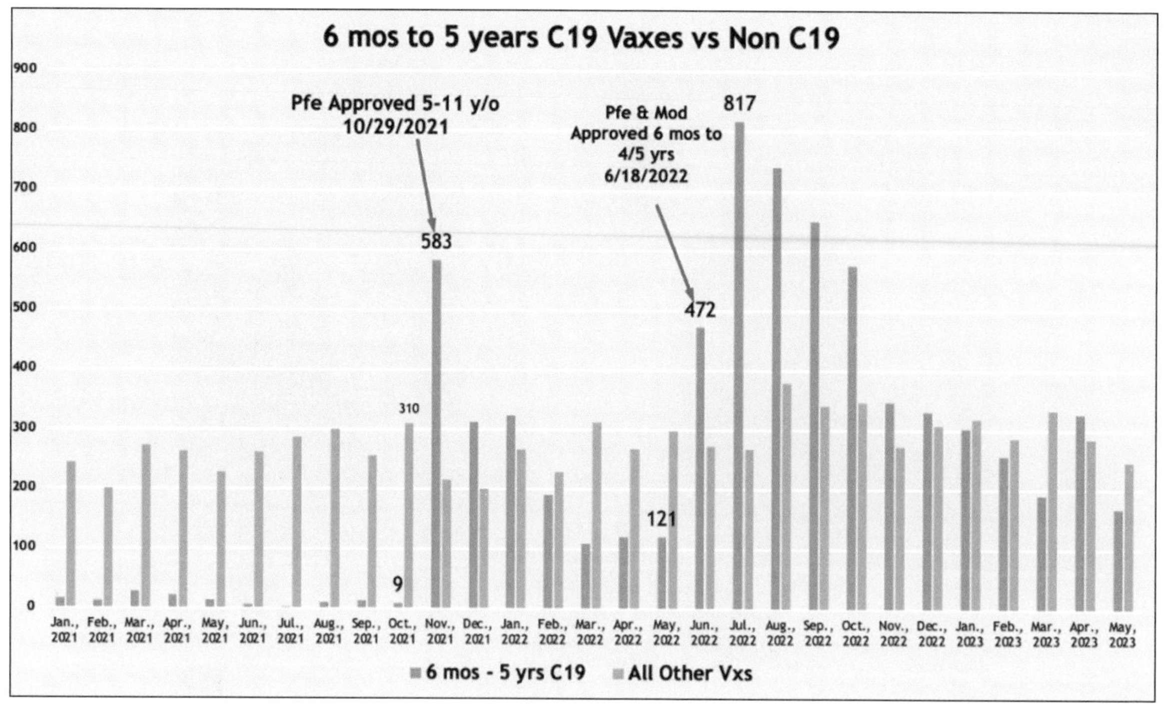

Eighty-five percent of autoimmune diseases following vaccination in the past 10 years have followed the public introduction of C19 gene therapy drugs. Females dominate the 18- to 29-year-old bracket.

Details for VAERS ID: 1486812-1

Event Information			
Patient Age	12.00	Sex	Female
State / Territory	Michigan	Date Report Completed	2021-07-20
Date Vaccinated	2021-06-01	Date Report Received	2021-07-20
Date of Onset	2021-07-20	Date Died	
Days to onset	49		
Vaccine Administered By	Private	Vaccine Purchased By	Not Applicable *
Mfr/Imm Project Number	NONE	Report Form Version	2
Recovered	No	Serious	No

* VAERS 2.0 Report Form Only
** VAERS-1 Report Form Only
"Not Applicable" will appear when information is not available on this report form version.

Event Categories	
Death	No
Life Threatening	No
Permanent Disability	No
Congenital Anomaly / Birth Defect *	No
Hospitalized	No
Days in Hospital	None
Existing Hospitalization Prolonged	No
Emergency Room / Office Visit **	N/A
Emergency Room *	No
Office Visit *	No

* VAERS 2.0 Report Form Only
** VAERS-1 Report Form Only
"N/A" will appear when information is not

Symptom
ALOPECIA
AUTOIMMUNE DISORDER
HYPERTHYROIDISM

Vaccine Type	Vaccine	Manufacturer	Lot	Dose	Route	Site
COVID19 VACCINE	COVID19 (COVID19 (PFIZER-BIONTECH))	PFIZER\BIONTECH	EW0191	2	SYR	LA

Adverse Event Description

Hyperthyroidism or some sort of autoimmune disorder. No one in our family on any side, has hyperthyroidism. We are trying to figure it out. It mimics hyperthyroidism. Has the same symptoms as Grave but Graves? disease was negative. Dr. said ?she?s not textbook?. He?s been doing this for about 30 years. Started losing massive amount of hair within a couple weeks of second dose. She has less than a third of the hair she had before the vaccine and is losing more daily. When put on methimazole for treatment, she reacts with extremely painful, deep mouth sores. Which isn?t typical. She?s never had mouth sores before. I?m wondering if it?s related to the vaccine. Her thyroid numbers were ? a little high but not bad? before the vaccine. But I?m wondering if the vaccine has something to do with it.

Lab Data	Current Illness	Adverse Events After Prior Vaccinations
Lots of labs june 10th, July 9th. She has an ultrasound of her thyroid and an two day scan of her thyroid on August 18 & 19.	Seasonal allergies	

Medications At Time Of Vaccination	History/Allergies
Onfi, gabapentin, respirdone	Asd,None

VAERS ID 1486812: This 12-year-old girl (above) developed autoimmune thyroiditis 49 days after dose two of BNT162b2. CoVax illnesses often resemble known illnesses, but, as in this case with "massive" hair loss, something is unique in many of them.

G. Multisystem Inflammatory Syndrome in Children (MIS-C)

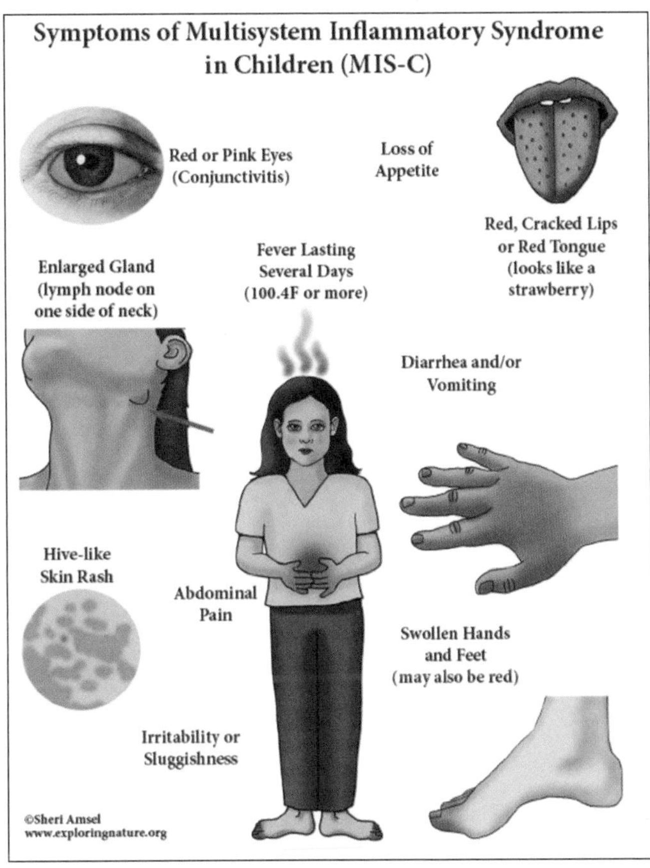

MIS-C is an inflammatory process that involves multiple organs and physiological systems simultaneously or in sequence, sometimes mild at first but proceeding rapidly at times to critical illness. MIS-C was discussed in Part 4C of this series. (https://dailyclout.io/report-70-part-4c-clinical-evidence-supporting-the-hypothesis-that-autoimmunity-is-one-of-the-principal-pathological-mechanisms-of-harm-from-covid-19-gene-therapy-drugs/)

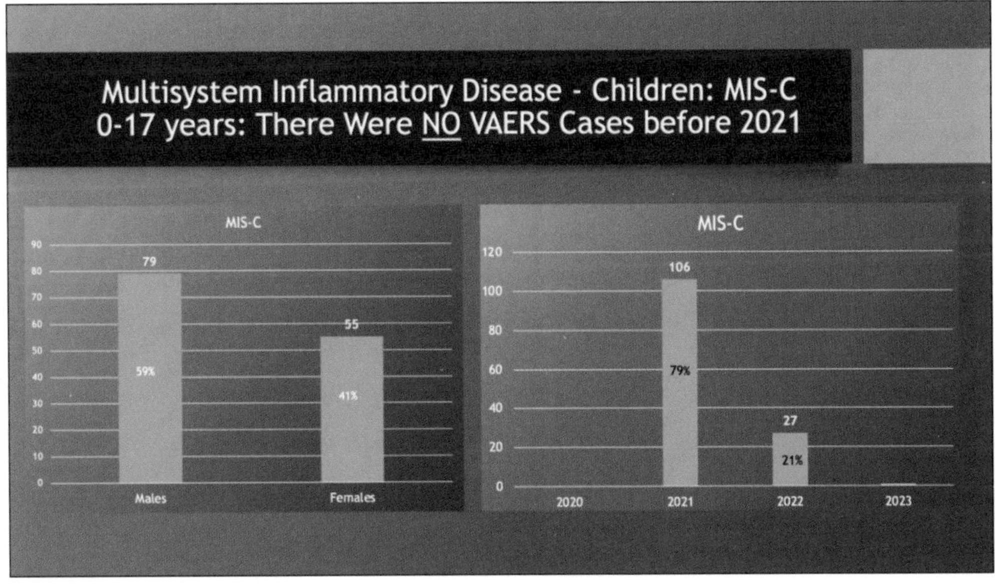

VAERS through 5/12/2023

Males are afflicted in 59% of reported cases (left histogram of previous image). There were no cases reports from 2020, but 106 were reported in 2021 and then, in 2022, the reports dropped to 27 as the spike drug use declined (right histogram).

As noted, prior to rollout of the C19 "vaccines," there were no MIS-C cases in VAERS. As the public's willingness to participate in the inoculation with C19 "vaccines" waned, so did the reports of MIS-C as shown in the right histogram above.

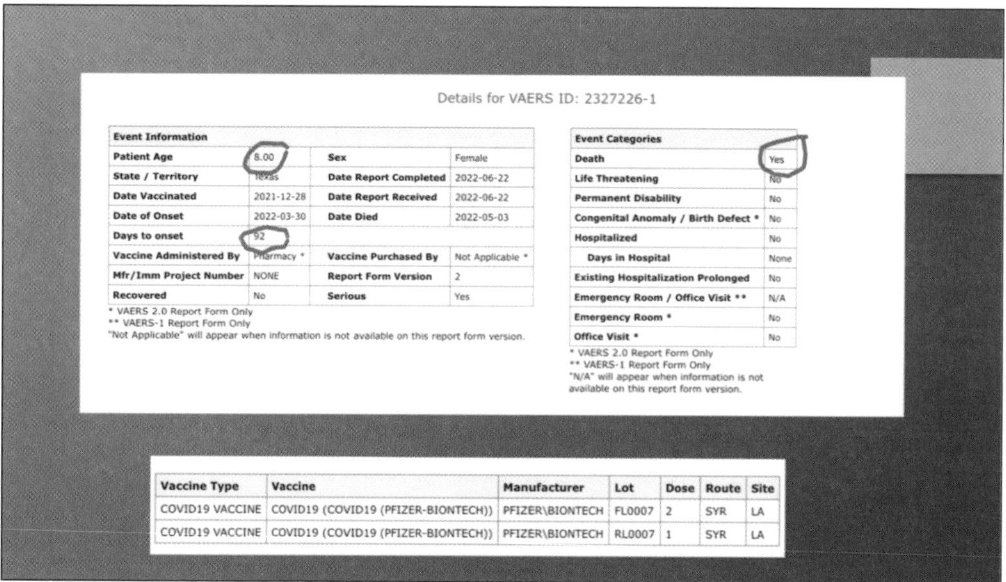

Details for VAERS ID: 2327226-1

Event Information			
Patient Age	8.00	Sex	Female
State / Territory	Texas	Date Report Completed	2022-06-22
Date Vaccinated	2021-12-28	Date Report Received	2022-06-22
Date of Onset	2022-03-30	Date Died	2022-05-03
Days to onset	92		
Vaccine Administered By	Pharmacy *	Vaccine Purchased By	Not Applicable *
Mfr/Imm Project Number	NONE	Report Form Version	2
Recovered	No	Serious	Yes

* VAERS 2.0 Report Form Only
** VAERS-1 Report Form Only
"Not Applicable" will appear when information is not available on this report form version.

Event Categories	
Death	Yes
Life Threatening	No
Permanent Disability	No
Congenital Anomaly / Birth Defect *	No
Hospitalized	No
Days in Hospital	None
Existing Hospitalization Prolonged	No
Emergency Room / Office Visit **	N/A
Emergency Room *	No
Office Visit *	No

* VAERS 2.0 Report Form Only
** VAERS-1 Report Form Only
"N/A" will appear when information is not available on this report form version.

Vaccine Type	Vaccine	Manufacturer	Lot	Dose	Route	Site
COVID19 VACCINE	COVID19 (COVID19 (PFIZER-BIONTECH))	PFIZER\BIONTECH	FL0007	2	SYR	LA
COVID19 VACCINE	COVID19 (COVID19 (PFIZER-BIONTECH))	PFIZER\BIONTECH	RL0007	1	SYR	LA

VAERS ID 2327226: This eight-year-old girl was not so fortunate as others and passed away four months following her second dose of BNT162b2. She was febrile for three weeks before being admitted to the hospital for multiple organ system failure.

Her organ involvement included lymph nodes, skin, heart, intestines, lungs, and liver.

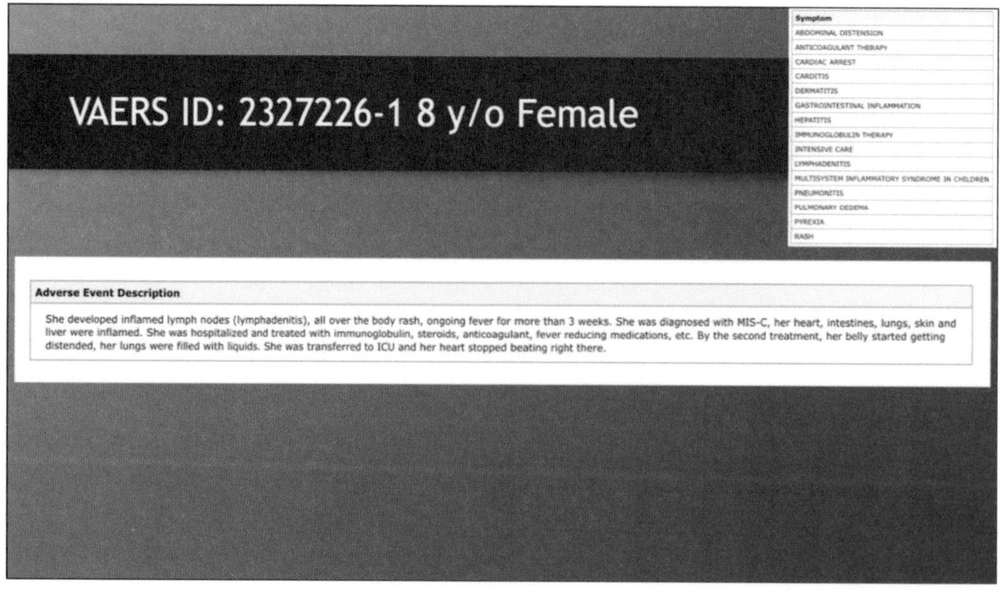

VAERS ID: 2327226-1 8 y/o Female

Symptom
ABDOMINAL DISTENSION
ANTICOAGULANT THERAPY
CARDIAC ARREST
CARDITIS
DERMATITIS
GASTROINTESTINAL INFLAMMATION
HEPATITIS
IMMUNOGLOBULIN THERAPY
INTENSIVE CARE
LYMPHADENITIS
MULTISYSTEM INFLAMMATORY SYNDROME IN CHILDREN
PNEUMONITIS
PULMONARY OEDEMA
PYREXIA
RASH

Adverse Event Description
She developed inflamed lymph nodes (lymphadenitis), all over the body rash, ongoing fever for more than 3 weeks. She was diagnosed with MIS-C, her heart, intestines, lungs, skin and liver were inflamed. She was hospitalized and treated with immunoglobulin, steroids, anticoagulant, fever reducing medications, etc. By the second treatment, her belly started getting distended, her lungs were filled with liquids. She was transferred to ICU and her heart stopped beating right there.

The final two sentences need emphasis: "By the second treatment, her belly started getting distended, her lungs filled with liquids. She was transferred to ICU and her heart stopped beating right there."

Six cases from VAERS are summarized below. All six of these youngsters survived. There has been no assessment and reporting of impairment and disability in survivors.

Age	Sex	Symptoms	VAERS ID	Adverse Event Description
6-17 years	Female	MULTI-ORGAN DISORDER	1713661-1	Atypical MIS-C: multiple organ involvement (skin, GI), elevated ESR, CRP, lymphopenia, neutrophilia Symptoms started approx 7 days prior to admission Treated w/ IVIG, steroids, ASA
6-17 years	Female	MULTI-ORGAN DISORDER	2271236-1	1/6 - 8 YR OLD PRESENTED TO THE ER AFTER 4 DAYS OF FEVER/MAX 103, GENERALIZED RASH, HA, SORE THROAT, ABDOMINAL PAIN, MALAISE, REDNESS OF HANDS. POSSIBLE COVID EXP 12/24/21, STREP/ COVID/FLU NEG AT DR OFFICE 1/3. SYMPTOMS PROGRESSED, SENT TO ER-PRIMARY MD CONCERNED FOR MIS-C. UPON EVAL PT WAS ILL APPEARING, FEBRILE, TACHYCARDIC, HYPOTENSIVE 90/40. CONCERN FOR MIS-C VS. KAWASASKI VS. VIRAL EXANTHEM VS. TICK BORNE ILLNESS. ADMITTED TO UNIT FLOOR. REC'D IV FLUIDS, EMPIRIC CEFTRIAXONE, IVIG AND LOW DOSE ASPIRIN PER MIS-C PROTOCOL. 1/7 ADDED IV STEROIDS & LOVENOX. Blood and urine cx NG >36 hours and at 5 days. Hospital course: Blood and urine cx NG >36 hours & at 5 days, - Tick studies negative, Pancytopenia improving -Received 2nd dose of IVIG due to being febrile and symptomatic within 36 hours after 1st infusion -Kept on mIVF for hydration -Appetite improved over time -Developed Abdominal pain; abd XR showed prominent liver, inc LFTS; abd US showed diffuse inflammation in the abdomen involving multiple organs -Abdomen pain and LFTs improved over following days, no N/V, constipation or diarrhea Discharge: -Follow up with physician within 48 hours of discharge -follow up with cardiology and rheumatology in the outaptient setting -continue with aspirin once a day, steroids once a day and famotidine for belly protection - rheumatology will help taper the steroids, for now continue taking them once a day.
6-17 years	Male	MULTI-ORGAN DISORDER	1399370-1	Patient was admitted from PCP for extreme tachycardia and tachypnea and developed multi organ involvement with tachycardia (HR to 140-150s), slight elevation in BNP (H of 490), Troponin (H of 0.244), mild proteinuria (50-70 proteins), respiratory distress with tachypnea (RR 50s) and hypoxia requiring escalation in O2 supplementation. Also with daily fevers until starting steroids. Laboratory findings concerning for slight hypertriglyceridemia, normal Ferritin, worsening thrombocytopenia, lymphopenia, hyponatremia, and hypoalbuminemia. CT with bibasilar atelectasis vs. consolidation, but no evidence of PE. Extensive ID and rheumatological evaluation performed and unremarkable so far. Received 2 days of Doxycycline. Was started on pulse dose steroids and began to show improvement in all markers.
6-17 years	Male	MULTI-ORGAN DISORDER	2001194-1	Patient received second dose of Pfizer Covid19 vaccine on 12/18/2021. Lot number FL0007. Patient first developed MIS-C symptoms on 12/24/2021, became febrile on 12/27/2021, and was subsequently hospitalized in the ICU on 12/29/2021. Patient experienced mild perihilar peribronchial cuffing, volume loss at lung bases, mildly suppressed systolic function,abnormal LV longitudinal strain. Evidence of cardiac, hematologic and gastrointestinal organ involvement (shock, hypotension, elevated troponin, chest pain/tightness, elevated d-dimers, abdominal pain, nausea, vomiting, diarrhea, elevated liver enzymes, conjunctival infection and periorbital edema.
6-17 years	Male	MULTI-ORGAN DISORDER	2151905-1	Case-patient had first Pfizer COVID-19 vaccine on 12/7/21, developed MIS-C symptoms on 1/5/2022, and received second Pfizer vaccine on 1/7/22. Case reported having an undiagnosed upper respiratory infection in late December/early January (specific date unknown). Case-patient met case definition for MIS-C, with evidence of clinically severe illness requiring hospitalization, fever, multisystem organ involvement (cardiac, hematologic, and GI). The case-patient experienced tachycardia, headaches, abdominal pain, vomiting, and various elevated inflammatory markers (see box 19 below). Case-patient was treated with IVIG, ASA, and steroids and survived after a 2 day hospitalization.
6-17 years	Male	MULTI-ORGAN DISORDER	2233852-1	Case-patient had first Pfizer COVID-19 vaccine on 12/18/2021 and illness onset for MIS-C was on the same day. Case had COVID-19 with mild symptoms approximately 4 weeks before MIS-C onset. Case-patient met case definition for MIS-C with evidence of clinically severe illness requiring hospitalization, fever, multisystem organ involvement (cardiac, hematologic, & GI). The case-patient experienced cardiac shock, chest pain, abdominal pain, vomiting, diarrhea, conjunctival injection, and various elevated inflammatory markers (see box 19 below). Case-patient was treated with IVIG, ASA, steroids, and epinephrine and survived after a 4 day hospitalization (1 day in PICU).

Note: Submitting a report to VAERS does not mean that healthcare personnel or the vaccine caused or contributed to the adverse event (possible side effect).

This category of CoVax Disease, MIS-C, is another example of complex and aggressive illness following GTPs.

H. Myocarditis/Pericarditis

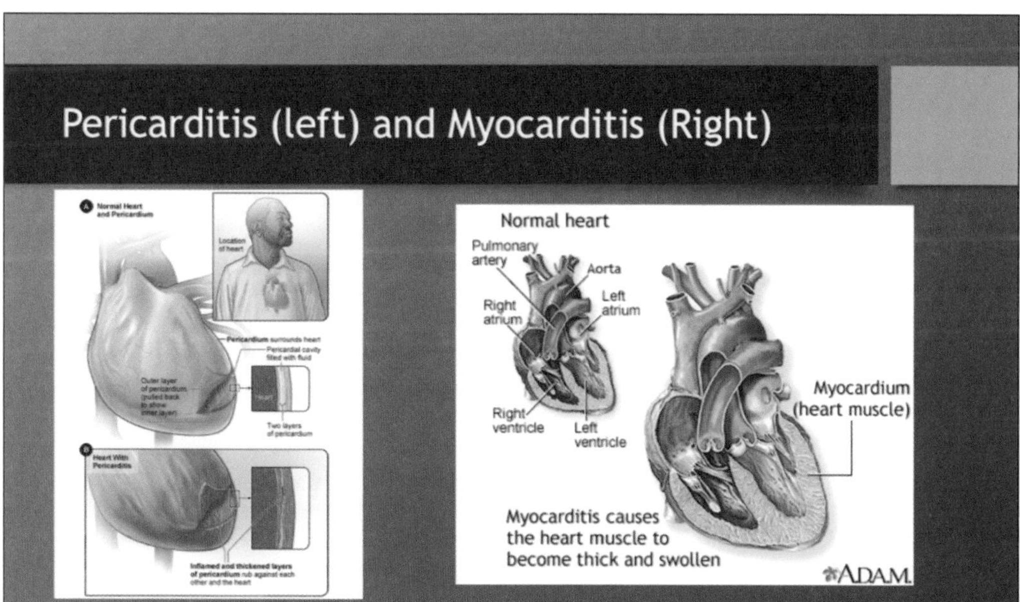

The pericardium is a fibrous tissue layer surrounding the heart. Inflammation of this sac-like structure can compromise cardiac function.

Myocarditis is inflammation of the heart muscle itself. Actual tissue destruction occurs to variable degrees; and, like any muscle, once the muscle cell dies, the muscle is replaced by rigid scar tissue (previous image).

Inflammation of the heart not only damages the muscle but can interfere with transmission of electrical signals to activate the heart muscle, causing an irregular rhythm that can be fatal.

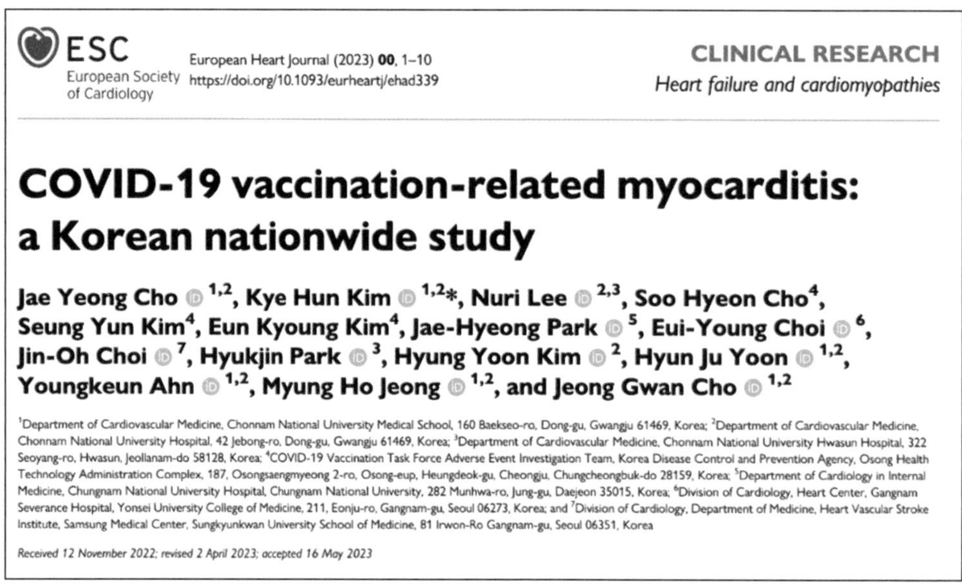

ESC European Society of Cardiology
European Heart Journal (2023) **00**, 1–10
https://doi.org/10.1093/eurheartj/ehad339
CLINICAL RESEARCH
Heart failure and cardiomyopathies

COVID-19 vaccination-related myocarditis: a Korean nationwide study

Jae Yeong Cho [1,2], Kye Hun Kim [1,2]*, Nuri Lee [2,3], Soo Hyeon Cho[4], Seung Yun Kim[4], Eun Kyoung Kim[4], Jae-Hyeong Park [5], Eui-Young Choi [6], Jin-Oh Choi [7], Hyukjin Park [3], Hyung Yoon Kim [2], Hyun Ju Yoon [1,2], Youngkeun Ahn [1,2], Myung Ho Jeong [1,2], and Jeong Gwan Cho [1,2]

[1]Department of Cardiovascular Medicine, Chonnam National University Medical School, 160 Baekseo-ro, Dong-gu, Gwangju 61469, Korea; [2]Department of Cardiovascular Medicine, Chonnam National University Hospital, 42 Jebong-ro, Dong-gu, Gwangju 61469, Korea; [3]Department of Cardiovascular Medicine, Chonnam National University Hwasun Hospital, 322 Seoyang-ro, Hwasun, Jeollanam-do 58128, Korea; [4]COVID-19 Vaccination Task Force Adverse Event Investigation Team, Korea Disease Control and Prevention Agency, Osong Health Technology Administration Complex, 187, Osongsaengmyeong 2-ro, Osong-eup, Heungdeok-gu, Cheongju, Chungcheongbuk-do 28159, Korea; [5]Department of Cardiology in Internal Medicine, Chungnam National University Hospital, Chungnam National University, 282 Munhwa-ro, Jung-gu, Daejeon 35015, Korea; [6]Division of Cardiology, Heart Center, Gangnam Severance Hospital, Yonsei University College of Medicine, 211, Eonju-ro, Gangnam-gu, Seoul 06273, Korea; and [7]Division of Cardiology, Department of Medicine, Heart Vascular Stroke Institute, Samsung Medical Center, Sungkyunkwan University School of Medicine, 81 Irwon-Ro Gangnam-gu, Seoul 06351, Korea

Received 12 November 2022; revised 2 April 2023; accepted 16 May 2023

(https://academic.oup.com/eurheartj/advance-article/doi/10.1093/eurheartj/ehad339/7188747)

Cho et al. reviewed the Korean national database reports of 1,533 cases of myo/pericarditis in the Korean vaccinated population of 44,276,704 persons, out of 51,349,116 (comprising the total Korean population) following review by the government-organized Expert Adjudication Committee on COVID-19 Vaccination Pericarditis/Myocarditis. After screening, the committee confirmed 480 cases of vaccine-associated heart disease.

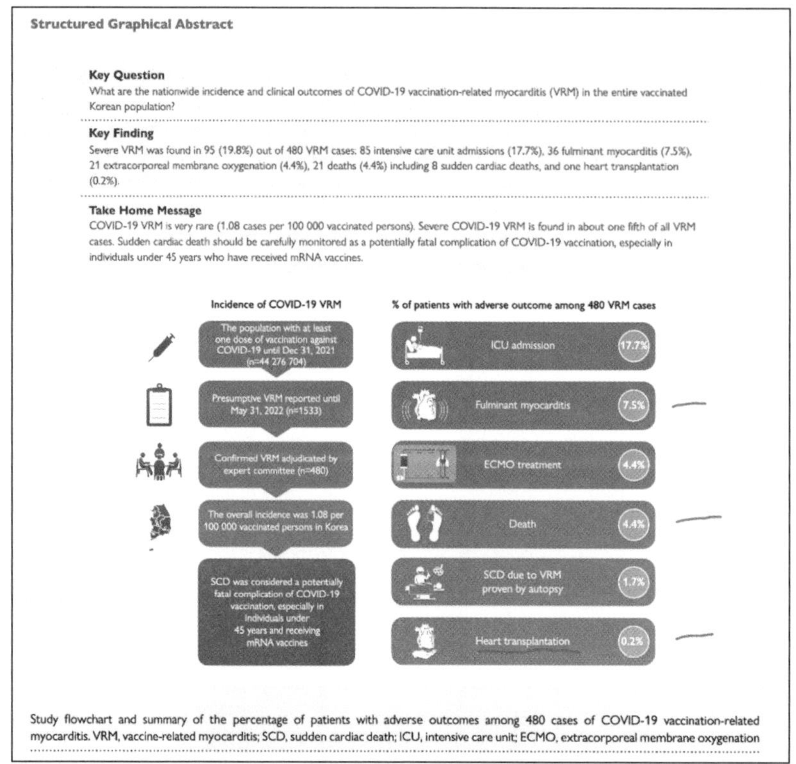

Structured Graphical Abstract

Key Question
What are the nationwide incidence and clinical outcomes of COVID-19 vaccination-related myocarditis (VRM) in the entire vaccinated Korean population?

Key Finding
Severe VRM was found in 95 (19.8%) out of 480 VRM cases: 85 intensive care unit admissions (17.7%), 36 fulminant myocarditis (7.5%), 21 extracorporeal membrane oxygenation (4.4%), 21 deaths (4.4%) including 8 sudden cardiac deaths, and one heart transplantation (0.2%).

Take Home Message
COVID-19 VRM is very rare (1.08 cases per 100 000 vaccinated persons). Severe COVID-19 VRM is found in about one fifth of all VRM cases. Sudden cardiac death should be carefully monitored as a potentially fatal complication of COVID-19 vaccination, especially in individuals under 45 years who have received mRNA vaccines.

Study flowchart and summary of the percentage of patients with adverse outcomes among 480 cases of COVID-19 vaccination-related myocarditis. VRM, vaccine-related myocarditis; SCD, sudden cardiac death; ICU, intensive care unit; ECMO, extracorporeal membrane oxygenation

Males accounted for 62% of cases with a median age of 30 years and a range of 20 to 45 years. ICU admission was required in almost 18%, and 0.2% underwent heart transplantation. Death occurred in 4.4%. The authors offer incidence estimates of 1.08 cases of vaccine-related cases of myo/pericarditis cases per 100,000 vaccinated persons. **National dosing data is not a suitable denominator for prevalence calculation.**

Cho et al. discussed under-reporting as a limitation of the study, ". . . thorough measurement of cardiac troponin levels and endomyocardial biopsy could minimize underreporting in the present study." Detailed prevalence studies using cardiac MRI (cMRI) with late gadolinium enhancement (LGE), echocardiography, and other diagnostic studies are necessary.

Recently, Barmada et al. evaluated 23 young patients with myo/pericarditis. The bulk of the article concerns analysis of the immunological aspects of these cases, but they also presented one of the larger series of myopericarditis cases with follow up cMRI data.

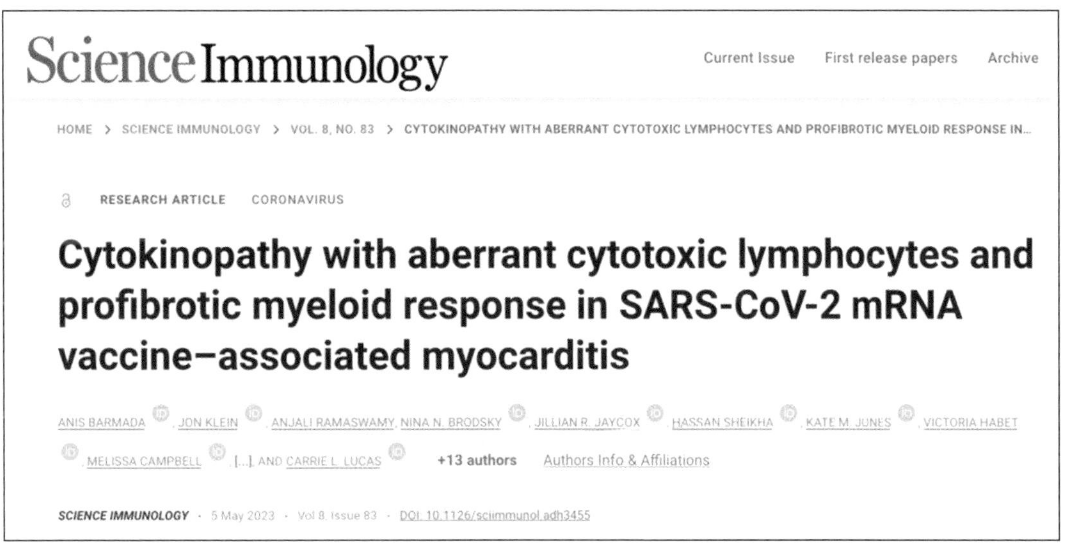
Science Immunology

Current Issue First release papers Archive

HOME > SCIENCE IMMUNOLOGY > VOL. 8, NO. 83 > CYTOKINOPATHY WITH ABERRANT CYTOTOXIC LYMPHOCYTES AND PROFIBROTIC MYELOID RESPONSE IN...

RESEARCH ARTICLE CORONAVIRUS

Cytokinopathy with aberrant cytotoxic lymphocytes and profibrotic myeloid response in SARS-CoV-2 mRNA vaccine–associated myocarditis

ANIS BARMADA, JON KLEIN, ANJALI RAMASWAMY, NINA N. BRODSKY, JILLIAN R. JAYCOX, HASSAN SHEIKHA, KATE M. JONES, VICTORIA HABET, MELISSA CAMPBELL, [...], AND CARRIE L. LUCAS +13 authors Authors Info & Affiliations

SCIENCE IMMUNOLOGY · 5 May 2023 · Vol 8, Issue 83 · DOI: 10.1126/sciimmunol.adh3455

(https://www.science.org/doi/full/10.1126/sciimmunol.adh3455)

Males comprised 87%, and the average age was 16.9 years, (range 13 to 21 years). Onset of symptoms ranged from a few days to over a week after the second dose of BNT162b2. Outcome data is contained in Supplement 1 found online but was lacking necessary clinical information including, but not limited to, age, follow-up time period, and functional recovery including athletics.

Six patients were excluded, but the authors admitted these cases might be worth studying as much as the others. The reasons for exclusion were a positive polymerase chain reaction (PCR) test for SC2 or greater than seven days between injection and onset of symptoms even though about a third of myocarditis cases in VAERS occur after seven days.

The six excluded patients were not differentiated from the six patients with incomplete cMRI data.

Blood studies were positive for myocardial injury and inflammation; elevated troponin, CRP, BNP, and NLR. (https://www.ahajournals.org/doi/full/10.1161/circulationaha.111.023697, https://www.mayoclinic.org/tests-procedures/c-reactive-protein-test/about/pac-20385228,https://my.clevelandclinic.org/health/diagnostics/22629-b-type-natriuretic-peptide, https://pubmed.ncbi.nlm.nih.gov/26878164/.)

Barmada et al. evaluated RVEF and LVEF, which stand for right and left ventricle ejection fraction—i.e., the percent of the blood ejected from the right and left ventricles during contraction (systole).

Data from initial hospitalization and follow-up after at least two months right and left ventricle ejection fraction data are presented with the right ventricle ejection fraction on top and the left ventricle data on bottom (available for 17 of 23 patients).

Values for healthy young males should be >55%, although this number varies from source to source. (https://doi.org/10.1161/CIRCIMAGING.113.000706 Circulation: Cardiovascular Imaging. 2013; 6:700–7)

	RVEF1	RVEF2	Change
P1	47	51	4
P2	53	49	-4
P3	58	48	-10
P4	55	54	-1
P5	50	50	0
P6	59	52	-7
P7	51	50	-1
P9	44	46	2
P11	38	43	5
P15	54	58	4
P16	51	46	-5
P17	50	50	0
P18	44	44	0
P19	48	51	3
P20	39	51	12
P21	59	61	2
P22	54	43	-11
	50.2	49.8	-0.41

	LVEF1	LVEF2	Change
P1	57	56	-1
P2	60	56	-4
P3	65	59	-6
P4	55	57	2
P5	58	57	-1
P6	61	58	-3
P7	52	50	-2
P9	53	46	-7
P11	45	48	3
P15	59	56	-3
P16	51	49	-2
P17	49	53	4
P18	44	48	4
P19	50	57	7
P20	44	58	14
P21	58	60	2
P22	53	47	-6
	53.8	53.8	1

From Barmada et al.

RVEF: Two patients had normal values at follow-up. Seven had lower EFs after "recovery," and seven improved but were below 55% at follow-up.

LVEF: Seven patients were in the normal range during hospitalization and after at least two months. Three patients improved, and nine declined.

Late gadolinium enhancement (LGE) on cMRI is an indication of ongoing myocardial inflammation/fibrosis. (https://radiopaedia.org/articles/late-gadolinium-enhancement-2?lang=us) Data are available for 17 patients (following image). LGE1 indicates results during hospitalization. LGE2 follow-up was performed at least two months later.

	LGE1	LGE2
P1	+	+
P2	-	+
P3	+	+
P4	-	+
P5	+	+
P6	+	+
P7	+	+
P9	+	+
P11	+	-
P15	+	+
P16	+	+
P17	+	+
P18	+	+
P19	+	-
P20	+	-
P21	+	+
P22	+	+
	15/17	14/17

Published 5/05/2023

Fifteen of 17 patients had LGE in the acute phase of their illness and *14 of 17 began or continued to have LGE after their "recovery."* Three patients went from positive to negative and two went from negative to positive. The long-term damage to these hearts was not emphasized by Barmada et al. We cannot determine how many of these young people will be needing heart transplants, pacemakers, and other future treatment for their damaged hearts.

The CDC has known that myo/pericarditis occurs after C19 gene therapy as has been made public. (https://www.cdc.gov/coronavirus/2019-ncov/vaccines/safety/adverse-events.html)

Myocarditis and pericarditis **after COVID-19 vaccination are rare.** Myocarditis is inflammation of the heart muscle, and pericarditis is inflammation of the outer lining of the heart. Most patients with myocarditis or pericarditis after COVID-19 vaccination responded well to medicine and rest and felt better quickly. Most cases have been reported after receiving Pfizer-BioNTech or Moderna (mRNA COVID-19 vaccines), particularly in male adolescents and young adults.

As of 6/16/2023

The following data are from VAERS using myocarditis "signal" for all vaccines.

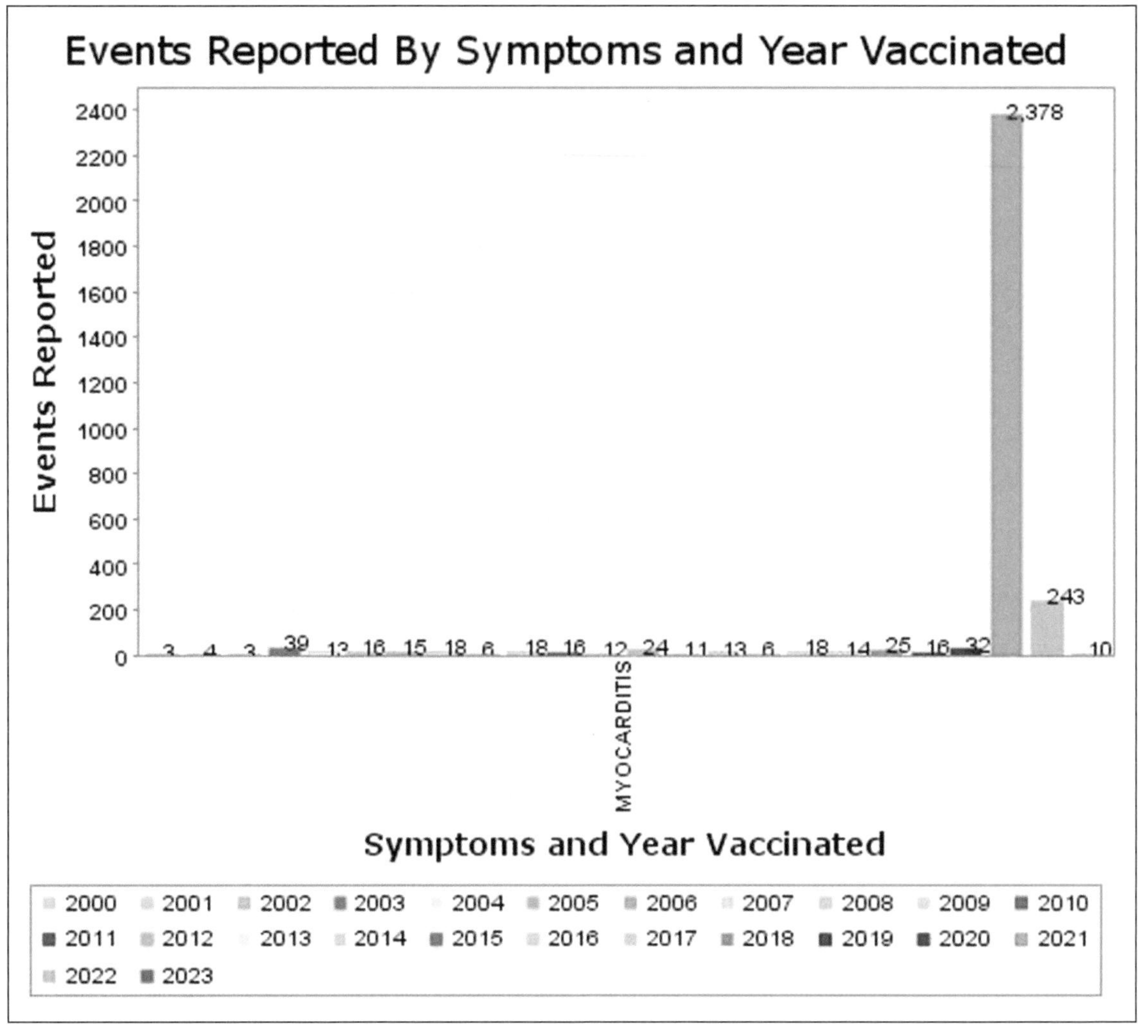

Of the 2,378 event reports in 2021, 2,345 were associated with C19 gene therapy products.

Messages:
- VAERS data in CDC WONDER are updated every Friday. Hence, results for the same query can change from week to week.
- These results are for 2,607 total events.

Symptoms	Year Vaccinated	Events Reported	Percent (of 2,607)
MYOCARDITIS	2020	24	0.92%
	2021	2,345	89.95%
	2022	232	8.90%
	2023	6	0.23%
	Total	**2,607**	**100.00%**
Total		**2,607**	**100.00%**

Males had 71.68% of these events, 21% occurred in ages 6 to 17, and 33% occurred in 18- to 29-year-olds.

As noted previously, VAERS numbers change over time, but this pattern is remarkably consistent with a low level of reporting in children under 18 years old in the VAERS database for 30 years including the COVID-19 year of 2020.

It was in 2021 that the number of myocarditis reports jumped, and then they fell back in 2022 as dosing with C19 "vaccines" tapered off. The mRNA products were authorized for 12- to 18-year-olds in May 2021 as the dosing program headed to its peak.

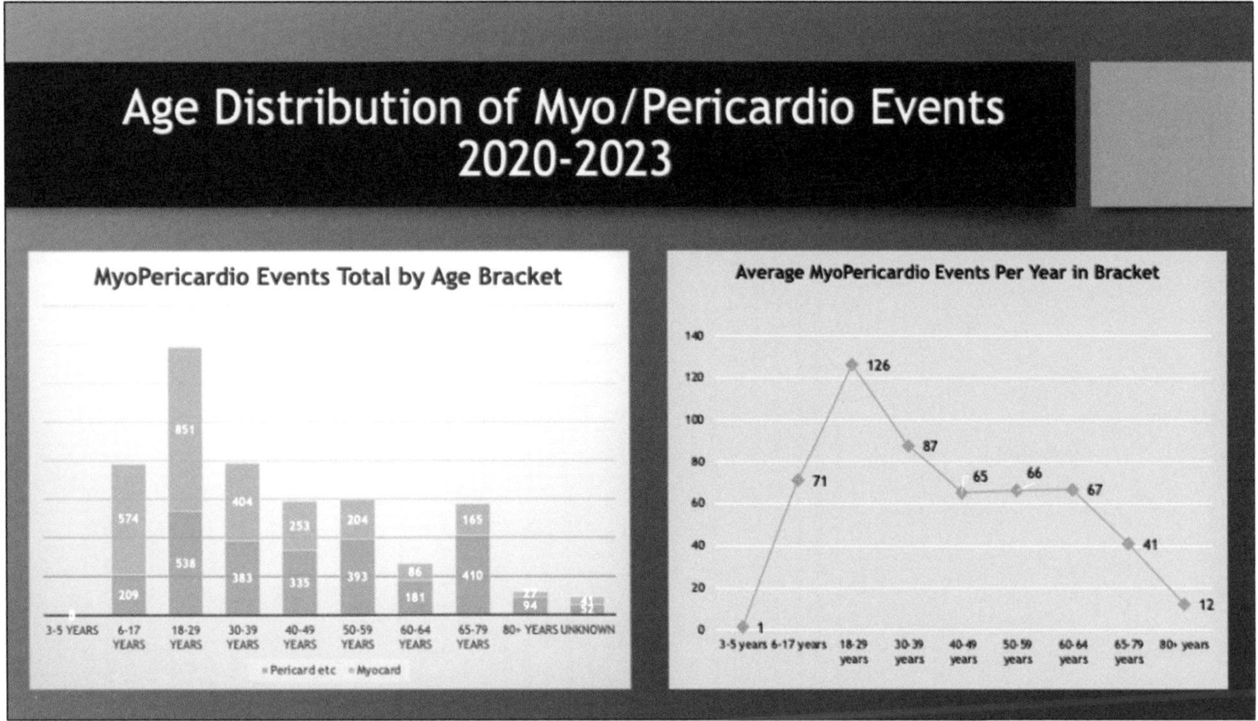

The left histogram above is a plot of the distribution of combined myo/pericardial event reports by age bracket with the average per year within the bracket on the right.

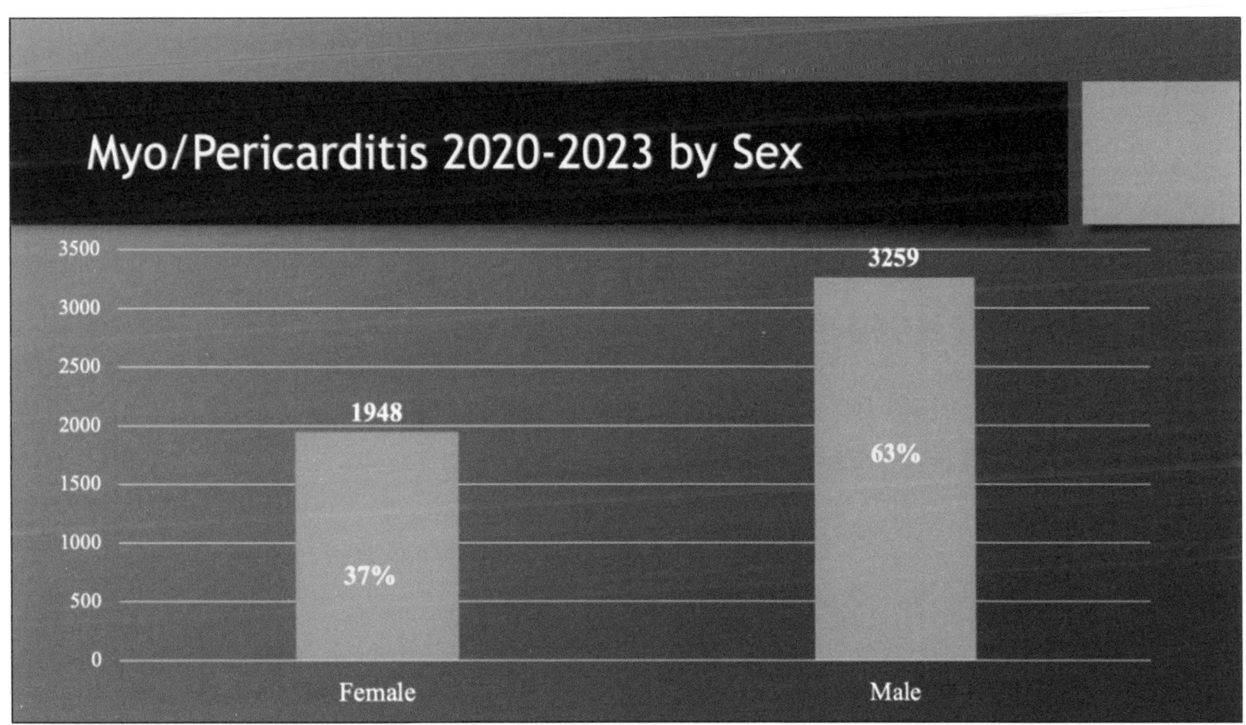

Males dominate the category of inflammatory cardiac disease.

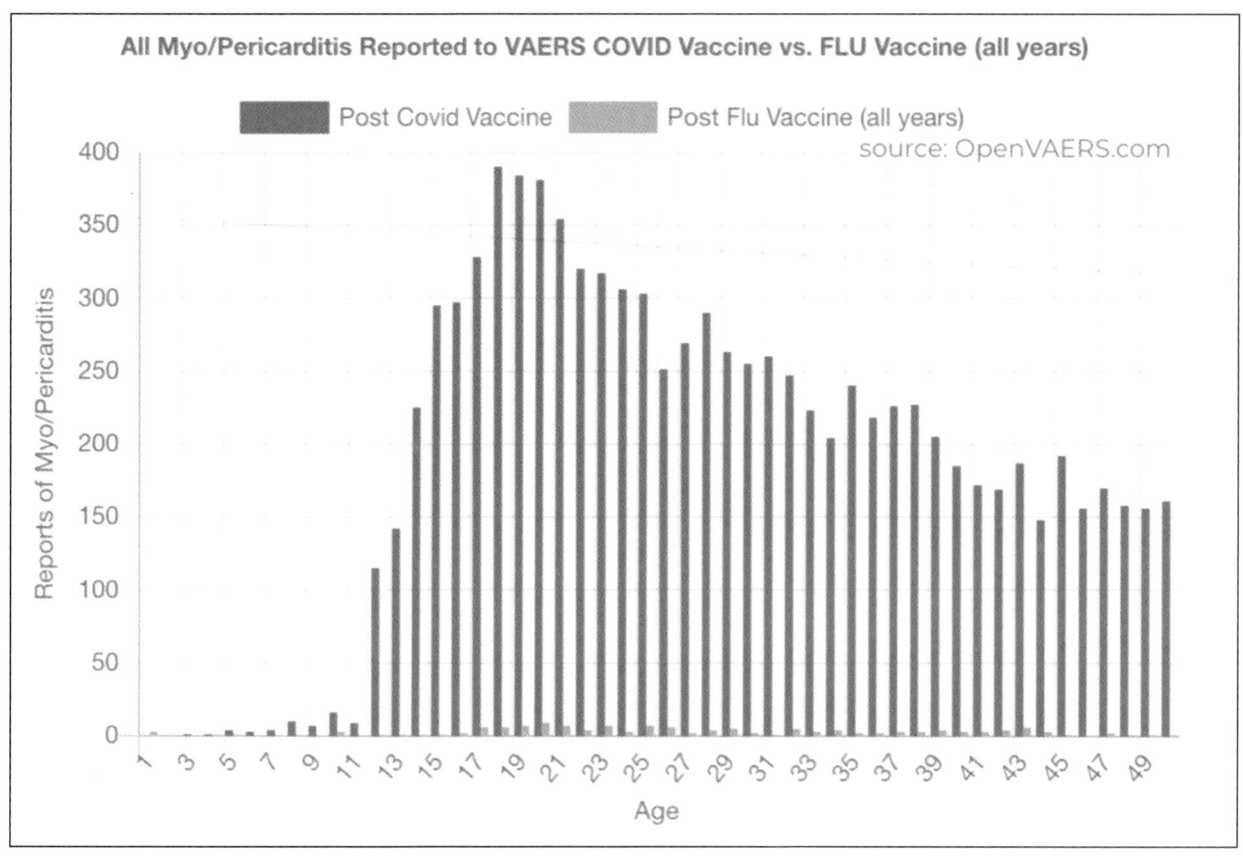

(https://openvaers.com/)

This chart from OpenVAERS shows the contrast between the reporting of myo/pericarditis after flu vaccine, in blue, compared with C19 gene therapy products, in red. Flu hardly shows up.

IV. Death, Sudden Death, and Sudden Cardiac Death
A. Tampering with VAERS Reports

Dr. Makis tracks morbidity and mortality reports in public media on his Substack. His June 10, 2023, article reports on the work of Alberto Benavidez who has been researching the veracity of VAERS for two years. (https://welcometheeagle.substack.com/p/vaers-jun-2–2023-wrap-up) Mr. Benavidez estimated the number of **fatalities in children listed in VAERS should be increased by at least 182.**

> There are at least 182 children who died from COVID-19 vaccines hidden in the VAERS database, that don't show up when you search for child vaccine deaths.
>
> How is VAERS doing this? It's creative and diabolical:
>
> **Some brilliant investigative work was done by TheEagle88, who publishes his work on his Substack, and who made this shocking discovery**.
>
> (https://makismd.substack.com/p/vaers-is-cleverly-hiding-182 child)

In addition to mislabeling, TheEagle88 cites the efforts of Jessica Rose to track the disappearance of reports. The cases entered into the system can be removed as new ones are added. (https://jessicar.substack.com/p/the-death-counts-been-slowing-down, https://public.tableau.com/app/profile/alberto.benavidez.)

Mr. Benavidez says he has specialized training and experience in investigating billing and health insurance matters. He presents the following list of ways VAERS is being manipulated.

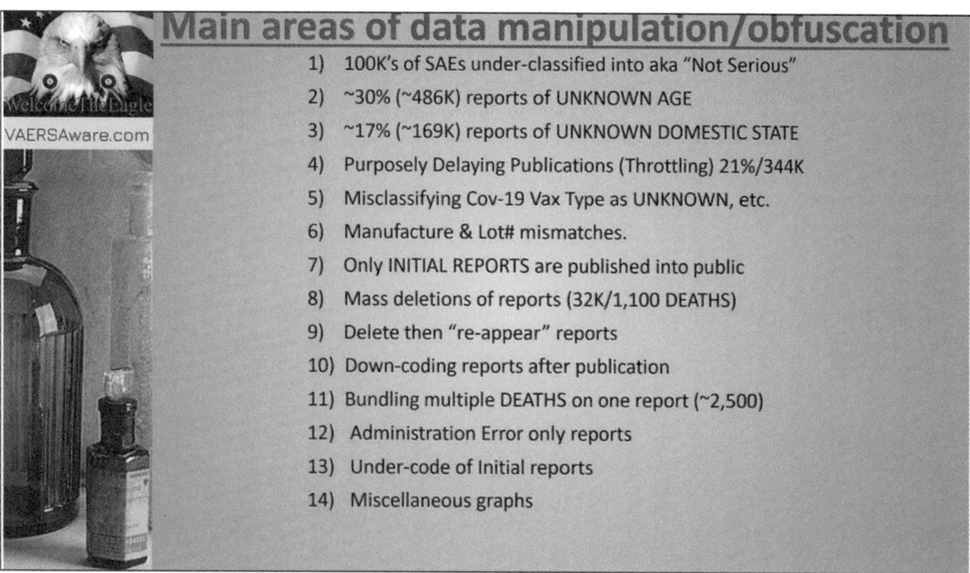

"Conclusion: VAERS is actively covering up catastrophic injury and to add insult, VAERS does not publish all legitimate reports received!"

Case 1:

VAERS ID 1952747, mentioned in TheEagle88's reporting, is reproduced below. It shows the clinical detail submitted for a 12-year-old male captured by Mr. Benavidez before it disappeared. No age is indicated, yet the clinical detail identifies the young man's age. (https://www.medalerts.org/vaersdb/findfield.php?IDNUMBER=1952747)

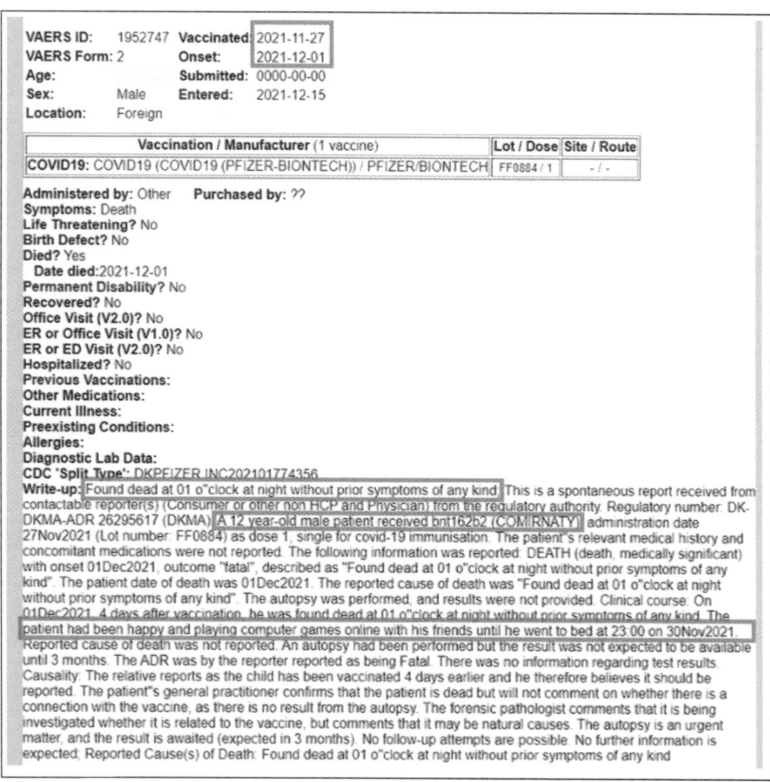

VAERS ID: 1952747 **Vaccinated:** 2021-11-27
VAERS Form: 2 **Onset:** 2021-12-01
Age: **Submitted:** 0000-00-00
Sex: Male **Entered:** 2021-12-15
Location: Foreign

Vaccination / Manufacturer (1 vaccine)	Lot / Dose	Site / Route
COVID19: COVID19 (COVID19 (PFIZER-BIONTECH)) / PFIZER/BIONTECH	FF0884 / 1	- / -

Administered by: Other **Purchased by:** ??
Symptoms: Death
Life Threatening? No
Birth Defect? No
Died? Yes
Date died:2021-12-01
Permanent Disability? No
Recovered? No
Office Visit (V2.0)? No
ER or Office Visit (V1.0)? No
ER or ED Visit (V2.0)? No
Hospitalized? No
Previous Vaccinations:
Other Medications:
Current Illness:
Preexisting Conditions:
Allergies:
Diagnostic Lab Data:
CDC 'Split Type': DKPFIZER INC202101774356
Write-up: Found dead at 01 o"clock at night without prior symptoms of any kind. This is a spontaneous report received from contactable reporter(s) (Consumer or other non HCP and Physician) from the regulatory authority. Regulatory number: DK-DKMA-ADR 26295617 (DKMA). A 12 year-old male patient received bnt162b2 (COMIRNATY), administration date 27Nov2021 (Lot number: FF0884) as dose 1, single for covid-19 immunisation. The patient"s relevant medical history and concomitant medications were not reported. The following information was reported: DEATH (death, medically significant) with onset 01Dec2021, outcome "fatal", described as "Found dead at 01 o"clock at night without prior symptoms of any kind". The patient date of death was 01Dec2021. The reported cause of death was "Found dead at 01 o"clock at night without prior symptoms of any kind". The autopsy was performed, and results were not provided. Clinical course: On 01Dec2021, 4 days after vaccination, he was found dead at 01 o"clock at night without prior symptoms of any kind. The patient had been happy and playing computer games online with his friends until he went to bed at 23:00 on 30Nov2021. Reported cause of death was not reported. An autopsy had been performed but the result was not expected to be available until 3 months. The ADR was by the reporter reported as being Fatal. There was no information regarding test results. Causality: The relative reports as the child has been vaccinated 4 days earlier and he therefore believes it should be reported. The patient"s general practitioner confirms that the patient is dead but will not comment on whether there is a connection with the vaccine, as there is no result from the autopsy. The forensic pathologist comments that it is being investigated whether it is related to the vaccine, but comments that it may be natural causes. The autopsy is an urgent matter, and the result is awaited (expected in 3 months). No follow-up attempts are possible. No further information is expected; Reported Cause(s) of Death: Found dead at 01 o"clock at night without prior symptoms of any kind

This entry for VAERS ID 1952747 is now blank. See below.

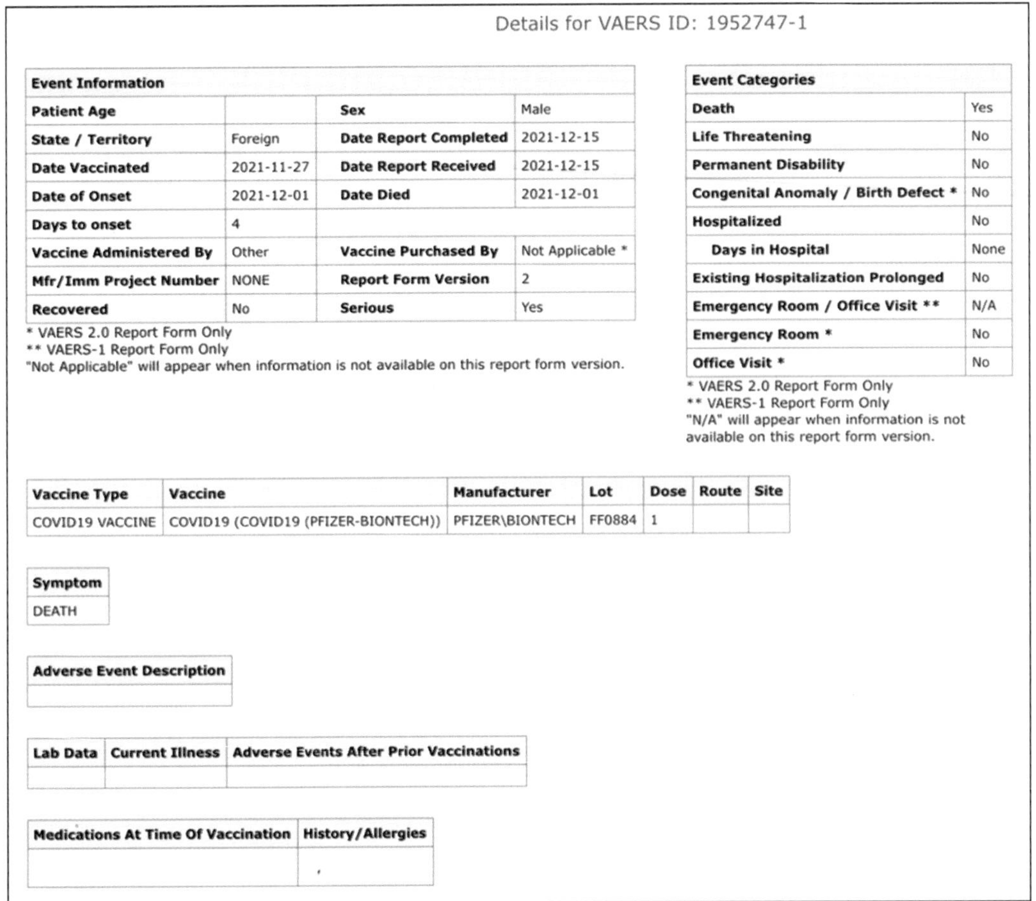

Details for VAERS ID: 1952747-1

Event Information			
Patient Age		**Sex**	Male
State / Territory	Foreign	**Date Report Completed**	2021-12-15
Date Vaccinated	2021-11-27	**Date Report Received**	2021-12-15
Date of Onset	2021-12-01	**Date Died**	2021-12-01
Days to onset	4		
Vaccine Administered By	Other	**Vaccine Purchased By**	Not Applicable *
Mfr/Imm Project Number	NONE	**Report Form Version**	2
Recovered	No	**Serious**	Yes

* VAERS 2.0 Report Form Only
** VAERS-1 Report Form Only
"Not Applicable" will appear when information is not available on this report form version.

Event Categories	
Death	Yes
Life Threatening	No
Permanent Disability	No
Congenital Anomaly / Birth Defect *	No
Hospitalized	No
Days in Hospital	None
Existing Hospitalization Prolonged	No
Emergency Room / Office Visit **	N/A
Emergency Room *	No
Office Visit *	No

* VAERS 2.0 Report Form Only
** VAERS-1 Report Form Only
"N/A" will appear when information is not available on this report form version.

Vaccine Type	Vaccine	Manufacturer	Lot	Dose	Route	Site
COVID19 VACCINE	COVID19 (COVID19 (PFIZER-BIONTECH))	PFIZER\BIONTECH	FF0884	1		

Symptom
DEATH

Adverse Event Description

Lab Data	Current Illness	Adverse Events After Prior Vaccinations

Medications At Time Of Vaccination	History/Allergies

VAERS 6/10/2023

Case 2:

VAERS ID 1887456 concerns a two-year-old male who received one dose of BNT162b2 on November 18, 2021, seven months before the drug was released to his age group by Health and Human Services (HHS) Directive on June 18, 2022. Within six hours of receiving the injection, the child began hemorrhaging from his eyes, ears, nose, and mouth and then he died. The report was received November 20, 2021.

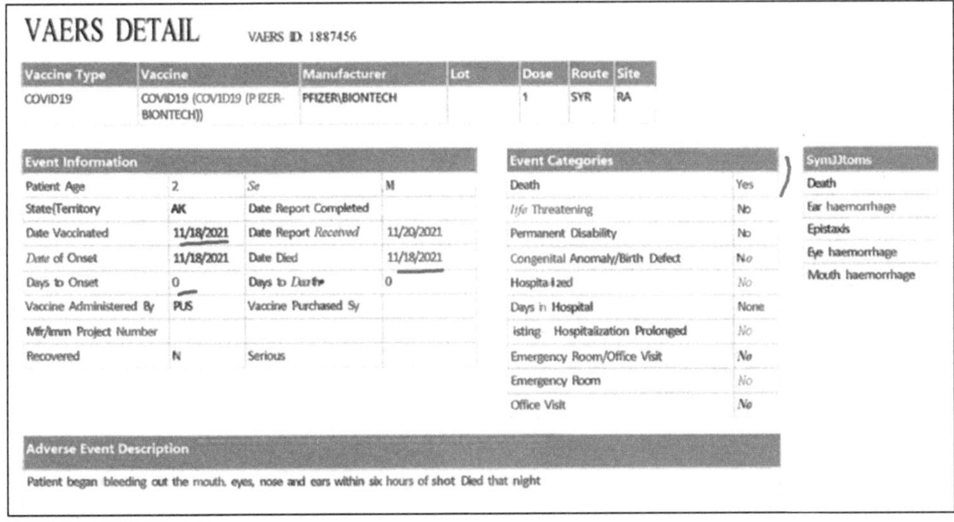

VAERS DETAIL VAERS ID: 1887456

Vaccine Type	Vaccine	Manufacturer	Lot	Dose	Route	Site
COVID19	COVID19 (COVID19 (PFIZER-BIONTECH))	PFIZER\BIONTECH		1	SYR	RA

Event Information			
Patient Age	2	Se	M
State(Territory	AK	Date Report Completed	
Date Vaccinated	11/18/2021	Date Report Received	11/20/2021
Date of Onset	11/18/2021	Date Died	11/18/2021
Days to Onset	0	Days to Death	0
Vaccine Administered By	PUS	Vaccine Purchased Sy	
Mfr/Imm Project Number			
Recovered	N	Serious	

Event Categories	
Death	Yes
Life Threatening	No
Permanent Disability	No
Congenital Anomaly/Birth Defect	No
Hospitalized	No
Days in Hospital	None
isting Hospitalization Prolonged	No
Emergency Room/Office Visit	No
Emergency Room	No
Office Visit	No

SymJJtoms
Death
Ear haemorrhage
Epistaxis
Eye haemorrhage
Mouth haemorrhage

Adverse Event Description
Patient began bleeding out the mouth, eyes, nose and ears within six hours of shot Died that night

The report has since gone down the CDC memory hole, as documented on June 12, 2023, below.

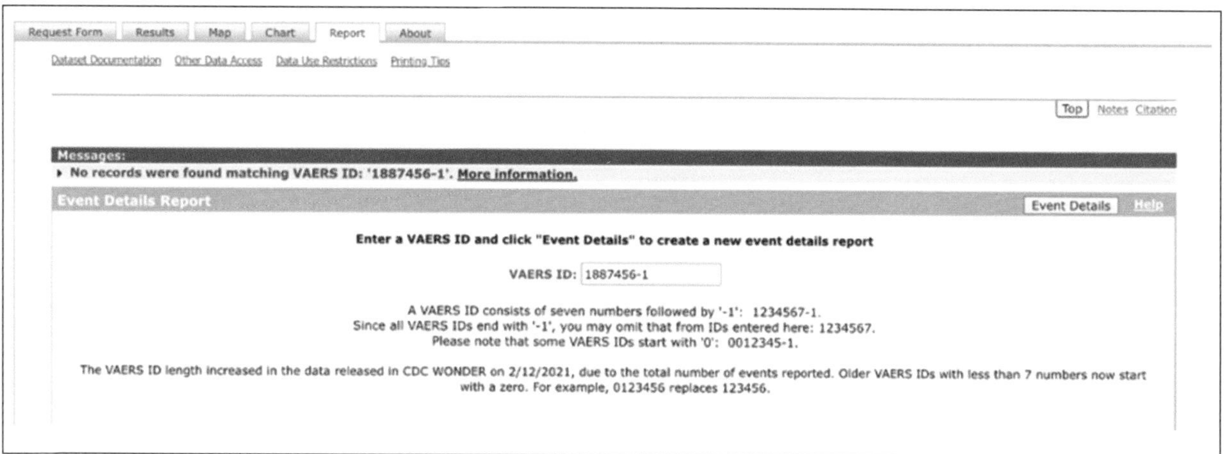

B. VAERS Data: Red Flag Alert

The following composite illustrates search results for "death" after "vaccination" in different VAERS age brackets.

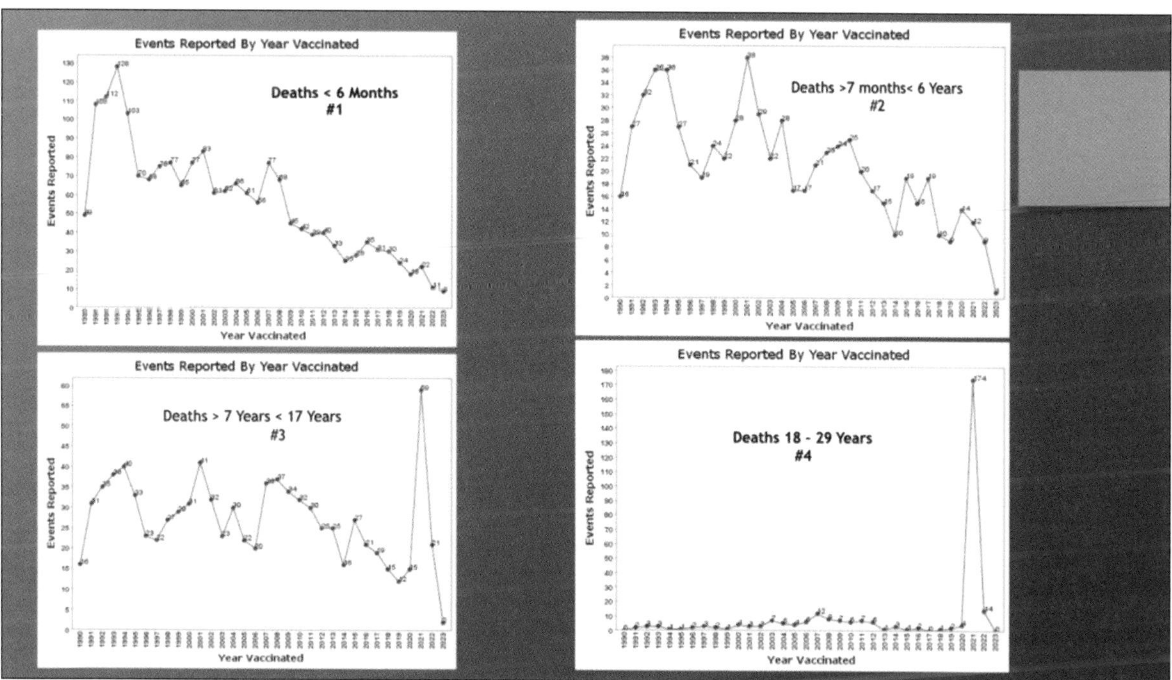

Charts #1 and #2 are reports of death in ages under seven years. Both plots show a 30-year downward trend in fatalities after vaccination. In 2021 there was a slight bump up in infant fatalities but within the range of past variation. The popularity of the injection waned before these children were included in the inoculation scheme.

Charts #3 and #4 show an entirely different pattern with a substantial "death" bump in 2021 with a 47x increase in fatalities for ages 7 to 17 years and a 43x increase in deaths in 2021 for ages 18 to 29 years.

The gene therapy products were authorized for ages 16 years and older on December 11, 2020, ages 12 to 15 years on May 10, 2021, and ages 5 to 11 on October 29, 2021. The age six months and older release date was June 17, 2022, supporting the hypothesis that these products played a role in the fatality spikes.

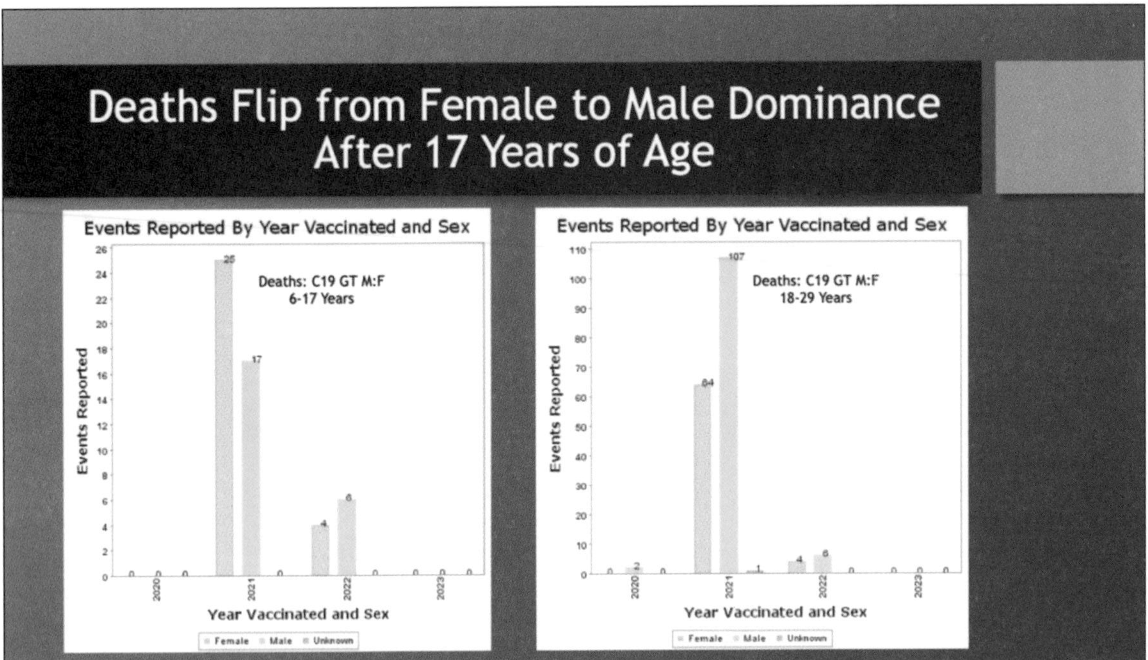

From 6 to 17 years of age, females had 60% of the fatalities contrasted with the next older age bracket, 18–29 years old, in which males had 63% of the fatalities. The total number went from 42 in the 6–17 years old bracket to 171 in the 18–29 years old bracket.

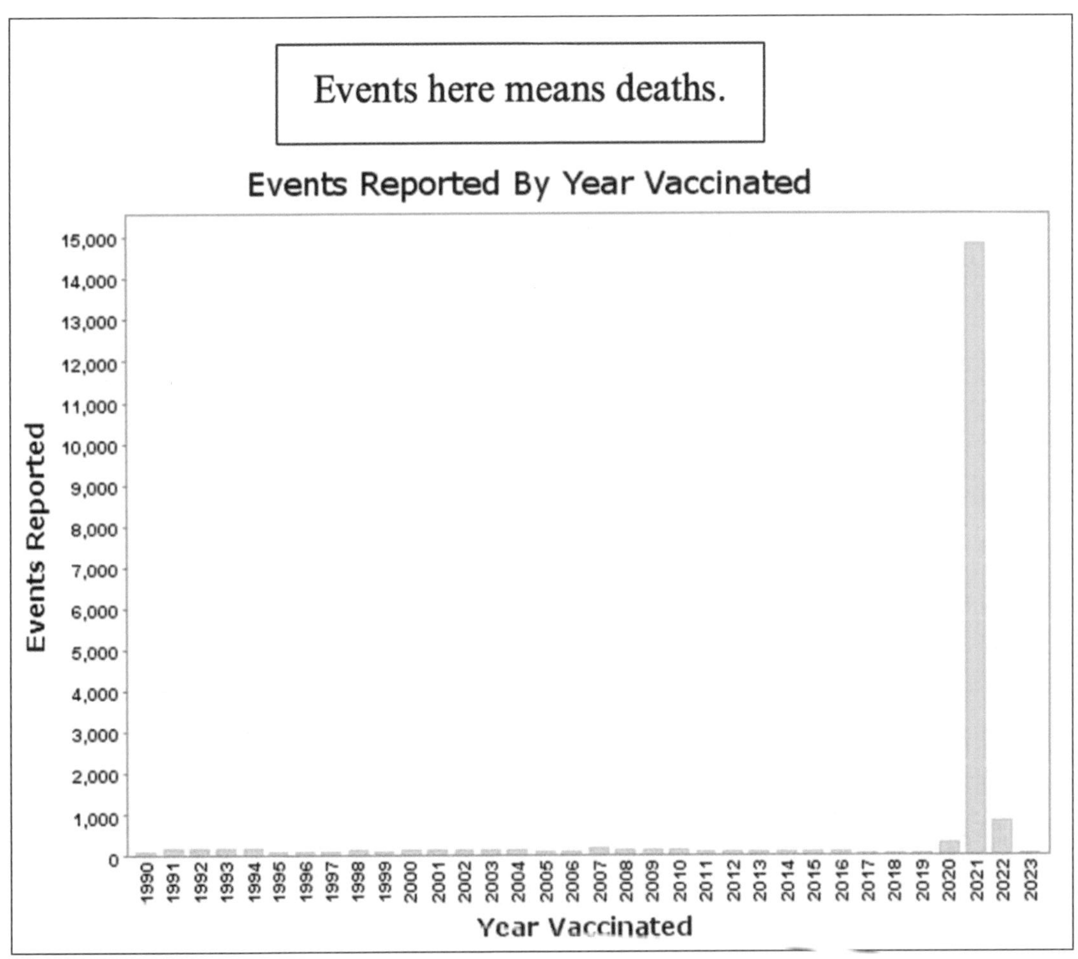

Total deaths (all ages) reported to VAERS for all "vaccine" products spiked in 2021, the year of the widescale C19 gene therapy products rollout (previous and following images).

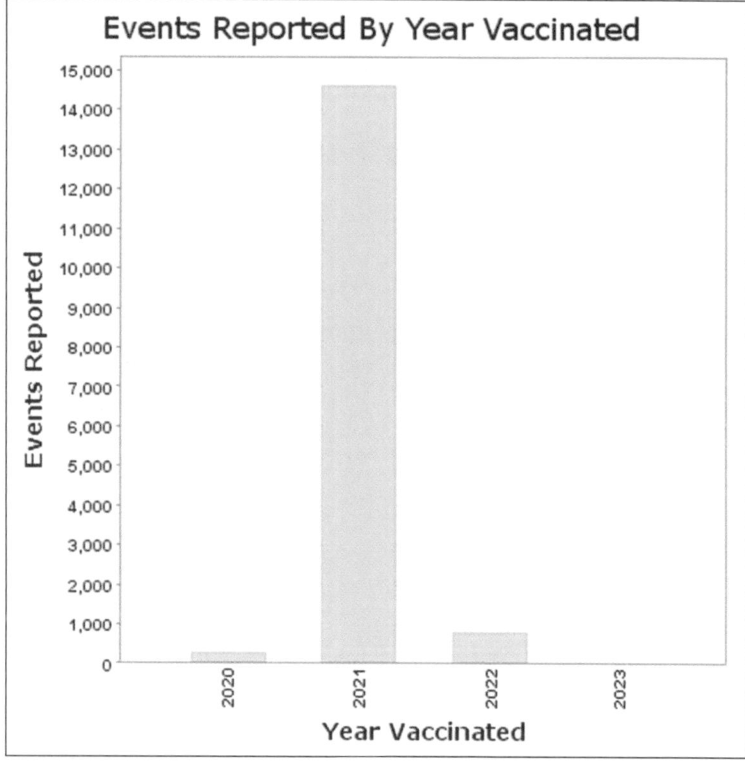

CoVax gene therapy products account for almost all deaths since the December 2020 emergency use authorization with very small adjustment for non-C19 drug products (compare this with preceding histogram).

VAERS ID 1913198 concerns a 13-year-old girl who experienced rapid onset of a very rare and aggressive epithelioid sarcoma of her heart one month after her first dose of BNT162b2.

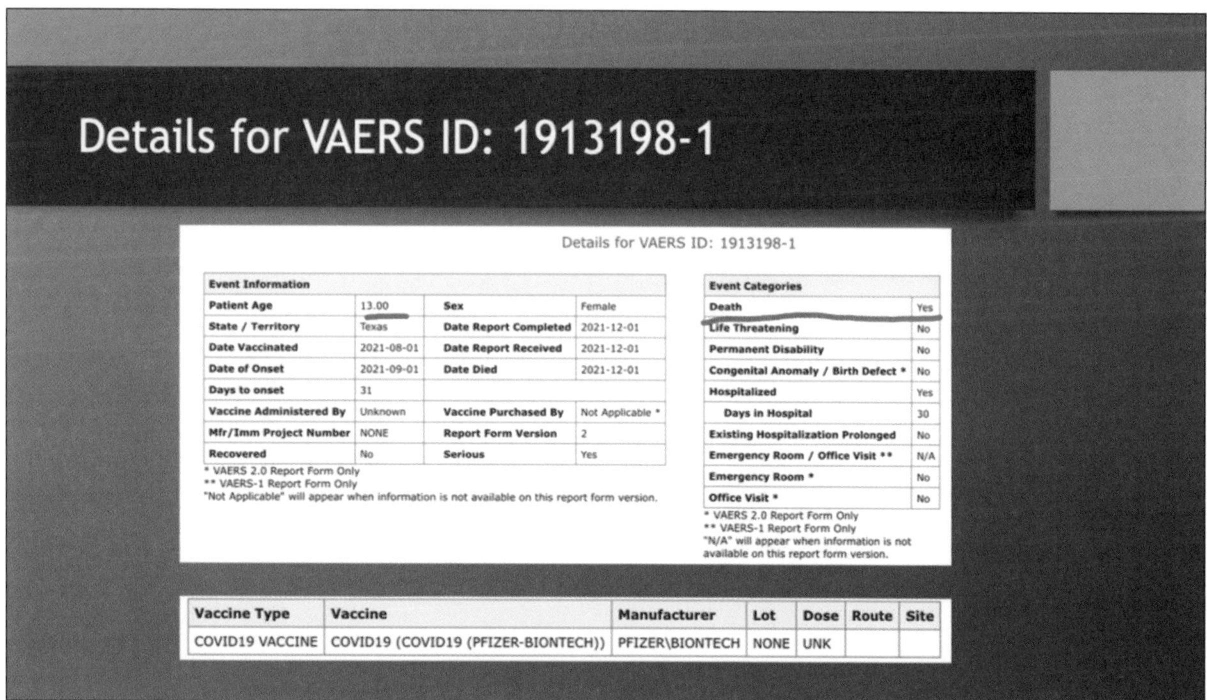

Details for VAERS ID: 1913198-1

Details for VAERS ID: 1913198-1

Event Information			
Patient Age	13.00	Sex	Female
State / Territory	Texas	Date Report Completed	2021-12-01
Date Vaccinated	2021-08-01	Date Report Received	2021-12-01
Date of Onset	2021-09-01	Date Died	2021-12-01
Days to onset	31		
Vaccine Administered By	Unknown	Vaccine Purchased By	Not Applicable *
Mfr/Imm Project Number	NONE	Report Form Version	2
Recovered	No	Serious	Yes

* VAERS 2.0 Report Form Only
** VAERS-1 Report Form Only
"Not Applicable" will appear when information is not available on this report form version.

Event Categories	
Death	Yes
Life Threatening	No
Permanent Disability	No
Congenital Anomaly / Birth Defect *	No
Hospitalized	Yes
Days in Hospital	30
Existing Hospitalization Prolonged	No
Emergency Room / Office Visit **	N/A
Emergency Room *	No
Office Visit *	No

* VAERS 2.0 Report Form Only
** VAERS-1 Report Form Only
"N/A" will appear when information is not available on this report form version.

Vaccine Type	Vaccine	Manufacturer	Lot	Dose	Route	Site
COVID19 VACCINE	COVID19 (COVID19 (PFIZER-BIONTECH))	PFIZER\BIONTECH	NONE	UNK		

She spent 30 days in the hospital before she expired.

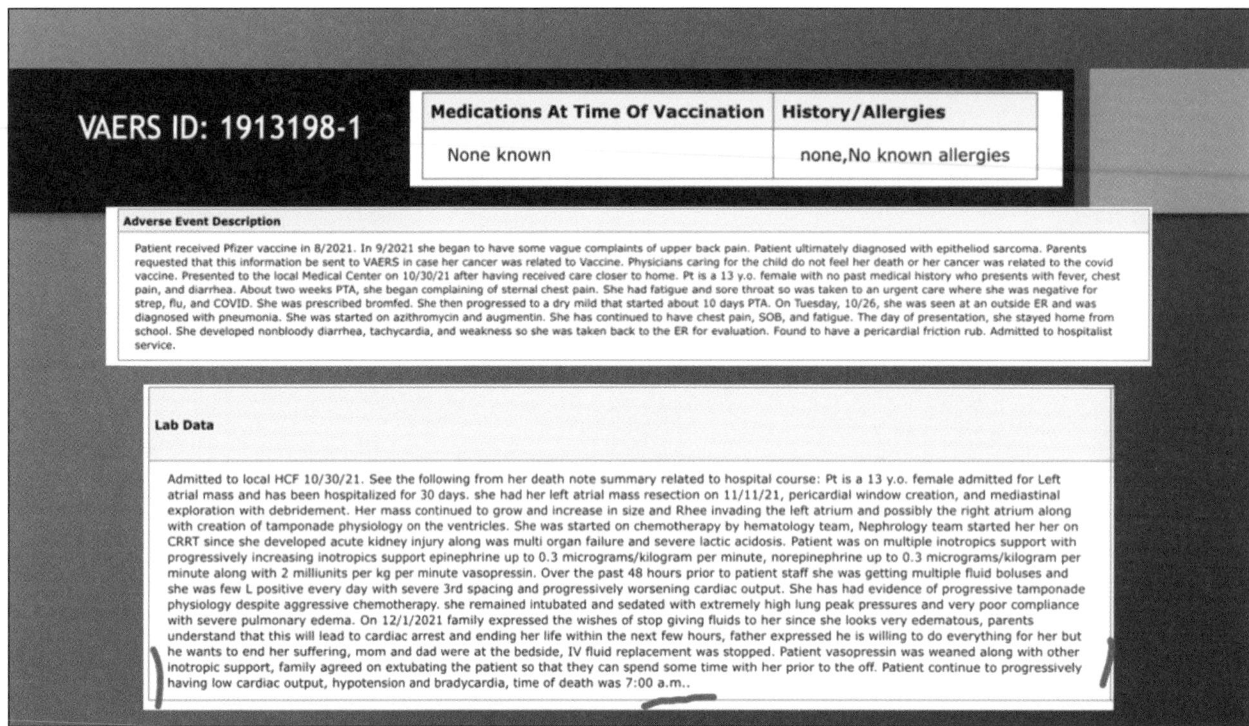

VAERS ID: 1913198-1

Medications At Time Of Vaccination	History/Allergies
None known	none,No known allergies

Adverse Event Description

Patient received Pfizer vaccine in 8/2021. In 9/2021 she began to have some vague complaints of upper back pain. Patient ultimately diagnosed with epitheliod sarcoma. Parents requested that this information be sent to VAERS in case her cancer was related to Vaccine. Physicians caring for the child do not feel her death or her cancer was related to the covid vaccine. Presented to the local Medical Center on 10/30/21 after having received care closer to home. Pt is a 13 y.o. female with no past medical history who presents with fever, chest pain, and diarrhea. About two weeks PTA, she began complaining of sternal chest pain. She had fatigue and sore throat so was taken to an urgent care where she was negative for strep, flu, and COVID. She was prescribed bromfed. She then progressed to a dry mild that started about 10 days PTA. On Tuesday, 10/26, she was seen at an outside ER and was diagnosed with pneumonia. She was started on azithromycin and augmentin. She has continued to have chest pain, SOB, and fatigue. The day of presentation, she stayed home from school. She developed nonbloody diarrhea, tachycardia, and weakness so she was taken back to the ER for evaluation. Found to have a pericardial friction rub. Admitted to hospitalist service.

Lab Data

Admitted to local HCF 10/30/21. See the following from her death note summary related to hospital course: Pt is a 13 y.o. female admitted for Left atrial mass and has been hospitalized for 30 days. she had her left atrial mass resection on 11/11/21, pericardial window creation, and mediastinal exploration with debridement. Her mass continued to grow and increase in size and Rhee invading the left atrium and possibly the right atrium along with creation of tamponade physiology on the ventricles. She was started on chemotherapy by hematology team, Nephrology team started her her on CRRT since she developed acute kidney injury along was multi organ failure and severe lactic acidosis. Patient was on multiple inotropics support with progressively increasing inotropics support epinephrine up to 0.3 micrograms/kilogram per minute, norepinephrine up to 0.3 micrograms/kilogram per minute along with 2 milliunits per kg per minute vasopressin. Over the past 48 hours prior to patient staff she was getting multiple fluid boluses and she was few L positive every day with severe 3rd spacing and progressively worsening cardiac output. She has had evidence of progressive tamponade physiology despite aggressive chemotherapy. she remained intubated and sedated with extremely high lung peak pressures and very poor compliance with severe pulmonary edema. On 12/1/2021 family expressed the wishes of stop giving fluids to her since she looks very edematous, parents understand that this will lead to cardiac arrest and ending her life within the next few hours, father expressed he is willing to do everything for her but he wants to end her suffering, mom and dad were at the bedside, IV fluid replacement was stopped. Patient vasopressin was weaned along with other inotropic support, family agreed on extubating the patient so that they can spend some time with her prior to the off. Patient continue to progressively having low cardiac output, hypotension and bradycardia, time of death was 7:00 a.m..

Her sarcoma recurred after excision and with chemotherapy.

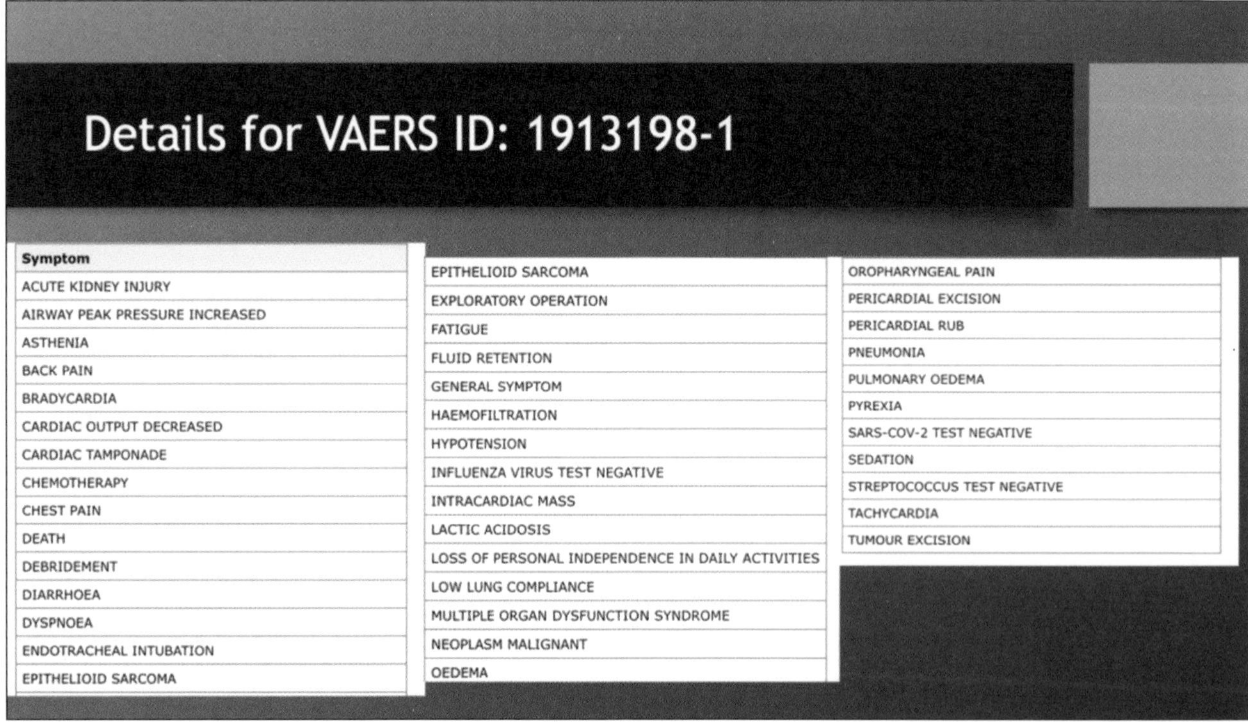

Details for VAERS ID: 1913198-1

Symptom
ACUTE KIDNEY INJURY
AIRWAY PEAK PRESSURE INCREASED
ASTHENIA
BACK PAIN
BRADYCARDIA
CARDIAC OUTPUT DECREASED
CARDIAC TAMPONADE
CHEMOTHERAPY
CHEST PAIN
DEATH
DEBRIDEMENT
DIARRHOEA
DYSPNOEA
ENDOTRACHEAL INTUBATION
EPITHELIOID SARCOMA
EPITHELIOID SARCOMA
EXPLORATORY OPERATION
FATIGUE
FLUID RETENTION
GENERAL SYMPTOM
HAEMOFILTRATION
HYPOTENSION
INFLUENZA VIRUS TEST NEGATIVE
INTRACARDIAC MASS
LACTIC ACIDOSIS
LOSS OF PERSONAL INDEPENDENCE IN DAILY ACTIVITIES
LOW LUNG COMPLIANCE
MULTIPLE ORGAN DYSFUNCTION SYNDROME
NEOPLASM MALIGNANT
OEDEMA
OROPHARYNGEAL PAIN
PERICARDIAL EXCISION
PERICARDIAL RUB
PNEUMONIA
PULMONARY OEDEMA
PYREXIA
SARS-COV-2 TEST NEGATIVE
SEDATION
STREPTOCOCCUS TEST NEGATIVE
TACHYCARDIA
TUMOUR EXCISION

In medicine, this is called a problem list. This is a long one. She could be placed in multiple diagnostic categories including turbo cancer, cardiac epithelioid sarcoma, MIS-C, and myopericardial disease. For more on turbo cancer see: https://makismd.substack.com/p/turbo-cancer-sarcomas-14-yo-jeremiah, https://makismd.substack.com/p/turbo-colon-cancer-diagnosis-to-death, and https://makismd.substack.com/p/turbo-lung-cancer-24-year-old-uk.

VAERS ID 2576556 was a 13-year-old girl who died on Christmas Eve, three weeks after her **third dose of Pfizer's BNT162b2**.

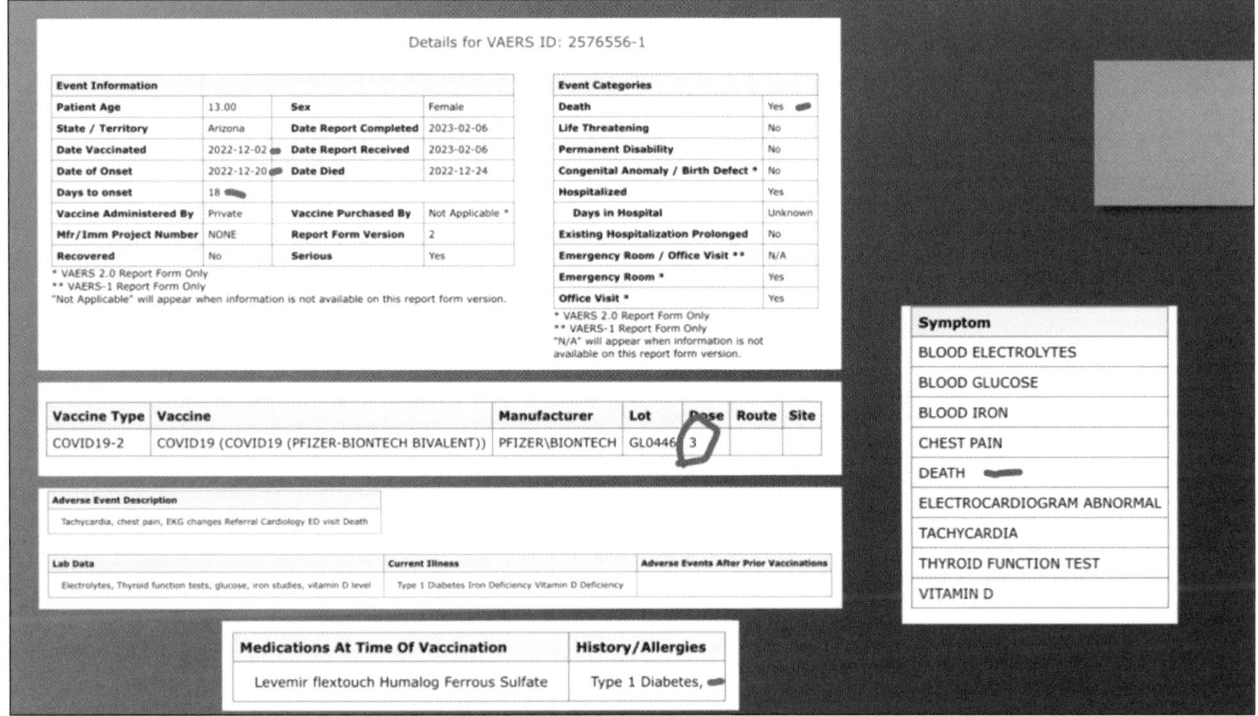

Details for VAERS ID: 2576556-1

Event Information			
Patient Age	13.00	Sex	Female
State / Territory	Arizona	Date Report Completed	2023-02-06
Date Vaccinated	2022-12-02	Date Report Received	2023-02-06
Date of Onset	2022-12-20	Date Died	2022-12-24
Days to onset	18		
Vaccine Administered By	Private	Vaccine Purchased By	Not Applicable *
Mfr/Imm Project Number	NONE	Report Form Version	2
Recovered	No	Serious	Yes

* VAERS 2.0 Report Form Only
** VAERS-1 Report Form Only
"Not Applicable" will appear when information is not available on this report form version.

Event Categories	
Death	Yes
Life Threatening	No
Permanent Disability	No
Congenital Anomaly / Birth Defect *	No
Hospitalized	Yes
Days in Hospital	Unknown
Existing Hospitalization Prolonged	No
Emergency Room / Office Visit **	N/A
Emergency Room *	Yes
Office Visit *	Yes

* VAERS 2.0 Report Form Only
** VAERS-1 Report Form Only
"N/A" will appear when information is not available on this report form version.

Symptom
BLOOD ELECTROLYTES
BLOOD GLUCOSE
BLOOD IRON
CHEST PAIN
DEATH
ELECTROCARDIOGRAM ABNORMAL
TACHYCARDIA
THYROID FUNCTION TEST
VITAMIN D

Vaccine Type	Vaccine	Manufacturer	Lot	Dose	Route	Site
COVID19-2	COVID19 (COVID19 (PFIZER-BIONTECH BIVALENT))	PFIZER\BIONTECH	GL0446	3		

Adverse Event Description
Tachycardia, chest pain, EKG changes Referral Cardiology ED visit Death

Lab Data	Current Illness	Adverse Events After Prior Vaccinations
Electrolytes, Thyroid function tests, glucose, iron studies, vitamin D level	Type 1 Diabetes Iron Deficiency Vitamin D Deficiency	

Medications At Time Of Vaccination	History/Allergies
Levemir flextouch Humalog Ferrous Sulfate	Type 1 Diabetes,

VAERS 5/12/202

This section will end with a case of what may someday be called fatal toxic vaccinosis:

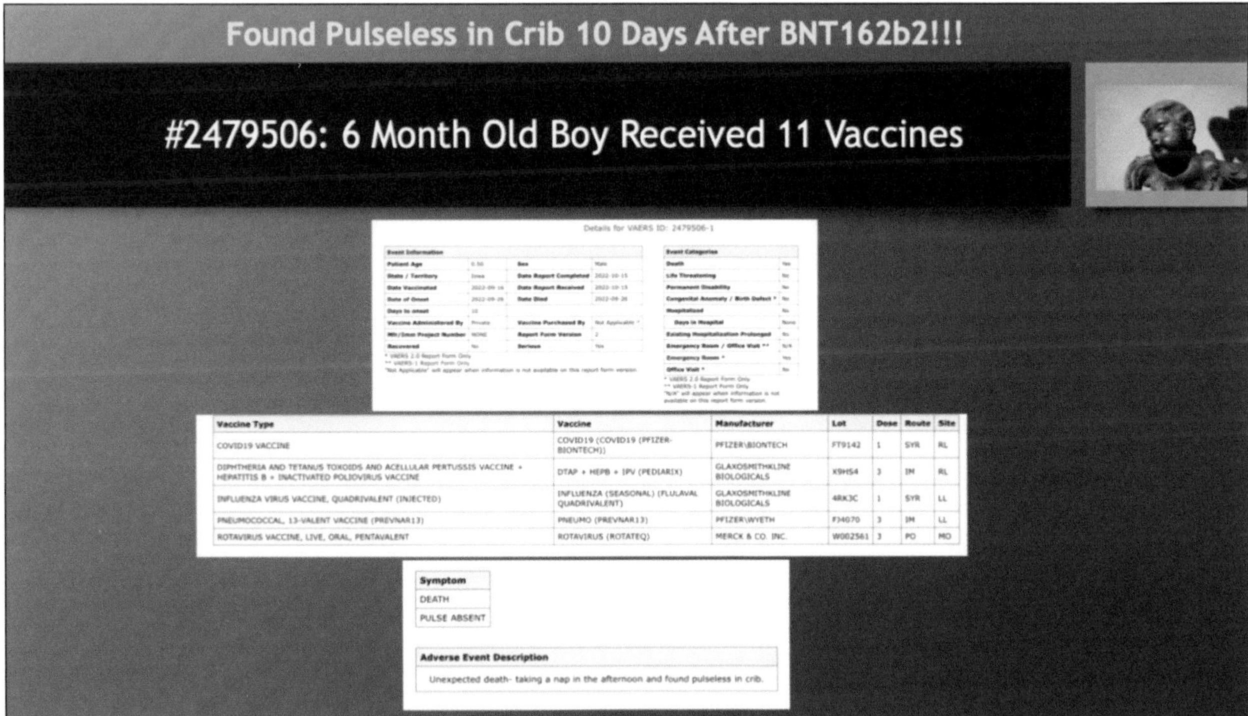

Vaccine Type	Vaccine	Manufacturer	Lot	Dose	Route	Site
COVID19 VACCINE	COVID19 (COVID19 (PFIZER-BIONTECH))	PFIZER\BIONTECH	FT9142	1	SYR	RL
DIPHTHERIA AND TETANUS TOXOIDS AND ACELLULAR PERTUSSIS VACCINE + HEPATITIS B + INACTIVATED POLIOVIRUS VACCINE	DTAP + HEPB + IPV (PEDIARIX)	GLAXOSMITHKLINE BIOLOGICALS	X9HS4	3	IM	RL
INFLUENZA VIRUS VACCINE, QUADRIVALENT (INJECTED)	INFLUENZA (SEASONAL) (FLULAVAL QUADRIVALENT)	GLAXOSMITHKLINE BIOLOGICALS	4RK3C	1	SYR	LL
PNEUMOCOCCAL, 13-VALENT VACCINE (PREVNAR13)	PNEUMO (PREVNAR13)	PFIZER\WYETH	FJ4070	3	IM	LL
ROTAVIRUS VACCINE, LIVE, ORAL, PENTAVALENT	ROTAVIRUS (ROTATEQ)	MERCK & CO. INC.	W002561	3	PO	MO

Symptom
DEATH
PULSE ABSENT

Adverse Event Description
Unexpected death- taking a nap in the afternoon and found pulseless in crib.

Was the cause of death 11 doses of "vaccines"? We cannot say, as VAERS has limitations, but VAERS was touted as being a resource to monitor adverse event signals. Unfortunately, the signals are being ignored.

C: Excess Mortality

Ethical Skeptic has looked at all-cause mortality in the zero- to 24-year-old age bracket (below). There was a big jump in 2021, a five-sigma event, at the time the spike-generating drugs were being distributed widely. The years covered have unique importance with 2018 and 2019 predating C19 and offering a baseline, 2020 reflecting C19 alone; 2021 reflecting C19 plus C19 gene therapy product mass inoculation, and 2022 reflecting tapering of both C19 and C19 gene therapy. Mortality took a big jump in 2021 as the mass inoculation program launched.

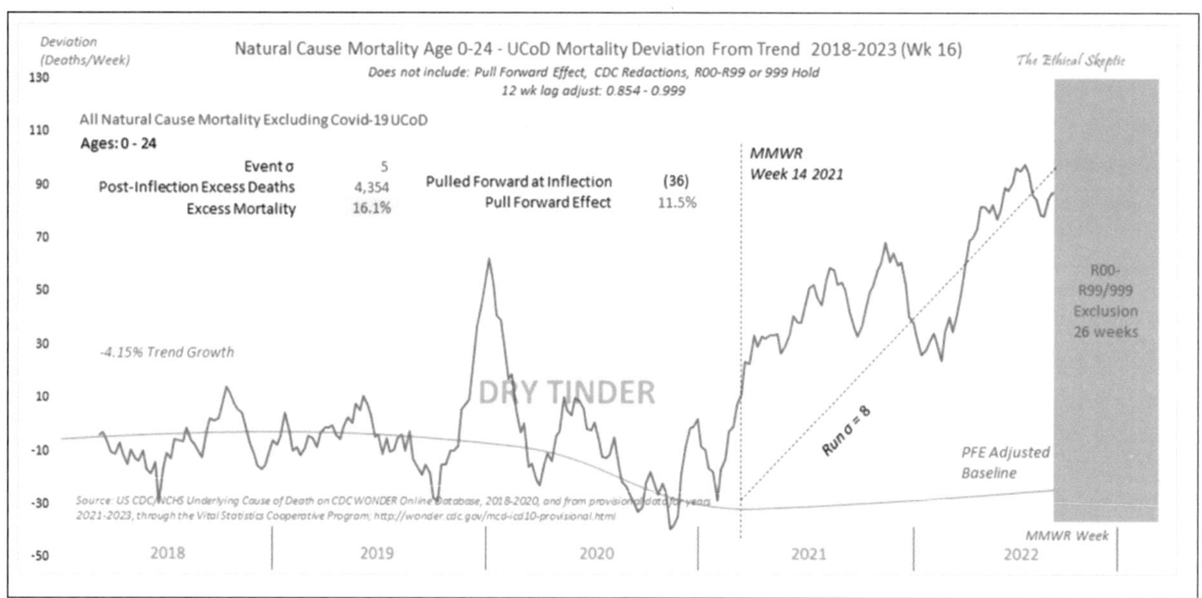

(https://theethicalskeptic.com/2022/08/20/houston-we-have-a-problem-part-1-of-3/) and (https://theethicalskeptic.com/2022/10/24/houston-the-cdc-has-a-problem-part-2-of-3/)

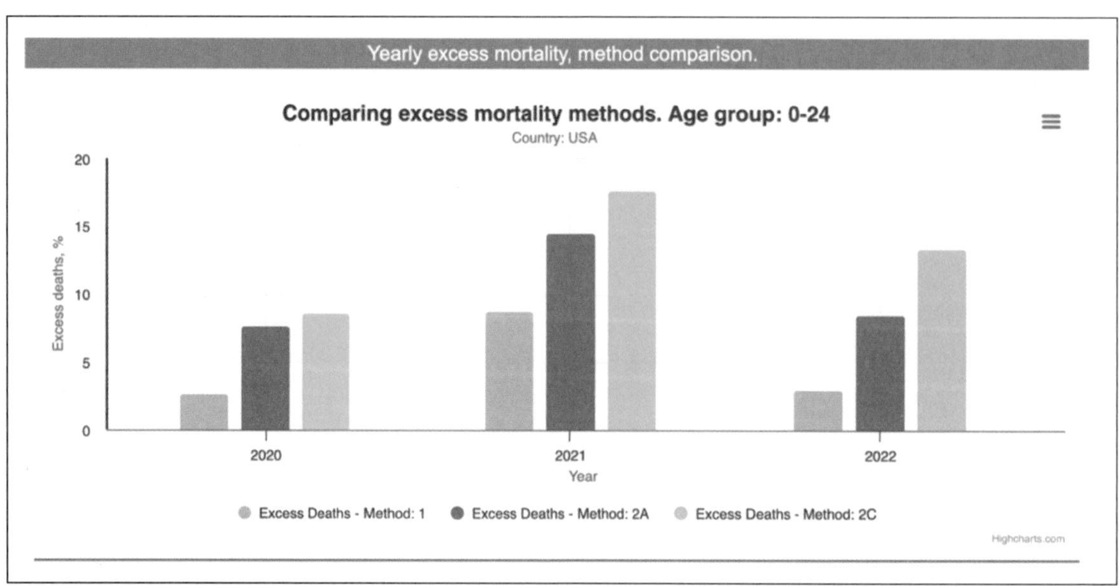

Ed Dowd and his group used three methods to estimate excess mortality in young people. (https://phinancetechnologies.com/HumanityProjects/Yearly%20Excess%20Death%20Rate%20Analysis%20-%20US.htm) The range was 2.7% to 8.6% in 2020, COVID-19 year one, and more than doubling in C19, "vaccine" year one, to 8.7% to 17.6% then dropping to 3% to 13% as the rate of "vaccination" dropped.

V. Adverse Events and Dosing Schedules: Correlation

The histogram below is a plot of annual VAERS reports data since 1990. Years 2021 and 2022 demonstrate VAERS reporting spikes as the GTP "vaccine" program picked up steam. Then, it began cooling off in 2022. From Pfizer Document 2.4 *(reporting on animal studies in the early phases of development beginning in early 2020)* forward, there has been documentation of dose-related adverse events associated with the LNP/mRNA products. (https://www.phmpt.org/wp-content/uploads/2022/03/125742_S1_M2_24_nonclinical-overview.pdf)

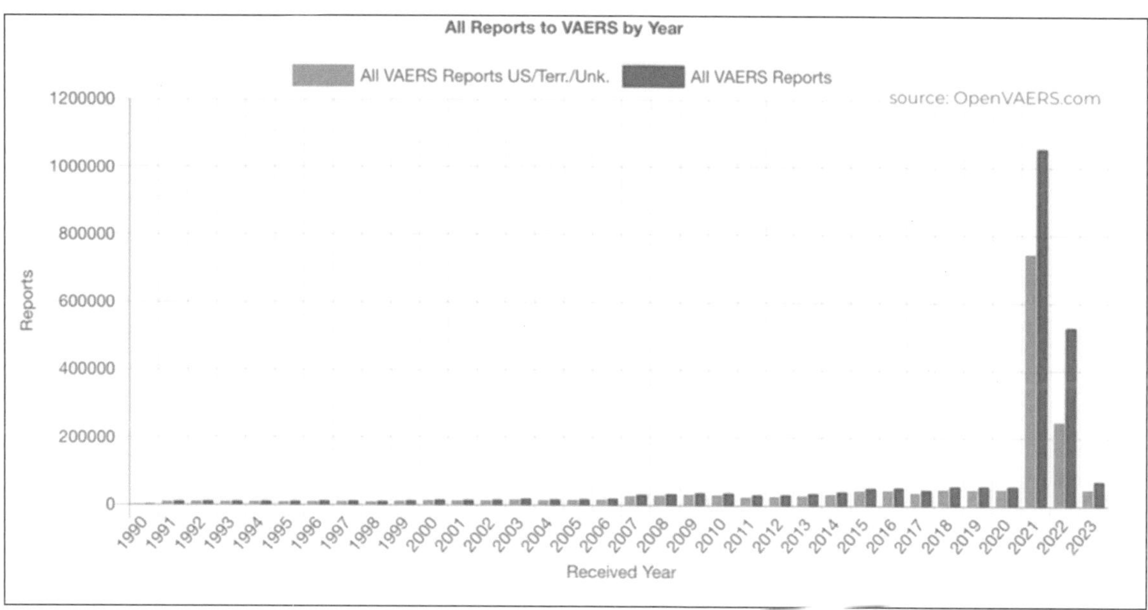

The almost two-and-a-half years of data support this observation. The more drug administered, the more complications.

Is this hypothesis supported with individual disease categories?

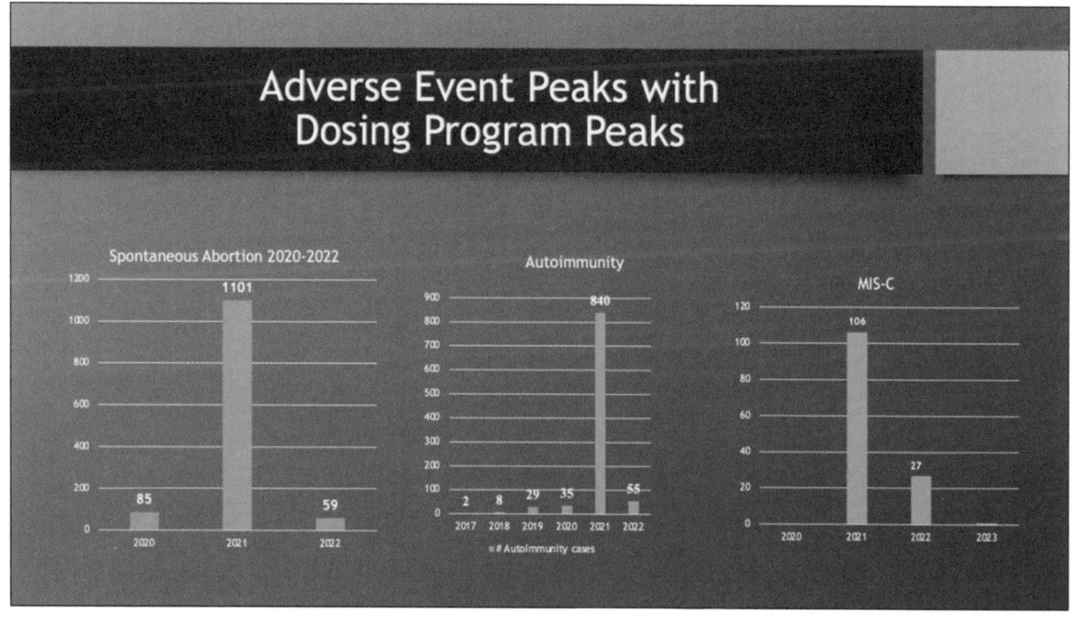

VAERS 5/12/2023

Spontaneous abortion, autoimmunity, and MIS-C patterns peaked in 2021 along with the peak in drug dosing.

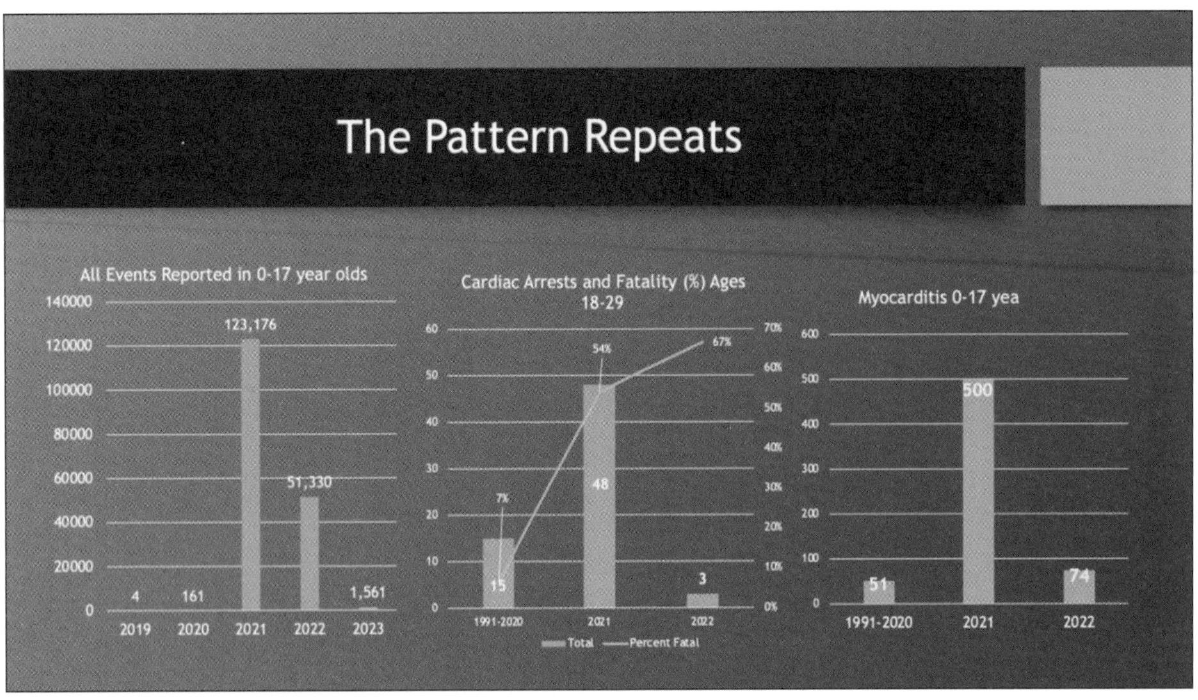

VAERS 5/12/2023

The same is true for all events (ages 0–17 years), cardiac arrests (ages 18–29 years), and myocarditis (0–17 years).

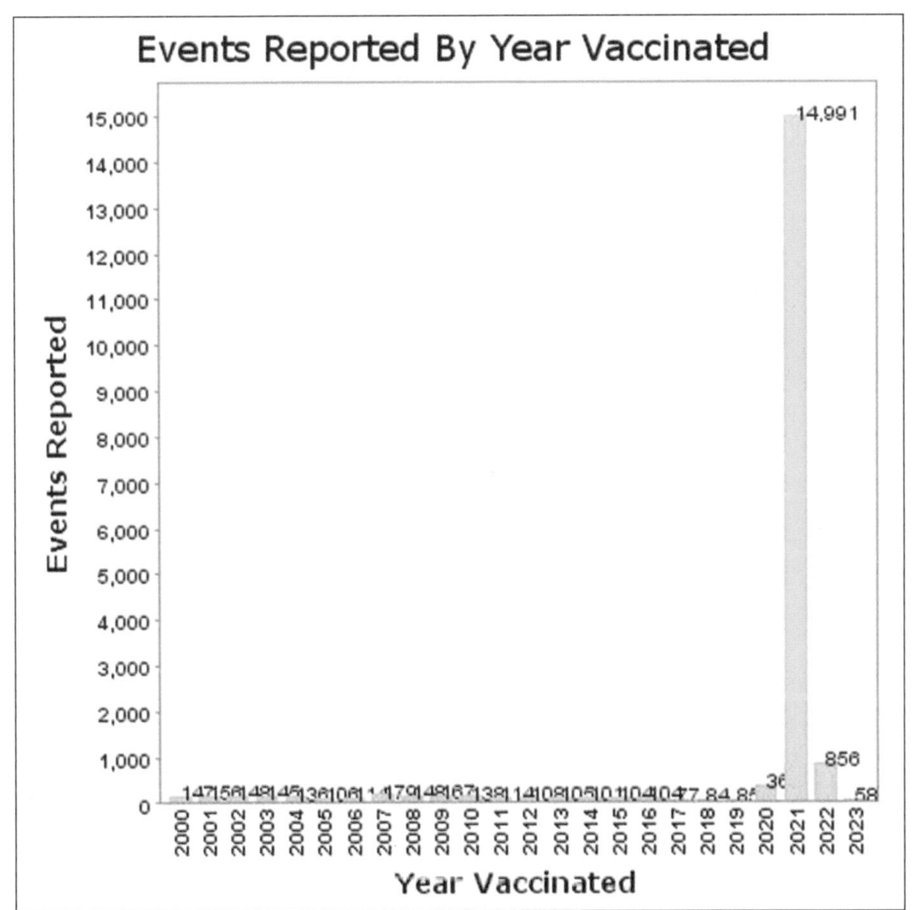

The plot above illustrates the enormous jump in mortality reporting events in VAERS in 2021, the year the gene therapy products were rolled out.

The next chart is a **plot of daily doses administered in the U.S. and Territories.** Note the primary peak in dosing during spring 2021 with a secondary peak beginning in late summer of 2021 and extending through January 2022. Two additional peaks are present.

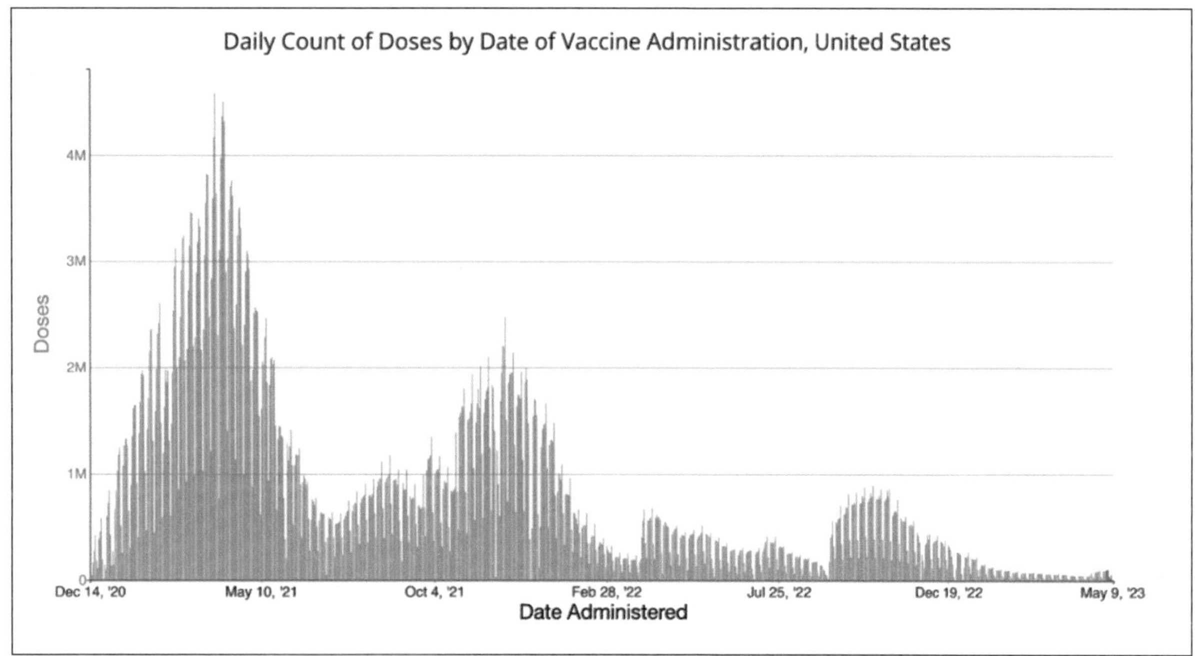

(https://covid.cdc.gov/covid data-tracker/#vaccination-trends)

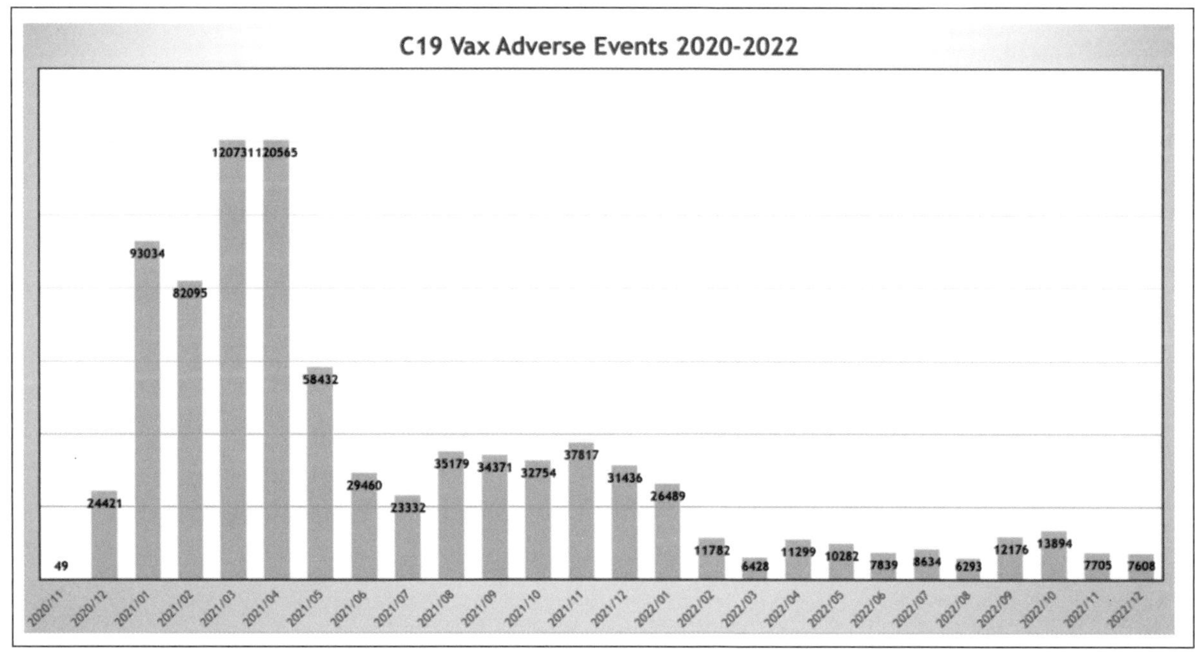

VAERS event reports followed a similar pattern with peaks in February and March 2021 and a secondary peak in August 2021 through January 2022.

Below is a plot of monthly people injected on the primary axis in orange, and monthly VAERS Event Reports are plotted on the secondary axis (scaled for illustration purposes).

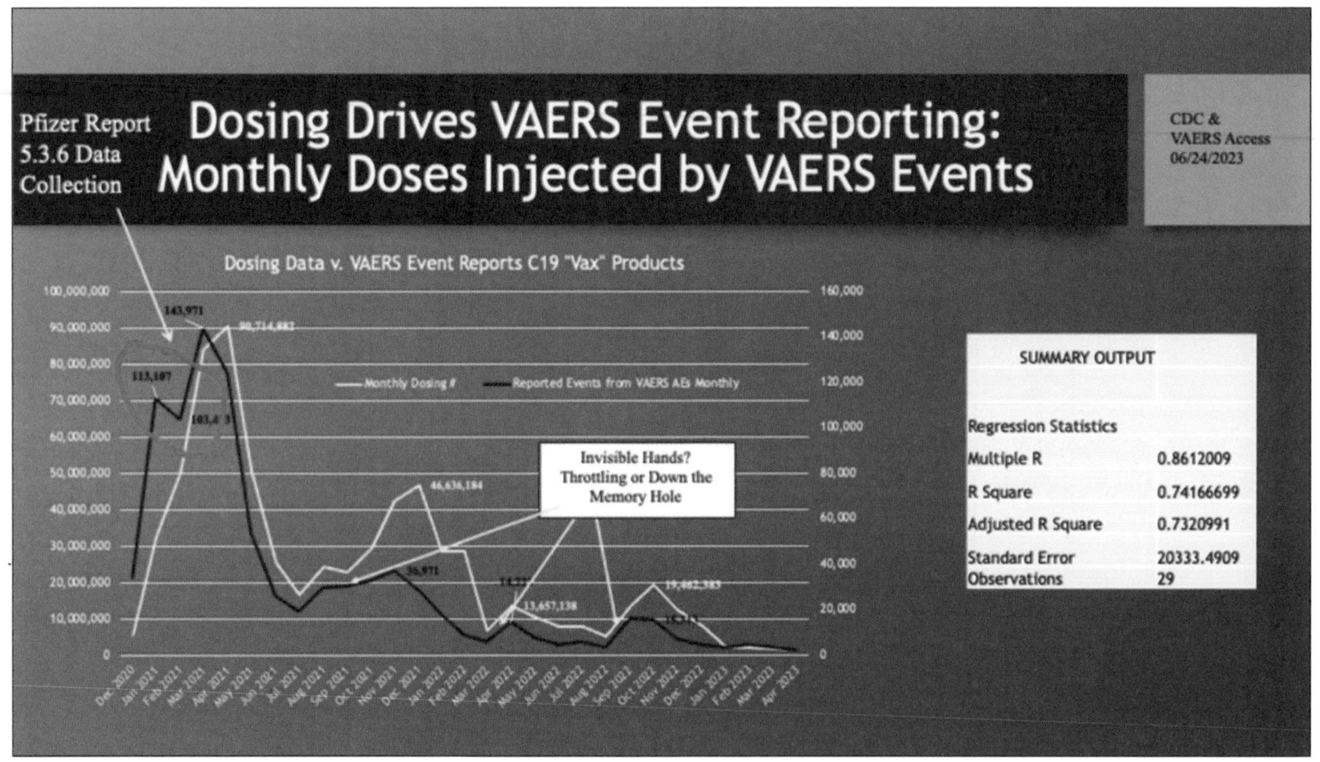

Data are restricted to U.S. and Territories. Data Accessed 06/24/2023.

There is a strong association between dosing and adverse events.

What accounts for the abrupt decline (red circle) in reporting of events in February 2021? At this point in time, the data collecting for Pfizer Confidential Post-Marketing Report 5.3.6 was in the late stage of its compilation. Pfizer had to add 2,400 employees to record over 140,000 adverse events during the first 10 weeks of widespread use of C19 gene therapy drugs. Did the CDC throttle posting of events? What are the alternative explanations of this anomaly? Did they throttle down the next three Event Report waves as well?

Superimposing monthly C19 gene therapy drug monthly doses (orange) on a plot of VAERS reports by month of injection (blue) shows a similar peak and valley pattern. When combined with the other plots, these are evidence of a possible causal relationship between the number of monthly doses and the number of adverse reports in the VAERS database.

The reduced amplitude of the secondary, tertiary, and quaternary peaks of reported events related to the dosing peaks are possibly explained by blanks and reformulations of the C19 gene therapy product or meddling with the numbers. Efficacy proved elusive, but toxicity could be reduced with removal of mRNA from some batches and reformulations of the "recipe." (https://dailysceptic.org/2023/06/28/pfizer-vaccine-batches-in-the-eu-were-placebos-say-scientists/) As dosing fell off in 2022, VAERS reporting fell in tandem (below).

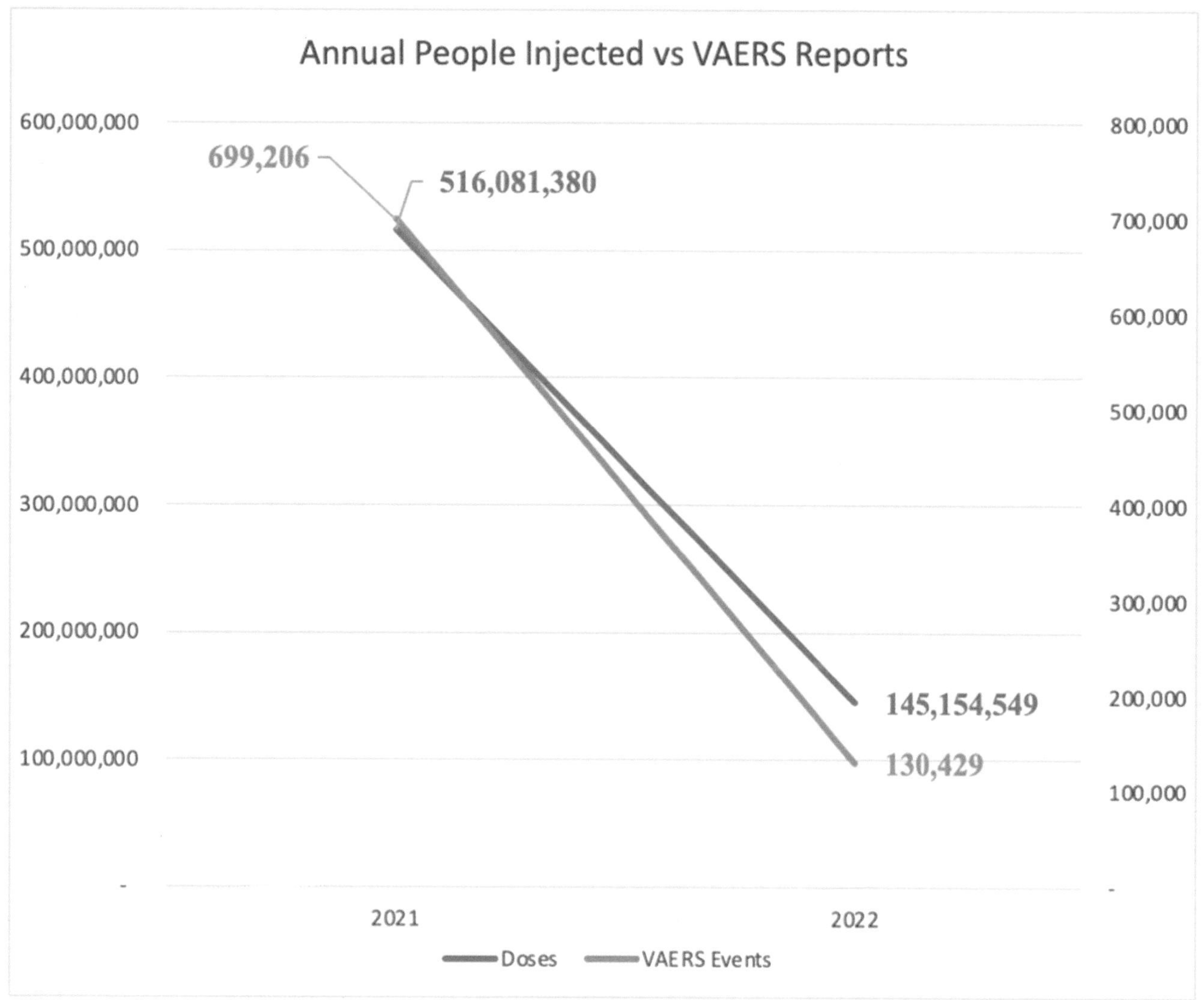

VI. Discussion

The SARS-CoV-2 virus emerged from a laboratory and spread around the world wreaking havoc in many ways. Some havoc was related to illness caused by Dr. Ralph Baric's chimera, but other than the elderly and those with certain co-morbidities, the virus posed little threat to healthy, young people. Recently, a Freedom of Information (FOI) request to the Ministry of Health in Israel asking for documentation of any deaths of persons less than 50 years of age without comorbidity brought the response, **there were none. (**https://www.illusionconsensus.com/p/new-israeli-report-no-covid-deaths)

As Dr. Ioannidis et al. recently reported, COVID-19 is an illness of Seniors.

> The COVID-19 death risk in people <65 years old during the period of fatalities from the epidemic was equivalent to the death risk from driving between 13 and 101 miles per day for 11 countries and 6 states, and was higher (equivalent to the death risk from driving 143–668 miles per day) for 6 other states and the UK. People <65 years old without underlying predisposing conditions accounted for only 0.7–2.6% of all COVID-19 deaths (data available from France, Italy, Netherlands, Sweden, Georgia, and New York City). (https://doi.org/10.1101/2020.04.05.20054361)

In response to COVID-19, an experimental gene therapy product was labeled a vaccine and, with massive funding for marketing from the U.S. government, was foisted on the peoples of the world. Like the virus itself, the new drugs had a negative impact on Seniors; but, unlike SARS-CoV-2, has had devastating effects on those less than 60 years old. Children with miniscule risk from the virus have died and been disabled by the spike-producing gene therapy drugs.

Beginning before conception, these drugs have negatively impacted reproduction. Women are disproportionately affected with adverse events and disability. Young men are disproportionately afflicted with heart damage, which may be permanent for many. Unlike the virus, the treatment of the virus has impacted all age groups, including children, healthy people, as well as those with no comorbidities.

We do not yet know whether these new drugs and the diseases that they cause will go away or are now part of the human condition. If, a big if, they are integrated into the human genome and transcribe and translate abnormal proteins, these new disease conditions may be propagated to future generations. Natural selection has never been set up against synthetic genetic materials and the host of new diseases produced by them.

Perhaps even more disturbing is the work of Kevin McKernan who identified bacterial DNA from the manufacturing process present in the "vaccines" at a level 1,000 times above the acceptable level. (https://sashalatypova.substack.com/p/kevin-mckernan-reports-on-plasmidgate) This bacterial DNA has unknown consequence at present but has the potential of producing resistance to an important class of antibiotics, the aminoglycosides.

For now, these negative thoughts are not supported by known science. The robustness and complexity of the human immune system may yet defeat this threat. Once governments around the world stop the coverup of these various man-made C19 gene therapy-related illnesses, doctors and researchers can begin the task of helping the injured and finding cures for their diseases.

Acknowledgements

Amy Kelly, COO of DailyClout.io and Project Director of the WarRoom/DailyClout Pfizer Documents Analysis Project, and Linnea Wahl, M.S., WarRoom/DailyClout Pfizer Documents Team Five Leader, provided invaluable editorial assistance in preparing this manuscript. Ed Clark and Tony Damian supplied data where noted, and Rudy Parrish critically reviewed the statistical analysis.

Twenty-One: "Women Suffered 94% of Dermatological Adverse Events Reported in First 90 Days of Pfizer COVID "Vaccine" Rollout. 80% of These Adverse Events Were Categorized As 'Serious.'"

—Barbara Gehrett, MD; Joseph Gehrett, MD; Chris Flowers, MD; and Loree Britt

The WarRoom/DailyClout Pfizer Documents Analysis Project Post-Marketing Group (Team 1)—Barbara Gehrett, MD; Joseph Gehrett, MD; Chris Flowers, MD; and Loree Britt—produced a report about Dermatological Adverse Events of Special Interest (AESIs) found in Pfizer document *5.3.6 Cumulative Analysis of Post-Authorization Adverse Event Reports of PF-07302048 (BNT162B2) Received Through 28-FEB-2021* (a.k.a., "*5.3.6*"). This category of AESIs contains two diagnoses and 20 adverse event reports.

It is important to note that the AESIs in the *5.3.6* document were reported to Pfizer for only a 90-day period starting on December 1, 2020, the date of the United Kingdom's public rollout of Pfizer's COVID-19 experimental mRNA "vaccine" product.

Key points of this report include:

- The two dermatological diagnoses in the report are erythema multiforme, a distinctive hypersensitivity reaction with target-like lesions involving the skin and mucous membranes, and chilblains, a localized form of vasculitis affecting mainly fingers and toes.
- There were 13 erythema multiforme adverse events, and seven chilblains adverse events.
- Eighteen of the adverse events occurred in adults and just one in the elderly age group. Age was not provided on the remaining case.
- Seventeen (94%) of the affected individuals were female. One was male, and one was of unknown sex.
- The two remarkable features of this small group of patients are the *overwhelming ratio of female to male* and the *occurrence in nonelderly adults.*
- The median onset was three days after the injection, ranging from under 24 hours to 17 days. Outcomes reported included seven "resolved/resolving," eight "not resolved," and six "unknown." Note: Pfizer's 5.3.6 document shows 20 cases (patients) and 20 adverse events; however, both sex and outcomes in Pfizer's document do not add up to 20. Rather, sex is fewer than 20 (19), and outcomes are more than 20 (21).
- Despite erythema multiforme being an immune-mediated reaction, Pfizer chose not to include it in the Immune-Mediated/Autoimmune AESI category, which had over 500 hypersensitivity adverse events.
- Chilblains is a type of vasculitis, yet these adverse events were not reported under Pfizer's "Vasculitis category" of AESIs.

- Sixteen (80%) of these adverse events were categorized as serious. According to the FDA, serious adverse events include, but are not limited to, patient outcomes such as death, life-threatening events, hospitalizations, and disability or permanent damage.
- Pfizer concludes: "This cumulative case review does not raise new safety issues. Surveillance will continue."

Please read the full report below.

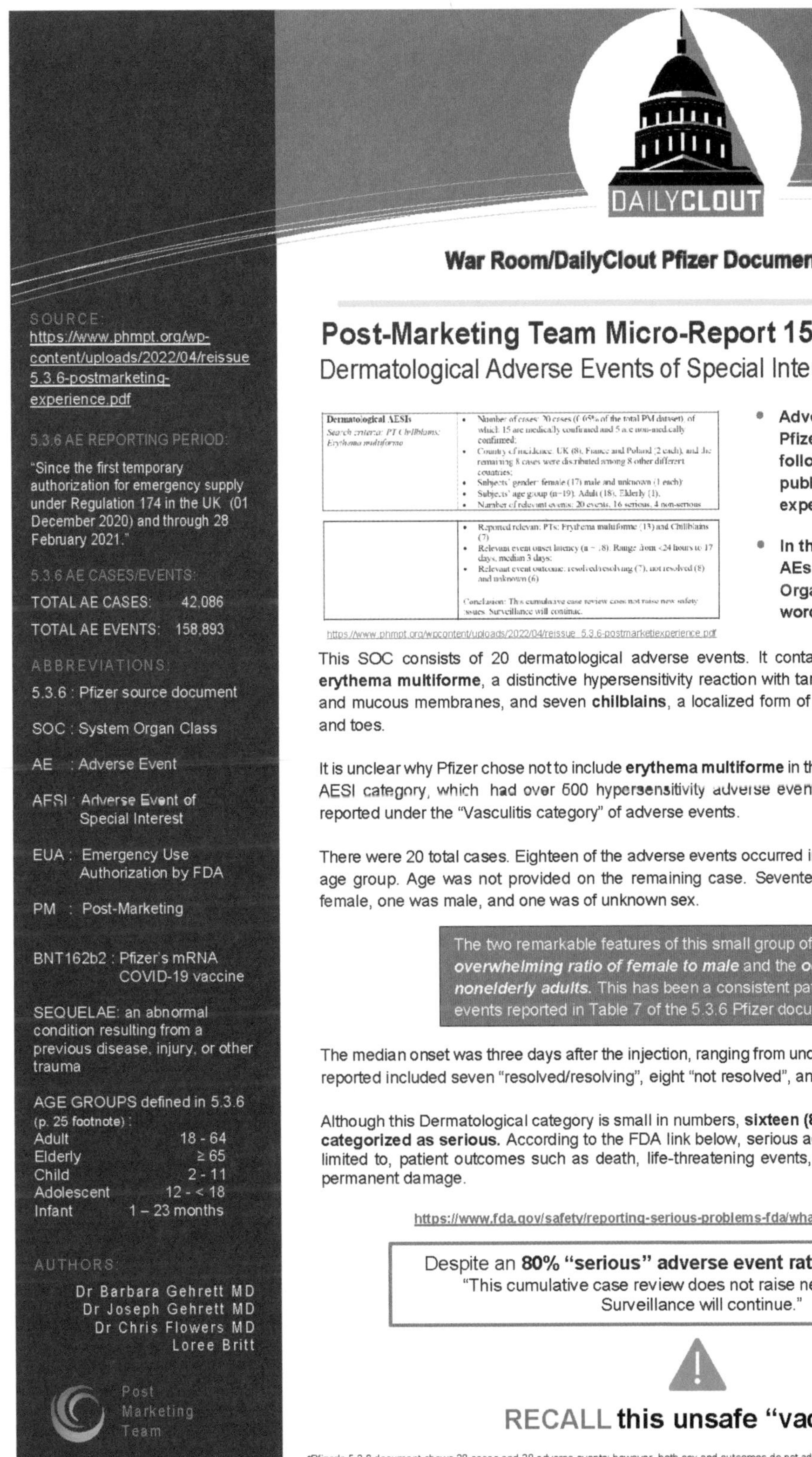

War Room/DailyClout Pfizer Document Analysis

SOURCE:
https://www.phmpt.org/wp-content/uploads/2022/04/reissue 5.3.6-postmarketing-experience.pdf

5.3.6 AE REPORTING PERIOD:

"Since the first temporary authorization for emergency supply under Regulation 174 in the UK (01 December 2020) and through 28 February 2021."

5.3.6 AE CASES/EVENTS:

TOTAL AE CASES: 42,086

TOTAL AE EVENTS: 158,893

ABBREVIATIONS:

5.3.6 : Pfizer source document

SOC : System Organ Class

AE : Adverse Event

AESI : Adverse Event of Special Interest

EUA : Emergency Use Authorization by FDA

PM : Post-Marketing

BNT162b2 : Pfizer's mRNA COVID-19 vaccine

SEQUELAE: an abnormal condition resulting from a previous disease, injury, or other trauma

AGE GROUPS defined in 5.3.6 (p. 25 footnote):

Adult	18 - 64
Elderly	≥ 65
Child	2 - 11
Adolescent	12 - < 18
Infant	1 – 23 months

AUTHORS:

Dr Barbara Gehrett MD
Dr Joseph Gehrett MD
Dr Chris Flowers MD
Loree Britt

Post Marketing Team

09Jul23

Post-Marketing Team Micro-Report 15:
Dermatological Adverse Events of Special Interest Review of 5.3.6

Dermatological AESIs *Search criteria: PT Chilblains; Erythema multiforme*	• Number of cases: 20 cases (0.05% of the total PM dataset), of which 15 are medically confirmed and 5 are non-medically confirmed; • Country of incidence: UK (8), France and Poland (2 each), and the remaining 8 cases were distributed among 8 other different countries; • Subjects' gender: female (17) male and unknown (1 each); • Subjects' age group (n=19): Adult (18), Elderly (1). • Number of relevant events: 20 events, 16 serious, 4 non-serious
	• Reported relevant PTs: Erythema multiforme (13) and Chilblains (7) • Relevant event onset latency (n = 18): Range from <24 hours to 17 days, median 3 days; • Relevant event outcome: resolved/resolving (7), not resolved (8) and unknown (6) Conclusion: This cumulative case review does not raise new safety issues. Surveillance will continue.

https://www.phmpt.org/wpcontent/uploads/2022/04/reissue_5.3.6-postmarketiexperience.pdf

- **Adverse Events were reported to Pfizer during a 90-day period, following the December 1, 2020, public rollout of its COVID-19 experimental "vaccine" product.**
- **In the Pfizer 5.3.6 document, these AEs were categorized by System Organ Classes (SOC) – in other words, by systems in the body.**

This SOC consists of 20 dermatological adverse events. It contains two separate diagnoses: 13 **erythema multiforme**, a distinctive hypersensitivity reaction with target-like lesions involving the skin and mucous membranes, and seven **chilblains**, a localized form of vasculitis affecting mainly fingers and toes.

It is unclear why Pfizer chose not to include **erythema multiforme** in the Immune-Mediated/Autoimmune AESI category, which had over 600 hypersensitivity adverse events. Similarly, **chilblains** were not reported under the "Vasculitis category" of adverse events.

There were 20 total cases. Eighteen of the adverse events occurred in adults and just one in the elderly age group. Age was not provided on the remaining case. Seventeen (94%) of the individuals were female, one was male, and one was of unknown sex.

> The two remarkable features of this small group of patients are the ***overwhelming ratio of female to male*** and the ***occurrence in nonelderly adults.*** This has been a consistent pattern of adverse events reported in Table 7 of the 5.3.6 Pfizer document.

The median onset was three days after the injection, ranging from under 24 hours to 17 days. Outcomes reported included seven "resolved/resolving", eight "not resolved", and six "unknown".*

Although this Dermatological category is small in numbers, **sixteen (80%) of the adverse events were categorized as serious.** According to the FDA link below, serious adverse events include, but are not limited to, patient outcomes such as death, life-threatening events, hospitalizations, and disability or permanent damage.

https://www.fda.gov/safety/reporting-serious-problems-fda/what-serious-adverse-event

> Despite an **80% "serious" adverse event rate,** Pfizer concludes:
> "This cumulative case review does not raise new safety issues.
> Surveillance will continue."

RECALL **this unsafe "vaccine."**

*Pfizer's 5.3.6 document shows 20 cases and 20 adverse events; however, both sex and outcomes do not add up to 20. Rather, sex is under (19), outcomes are over (21).

(https://dailyclout.io/wp-content/uploads/Post-Marketing-Dermatological-AESI-micro-report.pdf)

Twenty-Two: "Thirty-Two Percent of Pfizer's Post-Marketing Respiratory Adverse Event Patients Died, Yet Pfizer Found No New Safety Signals."

—Joseph Gehrett, MD; Barbara Gehrett, MD; Chris Flowers, MD; and Loree Britt

The WarRoom/DailyClout Pfizer Documents Analysis Project Post-Marketing Group (Team 1)—Joseph Gehrett, MD; Barbara Gehrett, MD; Chris Flowers, MD; and Loree Britt—wrote a compelling analysis of Respiratory Adverse Events of Special Interest (AESIs) found in Pfizer document *5.3.6 Cumulative Analysis of Post-Authorization Adverse Event Reports of PF-07302048 (BNT162B2) Received Through 28-FEB-2021* (a.k.a., "*5.3.6*"). This category of AESIs contains conditions of damaged lung structure or impaired oxygen or carbon dioxide exchange.

It is important to note that the AESIs in the *5.3.6* document were reported to Pfizer for **only a 90-day period** starting on December 1, 2020, the date of the United Kingdom's public rollout of Pfizer's COVID-19 experimental mRNA "vaccine" product.

Key information from this report:

- Of the 130 total patients in this SOC, **41 (32%) died.**
- Of the 137 relevant adverse events, **126, or 92%, were categorized as "serious,"** which, according to the Food and Drug Administration (FDA), includes patient outcomes such as death, life-threatening events, hospitalizations, and disability or permanent damage.
- Diagnoses included:
  - **"Severe acute respiratory syndrome" (SARS, also known as SARS-CoV-1).** According to the Centers for Disease Control and Prevention (CDC), **this disease still has not been seen in the world since 2004.** "*Since 2004, there have not been any known cases of SARS reported anywhere in the world. The content in this website was developed for the 2003 SARS epidemic. But some guidelines are still being used. Any new SARS updates will be posted on this website.*" The World Health Organization (WHO) agrees with that assessment.
  - **"Respiratory failure," 32% of the adverse events, includes all cases requiring a mechanical ventilator.**
  - **Acute respiratory distress syndrome (ARDS)**, a lung injury where fluid leaks from the blood vessels into the lung tissue as well as the air spaces (alveoli) resulting in stiffness of the lung. It causes a marked increase in the work required to breathe, as well as reduced oxygen and carbon dioxide exchange.
  - **Low blood oxygen** levels.
  - **High carbon dioxide** blood levels.

- For the adverse events that had a record of time from injection to onset, the range was from within 24 hours to 18 days, with **half of the events reported at one day**.
- These **AESIs occurred more in females (55.4%)** than in males (44.6%).
- Despite 32% of patients in this SOC dying, Pfizer concluded, "*This cumulative case review does not raise new safety issues. Surveillance will continue.*"

Please read this important report below.

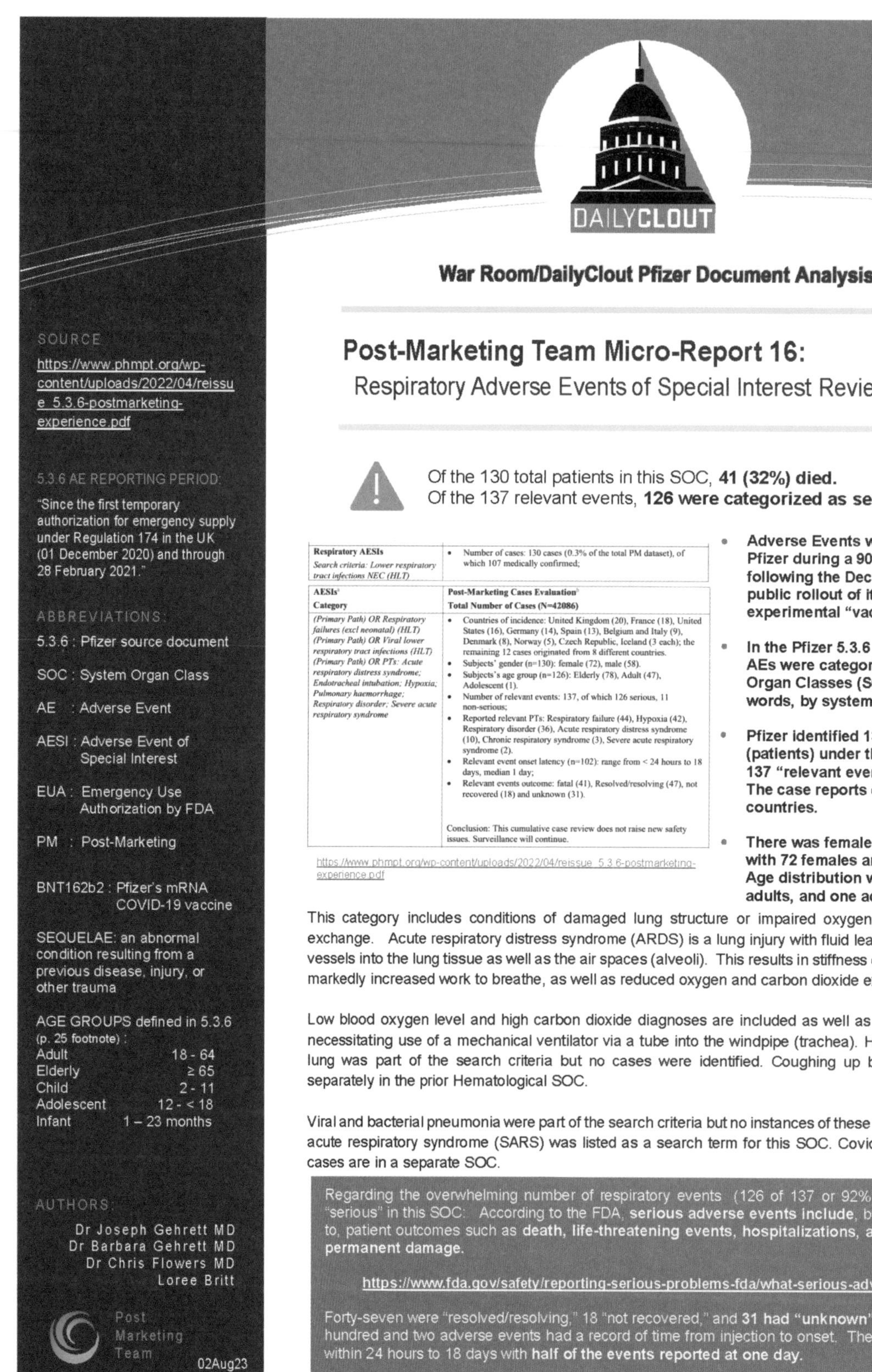

DAILYCLOUT

War Room/DailyClout Pfizer Document Analysis

Post-Marketing Team Micro-Report 16:

Respiratory Adverse Events of Special Interest Review of 5.3.6

SOURCE

https://www.phmpt.org/wp-content/uploads/2022/04/reissue_5.3.6-postmarketing-experience.pdf

5.3.6 AE REPORTING PERIOD:

"Since the first temporary authorization for emergency supply under Regulation 174 in the UK (01 December 2020) and through 28 February 2021."

ABBREVIATIONS:

5.3.6 : Pfizer source document

SOC : System Organ Class

AE : Adverse Event

AESI : Adverse Event of Special Interest

EUA : Emergency Use Authorization by FDA

PM : Post-Marketing

BNT162b2 : Pfizer's mRNA COVID-19 vaccine

SEQUELAE: an abnormal condition resulting from a previous disease, injury, or other trauma

AGE GROUPS defined in 5.3.6 (p. 25 footnote) :

Adult	18 - 64
Elderly	≥ 65
Child	2 - 11
Adolescent	12 - < 18
Infant	1 – 23 months

Of the 130 total patients in this SOC, **41 (32%) died.**
Of the 137 relevant events, **126 were categorized as serious.**

Respiratory AESIs *Search criteria: Lower respiratory tract infections NEC (HLT)*	• Number of cases: 130 cases (0.3% of the total PM dataset), of which 107 medically confirmed;
AESIs[a] Category	**Post-Marketing Cases Evaluation[b] Total Number of Cases (N=42086)**
(Primary Path) OR Respiratory failures (excl neonatal) (HLT) (Primary Path) OR Viral lower respiratory tract infections (HLT) (Primary Path) OR PTs: Acute respiratory distress syndrome; Endotracheal intubation; Hypoxia; Pulmonary haemorrhage; Respiratory disorder; Severe acute respiratory syndrome	• Countries of incidence: United Kingdom (20), France (18), United States (16), Germany (14), Spain (13), Belgium and Italy (9), Denmark (8), Norway (5), Czech Republic, Iceland (3 each); the remaining 12 cases originated from 8 different countries. • Subjects' gender (n=130): female (72), male (58). • Subjects's age group (n=126): Elderly (78), Adult (47), Adolescent (1). • Number of relevant events: 137, of which 126 serious, 11 non-serious; • Reported relevant PTs: Respiratory failure (44), Hypoxia (42), Respiratory disorder (36), Acute respiratory distress syndrome (10), Chronic respiratory syndrome (3), Severe acute respiratory syndrome (2). • Relevant event onset latency (n=102): range from < 24 hours to 18 days, median 1 day; • Relevant events outcome: fatal (41), Resolved/resolving (47), not recovered (18) and unknown (31). Conclusion: This cumulative case review does not raise new safety issues. Surveillance will continue.

https://www.phmpt.org/wp-content/uploads/2022/04/reissue_5.3.6-postmarketing-experience.pdf

- **Adverse Events were reported to Pfizer during a 90-day period, following the December 1, 2020, public rollout of its COVID-19 experimental "vaccine" product.**
- **In the Pfizer 5.3.6 document, these AEs were categorized by System Organ Classes (SOC) – in other words, by systems in the body.**
- **Pfizer identified 130 cases (patients) under these criteria with 137 "relevant events" (diagnoses). The case reports came from 19 countries.**
- **There was female predominance with 72 females and 58 males. Age distribution was 78 elderly, 47 adults, and one adolescent.**

This category includes conditions of damaged lung structure or impaired oxygen or carbon dioxide exchange. Acute respiratory distress syndrome (ARDS) is a lung injury with fluid leaking from the blood vessels into the lung tissue as well as the air spaces (alveoli). This results in stiffness of the lung requiring markedly increased work to breathe, as well as reduced oxygen and carbon dioxide exchange.

Low blood oxygen level and high carbon dioxide diagnoses are included as well as impaired breathing necessitating use of a mechanical ventilator via a tube into the windpipe (trachea). Hemorrhage into the lung was part of the search criteria but no cases were identified. Coughing up blood was reported separately in the prior Hematological SOC.

Viral and bacterial pneumonia were part of the search criteria but no instances of these were found. Severe acute respiratory syndrome (SARS) was listed as a search term for this SOC. Covid-19 (SARS CoV-2) cases are in a separate SOC.

Regarding the overwhelming number of respiratory events (126 of 137 or 92%) categorized as "serious" in this SOC: According to the FDA, **serious adverse events include**, but are not limited to, patient outcomes such as **death, life-threatening events, hospitalizations, and disability or permanent damage.**

https://www.fda.gov/safety/reporting-serious-problems-fda/what-serious-adverse-event

Forty-seven were "resolved/resolving," 18 "not recovered," and **31 had "unknown" outcome.** One hundred and two adverse events had a record of time from injection to onset. The range was from within 24 hours to 18 days with **half of the events reported at one day.**

AUTHORS:

Dr Joseph Gehrett MD
Dr Barbara Gehrett MD
Dr Chris Flowers MD
Loree Britt

Post Marketing Team

02Aug23

(https://dailyclout.io/wp-content/uploads/Post-Marketing-Team-Respiratory-MicroReport-p1.pdf)

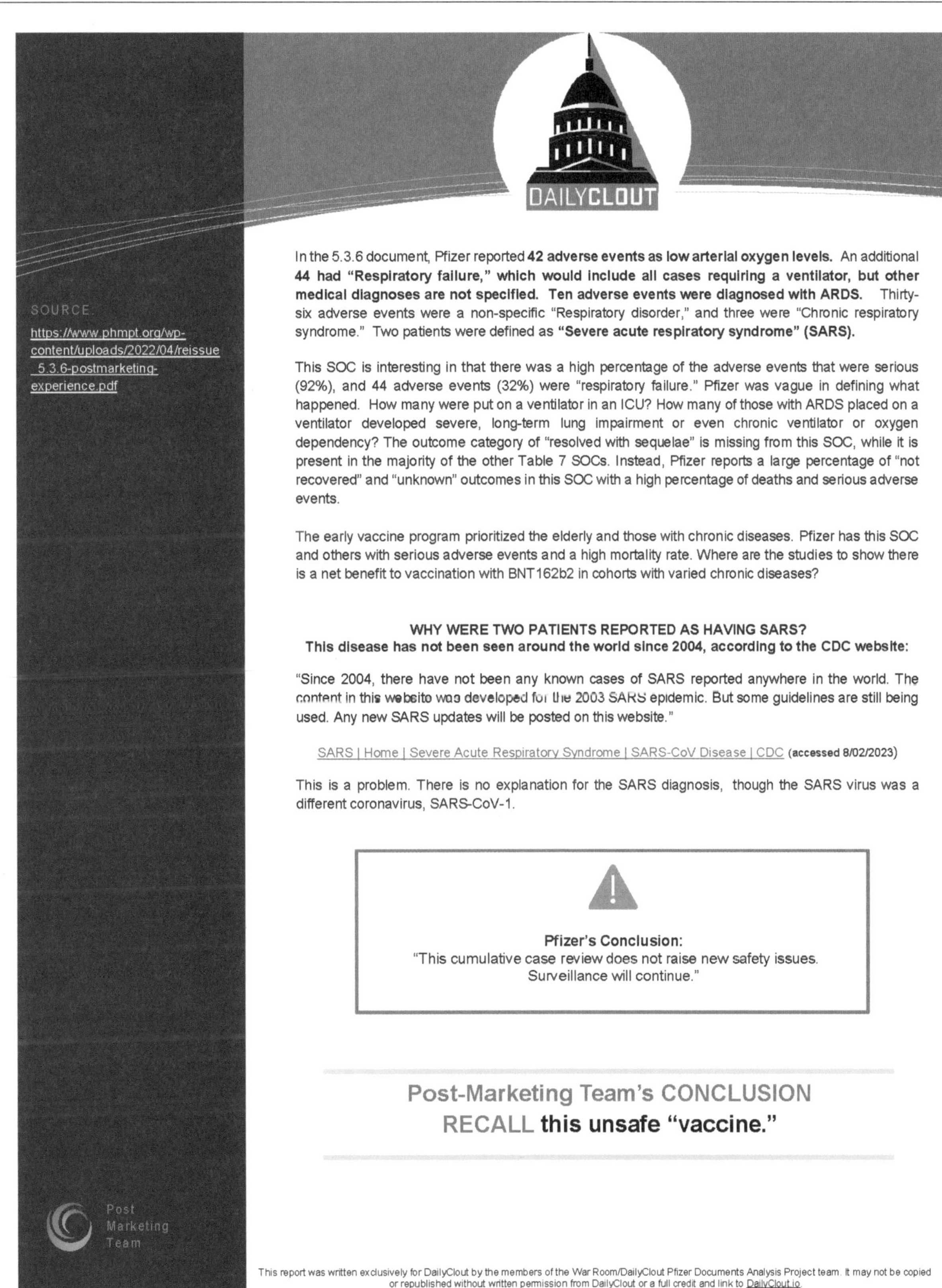

In the 5.3.6 document, Pfizer reported **42 adverse events as low arterial oxygen levels.** An additional **44 had "Respiratory failure," which would include all cases requiring a ventilator, but other medical diagnoses are not specified. Ten adverse events were diagnosed with ARDS.** Thirty-six adverse events were a non-specific "Respiratory disorder," and three were "Chronic respiratory syndrome." Two patients were defined as **"Severe acute respiratory syndrome" (SARS).**

This SOC is interesting in that there was a high percentage of the adverse events that were serious (92%), and 44 adverse events (32%) were "respiratory failure." Pfizer was vague in defining what happened. How many were put on a ventilator in an ICU? How many of those with ARDS placed on a ventilator developed severe, long-term lung impairment or even chronic ventilator or oxygen dependency? The outcome category of "resolved with sequelae" is missing from this SOC, while it is present in the majority of the other Table 7 SOCs. Instead, Pfizer reports a large percentage of "not recovered" and "unknown" outcomes in this SOC with a high percentage of deaths and serious adverse events.

The early vaccine program prioritized the elderly and those with chronic diseases. Pfizer has this SOC and others with serious adverse events and a high mortality rate. Where are the studies to show there is a net benefit to vaccination with BNT162b2 in cohorts with varied chronic diseases?

WHY WERE TWO PATIENTS REPORTED AS HAVING SARS?
This disease has not been seen around the world since 2004, according to the CDC website:

"Since 2004, there have not been any known cases of SARS reported anywhere in the world. The content in this website was developed for the 2003 SARS epidemic. But some guidelines are still being used. Any new SARS updates will be posted on this website."

SARS | Home | Severe Acute Respiratory Syndrome | SARS-CoV Disease | CDC **(accessed 8/02/2023)**

This is a problem. There is no explanation for the SARS diagnosis, though the SARS virus was a different coronavirus, SARS-CoV-1.

Pfizer's Conclusion:
"This cumulative case review does not raise new safety issues. Surveillance will continue."

Post-Marketing Team's CONCLUSION
RECALL this unsafe "vaccine."

(https://dailyclout.io/wp-content/uploads/Post-Marketing-Team-Respiratory-MicroReport-p2.pdf)

Twenty-Three: "mRNA COVID Vaccine-Induced Myocarditis at One Year Post-Injection: Spike Protein, Inflammation Still Present in Heart Tissue."

—Robert W. Chandler, MD, MBA

In spite of widespread censorship, the truth is coming out that myocarditis arising after injection with mRNA COVID "vaccines" is *not* rare, temporary, or mild.

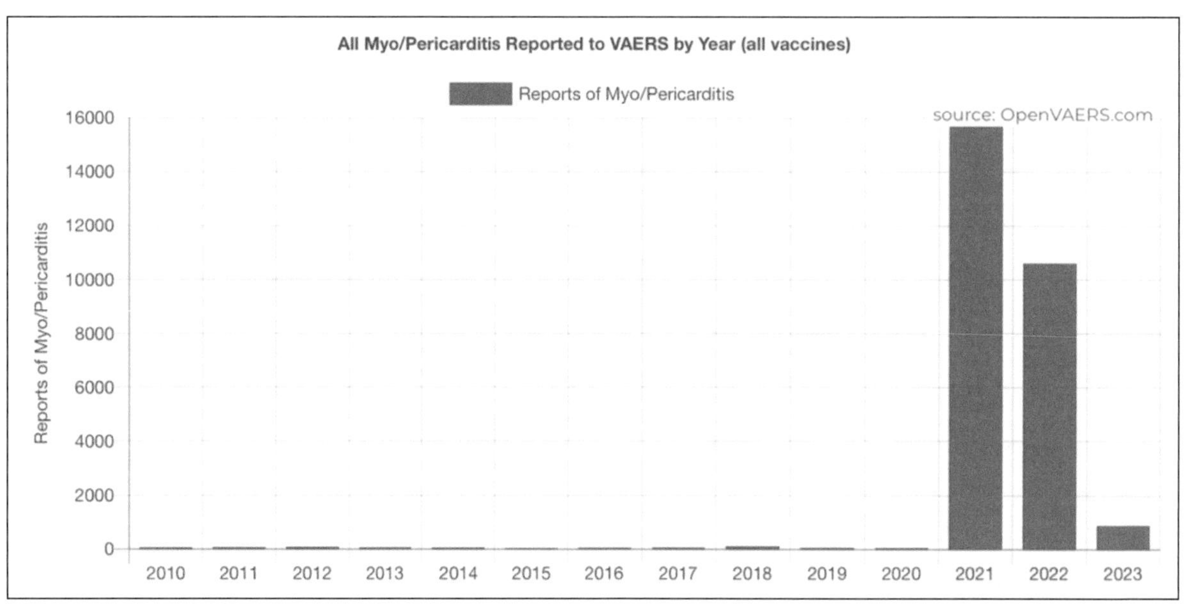

(https://openvaers.com/covid-data/myo-pericarditis)

Statement 1: Myocarditis after COVID-19 "vaccine" injection is rare.

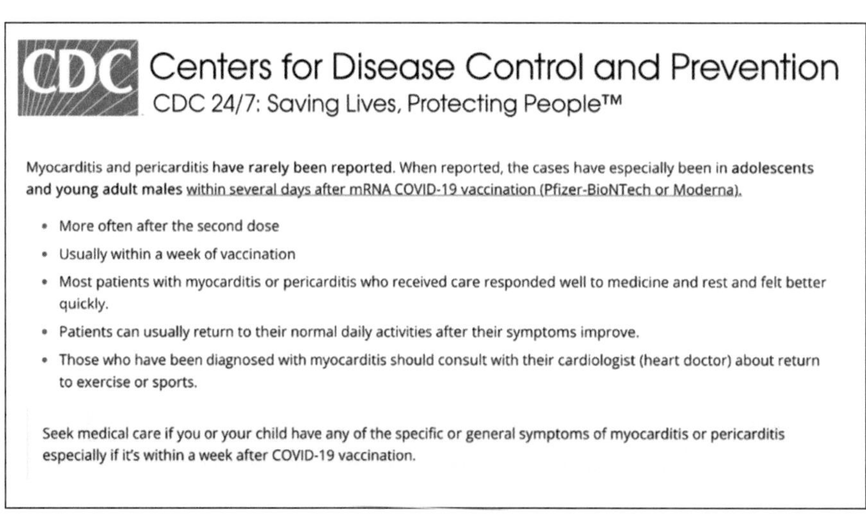

CDC Centers for Disease Control and Prevention
CDC 24/7: Saving Lives, Protecting People™

Myocarditis and pericarditis **have rarely been reported**. When reported, the cases have especially been in **adolescents and young adult males** within several days after mRNA COVID-19 vaccination (Pfizer-BioNTech or Moderna).

- More often after the second dose
- Usually within a week of vaccination
- Most patients with myocarditis or pericarditis who received care responded well to medicine and rest and felt better quickly.
- Patients can usually return to their normal daily activities after their symptoms improve.
- Those who have been diagnosed with myocarditis should consult with their cardiologist (heart doctor) about return to exercise or sports.

Seek medical care if you or your child have any of the specific or general symptoms of myocarditis or pericarditis especially if it's within a week after COVID-19 vaccination.

https://www.cdc.gov/coronavirus/2019-ncov/vaccines/safety/myocarditis.html)

The Centers for Disease Control and Prevention's (CDC) notice and its advice, above, ignore the following:

1. Myocarditis post-mRNA COVID vaccine is *not* rare: 2.8% of 777 subjects studied prospectively were diagnosed with myocarditis, with a median age of 37, which is outside of the high-risk years. (Buergin et al., https://doi.org/10.1002/ejhf.2978.)
2. According to the CDC's Vaccine Adverse Events Reporting Systems (VAERS), 30% of myocarditis reports occur one week post-mRNA COVID injection. There is no basis upon which to make assumptions about when heart disease will appear with a drug that has biologic activity, like the mRNA COVID vaccines, and has never been used before. Proper studies were not done nor are they in progress at any scale.
3. Return to normal activities after mRNA COVID vaccination and before cardiac diagnostic testing? NO! Just don't do it.

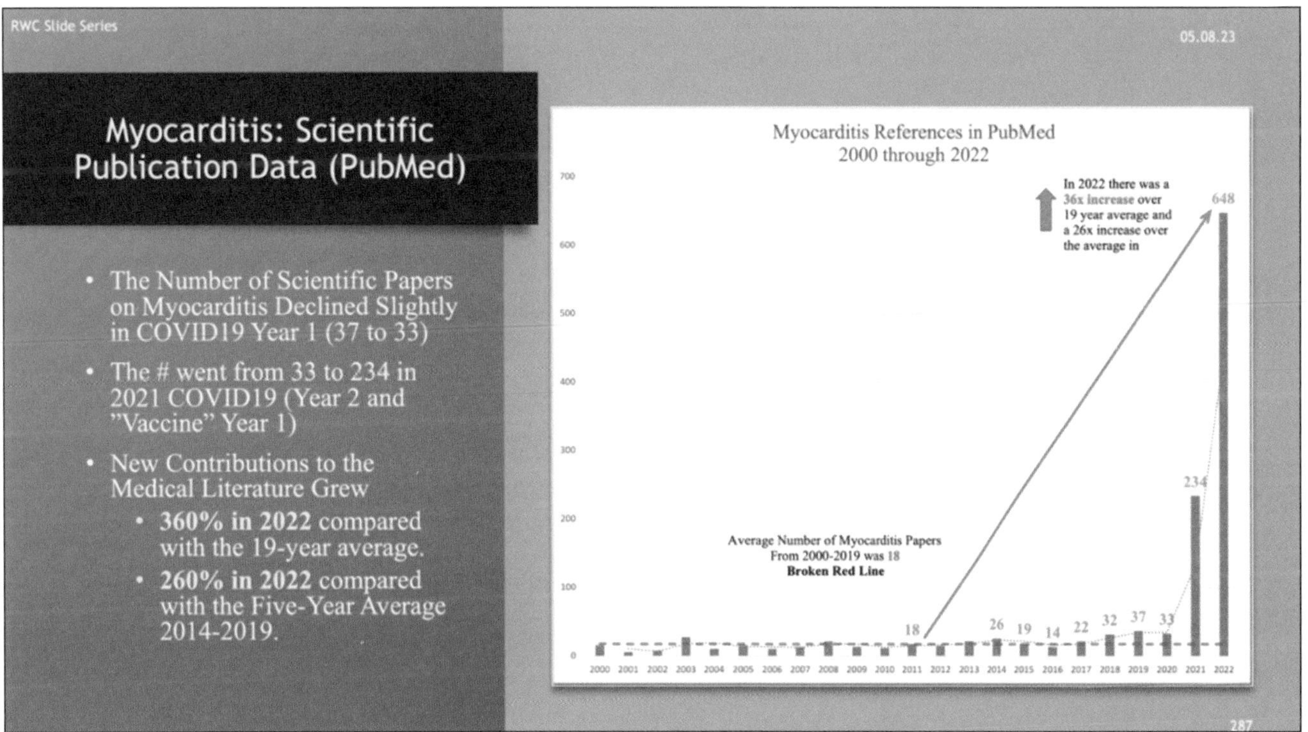

4. The publication volume of medical literature concerning myocarditis declined slightly in 2020, the first year of COVID-19, from 37 reports in 2019 to 33 in 2020.
   - In COVID "vaccine" Year 1/COVID-19 Year 2 (2021), the number of citations increased 13x compared with the 19-year average and 9x from the five-trailing-year average.
   - In "vaccine" Year 2/COVID-19 Year 3 (2022), the number of citations increased 2.77x over the prior year, *signaling a major medical event* that continued after COVID died down and the injections slowed to a trickle.
   - The cause can be sorted out in most myocarditis cases, and there is no reason to lump all myocarditis cases into the category of so-called "Long COVID." Unless, of course, you want to hide the real cause.

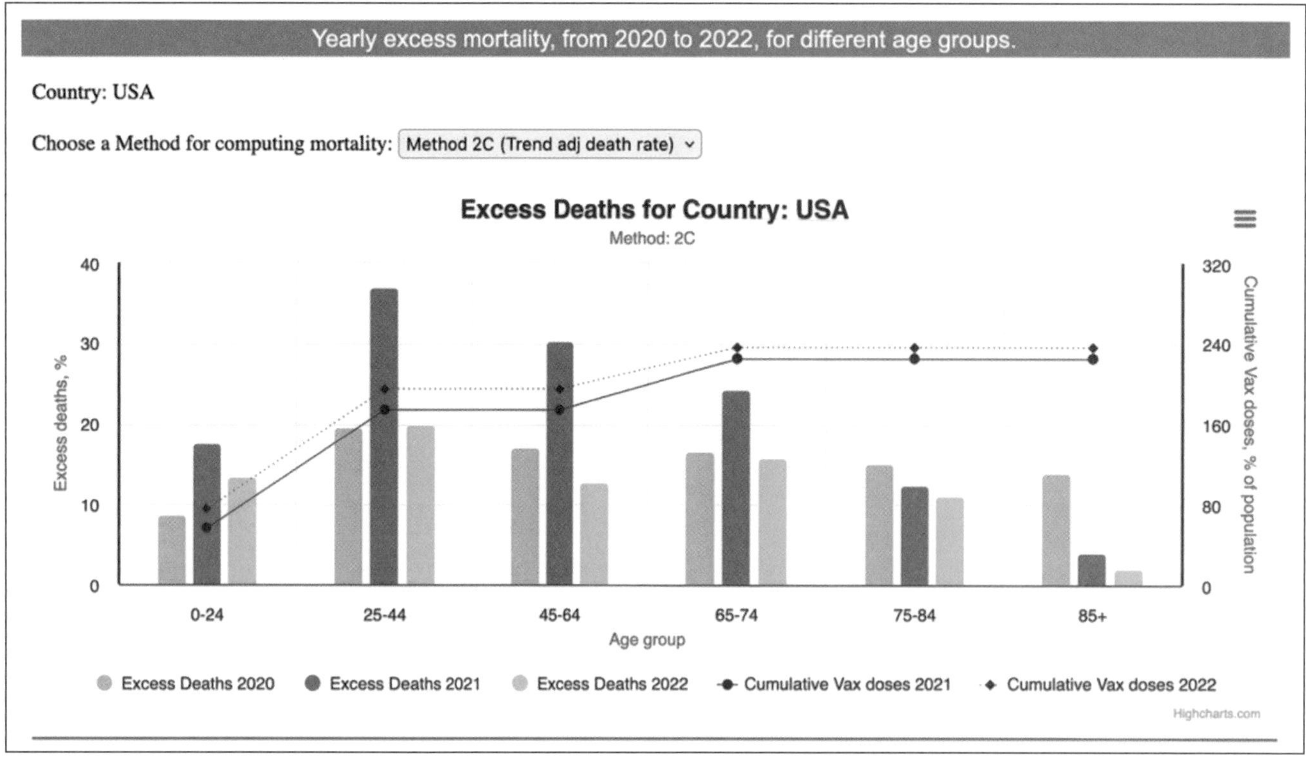

It is no surprise then to see the medical literature expand as excess deaths rise in both COVID and "vaccine" years, as indicated in Ed Dowd's graph above. (https://phinancetechnologies.com/HumanityProjects/Humanity%20Projects.asp)

These are crude indicators when it comes to sorting out etiology of the excess deaths, and more comprehensive analyses are needed from the perspective of population studies down to aggregation of individual patient data.

Statement 2: Myocarditis after Spike-generating drug injection is temporary.

This statement has no scientific support. The Barmada et al. study found myo/pericardial damage in 14 out of 17 young people averaging 16.9 years of age. Young men comprising 87% of the group. At two months or longer follow-up, some patients had signs of worsening on their follow-up cMRI studies. (DOI: 10.1126/sciimmunol.adh3455)

Long-term, widespread study is required to estimate the magnitude of the problem of myo/pericarditis on a population level. Screening protocols with now identifiable sensitive and specific diagnostic techniques need to be deployed in a scalable program with built-in directed treatment.

Sudden death cases and unexpected fatalities require specific autopsies following the guidelines prepared by Dr. Arne Burkhardt.

Notes and recommendations for conducting post-mortem examination (autopsy) of persons deceased in connection with COVID vaccination

doctors4covidethics.org

Authors:
Prof. Dr. A. Burkhardt[1]
Pathology Laboratory Reutlingen
Obere Wässere 3-7
72764 Reutlingen
Germany

[1] in collaboration with an international team of pathologists

Background and introduction

Dr. Burkhardt and colleagues recently carried out a series of 17 autopsies on persons deceased within days to months of vaccination. Initially, none of these deaths had been attributed to the vaccines. Nevertheless, Burkhardt and colleagues found characteristic lesions in multiple organs which led them to conclude that in most patients the vaccines were likely the cause of death. Key observations were widespread vasculitis with microthrombi as well as intense lymphocytic infiltration of multiple organs. A summary of these findings has been published before [1]. More recently, he and his colleagues have also demonstrated the expression of spike protein, induced by the vaccine, within the inflammatory lesions [2].

Here, Dr. Burkhardt gives guidance for conducting autopsies in similar circumstances. This document has been updated on March 17, 2022.

1 Conduct of autopsies

Autopsies should focus on the following phenomena:

- thromboembolic events (both macro- and microthrombi)
- vasculitis
- myocarditis
- lymphocytic alveolitis
- peculiar inflammatory reactions (autoimmune reactions?)
- foreign material

1.1 Inspection of the body, sampling of injection site and lymph nodes.

1. Carefully inspect the entire integument, paying special attention to discoloration due to allergic-exanthematous reactions, e.g. brown coloring indicating hemosiderosis in the context of leucoclastic vasculitis
2. Take tissue samples from the site of the vaccination (subcutaneous and muscle tissue)

3. Preserve the axillary lymph nodes on the side of injection, as well as enlarged lymph nodes from any other site
4. Check the veins of the lower legs for thrombi, and especially in bedridden persons also the plantar veins

1.2 Body cavities. Open up the three major body cavities according to standard practice. Take samples for histological examination from all organs and from any unusual lesions (infarctions, bleedings, thrombi etc.)

1.2.1 Thorax.

1. Check for thromboembolism by cutting open the major vessels
2. Check the lungs for focal lesions
3. Consider in-toto fixation of both lungs and preparation by serial section
4. Take histological samples from the heart muscle in several different locations
5. Optional: examine the heart's excitation conduction system, especially in cases of sudden cardiac death. Pay special attention to the region of the atrioventricular node

1.2.2 Abdomen.

1. Pay special attention to the spleen (histology) and to Peyer's Patches
2. Cut open the liver veins all the way to the periphery in order to check for veno-occlusive disease
3. Also examine the ovaries, which allegedly may contain deposits of foreign material

1.2.3 Brain, eyes, and ears.

1. Look for infarctions or bleedings. Pay special attention to the superior thalamostriate vein (vena terminalis)
2. Preserve the hypophyseal gland
3. If possible, carry out fixation in toto and subsequent neuropathological examination
4. Critical: examine the eyes in case the deceased had been suffering from impaired vision (e.g. recently developed visual field defects)
5. Examine the inner ear in patients with loss of hearing

1.3 Tissue sampling. Routine sampling from all organs, in addition to those specifically mentioned above:

1. Sample all recognizable lesions, especially thrombi, which should be preserved together with the surrounding vascular wall
2. Take samples from arteries even when they don't contain any thrombi, especially from the following:
    - aorta
    - coronary arteries

- carotid arteries
- the Circle of Willis (circulus arteriosus cerebri)
- arteries of the leg

3. Take samples of striated muscle from at least two locations, always including the lower leg muscles
4. Sample the bone marrow in at least two different sites with active hemopoiesis
5. Take samples from the thyroid gland and from the salivary glands (look for autoimmune phenomena)

1.4 General considerations.

1. Photographically document all relevant changes and important normal findings
2. Preserve organs until the histological samples have been assessed, for the purpose of possible further examinations
3. When embedding of the histological samples, ensure compatibility with subsequent immunohistological or PCR investigations (virus fragments)
4. If there is no significant autolysis yet, preserve samples for electron microscopy—search for virus particles or fragments, unusual materials etc.

2 Evaluation of organ samples from deceased or biopsies from living patients after COVID vaccination (microscopy, histology, immunohistochemistry)

In any case and on all organs:

- Search for birefringent material
- Stains: HE, PAS, iron

2.1 Immunohistochemical differentiation of inflammatory cells. In case of inflammation, further definition by immunohistochemistry, depending on the histological picture:

- CD 3 (T lymphocytes)
- CD 4 (T helper cells)
- CD 8 (cytotoxic T lymphocytes)
- CD 14 monocytes
- CD 20 B lymphocytes
- CD 56 cell adhesion (NK cells)
- CD 68 macrophages
- CD 31/D2-40 endothelium
- Complement deposits

2.2 Immunohistochemistry to detect vaccine-induced spike protein expression.

- Use anti-SARS-COV-2 spike protein/S1 antibodies to test for presence of spike protein in tissue samples. Always include myocardium and spleen tissue samples
- If spike protein is detected, use anti-nucleocapsid antibody to examine expression of SARS-COV-2 nucleocapsid: presence of nucleocapsid indicates viral "breakthrough" infection, absence of nucleocapsid supports vaccine-induced spike protein expression
- Perform positive and negative controls using vaccine-transfected and non-transfected cell cultures

2.3 Unidentified foreign material, other unusual findings. If histological examination reveals unidentified foreign material, preserve tissue samples using conditions suitable for

- electron microscopy
- Raman microscopy
- X-ray and laser microanalysis

In case of any unusual findings, carry out fixation for electron microscopy if possible.

3 Further considerations and measures

If the examinations detailed above provide evidence suggestive of vaccine-induced death, consider the following steps:

1. Preserve tissue samples of lesions, including the site of vaccine application
2. Obtain the consent of relatives, and if applicable the court prosecutor, for carrying out paraffin embedding and histological sections (HE, PAS, FE) of all organs
3. Depending on the findings, initiate further investigations by cooperating special laboratory or in a reference laboratory

References

[1] S. Bhakdi and A. Burkhardt: *On COVID vaccines: why they cannot work, and irrefutable evidence of their causative role in deaths after vaccination.* 2021. URL: https://doctors4covidethics.org/on-covid-vaccines-why-they-cannot-work-and-irrefutable-evidence-of-their-causative-role-in-deaths-after-vaccination/.

[2] A. Burkhardt and W. Lang: *First time detection of the vaccine spike protein in a person who died after vaccination against Covid-19.* 2022. URL: https://pathologie-konferenz.de/en/.

Few proper autopsies have been done. The slide below illustrates some of the findings of histopathological examination of the heart after mRNA COVID injection. Normal heart muscle is featured on the left and is juxtaposed with LNP/mRNA-damaged heart muscle on the right. The muscle structure is severely disrupted. There is infiltration of lymphocytes. The heart's muscle cells are dead.

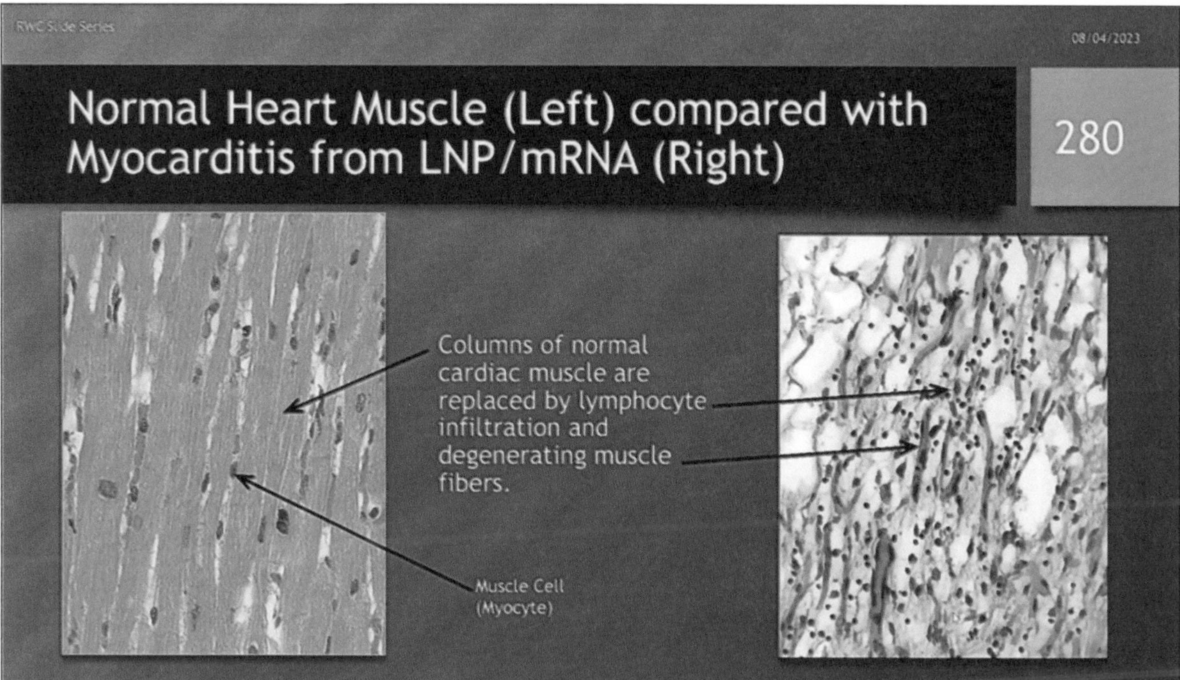

The exact process by which this destruction occurs has not been worked out at this point in time. Possibilities include an autoimmune reaction to modified host proteins, novel "vaccine"-induced non-self-proteins, molecular mimicry, or intense inflammatory reaction from mass cytokine release or complement cascade. Inflammation is a process that initiates healing, but it is also a process that can go out of control and destroy cells, tissues, and organs (image below).

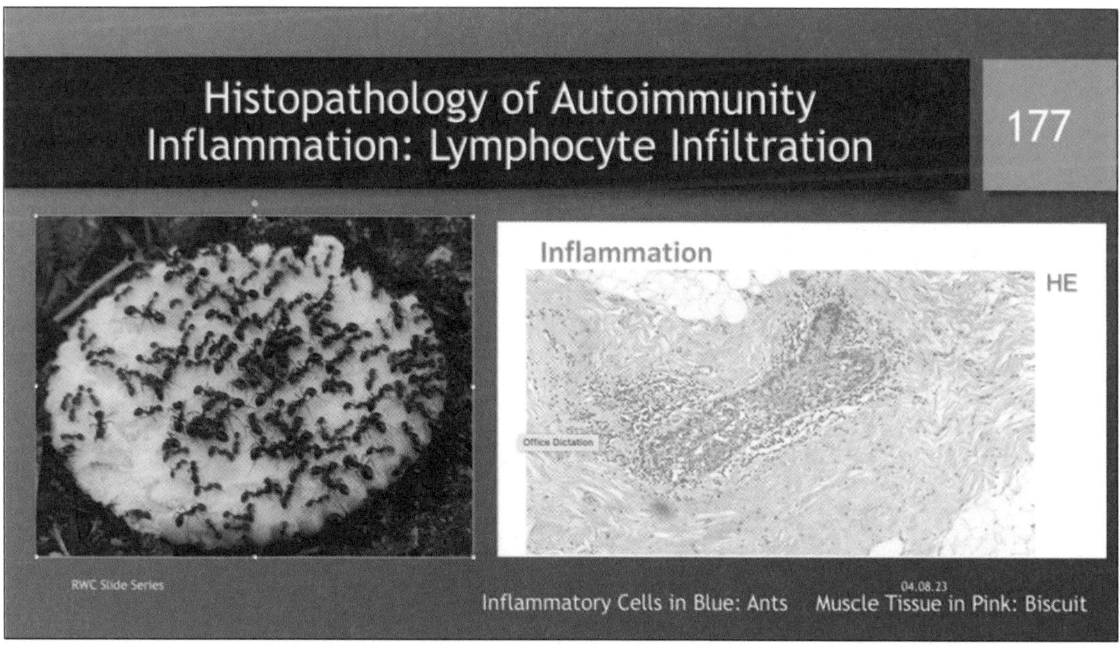

Early diagnosis and treatment are imperative. Acute phase diagnostics with inflammatory markers or signs of myocardial tissue damage, such as troponin, are necessary to make an accurate diagnosis. The graphic below is from Barmada et al. and shows how these indicators elevate with myocarditis.

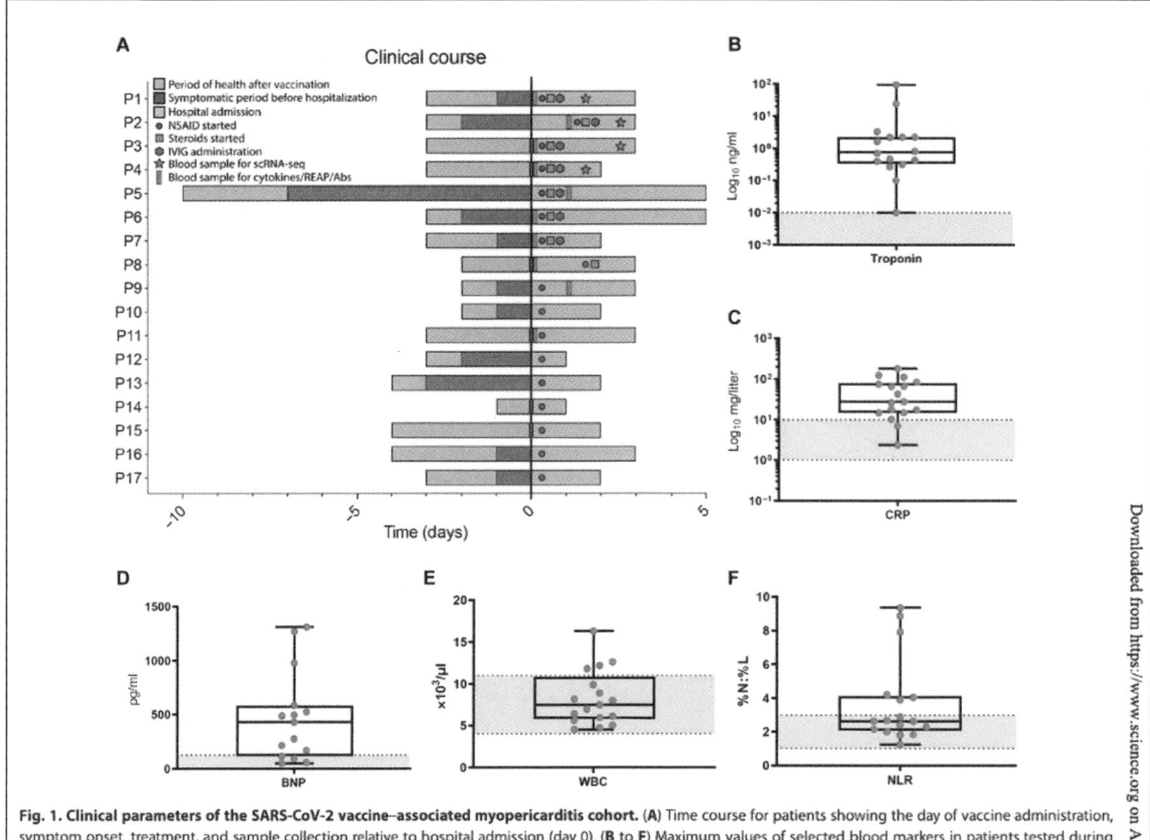

Fig. 1. Clinical parameters of the SARS-CoV-2 vaccine–associated myopericarditis cohort. (A) Time course for patients showing the day of vaccine administration, symptom onset, treatment, and sample collection relative to hospital admission (day 0). (B to F) Maximum values of selected blood markers in patients tested during hospital admission. Boxes depict the interquartile range (IQR), horizontal bars represent the median, whiskers extend to 1.5 × IQR, and red dots show the value of each patient. Dashed lines and gray area represent normal reference ranges used at Yale New Haven Hospital. CRP, C-reactive protein; BNP, B-type natriuretic peptide; WBC, white blood cell; NLR, neutrophil-to-lymphocyte ratio; REAP, rapid extracellular antigen profiling; Abs, antibodies; IVIG, intravenous immunoglobulin; NSAID, nonsteroidal anti-inflammatory drug.

DOI: 10.1126/sciimmunol.adh34

Cardiac MRI (cMRI) shows inflammatory change early and scarring once the inflammation subsides.

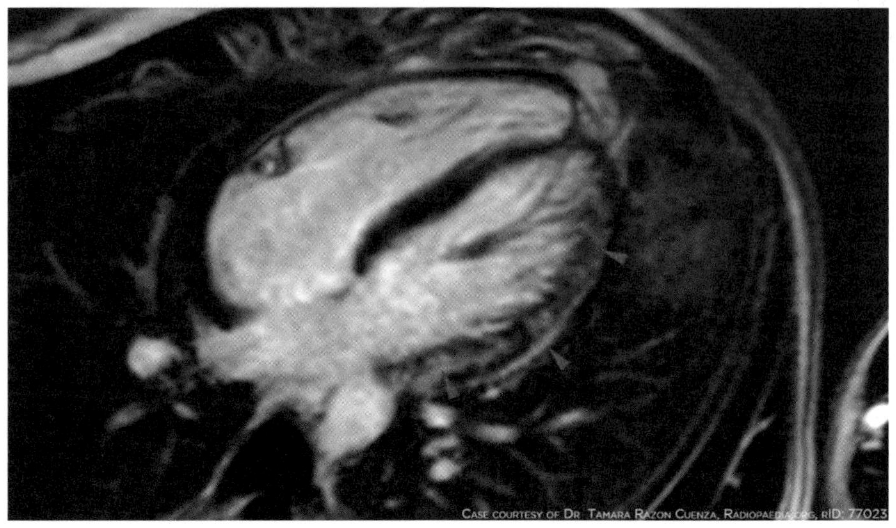

Statement Three: Myocarditis is mild.

Before he died, Dr. Burkhardt presented one of the few, perhaps only, cases of autopsy after one year in a Pfizer BNT162b2 myocardial and aorta injured 22-year-old male athlete who committed suicide because of his medical condition.

This case was presented to the Canadian COVID Care Alliance (https://rumble.com/user/CanadianCovidCareAlliance) shortly before Dr. Burkhardt's death on May 30, 2023.

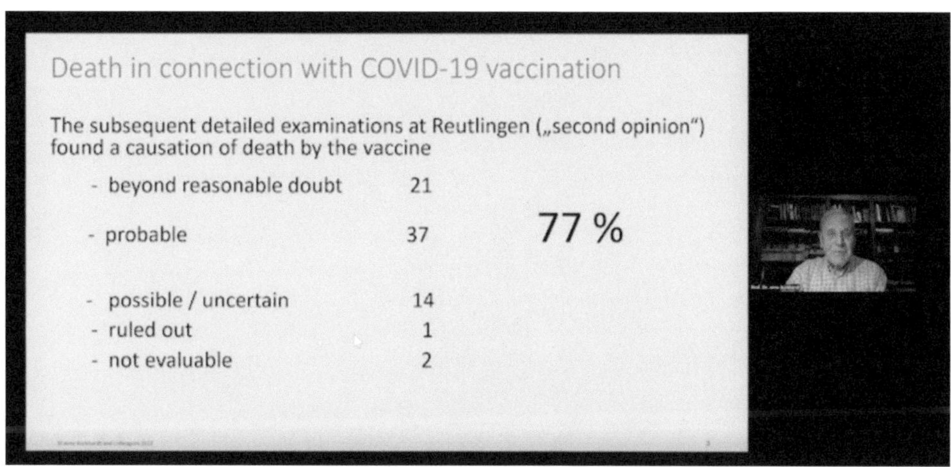

(https://rumble.com/v2jbj16-arne-burkhardt-presentation-to-the-ccca.html)

The inflamed and necrotic (dead) muscle seen here is replaced by extensive scar tissue that is almost full thickness across the wall of the heart (left, lower quadrant extending across to the right side).

Scar tissue replacement of normal cardiac muscle makes the heart stiffer with compromised contractility (decreasing ejection fraction). If this process affects the electrical conduction system, an irregularly beating heart (cardiac arrhythmia) can result.

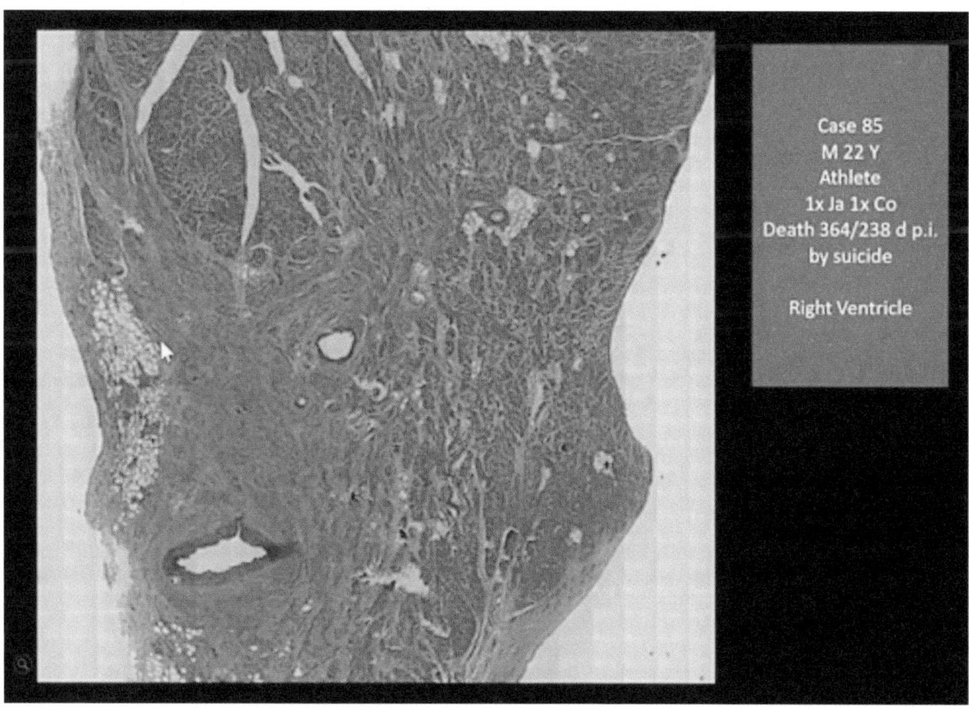

The section below illustrates some residual normal muscle in a matrix of rigid fibrous tissue that has replaced the necrotic muscle. There is a dense collection of lymphocytes with ongoing inflammation at 12 months post-injection.

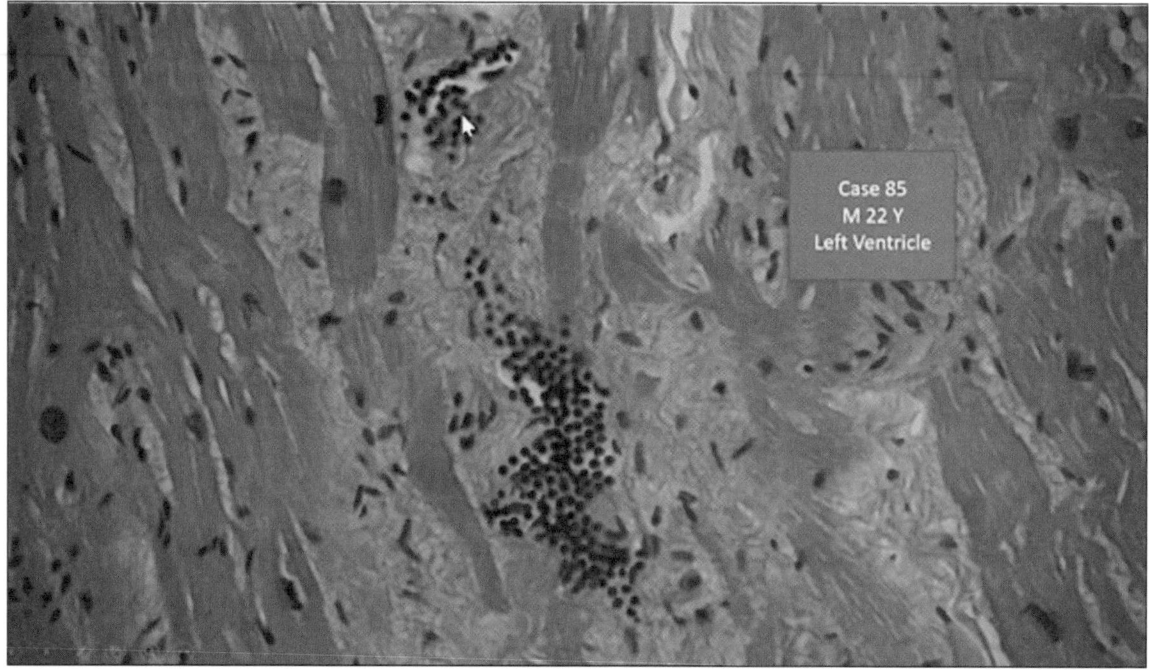

Dr. Burkhardt verified the cause of the diseased myocardium using one stain that is positive for spike from either the drug or the virus and a second that is positive only for cases caused by the virus. A year after injection, there is still spike protein in the heart, but there is no evidence that this heart was damaged by COVID.

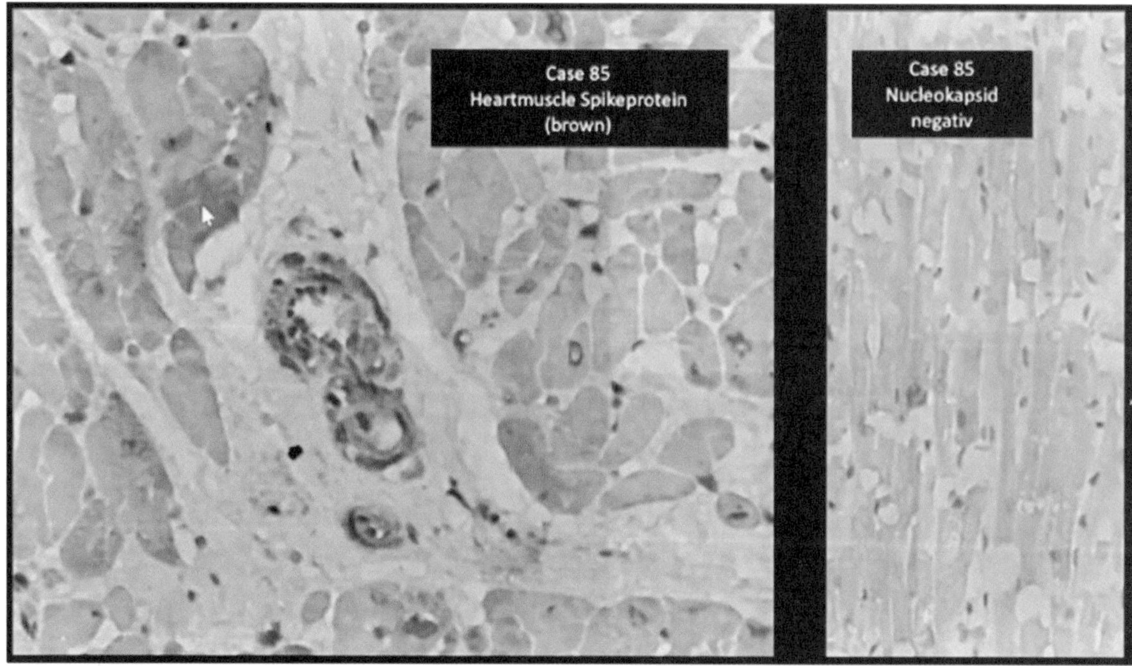

This unfortunate young man also had disruption of the muscular wall of his aorta that is forming an aneurysm as the inner layer of the aorta splits from the outer layer (below). Rupture of an aneurysm of the aorta can be rapidly fatal.

Spike proteins were designed to attach to a specific (ACE2) receptor on cells lining the inner wall of blood vessels called the endothelium. The slide below shows destruction of the endothelium in the aorta of the young athlete.

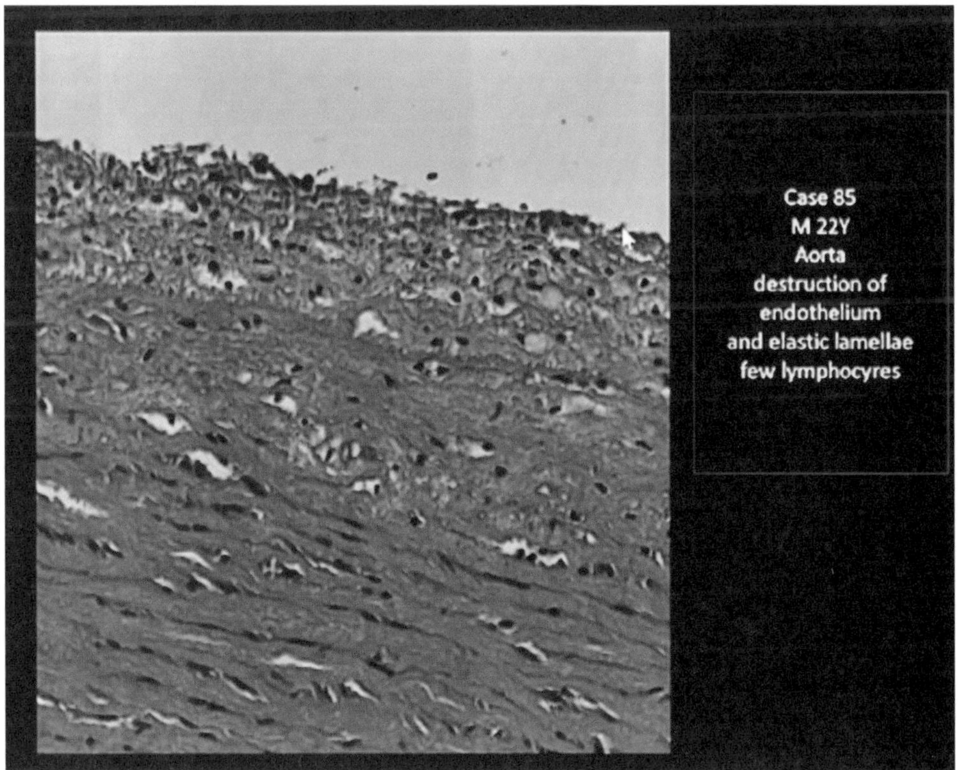

At this stage there is limited knowledge regarding the long-term prognosis of CoVax (i.e., LNP/mRNA) heart disease in a specific sense. CoVax myocarditis may behave differently than other types of myocarditis.

This case illustrates the tragic consequences of cutting corners and making assumptions instead of making proper science. Pfizer Confidential Document 2.4 provides the assumption made about proteins expressed by the RNA in BNT162b2.

5. *"The protein encoded by the RNA in BNT162b2 is expected to be proteolytically degraded like other endogenous proteins. RNA is degraded by cellular RNases and subjected to nucleic acid metabolism. Nucleotide metabolism occurs continuously within the cell, with the nucleoside being degraded to waste products and excreted or recycled for nucleotide synthesis. Therefore, no RNA or protein metabolism or excretion studies will be conducted." (p. 20, ¶3)*

Sadly, they simply skipped this piece of work. And at least 15 others. See the review below of the Pfizer Pre-clinical ("Nonclinical") Studies Confidential Document 2.4 to see the other 15 areas that were not studied.

From Australia

- Government Hearing: https://amgreatness.com/2023/08/04/pfizer-and-moderna-reps-put-on-the-hot-seat-in-fiery-senate-hearing-in-australia/

The Republican-led U.S. House Select Subcommittee on the Coronavirus Pandemic has not yet called any witnesses from Pfizer, Moderna, Johnson and Johnson, the CDC, FDA, Anthony Fauci, or Francis Collins to appear before the committee and has shown no interest in investigating the fraud that allegedly took place in the COVID vax clinical trials.

Class action lawsuit: https://rumble.com/v348nkr-covid-vaccine-injury-class-action.html

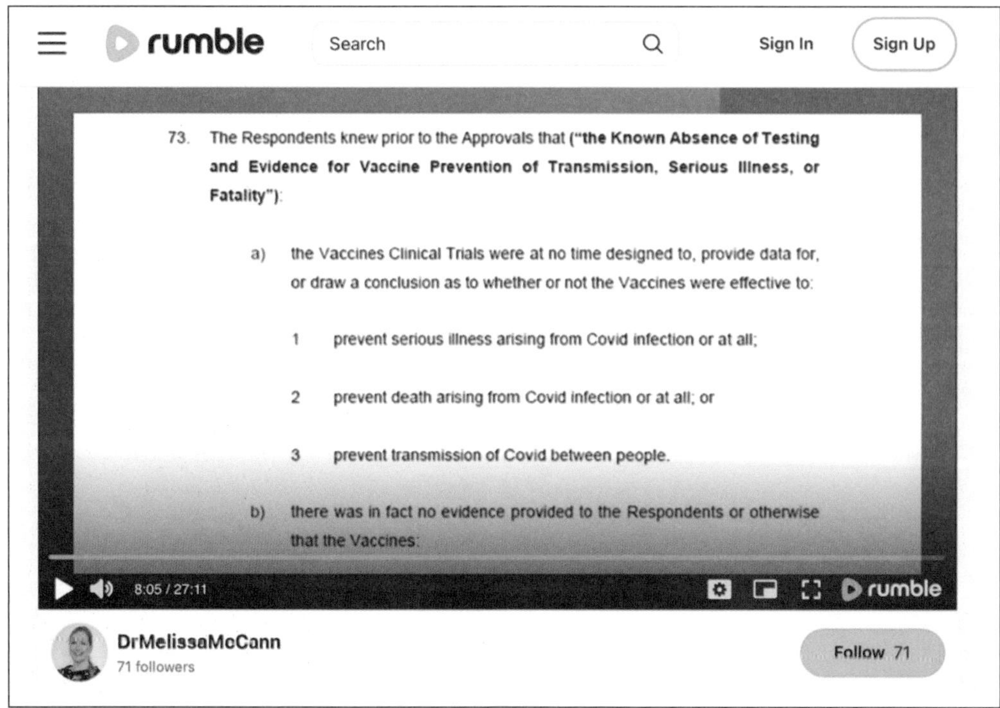

From Germany

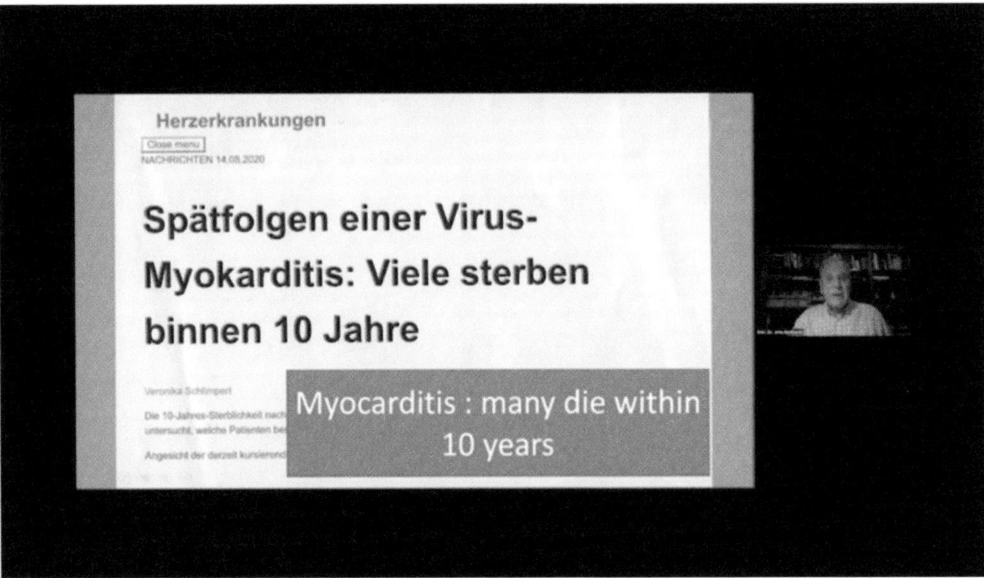

From All Over

- "RETIRED SUDDENLY—NFL Players and International soccer players injured after taking COVID-19 vaccines—pericarditis, arrhythmia, 'heart conditions', blood clots in legs & lungs—12 recent stories!" by William Makis, MD
- "Bronco KJ Hamler 'steps back'; LSU coach Jimmy Lindsey 'steps away'; Bears' Steve McMichael in ICU; ex-Buckeye Drue Chrisman in hospital; Sooners coach Brent Venables' wife has breast cancer," by Mark Crispin Miller

Conclusions

Heart disease following injection of COVID-19 mRNA "vaccine" products is:

1. Not rare.
2. Not temporary.
3. Not mild.

It is past time to seriously study this potentially catastrophic health problem, define its dimensions and severity, and then develop treatments to preserve life and function.

Twenty-Four: "Moderna mRNA COVID-19 Injection Damaged Mammals' Reproduction: 22% Fewer Pregnancies; Skeletal Malformations, Pain, Nursing Problems in Pups. FDA Knew, Yet Granted EUA."
—Linnea Wahl, MS—Team 5

Based on "A GLP intramuscular combined developmental and perinatal/postnatal reproductive toxicity study of mRNA-1273 in rats. FDA-CBER-2–22-4207–0015," Moderna's COVID-19 mRNA drug damaged the reproductive systems of female rats their estrous cycles (the recurring period of sexual receptivity and fertility in many female mammals) were disturbed, they were less likely to mate, and they were even less likely to get pregnant. When the mRNA-injected rats did get pregnant, their fetuses and live-born pups suffered malformations and distress. The results of Moderna's study on rats do not bode well for human reproduction.

About Rat Models

Moderna researchers used Sprague Dawley rats in their study. This rat model is commonly used to determine how a drug affects reproduction and development. In Moderna's study, injected female rats were mated with male rats who apparently were not injected.

According to Taconic Biosciences, a provider of Sprague Dawley rats, "the female rat accepts the male for mating only at the end of the 12-hour preliminary period of proestrus, and during the 12 hours of estrus. Ovulation occurs about 10 hours after the onset of estrus. Sperm migrate from the uterus to the oviduct about 15 minutes after copulation. At 1 hour post-copulation, sperm are found throughout the oviduct at 3 hours, 90% of the ova are fertilized" (p. 7). Researchers often determine if the female rat is in estrus by observing the appearance of the vagina or by taking a vaginal swab (Ajayi and Akhigbe 2022).

Similarly, researchers typically know if a female rat has mated by taking vaginal swabs and looking for sperm. By the thirteenth day after mating, a pregnant rat's abdomen begins to grow larger. This is followed the next day by mammary development, nipple enlargement, and increased eating and weight gain (Ypsilantis et al. 2009).

For rats that are indeed pregnant, their fetuses begin to develop organs about nine to 10 days into gestation. Fetuses develop bones and joints in the following few days. Normal gestation lasts about 21 to 23 days. An average litter consists of about 10 rat pups. (Taconic Biosciences)

Rat pups begin walking, using their forelimbs, and grooming themselves at about 10 to 20 days (Smirnov and Sitnikova 2019). Male rats are typically sexually mature at 40 to 60 days old (Fuochi et al. 2022); female rats experience proestrus at about 38 days old (Ajayi and Akhigbe 2020).

Moderna's Research on Developmental and Reproductive Toxicity

On June 16, 2020, Moderna began experiments on Sprague Dawley rats to find out whether Moderna's COVID-19 mRNA shot was toxic to pregnant females and their offspring (Moderna 2020). Researchers injected 44 female rats with mRNA-1273 (the experimental group) and 44 female rats with a buffer solution (the control group). Each rat got four doses: 28 days and 14 days before they mated, then 1 day and 13 days after they mated with male rats (who apparently were not injected).

When Moderna's female rats received their last injection 13 days after mating, they were likely just barely or not yet showing if they were pregnant. At this point, each group was divided equally into two subgroups. Half the experimental group (22 rats) and half the control group (22 rats) were sacrificed 21 days after they mated (at the end of the normal gestation period), and their fetuses were removed by Caesarean section. The other half (22 rats in each group) were allowed to deliver their pups naturally, then sacrificed along with their pups 21 days after birth (after the normal pup development period).

Developmental and Reproductive Toxicity Results

According to Moderna researchers, the mRNA-1273 shot "did not have any adverse effects on the F0 [injected females] or F1 [their offspring] generations" (Moderna 2020, p. 43). But the data in their report tell a different story.

Adverse effects in the rats who got the mRNA-1273 shot included thinning fur, swollen and partially paralyzed hind legs, weight gain, and abnormal eating patterns (Moderna 2020, pp. 37–38). Perhaps most alarming, the "mean number of [estrous] cycle lengths was statistically significantly higher" (longer) in mRNA-injected females than in the control group, meaning that the drug affected their reproductive systems (Moderna 2020, p. 38; Table 13, p. 67).

More evidence of reproductive harms to these rats is found in Table 14 (Moderna 2020, p. 68). Here the data show that only 39 of the 44 mRNA-injected females (88.6%) successfully mated, as compared to 42 of the 44 control group females (95.5%). This is a 7% decrease in mating in the experimental group; but, because the sample size is small, the difference is not statistically significant.

Pregnancy data, too, suggest harms to the rats' reproductive systems (Moderna 2020, p. 38). Only 37 of the 44 mRNA-injected females (84.1%) got pregnant, as compared to 41 out of 44 control group females (93.2%). That is a 9% decrease in pregnancies in the females who got the mRNA shot; but again, because the sample size is small, the difference is not statistically significant.

The data are even more discrepant in the rats that were designated for natural delivery. Recall that 22 control and 22 experimental rats were chosen, likely before researchers could tell if they were pregnant, to deliver their pups naturally. Just 15 of the 22 mRNA-injected females (68.2%) got pregnant and delivered their pups naturally, as compared to 20 out of 22 control group females (90.9%). This appears to be an alarming 22.7% decrease in pregnancies in the natural delivery, mRNA-injected female rats (Moderna 2020, Table 24, p. 103). In this analysis, the sample size of each group is even smaller; so, once again, the difference is not statistically significant.

The offspring, as fetuses (sacrificed by Caesarean section 21 days after their mothers mated) and as live-born pups, were affected by the mRNA shot as well. In both, Moderna researchers reported malformed bones: "mRNA-1273-related common skeletal variations consisting of wavy ribs and increased nodules were observed . . . the fetal and litter incidence of wavy ribs exceeded the range observed historically . . ." (Moderna 2020, p. 16). These skeletal malformations (wavy ribs) may suggest that vital serum proteins are not getting to the fetus (Kast 1994).

In the live-born pups, some whose mothers got the mRNA shot showed signs of pain (ungroomed fur is an indicator of pain in research rats; Carstens and Moberg 2000) and some had kidney abnormalities (renal papilla) (Moderna 2020, p. 41). Inexplicably, researchers noted that these pups had "statistically significant increases in mean pup body weights," while some were starving (no milk band present, meaning they were not getting milk through nursing; Moderna 2020, p. 41). In spite of these findings, the report declares that "there were no mRNA-1273-related effects on any natural delivery or litter observation parameters" (Moderna 2020, p. 40).

Alternate Conclusions

Clearly, Moderna's rat study showed that the mRNA-1273 shot had toxic effects on the female rats' reproductive systems and on their offspring. Unfortunately, Moderna researchers used too few rats in their study, limiting our ability (deliberately?) to use statistics to draw definitive conclusions.

Further, the study was criticized by the European Medicines Agency (EMA) because "no vaccine dose was administered during the early organogenesis, to address the direct embryotoxic effect of the components of the vaccine formulation" (European Medicines Agency 2021, p. 51). By not giving the mRNA shot throughout the female rats' pregnancies, Moderna researchers missed (deliberately?) an important opportunity to determine the drug's toxic effects on offspring.

The Moderna rat study ended on September 14, 2020. The US Food and Drug Administration (FDA) had the study and its alarming results in December 2020 when it approved the Moderna mRNA-1273 shot for emergency use in humans. (https://www.nih.gov/news-events/news-releases/statement-nih-barda-fda-emergency-use-authorization-moderna-covid-19-vaccine) Thus, Moderna and United States government agencies have known for years that mRNA-1273 is toxic to pregnant females and their babies, yet they continue to this day to recommend the shot to women "who are pregnant, breastfeeding, trying to get pregnant now, or those who might become pregnant in the future" (US Centers for Disease Control and Prevention).

When will the FDA alert the world to the fact that Moderna's mRNA-1273 shot is dangerous to pregnant women and their babies?

References

Ajayi AF, Akhigbe RE. 2020. Staging of the estrous cycle and induction of estrus in experimental rodents: an update. Fertility Research and Practice, 6:5. https://doi.org/10.1186%2Fs40738–020-00074–3.

Carstens E, Moberg GP. 2000. Recognizing pain and distress in laboratory animals. ILAR Journal, 41:2; 62–71. https://doi.org/10.1093/ilar.41.2.6.

European Medicines Agency. 2021. Assessment report: COVID-19 vaccine Moderna. EMA/15689 /2021, correction 1. https://www.ema.europa.eu/en/documents/assessment-report/spikevax-previously -covid-19-vaccine-moderna-epar-public-assessment-report_en.pdf.

Fuochi S, et al. 2022. Puberty onset curve in CD (Sprague Dawley) and Long Evans outbred male rats. Laboratory Animals, 56:5; 471–475. https://doi.org/10.1177/00236772221078725.

Kast A. 1994. "Wavy ribs" A reversible pathologic finding in rat fetuses. Experimental and Toxicologic Pathology. 46:3; 203–210. https://doi.org/10.1016/S0940–2993(11)80082–5.

Moderna. 2020. A GLP intramuscular combined developmental and perinatal/postnatal reproductive toxicity study of mRNA-1273 in rats. FDA-CBER-2–22-4207–0015. https://defendingtherepublic.org/wp-content/uploads/2023/07/6_Moderna-Study-A-GLP-Intramuscular-Combined-Developmental-and-PerinatalPostnatal-Reproductive-Toxicity-Study-of-mRNA-1273-in-Rats-814-pages.pdf.

Smirnov K, Sitnikova E. 2019. Developmental milestones and behavior of infant rats: the role of sensory input from whiskers. Behavioral Brain Research, 374; 112143. https://doi.org/10.1016/j.bbr.2019.112143

Taconic Biosciences. Sprague Dawley rat. https://www.taconic.com/pdfs/sprague-dawley-rat.pdf.

US Centers for Disease Control and Prevention. Safety and effectiveness of COVID-19 vaccination during pregnancy. https://www.cdc.gov/coronavirus/2019-ncov/vaccines/recommendations/pregnancy.html#anchor_1628692520287.

Ypsilantis P., et al. 2009. Ultrasonographic diagnosis of pregnancy in rats. Journal of the American Association for Laboratory Animal Science, 48:6; 734–739. https://www.ncbi.nlm.nih.gov/pmc/articles/PMC2786927/.

Twenty-Five: "23% of Vaccinated Mothers' Fetuses or Neonates Died. Suppressed Lactation and Breast Milk Discoloration Reported."

—Barbara Gehrett, MD; Joseph Gehrett, MD; Chris Flowers, MD; and Loree Britt

The WarRoom/DailyClout Pfizer Documents Analysis Project Post-Marketing Group (Team 1)—Barbara Gehrett, MD; Joseph Gehrett, MD; Chris Flowers, MD; and Loree Britt—wrote a shocking analysis of the "Use in Pregnancy and Lactation" section found in Pfizer document *5.3.6 Cumulative Analysis of Post-Authorization Adverse Event Reports of PF-07302048 (BNT162B2) Received Through 28-FEB-2021* (a.k.a., "*5.3.6*"). This section includes three types or groups of cases:

- Pregnancy cases
- Breastfeeding baby cases
- Breastfeeding mother cases

In all, 274 pregnancy cases were reported, which includes 270 mother cases and four fetus/baby cases.

It is important to note that the information in the *5.3.6* document was reported to Pfizer for **only a 90-day period** starting on December 1, 2020, the date of the United Kingdom's public rollout of Pfizer's COVID-19 experimental mRNA "vaccine" product.

Important points from this report include:

- **Pfizer's BNT162b2 mRNA COVID vaccine was *not* recommended for use in pregnancy or with lactation during the time of this post-marketing data set.**
  - The Centers for Disease Control and Prevention (CDC) did recommend COVID vaccination for pregnant and lactating women until April 23, 2021. [https://www.verywellhealth.com/pregnant-women-covid-vaccine-5092509]
- Two hundred and seventy (270) pregnant women reported either **"exposure" (146) or "vaccination" (124).**
  - Exposure can mean unvaccinated women **exposed to a vaccinated partner or exposed via inhalation or skin exposure to the vaccine.**
  - Only a few "exposure cases" noted the timing of exposure during their pregnancies: 15 in the first trimester, seven in the second trimester, and two in the third trimester.
  - The timing of **vaccination during the pregnancy** was reported in just 22 of the 124 cases—19 during the first trimester, one in the second trimester, and two in the third trimester.
    - In this group, 49 cases were rated non-serious, and **75 rated serious.**

    - One mother had uterine contraction during pregnancy, and another had **premature rupture of membranes.**
- Twenty-eight deaths of either a fetus or neonate happened to women in the *vaccinated* group (124 women). So, **23% of the vaccinated mothers had fetuses or newborns who died.**
    - These "losses" were described as **spontaneous abortion (miscarriage)** or various other terms which mean death of the fetus or baby.
- No outcome or "outcome pending" was reported for 243 of the 274, or 88.7%, of the pregnancy cases.
    - Because of this, it is unknown if 243 of the pregnancy cases resulted in normal or abnormal outcomes.
    - **Only 11.3% of pregnancies had known outcomes.**
- There were **five serious clinical events in four babies: two fetal growth retardation, two premature babies, and one neonatal death.**
- **Four breast feeding mothers reported suppressed lactation**, one of which was considered serious and also involved "paresis," a weakness (less than complete paralysis) usually of an arm or leg.
- **Breast milk discoloration** was also reported by breastfeeding mothers.
- **Clinical events were only listed if they occurred in more than five cases.** *How many in this important section went unreported?*
    - Many non-serious and serious were not separated but reported together, so it is unknown which of the following symptoms or combination of symptoms were responsible for the remaining 40+ serious clinical cases:
        - Thirty-three (33) headache, 24 vaccination site pain, 22 pain in extremity, 22 fatigue, 16 myalgia (muscle aches), 16 pyrexia (fever), 13 chills, 12 nausea, 11 pain, nine arthralgia (joint aches), seven lymphadenopathy (swollen lymph nodes), **six chest pain**, six dizziness, six asthenia (weakness or lack of energy), and five malaise.
        - Two other clinical events were included: seven **drug ineffective** (defined as getting COVID between 14 days after the first shot and six days after the second shot) and **five COVID-19** (presumably infection more than a week after the second injection, in other words failure of full immunization).
- One hundred and thirty-three **(133) breastfeeding baby cases** were reported.
    - One hundred and sixteen (116) simply reported exposure but no adverse reaction.
    - **Seventeen (17) adverse reactions were reported**, three classified as serious and 14 as non-serious.
        - Breastfeeding babies' reactions included:
            - **Pyrexia (fever), rash, infant irritability, infantile vomiting, diarrhea, insomnia, illness, poor feeding infant, lethargy, abdominal discomfort, vomiting, *allergy to vaccine*, increased appetite, anxiety, crying, poor quality sleep, eructation (burping), agitation, pain, and urticaria (hives).**
- There is a stark contrast in the cases shown in Pfizer's Pregnancy and Lactation Cumulative Review and the pregnancy and lactation cases reported in *5.3.6*, which is surprising given that the bulk of inoculations reflected in the Cumulative Review likely occurred during the period of time included in *5.3.6*.

- Pfizer's Cumulative Report documents 53 spontaneous abortions and two premature births with neonatal death, compared to 26 and two, respectively, documented in 5.3.6.
- Pfizer's Cumulative Report reported 41 baby/infant cases exposed via breastmilk who had adverse events, with 10 of the cases experiencing serious adverse events. Yet, the comparable figures from *5.3.6* Table 6 were 17 cases and three serious cases.
- *Why are there twice as many spontaneous abortions in Pfizer's Cumulative Report? Why does the Cumulative Report have so many more baby cases with adverse and serious adverse events?*

Please read the full report below.

DAILYCLOUT

War Room/DailyClout Pfizer Document Analysis

Post-Marketing Team Micro-Report 17:

Use in Pregnancy and Lactation Review of 5.3.6

SOURCE

https://www.phmpt.org/wp-content/uploads/2022/04/reissue_5.3.6-postmarketing-experience.pdf

5.3.6 AE REPORTING PERIOD:

"Since the first temporary authorization for emergency supply under Regulation 174 in the UK (01 December 2020) and through 28 February 2021."

Adverse Events were reported to Pfizer during a 90-day period, following the December 1, 2020, public rollout of its COVID-19 experimental "vaccine" product.

In the Pfizer 5.3.6 document, these AEs were categorized by System Organ Classes (SOC) – in other words, by systems in the body.

ABBREVIATIONS:

5.3.6 : Pfizer source document

SOC : System Organ Class

AE : Adverse Event

AESI : Adverse Event of Special Interest

EUA : Emergency Use Authorization by FDA

PM : Post-Marketing

BNT162b2 : Pfizer's mRNA COVID-19 vaccine

AUTHORS:

Dr Barbara Gehrett MD
Dr Joseph Gehrett MD
Dr Chris Flowers MD
Loree Britt

Post Marketing Team

30Aug23

Pfizer identified three categories of **Safety Concerns** in Table 3. These were:

1) Important identified risks: Anaphylaxis (Table 4)
2) Important potential risks: Vaccine-Associated Enhanced Disease (VAED), including Vaccine-Associated Enhanced Respiratory Disease (VAERD) (Table 5)
3) Missing Information: **Use in Pregnancy and Lactation**, Use in Paediatric Individuals < 12 years of Age, and Vaccine Effectiveness (Table 6)

Pregnancy Related AESIs	For relevant cases, please refer to Table 6, Description of Missing Information, *Use in Pregnancy and While Breast Feeding*
Search criteria: PTs Amniotic cavity infection; Caesarean section; Congenital anomaly; Death neonatal; Eclampsia; Foetal distress syndrome; Low birth weight baby; Maternal exposure during pregnancy; Placenta praevia; Pre-eclampsia; Premature labour; Stillbirth; Uterine rupture; Vasa praevia	

https://www.phmpt.org/wp-content/uploads/2022/04/reissue_5.3.6-postmarketing-experience.pdf

Table 6 reports 413 pregnancy and lactation cases, making up 0.98% of the 42,086 cases in the post-marketing data set. Of the 413 cases, 84 were considered serious and 329 non-serious. Of the non-serious adverse effects reported, 262 were exposure to the vaccine. **(It should be noted that BNT162b2 was not recommended for use in pregnancy or with lactation at the time of the post-marketing data set.)**

Three groups are represented in this report:
Pregnancy cases
Breast feeding baby cases
Breast feeding mother cases

Pregnancy cases: 274 pregnancy cases were reported, 270 mothers and four fetus/baby cases.

What happened to these 270 pregnancies?

There were 28 deaths of either a fetus or neonate. These were described as spontaneous abortion (miscarriage) or various other terms which mean loss of the fetus/baby. **No outcome was reported for 238 pregnancies (88%)!** In addition to the 28 deaths, there were five "outcome pending" and one "normal outcome." The numbers add up to more than 270 because a set of twins had different outcomes.

Under Pfizer's Clinical Protocol Document of 11/2020, C4591001, section 8.3.5, "exposure" can mean inhalation or skin contact with the vaccine, contact around the time of conception with a partner who was vaccinated or exposed to inhalation or skin contact, or vaccination. It is not known from 5.3.6 whether these 146 mothers were vaccinated or had other forms of exposure.

One hundred and forty-six of the 270 pregnant women simply noted their exposure to the vaccine. Only a few of these noted the timing during the pregnancy: 15 in the first trimester, seven in the second trimester, and two in the third trimester. Table 6 does not clarify whether "exposure" is the same as vaccination.

There were 28 deaths of either fetus or neonate.

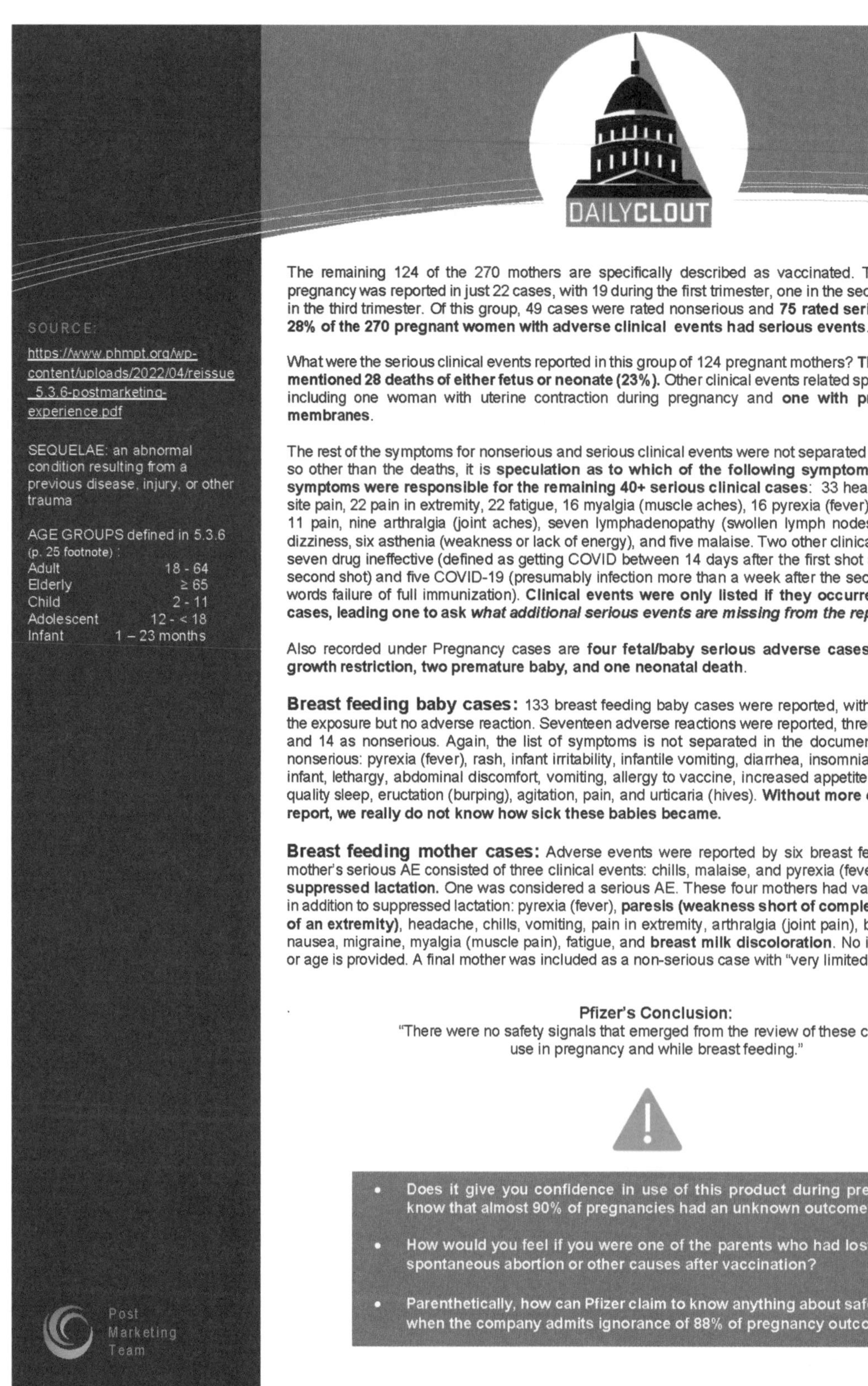

SOURCE:

https://www.phmpt.org/wp-content/uploads/2022/04/reissue_5.3.6-postmarketing-experience.pdf

SEQUELAE: an abnormal condition resulting from a previous disease, injury, or other trauma

AGE GROUPS defined in 5.3.6 (p. 25 footnote):

Adult	18 - 64
Elderly	≥ 65
Child	2 - 11
Adolescent	12 - < 18
Infant	1 – 23 months

The remaining 124 of the 270 mothers are specifically described as vaccinated. The timing during the pregnancy was reported in just 22 cases, with 19 during the first trimester, one in the second trimester and two in the third trimester. Of this group, 49 cases were rated nonserious and **75 rated serious.** This means that **28% of the 270 pregnant women with adverse clinical events had serious events**.

What were the serious clinical events reported in this group of 124 pregnant mothers? **There were the above-mentioned 28 deaths of either fetus or neonate (23%).** Other clinical events related specifically to pregnancy including one woman with uterine contraction during pregnancy and **one with premature rupture of membranes**.

The rest of the symptoms for nonserious and serious clinical events were not separated but reported together; so other than the deaths, it is **speculation as to which of the following symptoms or combination of symptoms were responsible for the remaining 40+ serious clinical cases**: 33 headache, 24 vaccination site pain, 22 pain in extremity, 22 fatigue, 16 myalgia (muscle aches), 16 pyrexia (fever), 13 chills, 12 nausea, 11 pain, nine arthralgia (joint aches), seven lymphadenopathy (swollen lymph nodes), six chest pain, six dizziness, six asthenia (weakness or lack of energy), and five malaise. Two other clinical events are included: seven drug ineffective (defined as getting COVID between 14 days after the first shot and six days after the second shot) and five COVID-19 (presumably infection more than a week after the second injection, in other words failure of full immunization). **Clinical events were only listed if they occurred in more than five cases, leading one to ask *what additional serious events are missing from the report.***

Also recorded under Pregnancy cases are **four fetal/baby serious adverse cases including two fetal growth restriction, two premature baby, and one neonatal death**.

Breast feeding baby cases: 133 breast feeding baby cases were reported, with 116 simply reporting the exposure but no adverse reaction. Seventeen adverse reactions were reported, three classified as serious and 14 as nonserious. Again, the list of symptoms is not separated in the document into serious versus nonserious: pyrexia (fever), rash, infant irritability, infantile vomiting, diarrhea, insomnia, illness, poor feeding infant, lethargy, abdominal discomfort, vomiting, allergy to vaccine, increased appetite, anxiety, crying, poor quality sleep, eructation (burping), agitation, pain, and urticaria (hives). **Without more details or a narrative report, we really do not know how sick these babies became.**

Breast feeding mother cases: Adverse events were reported by six breast feeding mothers. One mother's serious AE consisted of three clinical events: chills, malaise, and pyrexia (fever). **Four women had suppressed lactation.** One was considered a serious AE. These four mothers had various other symptoms in addition to suppressed lactation: pyrexia (fever), **paresis (weakness short of complete paralysis, usually of an extremity)**, headache, chills, vomiting, pain in extremity, arthralgia (joint pain), breast pain, scar pain, nausea, migraine, myalgia (muscle pain), fatigue, and **breast milk discoloration**. No information on latency or age is provided. A final mother was included as a non-serious case with "very limited information."

Pfizer's Conclusion:
"There were no safety signals that emerged from the review of these cases of use in pregnancy and while breast feeding."

- **Does it give you confidence in use of this product during pregnancy to know that almost 90% of pregnancies had an unknown outcome?**
- **How would you feel if you were one of the parents who had lost a baby to spontaneous abortion or other causes after vaccination?**
- **Parenthetically, how can Pfizer claim to know anything about safety signals when the company admits ignorance of 88% of pregnancy outcomes?**

Post Marketing Team

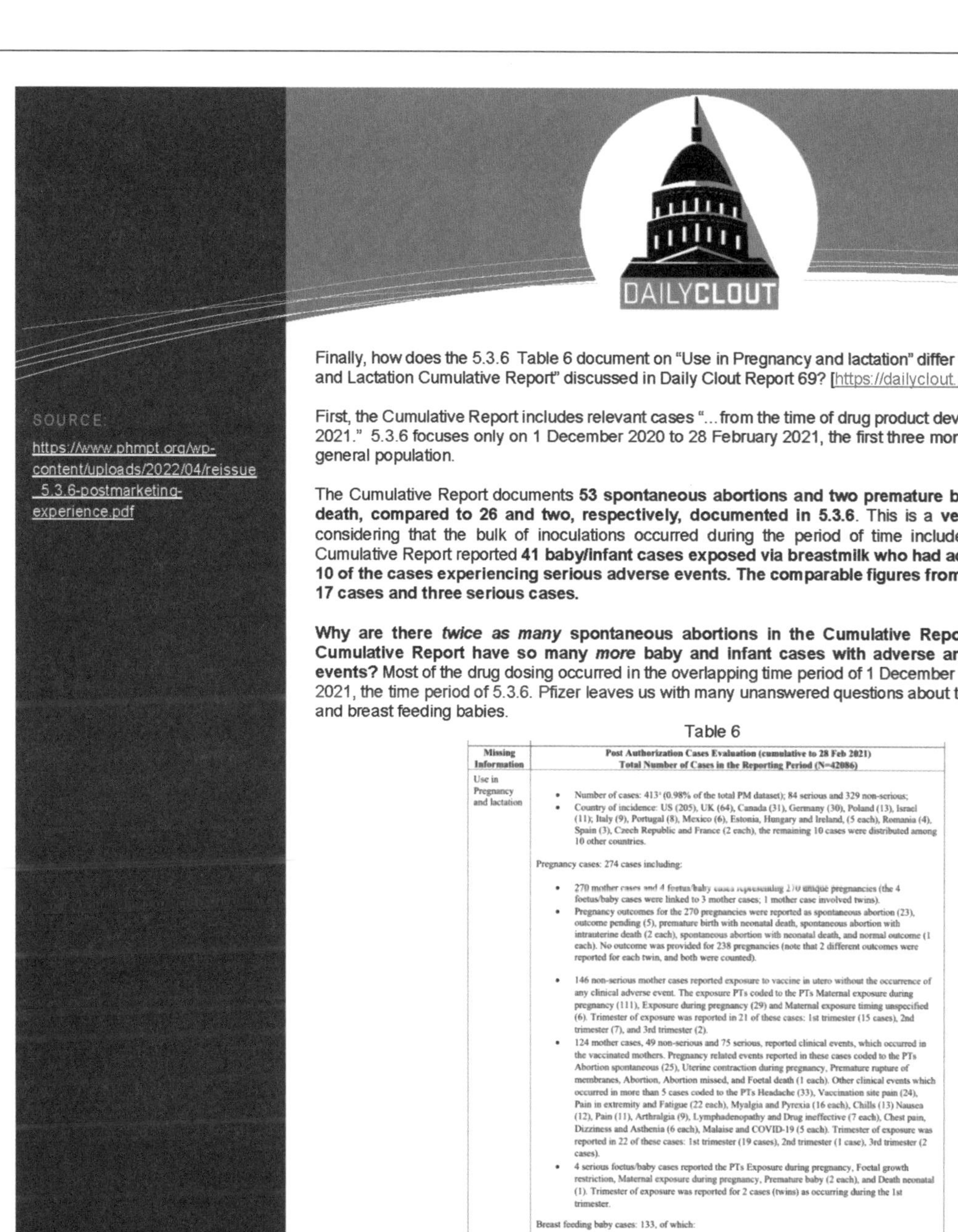

SOURCE:
https://www.phmpt.org/wp-content/uploads/2022/04/reissue_5.3.6-postmarketing-experience.pdf

Finally, how does the 5.3.6 Table 6 document on "Use in Pregnancy and lactation" differ from the "Pregnancy and Lactation Cumulative Report" discussed in Daily Clout Report 69? [https://dailyclout.io/?s=report+69]

First, the Cumulative Report includes relevant cases "…from the time of drug product development to 28-Feb-2021." 5.3.6 focuses only on 1 December 2020 to 28 February 2021, the first three months of release to the general population.

The Cumulative Report documents **53 spontaneous abortions and two premature births with neonatal death, compared to 26 and two, respectively, documented in 5.3.6**. This is a **very large difference** considering that the bulk of inoculations occurred during the period of time included in 5.3.6. Pfizer's Cumulative Report reported **41 baby/infant cases exposed via breastmilk who had adverse events, with 10 of the cases experiencing serious adverse events. The comparable figures from 5.3.6 Table 6 were 17 cases and three serious cases.**

Why are there *twice as many* spontaneous abortions in the Cumulative Report? Why does the Cumulative Report have so many *more* baby and infant cases with adverse and serious adverse events? Most of the drug dosing occurred in the overlapping time period of 1 December 2020 to 28 February 2021, the time period of 5.3.6. Pfizer leaves us with many unanswered questions about the risk to pregnancy and breast feeding babies.

Table 6

Missing Information	Post Authorization Cases Evaluation (cumulative to 28 Feb 2021) Total Number of Cases in the Reporting Period (N=42086)
Use in Pregnancy and lactation	• Number of cases: 413[a] (0.98% of the total PM dataset); 84 serious and 329 non-serious; • Country of incidence: US (205), UK (64), Canada (31), Germany (30), Poland (13), Israel (11); Italy (9), Portugal (8), Mexico (6), Estonia, Hungary and Ireland, (5 each), Romania (4), Spain (3), Czech Republic and France (2 each), the remaining 10 cases were distributed among 10 other countries. Pregnancy cases: 274 cases including: • 270 mother cases and 4 foetus/baby cases representing 270 unique pregnancies (the 4 foetus/baby cases were linked to 3 mother cases; 1 mother case involved twins). • Pregnancy outcomes for the 270 pregnancies were reported as spontaneous abortion (23), outcome pending (5), premature birth with neonatal death, spontaneous abortion with intrauterine death (2 each), spontaneous abortion with neonatal death, and normal outcome (1 each). No outcome was provided for 238 pregnancies (note that 2 different outcomes were reported for each twin, and both were counted). • 146 non-serious mother cases reported exposure to vaccine in utero without the occurrence of any clinical adverse event. The exposure PTs coded to the PTs Maternal exposure during pregnancy (111), Exposure during pregnancy (29) and Maternal exposure timing unspecified (6). Trimester of exposure was reported in 21 of these cases: 1st trimester (15 cases), 2nd trimester (7), and 3rd trimester (2). • 124 mother cases, 49 non-serious and 75 serious, reported clinical events, which occurred in the vaccinated mothers. Pregnancy related events reported in these cases coded to the PTs Abortion spontaneous (25), Uterine contraction during pregnancy, Premature rupture of membranes, Abortion, Abortion missed, and Foetal death (1 each). Other clinical events which occurred in more than 5 cases coded to the PTs Headache (33), Vaccination site pain (24), Pain in extremity and Fatigue (22 each), Myalgia and Pyrexia (16 each), Chills (13) Nausea (12), Pain (11), Arthralgia (9), Lymphadenopathy and Drug ineffective (7 each), Chest pain, Dizziness and Asthenia (6 each), Malaise and COVID-19 (5 each). Trimester of exposure was reported in 22 of these cases: 1st trimester (19 cases), 2nd trimester (1 case), 3rd trimester (2 cases). • 4 serious foetus/baby cases reported the PTs Exposure during pregnancy, Foetal growth restriction, Maternal exposure during pregnancy, Premature baby (2 each), and Death neonatal (1). Trimester of exposure was reported for 2 cases (twins) as occurring during the 1st trimester. Breast feeding baby cases: 133, of which: • 116 cases reported exposure to vaccine during breastfeeding (PT Exposure via breast milk) without the occurrence of any clinical adverse events; • 17 cases, 3 serious and 14 non-serious, reported the following clinical events that occurred in the infant/child exposed to vaccine via breastfeeding: Pyrexia (5), Rash (4), Infant irritability (3), Infantile vomiting, Diarrhoea, Insomnia, and Illness (2 each), Poor feeding infant, Lethargy, Abdominal discomfort, Vomiting, Allergy to vaccine, Increased appetite, Anxiety, Crying, Poor quality sleep, Eructation, Agitation, Pain and Urticaria (1 each). Breast feeding mother cases (6): • 1 serious case reported 3 clinical events that occurred in a mother during breast feeding (PT Maternal exposure during breast feeding); these events coded to the PTs Chills, Malaise, and Pyrexia • 1 non-serious case reported with very limited information and without associated AEs.
	• In 4 cases (3 non-serious; 1 serious) Suppressed lactation occurred in a breast feeding women with the following co-reported events: Pyrexia (2), Paresis, Headache, Chills, Vomiting, Pain in extremity, Arthralgia, Breast pain, Scar pain, Nausea, Migraine, Myalgia, Fatigue and Breast milk discolouration (1 each). Conclusion: There were no safety signals that emerged from the review of these cases of use in pregnancy and while breast feeding.

https://www.phmpt.org/wp-content/uploads/2022/04/reissue_5.3.6-postmarketing-experience.pdf

Post Marketing Team

Post-Marketing Team's CONCLUSION

RECALL this unsafe "vaccine."

Twenty-Six: "WarRoom/DailyClout Research Team Breaks Huge Story: More Cardiovascular Deaths in Vaxxed Than Unvaxxed; Pfizer Did Not Report Adverse Event Signal; Death Reporting Delays Favored Pfizer/Vaccinated."

—Amy Kelly

WarRoom/DailyClout Pfizer Documents Analysis team members Corinne Michels, PhD; Daniel Perrier; Jeyanthi Kunadhasan, MD; Ed Clark, MSE; Joseph Gehrett, MD; Barbara Gehrett, MD; Kim Kwiatek, MD; Sarah Adams; Robert Chandler, MD; Leah Stagno; Tony Damian; Erika Delph; and Chris Flowers, MD have published a bombshell study titled, "Forensic Analysis of the 38 Subject Deaths in the 6-Month Interim Report of the Pfizer/BioNTech BNT162b2 mRNA Vaccine Clinical Trial."

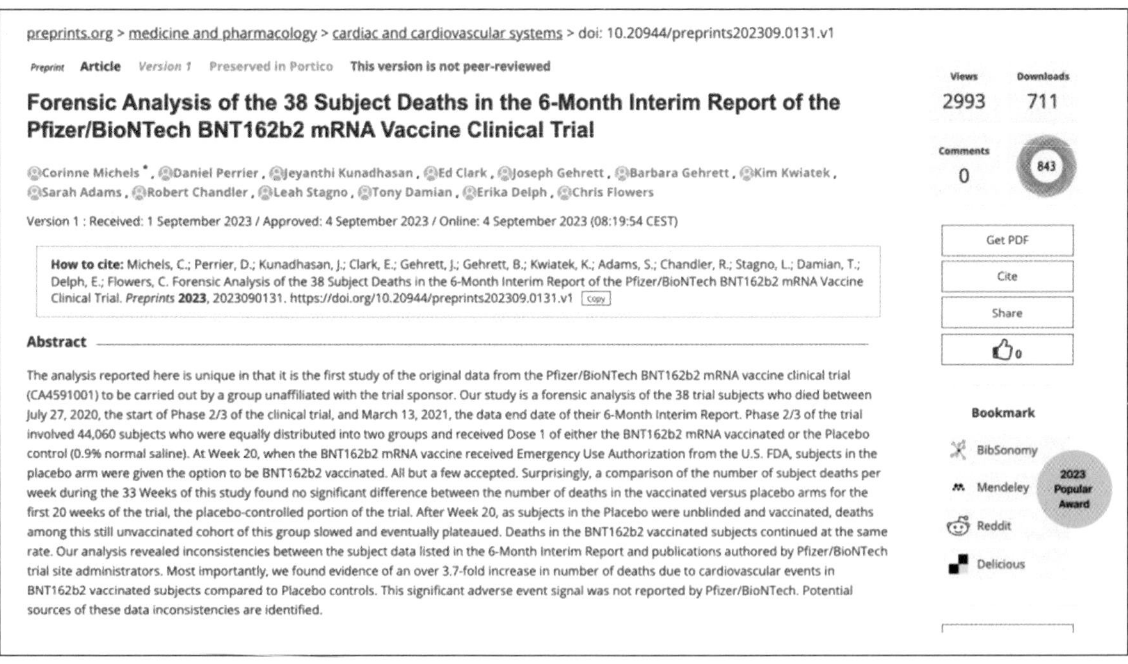

preprints.org > medicine and pharmacology > cardiac and cardiovascular systems > doi: 10.20944/preprints202309.0131.v1

Preprint Article Version 1 Preserved in Portico This version is not peer-reviewed

Forensic Analysis of the 38 Subject Deaths in the 6-Month Interim Report of the Pfizer/BioNTech BNT162b2 mRNA Vaccine Clinical Trial

Corinne Michels *, Daniel Perrier, Jeyanthi Kunadhasan, Ed Clark, Joseph Gehrett, Barbara Gehrett, Kim Kwiatek, Sarah Adams, Robert Chandler, Leah Stagno, Tony Damian, Erika Delph, Chris Flowers

Version 1 : Received: 1 September 2023 / Approved: 4 September 2023 / Online: 4 September 2023 (08:19:54 CEST)

How to cite: Michels, C.; Perrier, D.; Kunadhasan, J.; Clark, E.; Gehrett, J.; Gehrett, B.; Kwiatek, K.; Adams, S.; Chandler, R.; Stagno, L.; Damian, T.; Delph, E.; Flowers, C. Forensic Analysis of the 38 Subject Deaths in the 6-Month Interim Report of the Pfizer/BioNTech BNT162b2 mRNA Vaccine Clinical Trial. *Preprints* **2023**, 2023090131. https://doi.org/10.20944/preprints202309.0131.v1 Copy

Abstract

The analysis reported here is unique in that it is the first study of the original data from the Pfizer/BioNTech BNT162b2 mRNA vaccine clinical trial (CA4591001) to be carried out by a group unaffiliated with the trial sponsor. Our study is a forensic analysis of the 38 trial subjects who died between July 27, 2020, the start of Phase 2/3 of the clinical trial, and March 13, 2021, the data end date of their 6-Month Interim Report. Phase 2/3 of the trial involved 44,060 subjects who were equally distributed into two groups and received Dose 1 of either the BNT162b2 mRNA vaccinated or the Placebo control (0.9% normal saline). At Week 20, when the BNT162b2 mRNA vaccine received Emergency Use Authorization from the U.S. FDA, subjects in the placebo arm were given the option to be BNT162b2 vaccinated. All but a few accepted. Surprisingly, a comparison of the number of subject deaths per week during the 33 Weeks of this study found no significant difference between the number of deaths in the vaccinated versus placebo arms for the first 20 weeks of the trial, the placebo-controlled portion of the trial. After Week 20, as subjects in the Placebo were unblinded and vaccinated, deaths among this still unvaccinated cohort of this group slowed and eventually plateaued. Deaths in the BNT162b2 vaccinated subjects continued at the same rate. Our analysis revealed inconsistencies between the subject data listed in the 6-Month Interim Report and publications authored by Pfizer/BioNTech trial site administrators. Most importantly, we found evidence of an over 3.7-fold increase in number of deaths due to cardiovascular events in BNT162b2 vaccinated subjects compared to Placebo controls. This significant adverse event signal was not reported by Pfizer/BioNTech. Potential sources of these data inconsistencies are identified.

Views 2993 Downloads 711

Comments 0

843

Get PDF

Cite

Share

0

Bookmark

BibSonomy

Mendeley

Reddit

Delicious

2023 Popular Award

https://www.preprints.org/manuscript/202309.0131/v1 and https://doi.org/10.20944/preprints202309.0131.v1

Highlights from the Study

- This is the **first study of the original data from the Pfizer/BioNTech BNT162b2 vaccine clinical trial to be carried out by a group *unaffiliated with the trial sponsor*.**
- **Thirty-eight (38) trial subjects died** between July 27, 2020, the start of Pfizer's Phase 2/3 of its clinical trial, and March 13, 2021, the data end-date of Pfizer/BioNTech's Six-Month Interim Report.
- At Week 20 of the trial, Pfizer's BNT162b2 mRNA vaccine received Emergency Use Authorization from the FDA, and subjects in the placebo arm were given the option to get vaccinated. Most accepted.

- For the first 20 weeks of trial, before the placebo cohort began receiving vaccines, there was no significant difference in the number of deaths in the vaccinated versus placebo arms of the trial.
  - **Once the placebo cohort was unblinded and began receiving the Pfizer's vaccine, following Week 20, deaths among the vaccinated subjects continued at the same rate, while deaths among the unvaccinated slowed and even plateaued.**
- Evidence found of an **over 3.7-fold increase in number of deaths due to cardiovascular events in vaccinated subjects compared to placebo subjects. This significant adverse event signal was not reported by Pfizer/BioNTech.** (In other words, Pfizer knew about this safety signal by 3/13/21 and hid it.)
- Three hundred and ninety-five (395) subjects were "lost to follow-up."
- **Patterns of delay:**
  - "Of the 8 [Pfizer] BNT162b2 vaccinated subjects that should have been reported to the VRBPAC [Vaccines and Related Biological Products Advisory Committee] on December 10, [2021], the median reporting delay was 18 days (average of 17.5 days). Among the 8 Placebo subjects, the median delay was 5 days (average of 5.9 days) "
  - "When the recording delay after December 11 is analyzed, we found a dramatic decrease in both arms of the trial. The median delay in the BNT162b2 arm of the trial was 7 days (average 9.8 days) and in the Placebo arm the delay was 3 days (average of 15.9 days)."
  - "These results are a clear demonstration that **the long official recording delays are not distributed equally between the two arms of the trial but are clustered in the BNT162b2 vaccinated arm, particularly before FDA [Food and Drug Administration] approval of the EUA [Emergency Use Authorization]. Once the EUA was approved, Pfizer/BioNTech reported the date of death in a timelier fashion, although delays were still longer among vaccinated subjects**." [Emphasis added.]
  - "Our analysis of the data in Table 3 [below] . . . showed that **Pfizer/BioNTech used the date that the death was officially recorded in the CRF to determine which time period to report the death NOT the actual date of death, although both were available to them**." [Emphasis added.]
    - According to the CA4591001 Protocol, Pfizer/BioNTech was to be notified of a subject death immediately.
    - Access is not available to records that would have confirmed that the trial sites were diligent regarding death notifications; however, the existence of other steps in the death notification process are alluded to in the Case Report Forms [CRFs], which could have played a role in delaying entries into the CRF.
    - Preliminary databases, such as a Death Details Form, are suggested in interactions logged into the CRFs.
      - Public access is not available to any of these.
      - Completion of the Death Details Form, and perhaps other requirements, appears to be partly computerized and automatic.

Table 3. Delay in recording subject death.

Period	Subject ID	Date of Death	Officially Recorded Date (from CRF)	Delay Recording Death (Days)
BNT162b2 mRNA vaccinated subjects (Subjects with available CRF)				
*P-C	11621327	13Sept2020	24Sept2020	11
P-C	11141050	19Oct2020	25Nov2020	37
*P-C	10071101	21Oct2020	5Nov2020	15
P-C	11201050	07Nov2020	3Dec2020	26
P-C	11521497	11Nov2020	18Nov2020	7
P-C	10891073	12Nov2020	4Dec2020	22
P-C	10391010	18Nov2020	9Dec2020	21
P-C	11271112	04Dec2020	05Dec2020	1
O-L, F	11361102	19Dec2020	22Jan2021	34
O-L, F	10211127	19Dec2020	30Dec2020	11
O-L, F	10971023	21Dec2020	28Dec2020	7
O-L, F	11561160	24Dec2020	14Jan2021	21
O-L, F	12521010	26Dec2020	29Dec2020	3
O-L, F	11401117	29Dec2020	05Jan2021	7
O-L, F	10841266	12Jan2021	15Jan2021	3
O-L, F	11201266	19Jan2021	25Jan2021	6
O-L, O	11351033	29Jan2021	24Feb2021	26
O-L, O	11291166	03Feb2021	5Feb2021	2
O-L, O	10361140	10Feb2021	22Feb2021	12
O-L, O	11311204	15Feb2021	18Feb2021	3
O-L, O	10881139	06Mar2021	08Mar2021	2
Placebo subjects (Subjects with available CRF)				
*P-C	11521085	26Aug2020	27Aug2020	1
*P-C	12313972	28Sept2020	1Oct2020	3
P-C	11561124	02Nov2020	19Nov2020	17
*P-C	10661350	03Nov2020	10Nov2020	7
*P-C	10811194	04Nov2020	11Nov2020	7
O-L, F	11681083	18Nov2020	19Nov2020	1
O-L, F	11281009	28Nov2020	8Dec2020	10
O-L, F	10881126	01Dec2020	11Feb2021	72
O-L, F	12314987	06Dec2020	7Dec2020	1
O-L, F	10191146	17Dec2020	5Feb2021	50
O-L, F	10941112	18Dec2020	21Dec2020	3
O-L, F	10891088	30Dec2020	04Jan2021	5
O-L, F	12291083	05Jan2021	5Jan2021	0
O-L, F	10841470	11Jan2021	19Jan2021	8
O-L, O	12315324	31Jan2021	3Feb2021	3
O-L, O	12071055	09Feb2021	11Feb2021	2
O-L, O	10271191	13Feb2021	16Feb2021	3

Table 3: Delay in recording subject death. Subjects who died during the period July 27, 2020 to March 13, 2021 are listed. Subjects receiving BNT162b2 vaccine are listed separately from those who received the Placebo and in order of the true date of death. *Indicates those subjects included whose death was reported in the Pfizer/BioNTech EUA application [5] and Polack *et al.* [6]. The rows shaded in gray highlight those individuals whose death was officially recorded in their CRF between November 14 and December 10, 2020 indicating that Pfizer/BioNTech knew the subject died during this time period. Periods of the trial: P-C is Placebo-controlled,

Table 3 from "Forensic Analysis of the 38 Subject Deaths in the 6-Month Interim Report of the Pfizer/BioNTech BNT162b2 mRNA Vaccine Clinical Trial."

- **Additional Commentary:**
  - As was also shown in the timing of test results in WarRoom/DailyClout Report 76 [https://dailyclout.io/report-76-pfizer-had-necessary-data-to-announce-its-covid-19-vaccines-alleged-efficacy-in-october-2020-why-did-pfizer-delay/] around the United States' 2020 presidential election, **there appears to be a consistent pattern of delay which always favors Pfizer's interests.** A release of or subpoena for the records in the "Death Details Form" and other related data would be required to show that the timing delays were not intentional.

Twenty-Seven: "The Underlying Pathology of Spike Protein Biodistribution in People That Died Post COVID-19 Vaccination"

—Dr. Arne Burkhardt—Compiled and Edited by Robert W. Chandler, MD, MBA* and Michael Palmer, MD

The Scientific and Medical Advisory Committee hosted a meeting with special guest speaker Professor Arne Burkhardt.

**A transcript was produced then edited with a diligent effort to leave the meaning unchanged. My additions are in italics.*

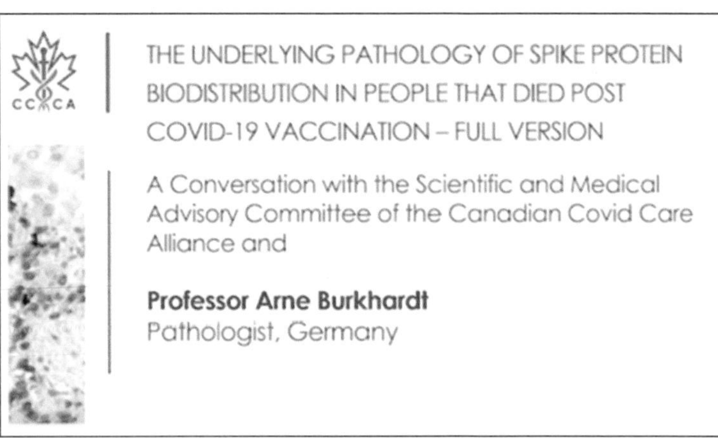

Professor Burkhardt was a German pathologist and researcher. He presented the findings of the ongoing work of an international team of pathologists. They have reviewed pathology specimens from people with new onset of symptoms or who have died following COVID-19 genetic vaccinations. He explained how to differentiate damage following natural infection with that following vaccination. Additionally, he presented tissue samples to illustrate the distribution of the spike protein following COVID-19 genetic vaccination and its associated damage, amyloid production, and clot formations.

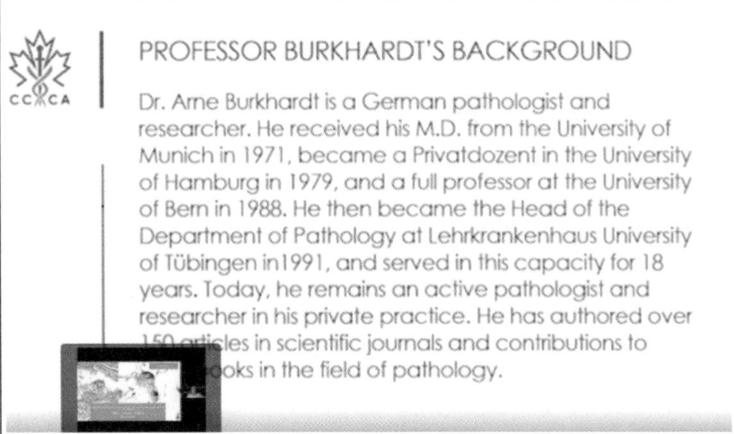

Original Lecture—"Arne Burkhardt Presentation to the CCCA"

(https://rumble.com/v2jbj16-arne-burkhardt-presentation-to-the-ccca.html)

Professor Arne Burkhardt

Transcript

I will tell you that actually very soon after the vaccine campaign started in Germany, there were relatives of deceased persons who went to me and said, well, we want to, we want that a pathologist takes a second opinion. We don't believe that our relative died of natural causes. And I thought this was an easy task.

I said, well, of course, I will look at the slides or the specimens that were taken during autopsy, and I will give a second opinion. But very soon with the first three or four cases, I realized that this was really a task that I could not handle just by myself alone. And fortunately, I found a pathologist who worked with me, Professor Walter Lang.

Professor Walter Lang

He joined me in these endeavors and very soon we got many more scientists and some pathologists, especially from outside of Germany, who joined us in these endeavors. And if I can show my first slide now. So, at this moment we are actually an international team.

It's about 10 pathologists, coroners, biologists, chemists, and physicists, not only from Germany, but actually many European countries, especially Austria, Switzerland, Italy, and Sweden.

We have completed the review of 75 autopsy studies, and we are right now having 41 biopsies from living persons. There are 40 men, 35 women of the deceased. The age range was from 21 to 94 ages.

Autopsy and histopathology studies on adverse events and deaths due to COVID-19 vaccinations, conducted at Reutlingen

- International team of 10 pathologists, coroners, biologists, chemists, physicists
- Histopathology studies on autopsy materials of 75 deceased patients, and on biopsies of 41 living patients
  - 40 men, 35 women
  - Age range: 21,2 – 94,7 years: median 65,7 years
  - Death occurred 1 day to 6 months after the most recent injection
  - Vaccines: Pfizer/BioNTech 57, Moderna 6, Janssen 3, AstraZeneca 1, unknown 8

Deaths occurred one day or to six months after the most recent injection. And the vaccines are those that are commonly used in Germany. Pfizer-BioNTech vaccine was the most common, but sometimes it was in connection with other vaccinations.

The autopsies were done by pathologists, by coroners, and one by both. So, you see it's about equal distribution. The primary diagnosis was natural death in 63 cases. And, in five cases, it was termed uncertain. So that makes 91% not related to the vaccination.

Only in seven cases it was stated that there might be a correlation with the vaccination and the disease. We did the second opinion, and these are our results. I will show you histological findings.

Death in connection with COVID-19 vaccination

Causes of death determined at autopsy

- In all but 2 cases, cause of death reported as "uncertain" or "natural"
- 1 case reported as „probably caused by vaccination"

The subsequent detailed examinations at Reutlingen found a causation of death by the vaccine

- Beyond reasonable doubt 21
- Probable 37
- Possible / uncertain 14
- Ruled out 1
- Not evaluable 2

77%

We saw a correlation of vaccination and the death occurrence in 21 cases, and in 37 cases it was probable. So, that is 77%. In another 14 (19%), we said it was uncertain or possible, ruled out in only one case, and two cases were not evaluable.

Sudden Adult Death Syndrome

Places of death (19 cases) **15 of 19 SADS**

Place	Number of cases	Individual cases no.
At home	5	1, 5, 15, 17, 20
In the street	1	11
In the car	1	2
At work	1	13
Assisted living	1	3
Hospital / ICU	4	4, 9, 14, 19
Hospital / short term (≤ 2 days)	4	6, 10, 12, 18
Not yet known	2	7, 8

This is the evaluation of 19 cases where they died. This is one important thing that most of the diseased we evaluated died suddenly, mostly at home, in the street, in the car, at work. So, we can rule out any post-mortem changes of the organs. There's no artificial respiration or anything like that.

So, of these 19 cases, 15 (79%) fulfill the criteria for what is now called the Sudden Adult Death Syndrome. And this is very important because I will come to the possible causes of this syndrome later.

COVID-19 vs. Modified mRNA (modRNA)

Now, at the beginning it was clear that there is a difference between the natural COVID-19 infection and the messenger modified RNA vaccination, but in both sets a spike protein that is probably the most important action that does harm to the tissues.

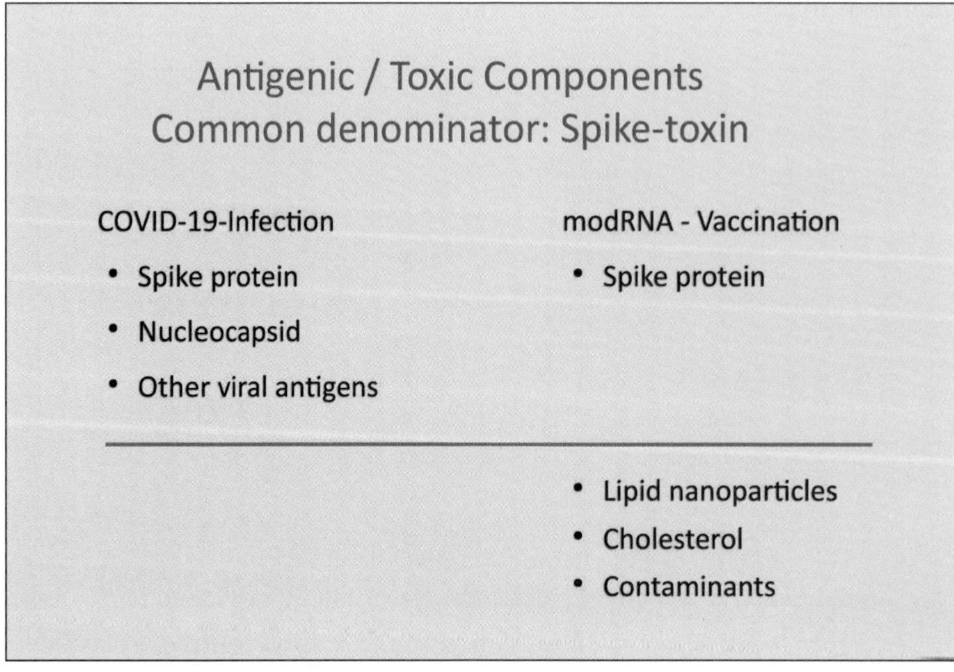

Now the COVID-19 infection, of course, has, besides a spike protein, other antigenic and possibly toxic agents like the nucleocapsid and others, and in the vaccination beside the spike protein, which of course is the leading mechanism. We have the lipid nanoparticles, we have cholesterol, and in some cases, contamination.

From the beginning, we realized that there's a different entry or primary target of these toxic and antigenic agents and the infection. It's the epithelium, starting from the nose, from the eye, pharynx, airways, lung. And these are immunocompetent linings of the body. And also, there's some protection by mucus and keratin.

Now, in the vaccination we do not use this natural entrance, but it is directly shot into the interstitial tissue, the basal tissue and the endothelium. And these are non-immunocompetent.

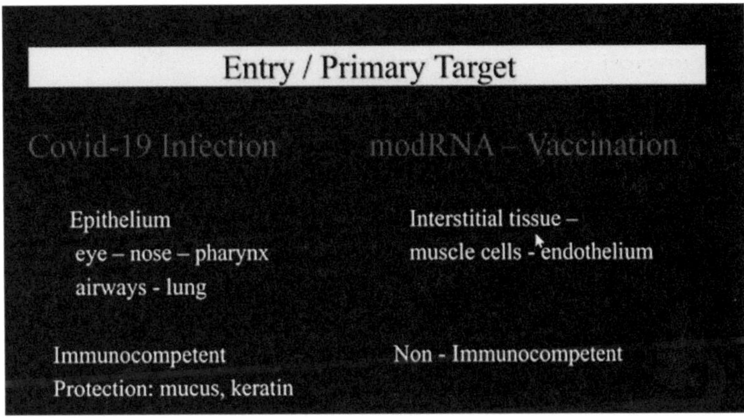

Methodology

We used normal histology (*hematoxylin and eosin stain)*, special stains, immunohistochemistry, and in some cases advanced physical chemical methods.

From the beginning on, it was clear to us that this was a novel examination because we had to demonstrate a toxin that was produced by the body itself, the spike protein. And this had to be differentiated from lesions induced by true viral infections.

And, as the body produces these toxins, we have to look for the toxin in the organs and tissues of the organ itself, which produce it, and not in body fluids or in the gastric contents.

Methods

- Histology
- Special stains
- Immunohistochemistry
- Advanced physico-chemical methods

Special toxicological problem:

- Demonstration of a toxin produced by the body itself (i.e., spike protein)
- Differentiation of lesions induced by vaccination vs. true viral infection

Immunohistochemistry: Spike (COVID-19 + modRNA) vs. Nucleocapsid (COVID-19)

So, we applied the method of immunochemistry to show the spike protein on the one hand and nucleocapsid on the other hand. Maybe most of you probably know the mechanism of immunochemistry is to have some stained material that binds to the protein that you are looking for. You see here some positively stained cells for spike protein from a nasal swab.

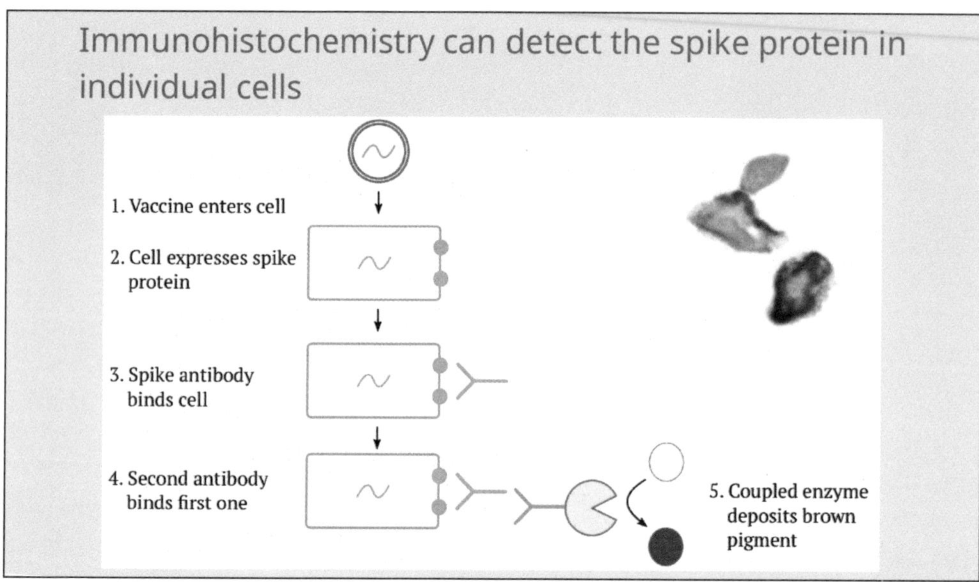

And this method was used to demonstrate the spike protein and, on the other hand, in some cases we excluded or found nucleocapsid.

General Lesions Affecting More Than One Organ (See Supplemental Resources SR 1- SR 5.)

General lesions affecting more than one organ

a. Expression of Spike protein S1
b. "Endothelialitis" /destruction and inflammation of vascular endothelium
c. Disturbance of the texture with destruction of elastic fibers of the aorta and larger vessels

d. "Displaced" (unidentified (?)) vacuolar and crystalline particles (cholesterol?)
e. Proteinaceous deposits (functional amyloidosis, fibrin amyloid)
f. Unusual Cancer manifestations – „turbo cancer"

g. "Clot"-formation
h. True foreign bodies from contaminated vaccine

First of all, we have general lesions affecting more than one organ. And, this is the **expression of the spike protein S1**. **(See SR 1 and SR 2.)**

Then, we have especially the detrimental causes on the vessels because, no matter how you inject the vaccine, it will always go into the blood and lymph vessels. We found what is called an **endotheliitis.** It's an inflammation and destruction of endothelium. And also, the larger vessels and the aorta are an aim of this toxic action. It's a **disturbance of the texture** and the **destruction of elastic fibers** in the larger vessels.

When we started, it was put forward by the pharmaceutical companies that it stays in the deltoid muscles where it was injected and that the muscle cells and maybe some other interstitial cells would produce the spike protein and cause the immunological reaction that they planned to.

Expression of Spike Protein

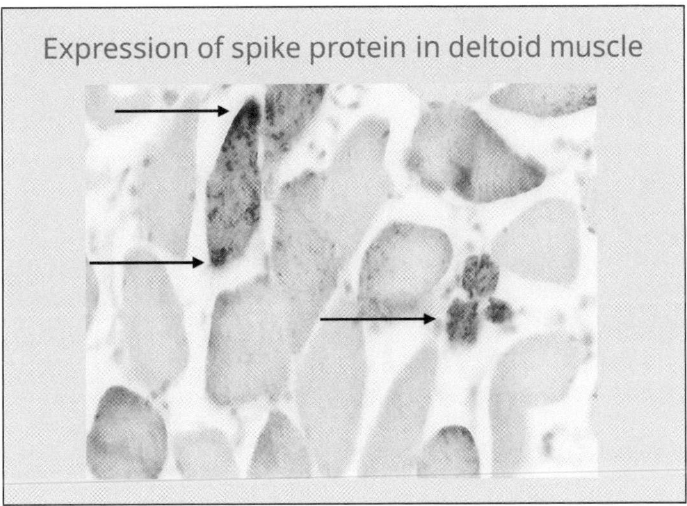

Here we examined the deltoid muscle in one case, and you can see this granular markation *(black arrows)* of the muscle cells. **Spike protein was produced at the spot of injection.**

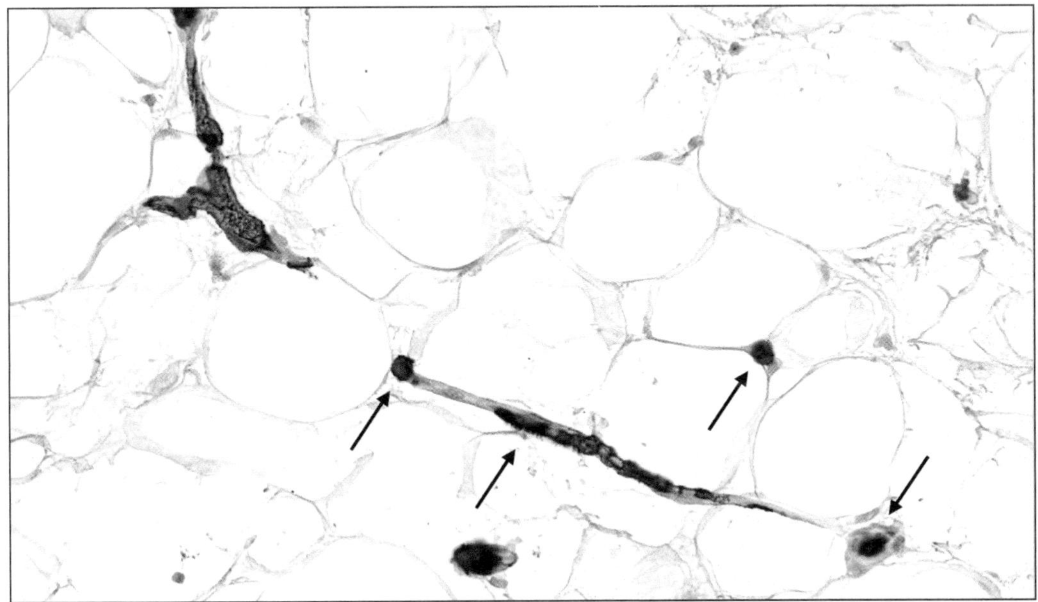

Then we also found it in capillaries *(black arrows)*. Here you see in fat tissue, you see the vacuoles and you see this capillary. Here it is out of the planar section, and here you can see it is cut vertically and you can see the endothelium strongly expresses a spike protein.

Endotheliitis

The spike protein, as you know, as we know now, it's a toxic agent and it harms the endothelium as we will see later.

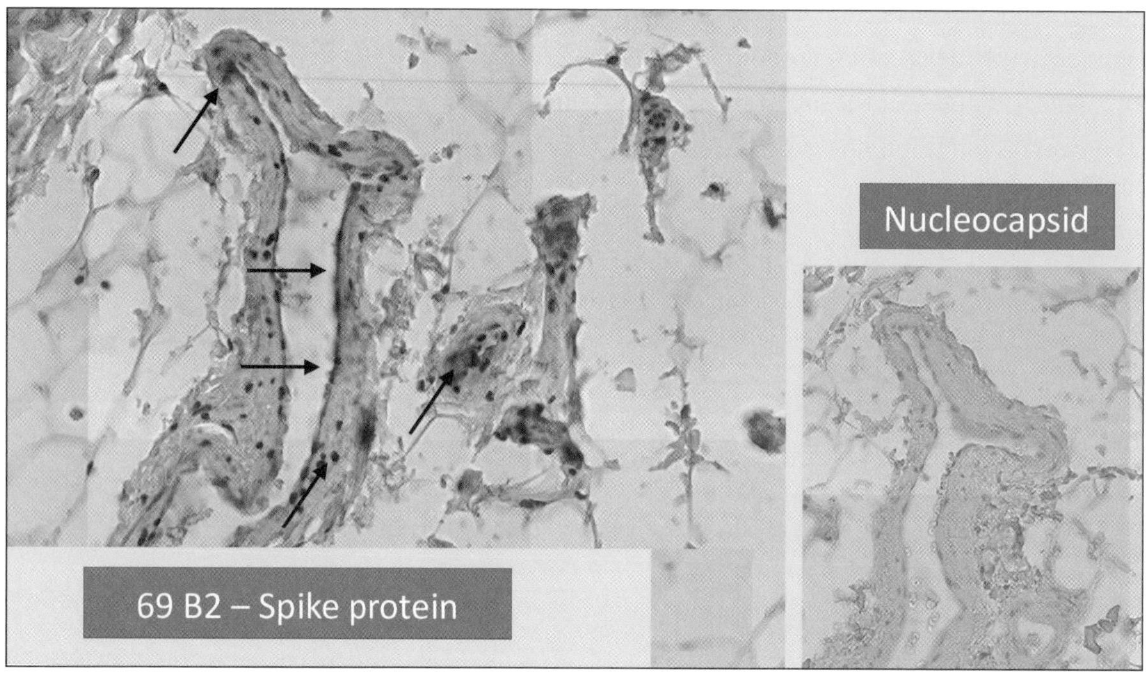

It is also found in other vessels in the surrounding, and you can see here an arteriole with a clear markation of the endothelium and some also in cells of the vessel walls. And this is a nucleocapsid stain of the same vessel. As you may see, it's a completely negative reaction.

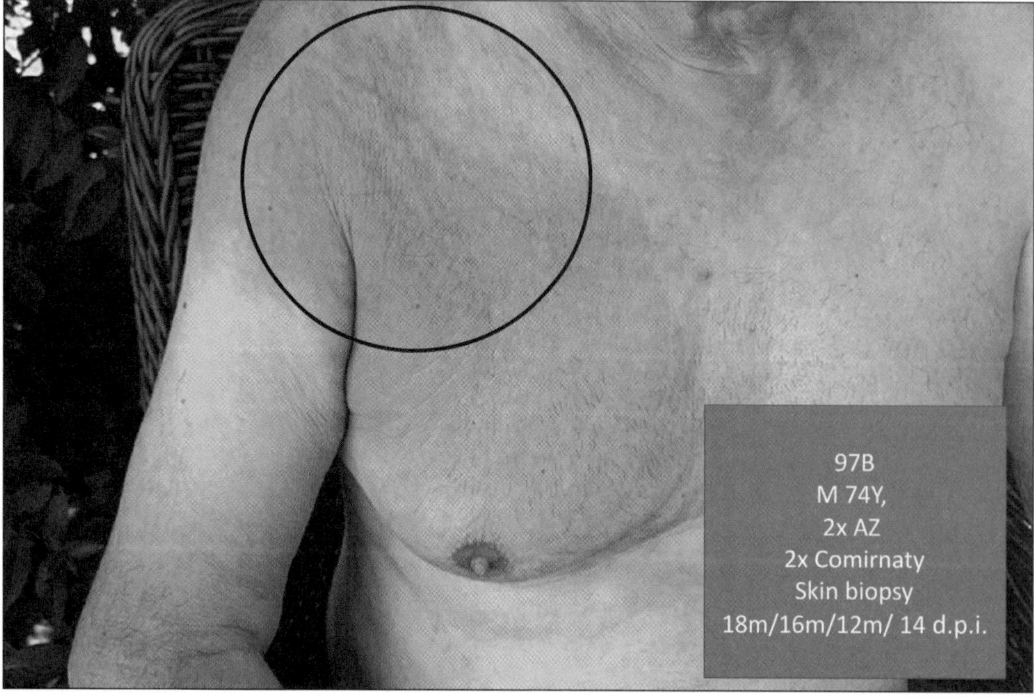

Here you can see at the site where the injection was performed. We can see still, for a very long time, inflammatory reaction in some cases. **(See SR 3 and SR 4.)**

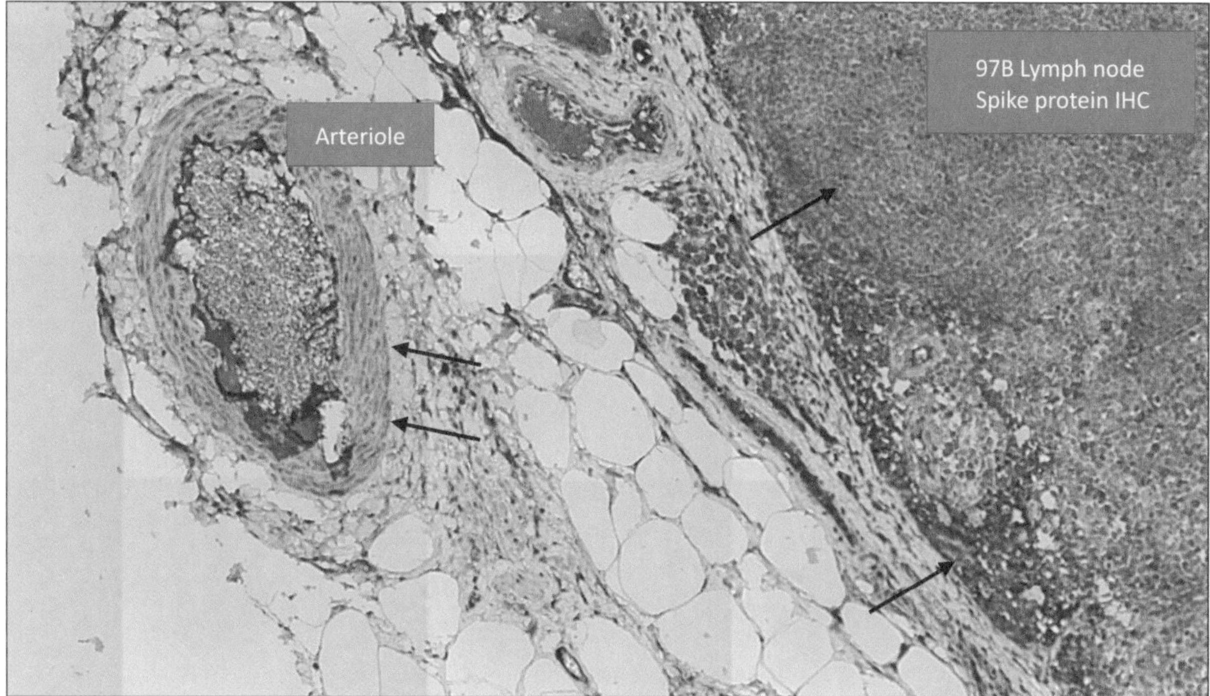

This is a biopsy from this person that you've just seen. You can see that in the lymph node a strong demonstration of the spike protein.

After that, the spike protein is distributed or can be demonstrated in many organs, almost all organs, more or less.

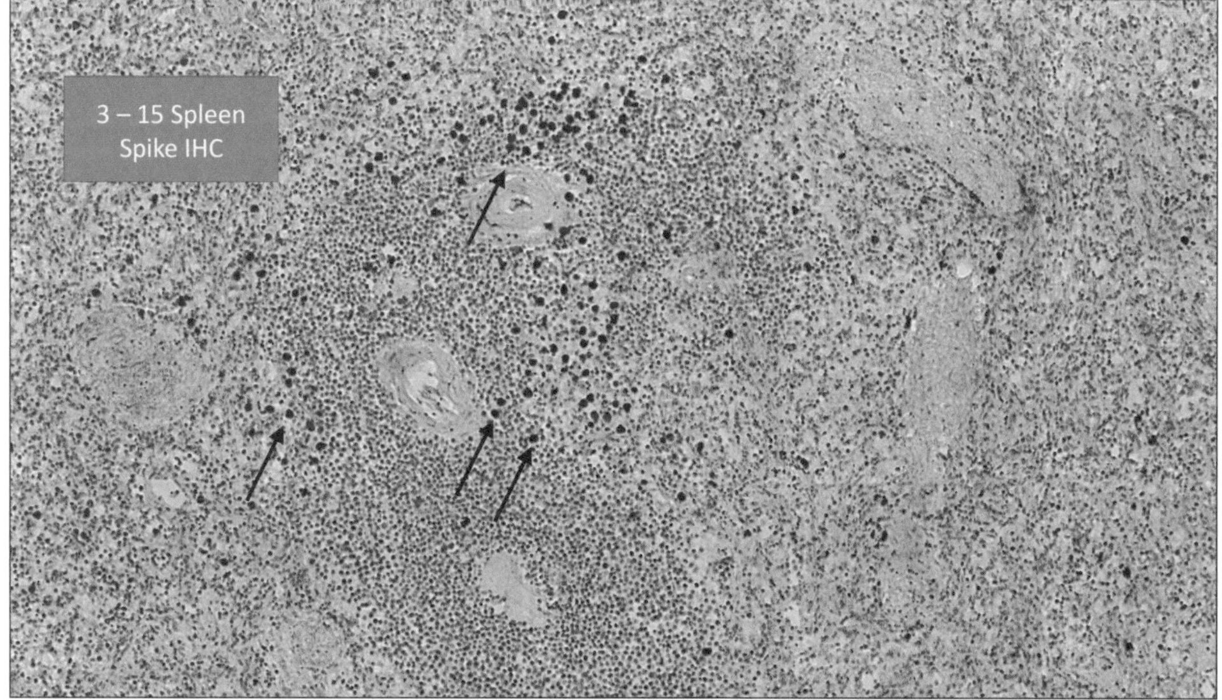

Spike in the spleen.

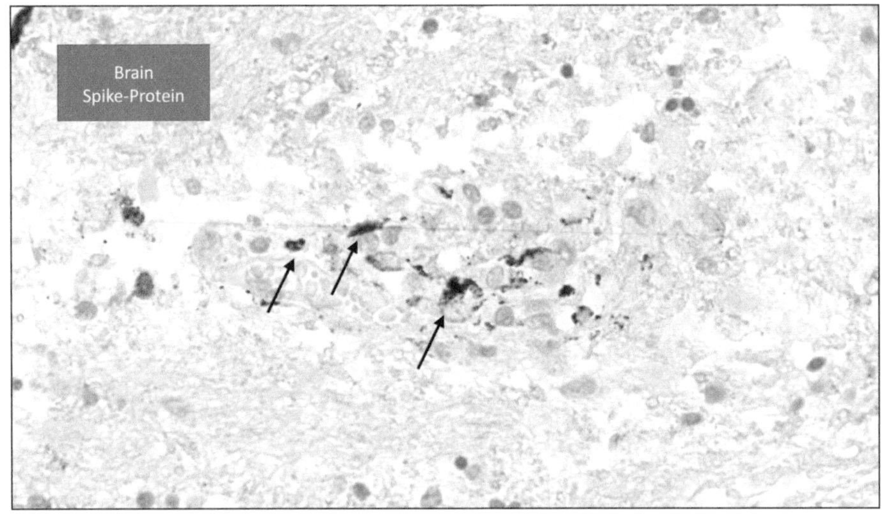

Spike protein in brain tissue.

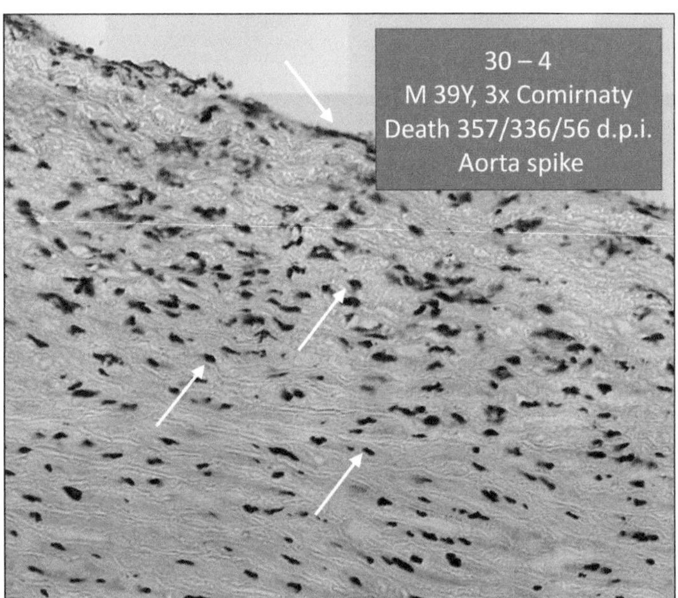

Spike protein in aorta.

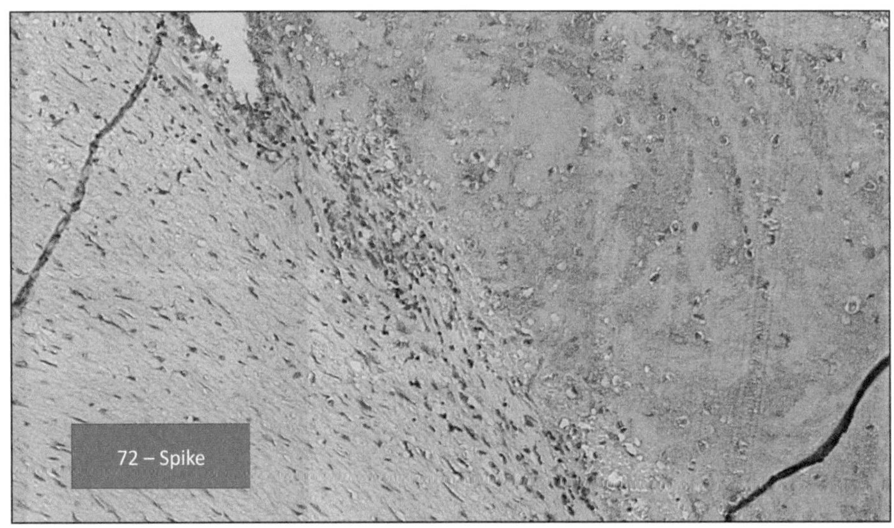

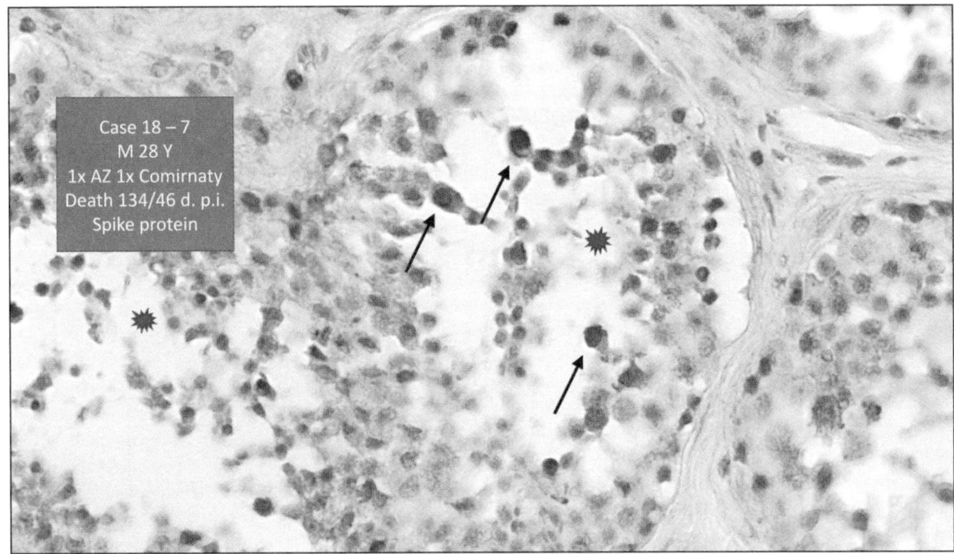

Spike in testis.

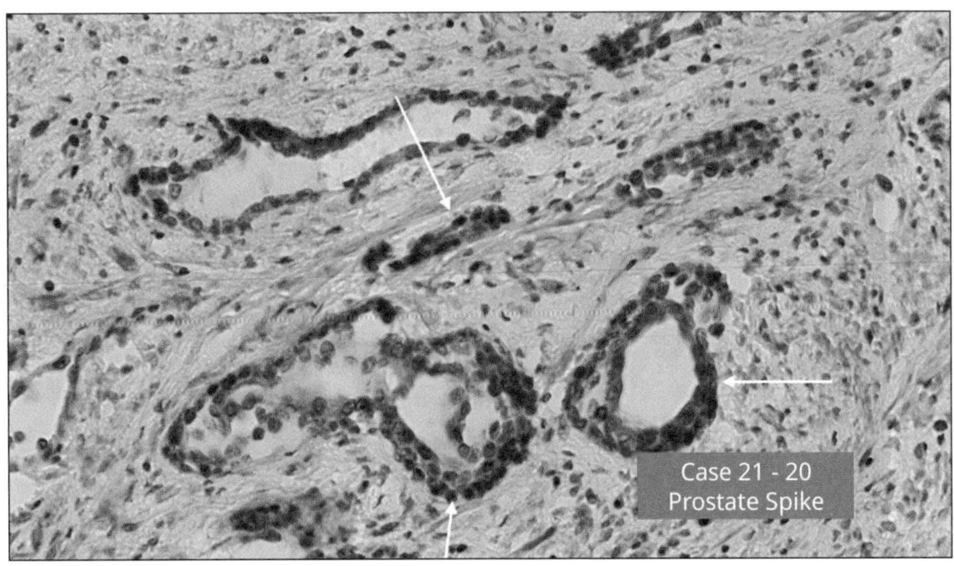

Spike in prostate.

There's some publication now about the effects on male fertility.

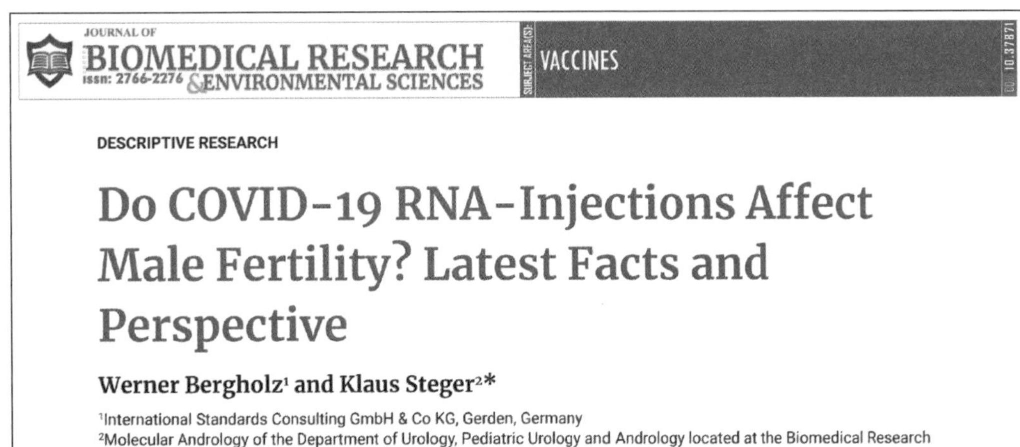

JOURNAL OF BIOMEDICAL RESEARCH & ENVIRONMENTAL SCIENCES
ISSN: 2766-2276
SUBJECT AREA(S): VACCINES

DESCRIPTIVE RESEARCH

Do COVID-19 RNA-Injections Affect Male Fertility? Latest Facts and Perspective

Werner Bergholz[1] and Klaus Steger[2]*

[1]International Standards Consulting GmbH & Co KG, Gerden, Germany
[2]Molecular Andrology of the Department of Urology, Pediatric Urology and Andrology located at the Biomedical Research Center of the Justus-Liebig University, Giessen, Germany

Conclusion: There remain still far more questions than answers. Due to the principle "primum non nocere," any new medical therapy must be banned until harmlessness beyond doubt has been proven. Most importantly, it must be realized that the active ingredient of RNA-based vaccines is not simply mRNA promoting the synthesis of a nota bene viral specific protein, but modRNA specifically designed for longevity and encapsulated in LNPs to bypass biological barriers and get access to all cells, possibly also germ cells. As mRNA is involved in regulation of gene expression, cells have mechanisms at hand to silence mRNA species not required, however, theses protective mechanisms will not work with modRNA.

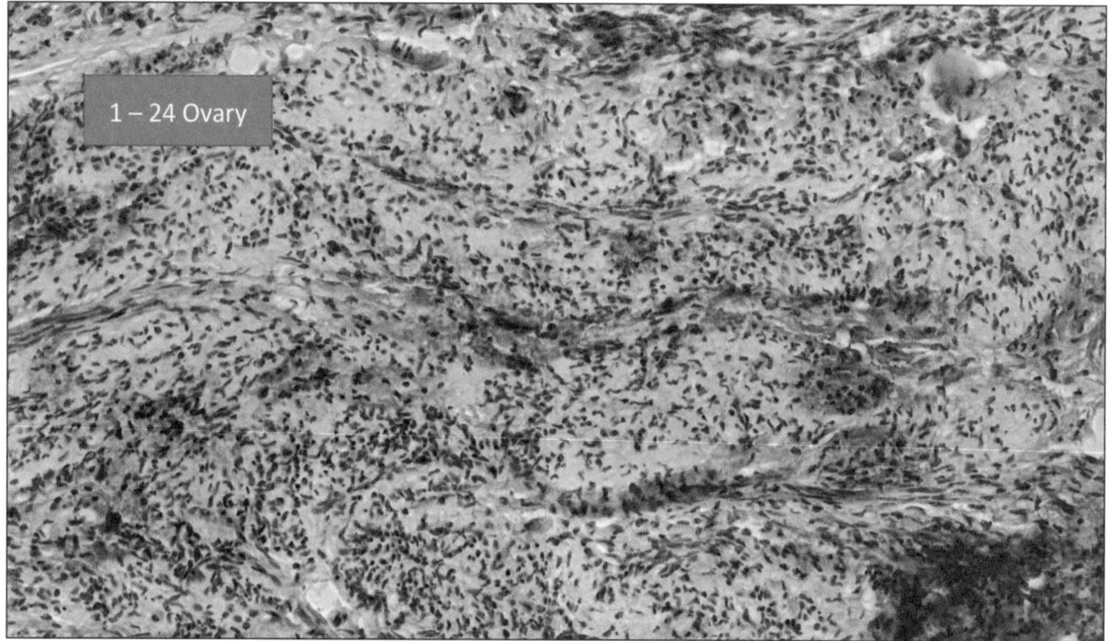

Post-modRNA lymphocytic infiltration in ovary.

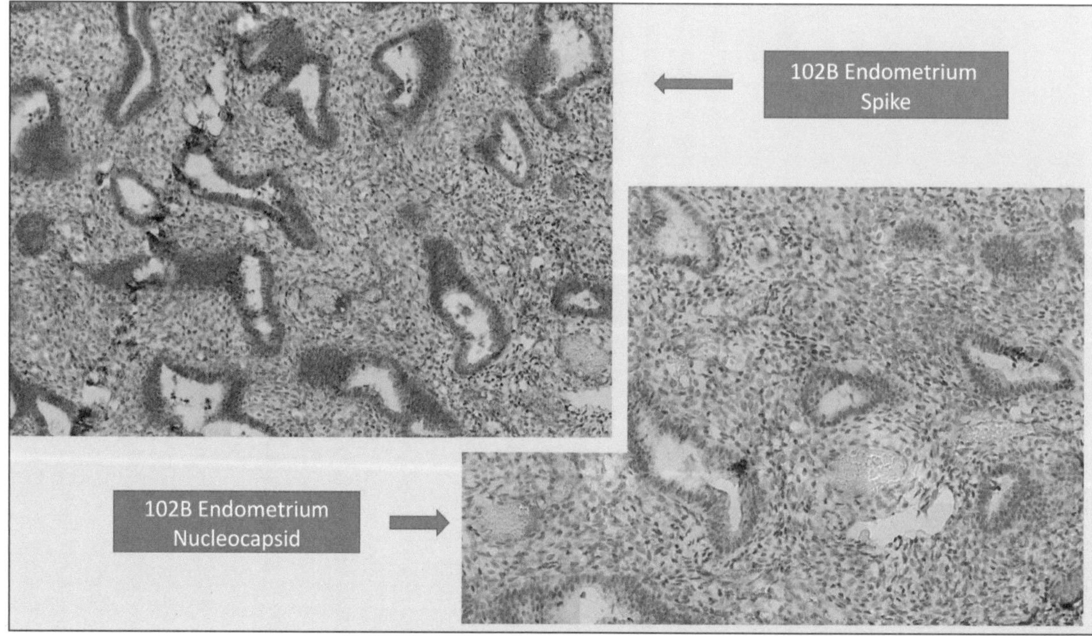

Post-modRNA lymphocytic infiltration in endometrium.

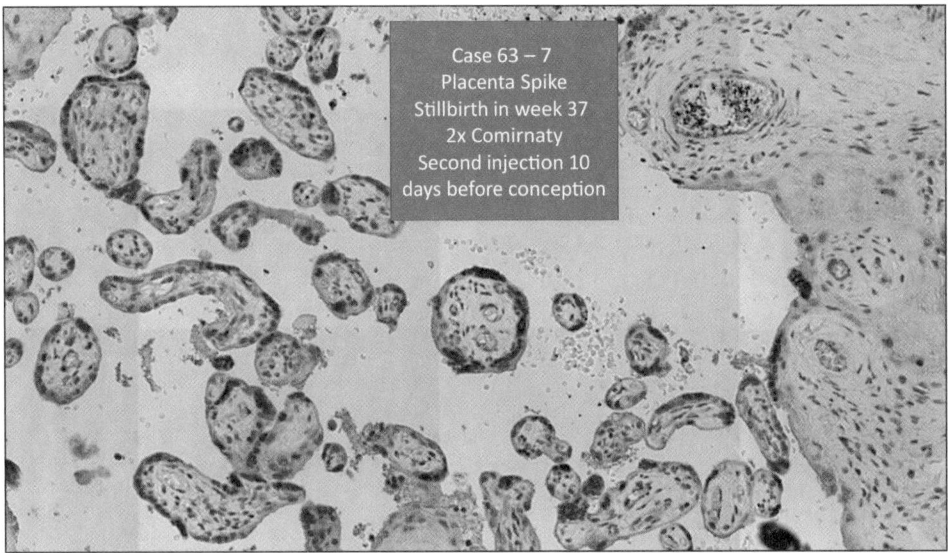

Post-modRNA lymphocytic infiltration in placenta.

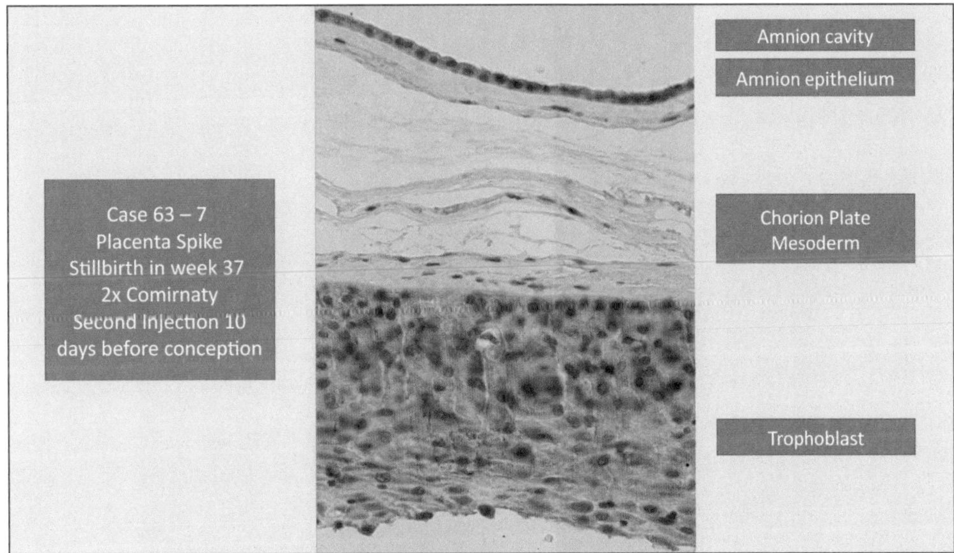

Post-modRNA lymphocytic infiltration in placenta and spike.

So, this is the first round about the general regions that we found.

General lesions affecting more than one organ

a. Expression of Spike protein S1
b. "Endothelialitis" /destruction and inflammation of endothelium
c. Disturbance of the texture with destruction of elastic fibers of the aorta and larger vessels

d. "Displaced" (unidentified (?)) vacuolar and crystalline particles (cholesterol?)
e. Proteinaceous deposits (functional amyloidosis, fibrin amyloid)

f. "Clot"-formation
g. True foreign bodies from contaminated vaccine

Displaced Unidentified Vacuolar and Crystal Particles

And we come now to the displaced vacuole and crystal particles at the proteinaceous deposits.

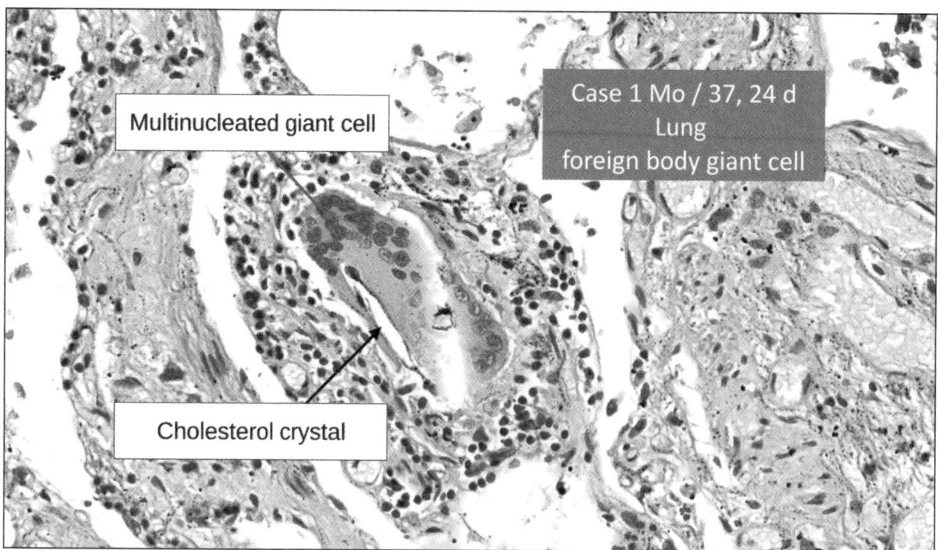

First of all, we found in some cases in the lung and also in other organs, these foreign body giant cells with these needle-like in inclusions *(black arrow)*. Now we know these inclusions to be so-called "**cholesterol needles**."

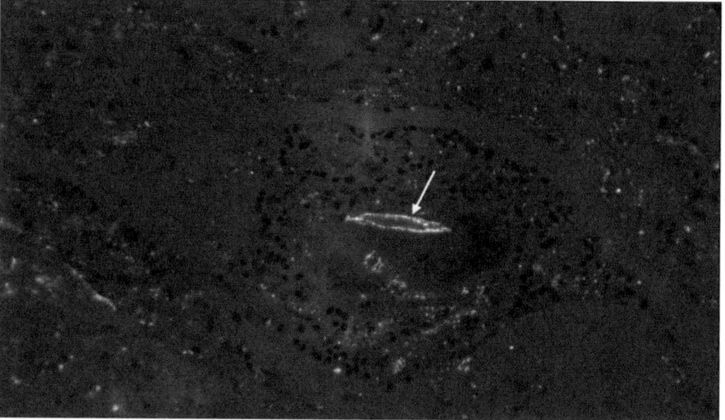

In addition, we found some other material there, and you can see that in the periphery you can see a birefringence of these materials *(white arrow)*.

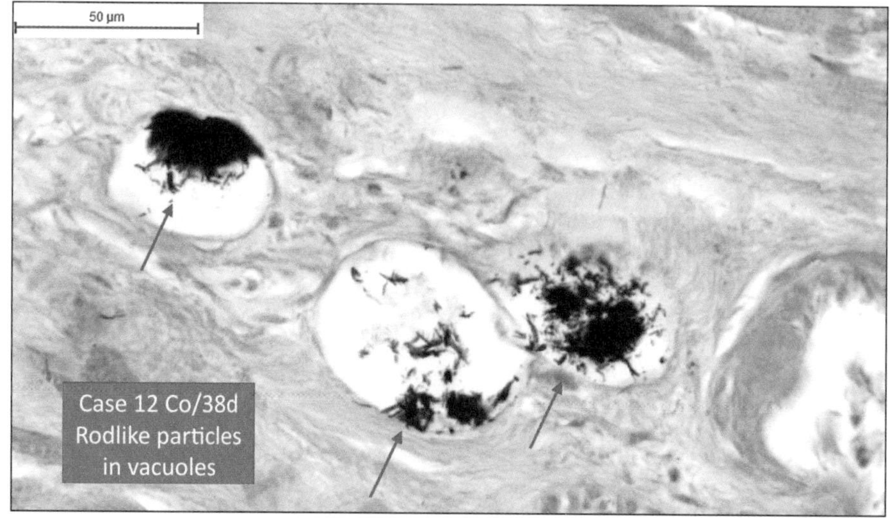

And then we found, in many organs, these rod-like particles *(red arrows)* which are in vacuoles (previous image). Now, these probably don't form vacuoles, but they contained lipids, which like in the fat tissue are extracted during preparation.

First of all, we overlooked these particles, because we thought it was so called formalin pigment, which is very well known to pathologists to be an artifact.

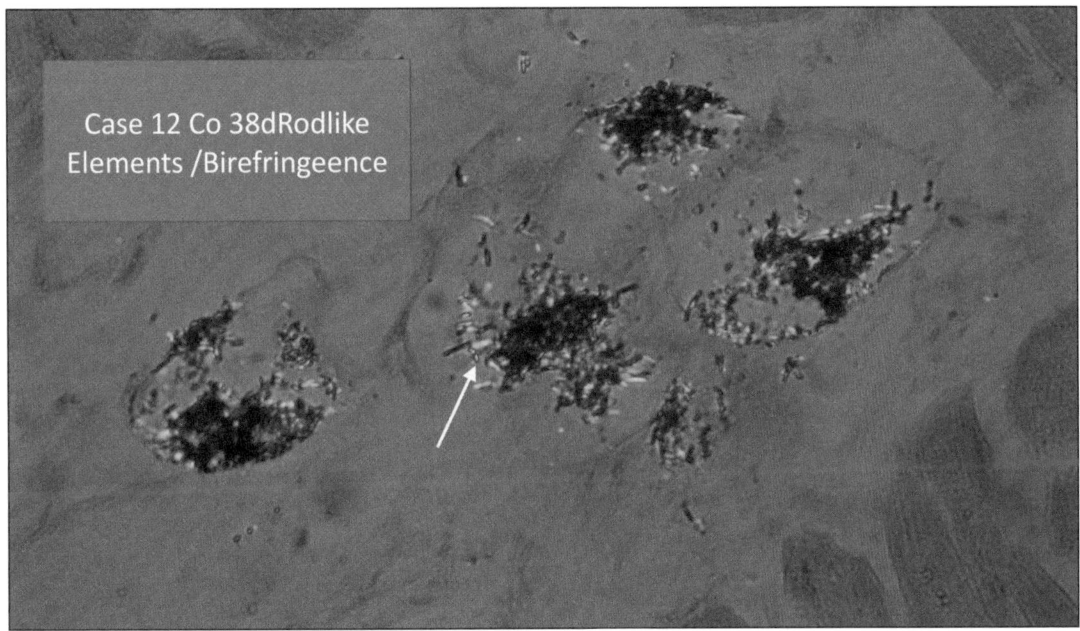

But we came to the conclusion that this is definitely not an artifact but that they must be some kind of material which is formed in the process of this vaccination. You can see some of these are also birefringent *(white arrow)*.

We thought a lot about where this comes from. It could come from the injection itself, but the mass that we saw was not compatible.

It's just I think it would not be possible that so much material is injected.

Also, the production of some substances could be the cause for these; but from, **Raman Spectrometry,** we had the notion that it could be cholesterol. And then we figured out that where cholesterol is located in the human body.

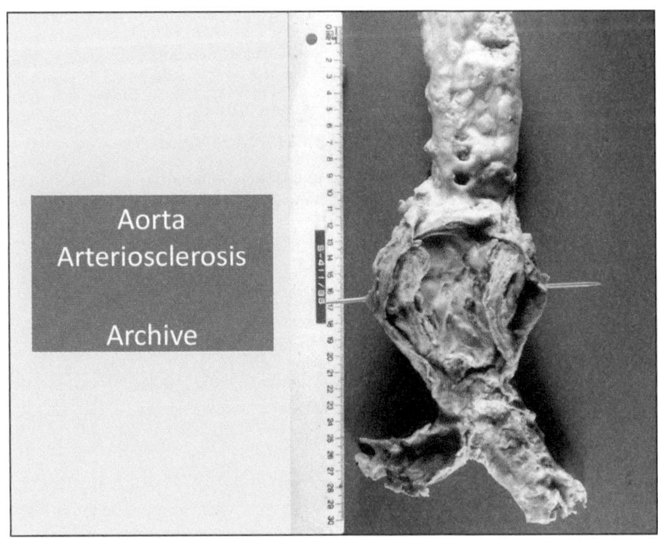

And, as you may know, now, this is not a vaccination victim. This is a normal person who died of natural causes, but he had this large arteriosclerotic plaque with perforation, and this contains, of course, masses of cholesterol.

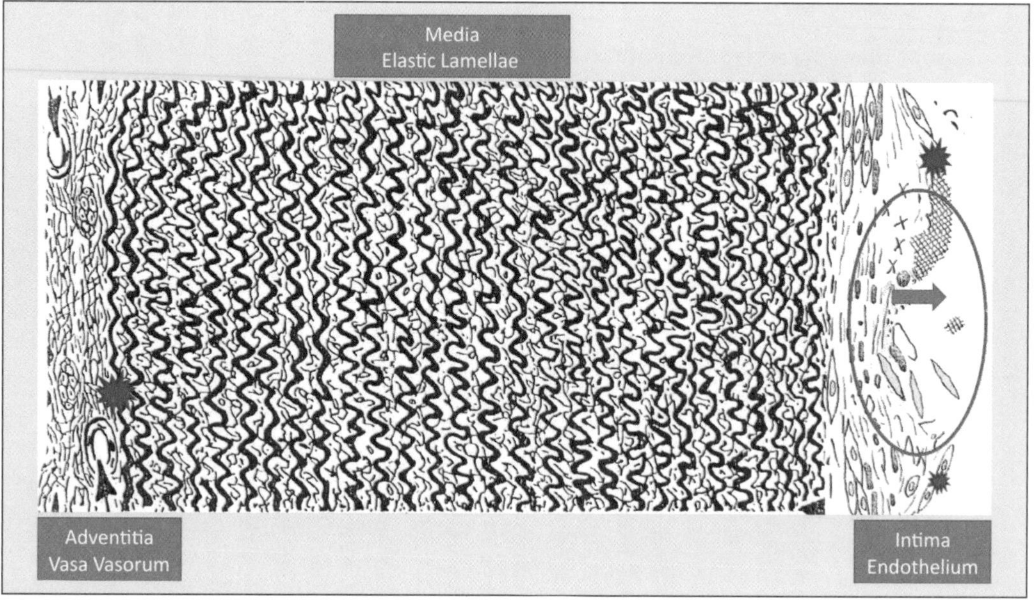

And just to show you the scheme of the aorta—you can see that the atheroma contains these needle-like cholesterols. And, if the epithelium is destroyed, it might come into the blood circulation. *(Cholesterol embolus after erosion of the inner lining of the artery overlying the plaque.)*

And secondary destruction could be due to the endothelial lesions of the vasa vasorum.

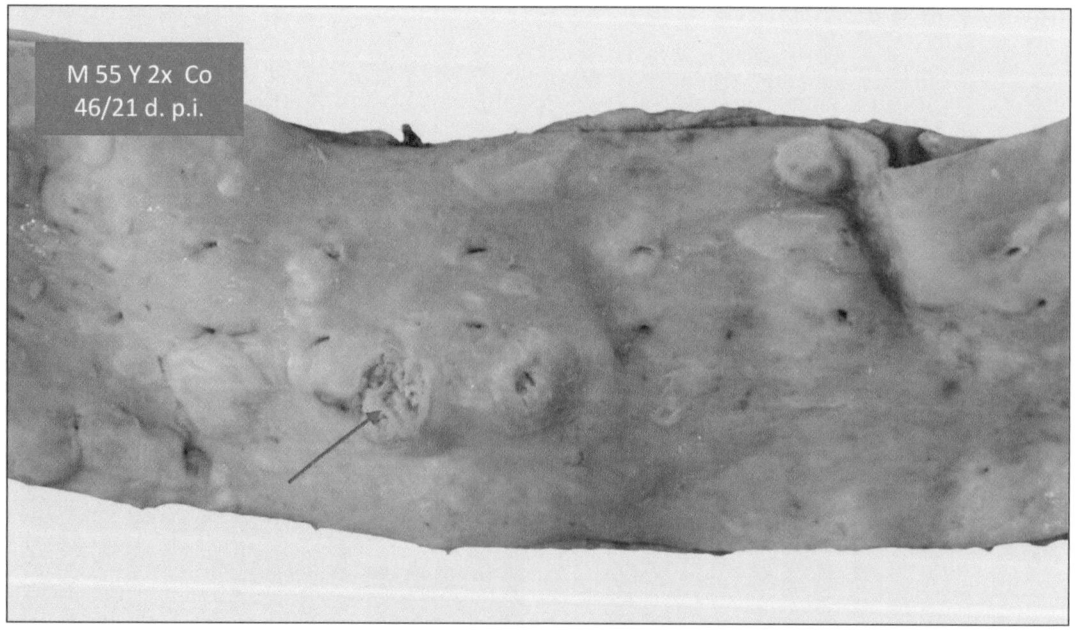

And here you can see a 55-year-old man, the aorta, and you can see that he has some atheromatous plaques; and you can see that one of them has broken open *(red arrow)*, and the contents of these atheromatous plugs went into the blood.

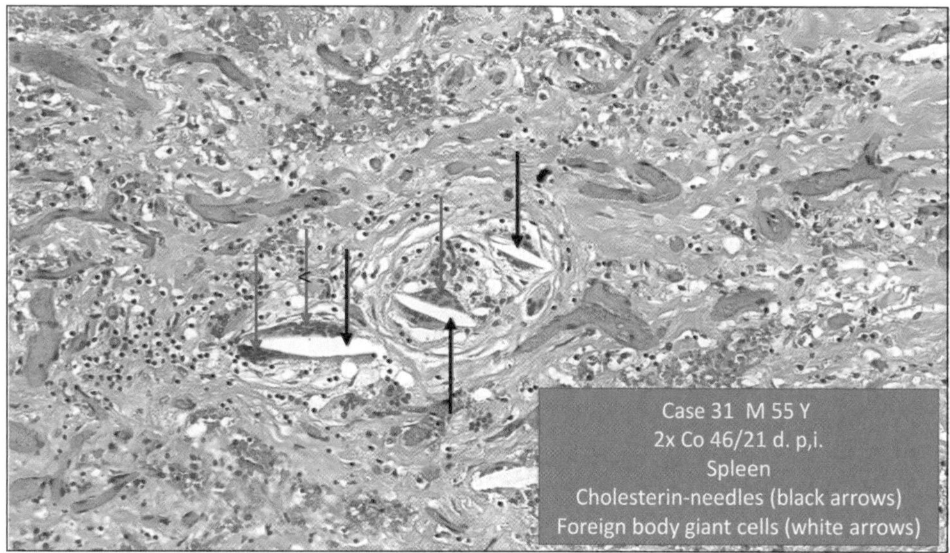

And we can show this because, in the spleen of this person, we found these foreign body giant cells with cholesterol needles *(black arrows)*, and this seems to be very reasonable that this is one of the sources of cholesterol in the in the organs of the deceased after vaccination.

Amyloid Formation (See SR 6.)

We come to the so-called "**amyloid formation**."

It was Swedish authors which found that the peptide sequences of the spike protein might be similar to amyloid and that amyloid-like substances may be formed.

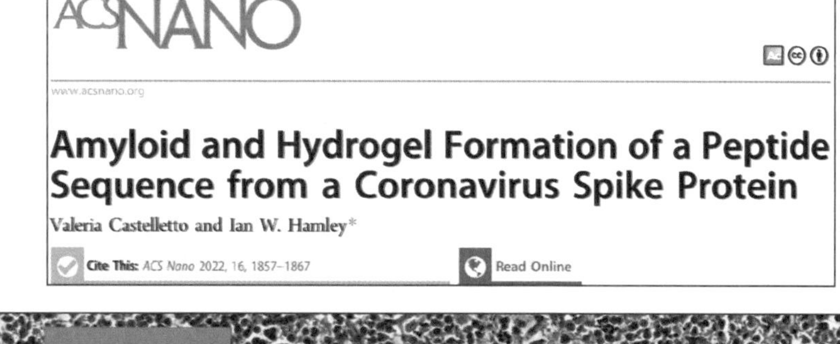
ACS NANO

www.acsnano.org

Amyloid and Hydrogel Formation of a Peptide Sequence from a Coronavirus Spike Protein

Valeria Castelletto and Ian W. Hamley*

Cite This: *ACS Nano* 2022, 16, 1857–1867 Read Online

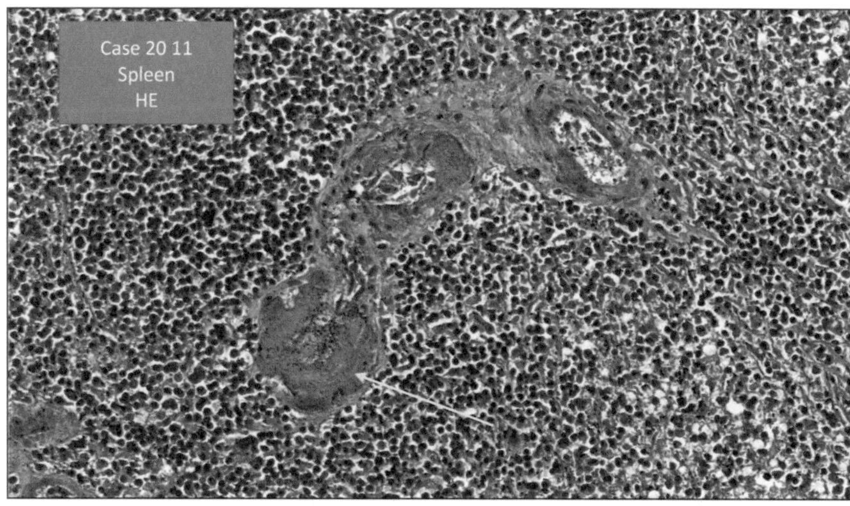

(The yellow arrow points to amyloid in the vessel wall.)

Very early we recognized that here was a strange extracellular deposit of strongly **eosinophilic proteinaceous material in the vessel** walls of many of the deceased that we looked at (previous image).

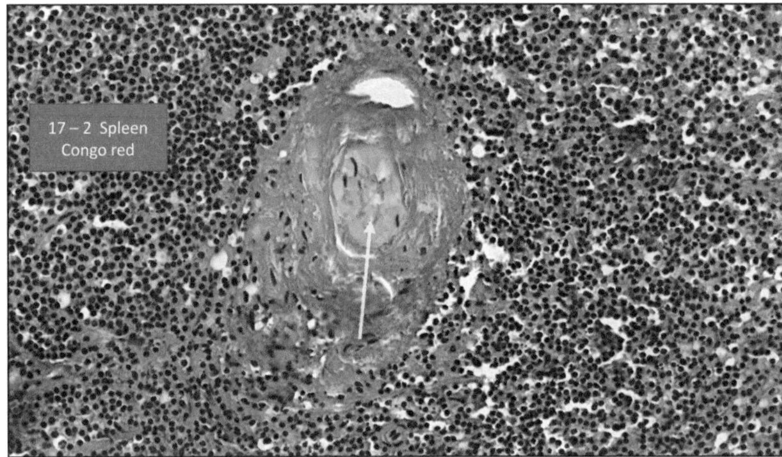

And, in some cases, even the vessel wall, here in the spleen, was occluded, and with the **Congo Red** stain these vessel walls are strongly positive.

So, we confirmed that this is an **amyloid deposition** *(yellow arrow)*.

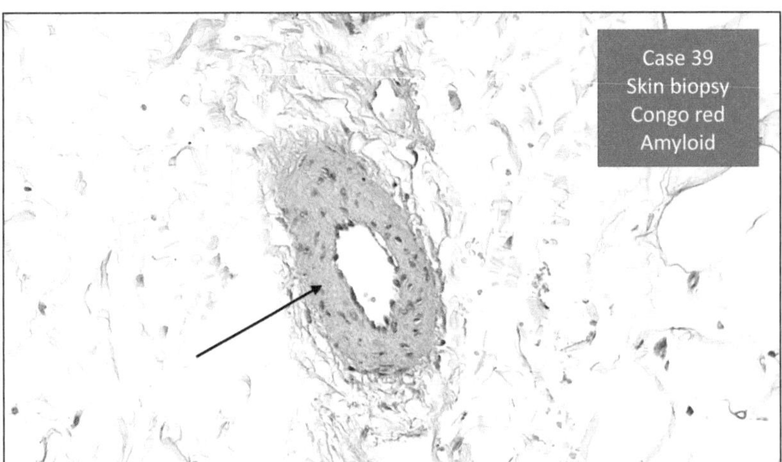

And this is a skin biopsy from a living person. She had problems of perfusion, blood perfusion. You can see also that a small subcutaneous vessel has some amyloid-like material *(black arrow)*.

General lesions affecting more than one organ

a. Expression of Spike protein S1
b. "Endothelialitis" /destruction and inflammation of endothelium
c. Disturbance of the texture with destruction of elastic fibers of the aorta and larger vessels

d. "Displaced" (unidentified (?)) vacuolar and crystalline particles (cholesterol?)
e. Proteinaceous deposits (functional amyloidosis)

f. "Clot"-formation
g. True foreign bodies from contaminated vaccine

Clot Formation

Now we come to the so-called "clot formation."

You may all be aware of these reports from undertakers that they found something that they never saw before; that there were some clots that were of elastic property.

And, the only thing they said is that it was not a **thrombus**, but some *(inaudible)*. That's why they are called it a clot.

And, until now, we do not know exactly what it is, but it's definitely not a thrombotic event. Because no person would survive these clot formations.

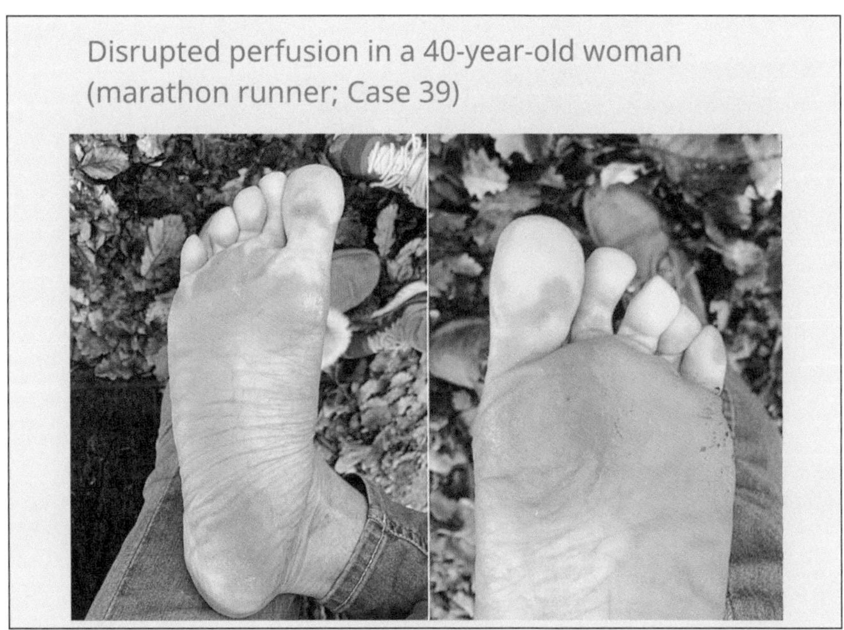

But we have a very interesting observation. We have this 40-year-old woman of around 40, which was an active marathon *runner.*

And after one vaccination with Comirnaty, she was hardly able to walk because she had this very severe disturbance of a perfusion of her lower legs.

And, in the radiograph, there was a dissection of the arteries of the lower leg.

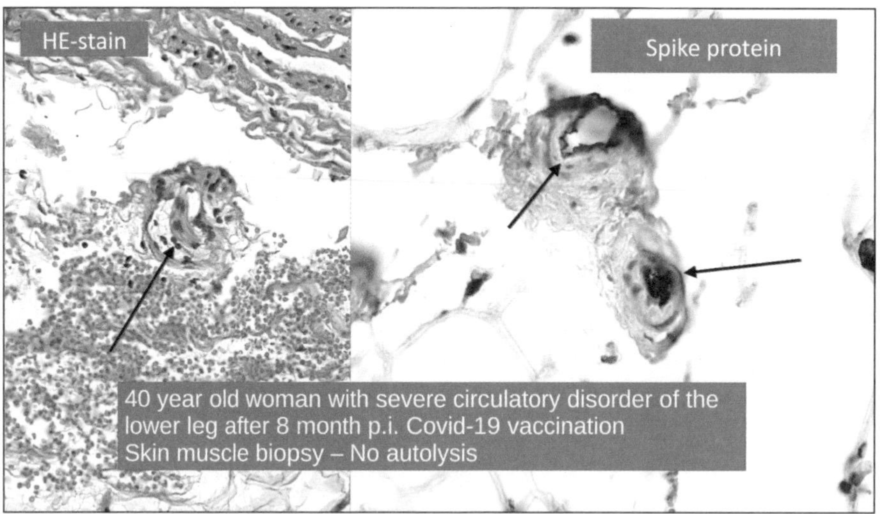

And, in the skin biopsy that was taken from her, we found these lesions of the small vessels, and you can see that the endothelium is swollen. It is detached from the basal membrane. And **in these vessels, we could clearly demonstrate the spike protein**.

And this is from a living person. It's definitely not an autolytic artifact. And, in this person, and this is the interesting part of it, blood samples were taken; and, after centrifugation and cooling, these clots were formed. There they are definitely separated from this part of the centrifugation.

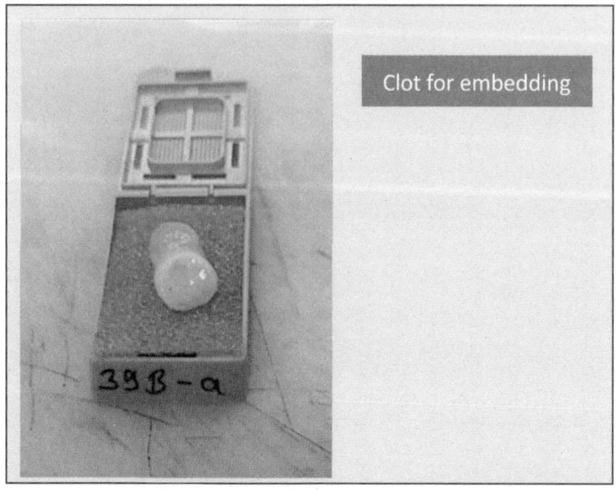

And we were asked, "Well, will you examine these clots formed by this lady after the cooling?" and we did this.

Here you can see this is the clot that we took, and you can see the histology, it's a proteinaceous-fibrous material which seems to be gross outgrowths on the periphery.

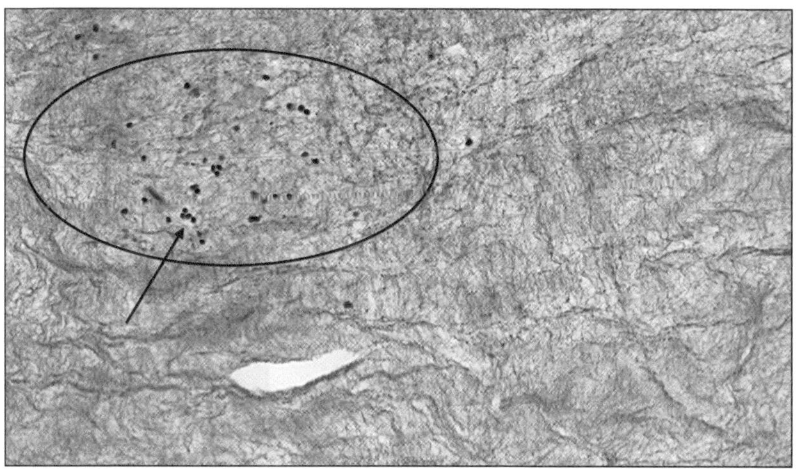

And there are some lymphocytes in there. It's not mature fibrin but fibrinogen, and we did some examination on it.

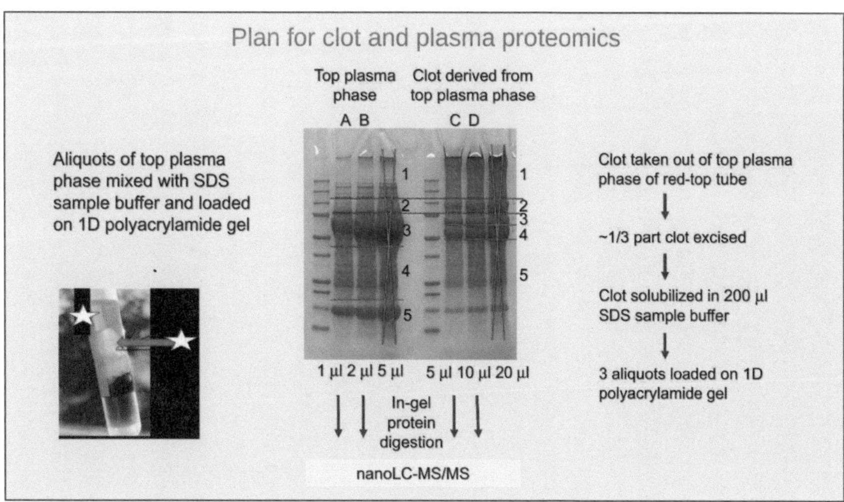

And you can see the top plasma phase and the clot derived from the plasma phase (previous image).

137 clot-enriched proteins

GO 0005201	Extracellular matrix structural constituent
GO 0062023	Collagen-containing extracellular matrix
GO 0005518	Collagen binding
GO 0043236	Laminin binding
GO 0071953	Elastic fiber
GO 0050839	Cell adhesion molecule binding
GO 2001027	Negative regulation Endothelial cell chemotaxis
GO 1900024	Regulation of substrate adhesion-dependent cell spreading
GO 2000352	Negative regulation of endothelial cell apoptotic process
GO 0045907	Positive regulation of vasoconstriction
GO 0036002	pre-mRNA binding
GO 0035198	miRNA binding

And there were **137 clot enriched proteins** that were **not found** in the serum. These are all elements from the:

1. Vessel walls,
2. Extracellular matrix collagen,
3. Collagen-containing extracellular matrix,
4. Collagen binding,
5. Laminin binding,
6. Elastic fiber, and
7. Cell adhesion binding molecule.

So, the conclusions that were drawn from this finding is that the **endothelial damage persisted** in this unlucky lady.

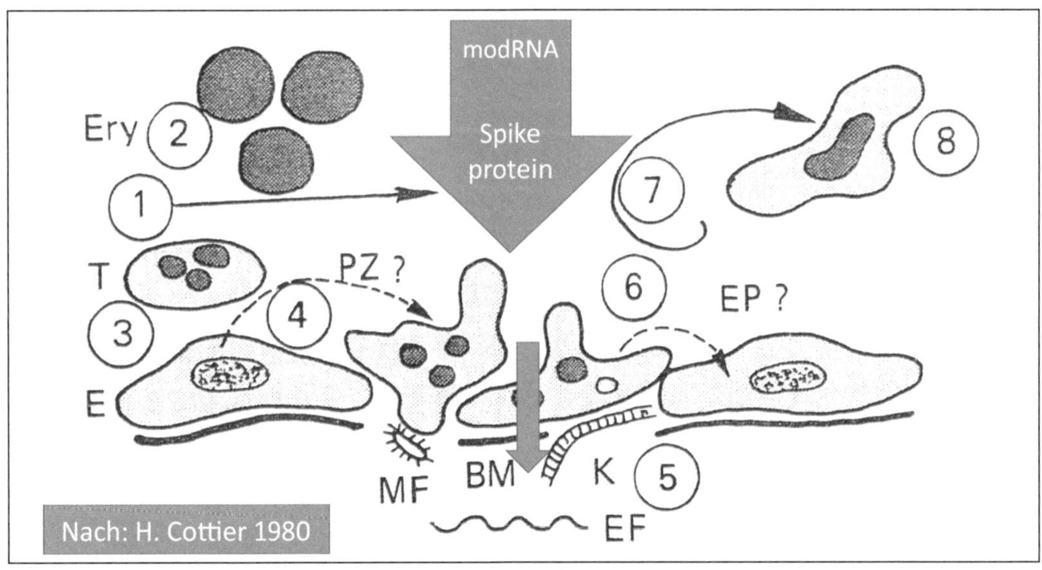

In the previous image, you can see a scheme. This is the endothelium, and, **if the endothelium is destroyed by the spike protein, the spike protein goes to the deeper layer.** And these constituents reach into the blood.

So, I think this is demonstration that the **endothelial damage is very important and may persist for a very long time**.

Specific Organ and Tissue Lesions (See SR 7.)

We come to the specific organ and tissue lesions, and we start with the main pathological findings in the blood vessels, the small vessels.

Specific organ- and tissue lesions

Main pathological findings on blood vessels

Small vessels

- **Endothelialitis, most prominently in heart, lungs and brain**
- **Aggregation of erythrocytes, bleeding, hemosiderosis within vessel wall**
- **Complex formation of amyloid-spike protein-fibrin in vessels - amyloidosis**
- **Thrombocyte aggregates and microthrombi**
- **Obliteration**

I already showed you the

1. Endothcliitis most prominent in the heart, lungs and brain,
2. Aggregation of erythrocytes,
3. Hemorrhage and bleeding,
4. Hemosiderosis into the vessel wall,
5. Complex formation of amyloid-spike protein-fibrin in the vessel walls,
6. Amyloid-like deposits,
7. Thrombocyte aggregates, and, finally,
8. Obliteration of small vessels.

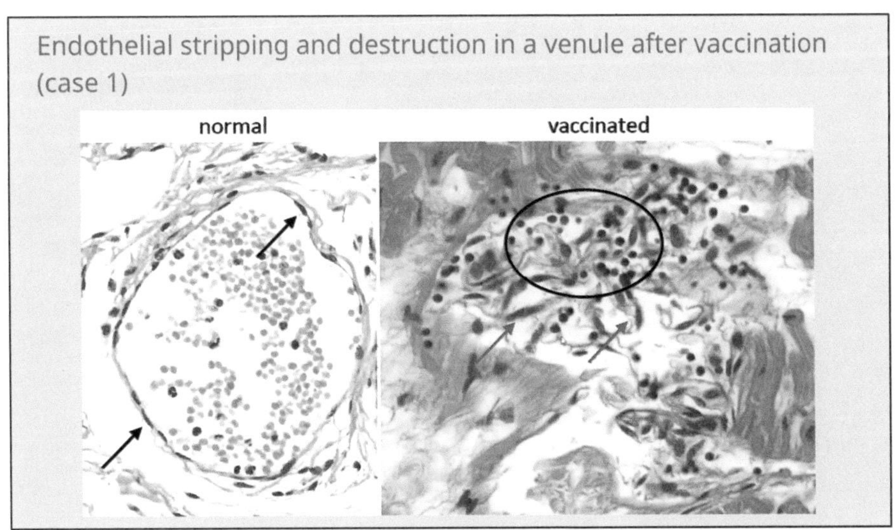

And here (previous image) you can see the normal small vessels with the very thin endothelium, left. And here you can see a destroyed small vessel in the heart muscle, right. And you can see these spindle-like cells, these you can find here, these are the destroyed endothelium (arrow); and you can see the lymphocytic infiltration (oval), which clearly marks this as an intravital process.

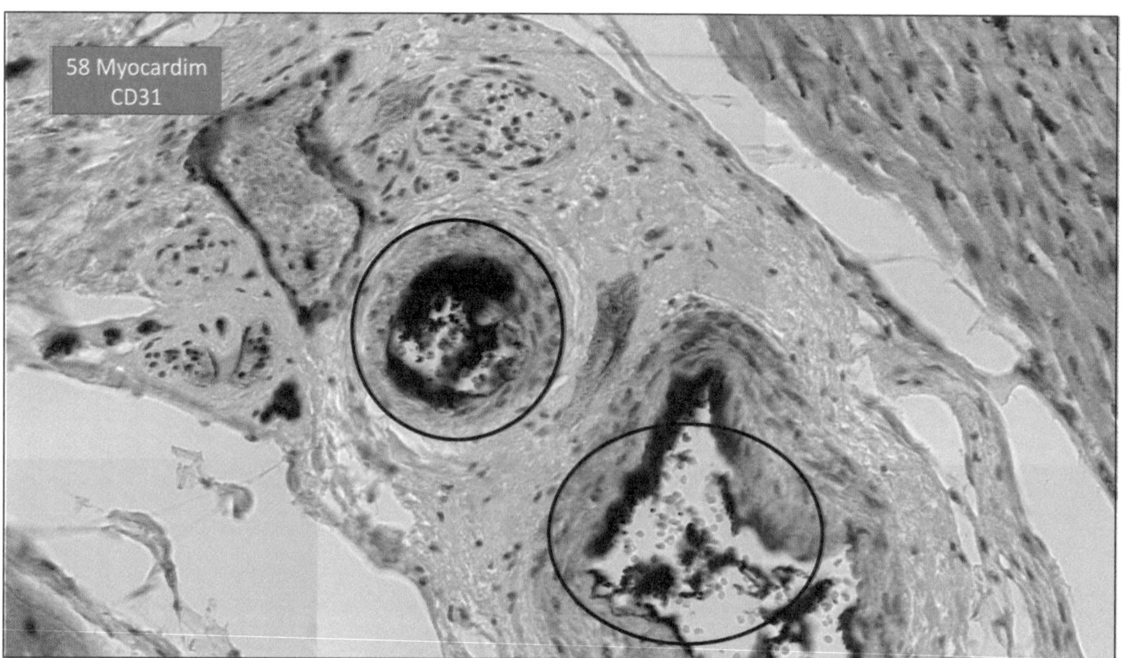

And we can show that these swollen and detached elements are endothelium by the **CD 31** marker. (*Cell adhesion molecule required for leukocyte transendothelial migration under most inflammatory conditions https://www.pathologyoutlines.com/topic/cdmarkerscd31.html.*)

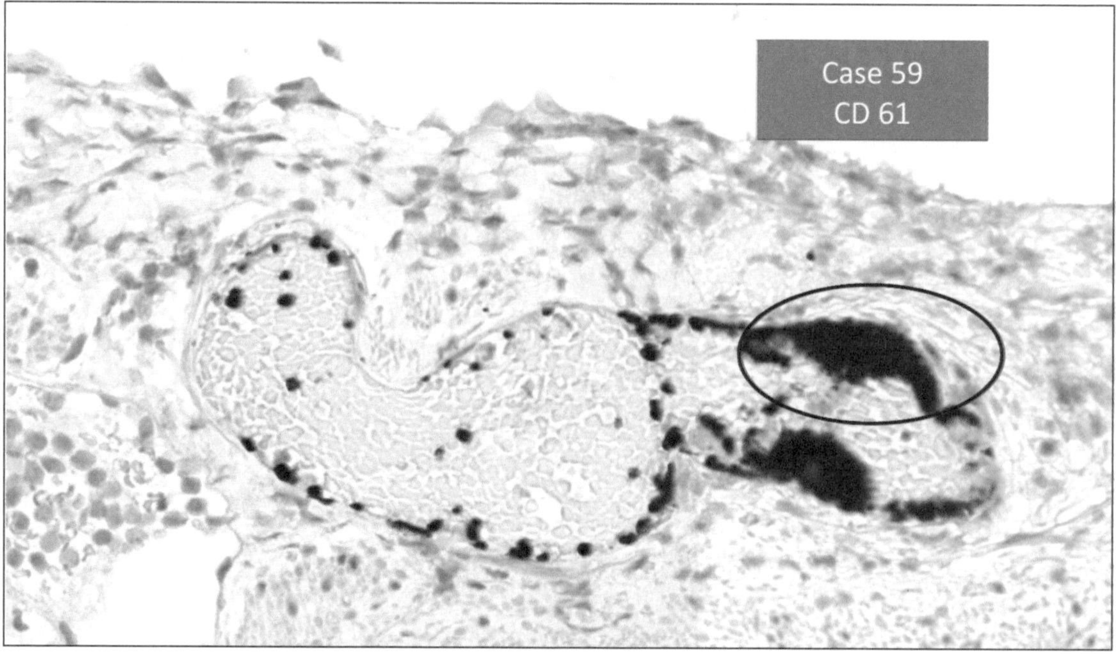

And then we can see that in these areas where the endothelium is destroyed there is attachment of platelets shown by the **CD 61**. (*This glycoprotein complex (GPIIb-IIIa) binds plasma proteins, such as fibrinogen, fibronectin, von Willebrand factor, and vitronectin, and plays a critical role in platelet aggregation.*)

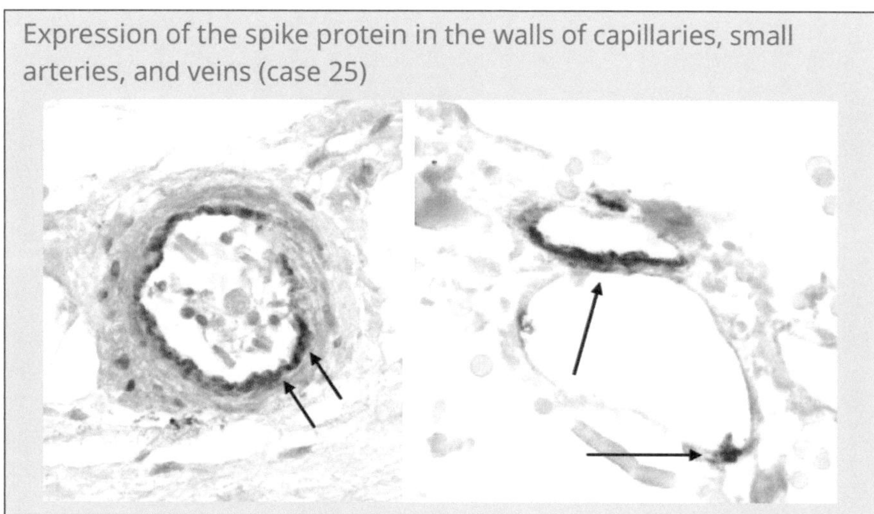
Expression of the spike protein in the walls of capillaries, small arteries, and veins (case 25)

Of course, again, we can show the spike protein in these capillaries and in the small arterial vessels, not only in the endothelium, but also in the near inner most of the vessel walls.

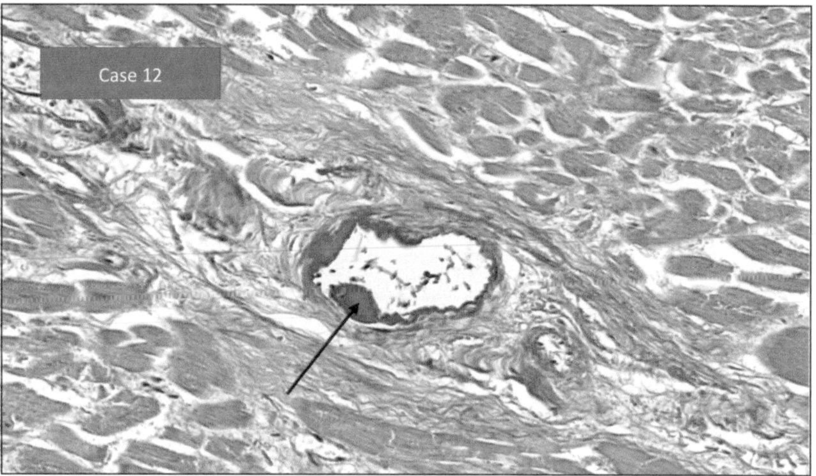

And, this can lead to these deposits here *(black arrow)*.

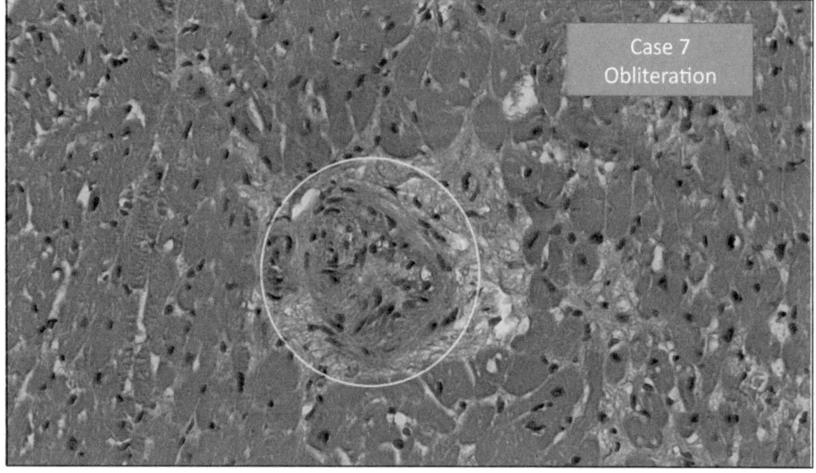

And finally, into obliteration here in the cardiac vessel . . .

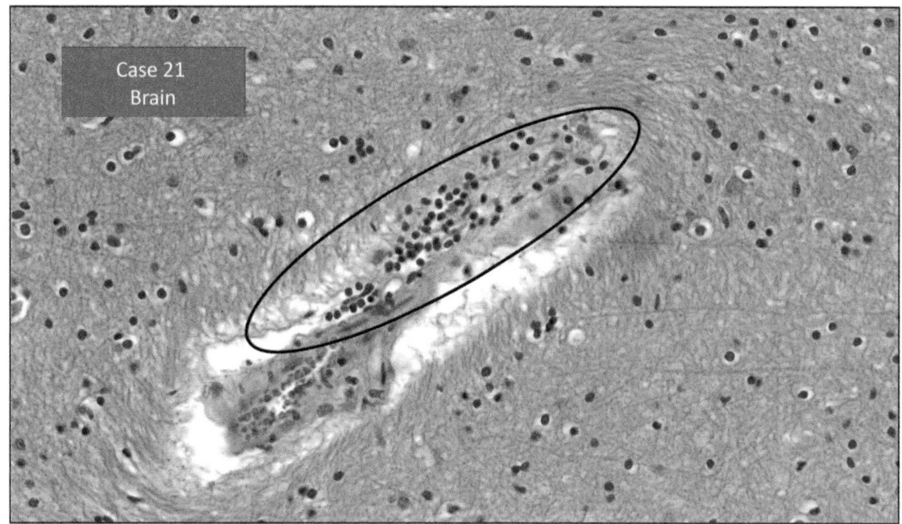

. . . also, inflammation of the small vessels in the brain. I will go to the brain later in detail, but this is in this context and . . .

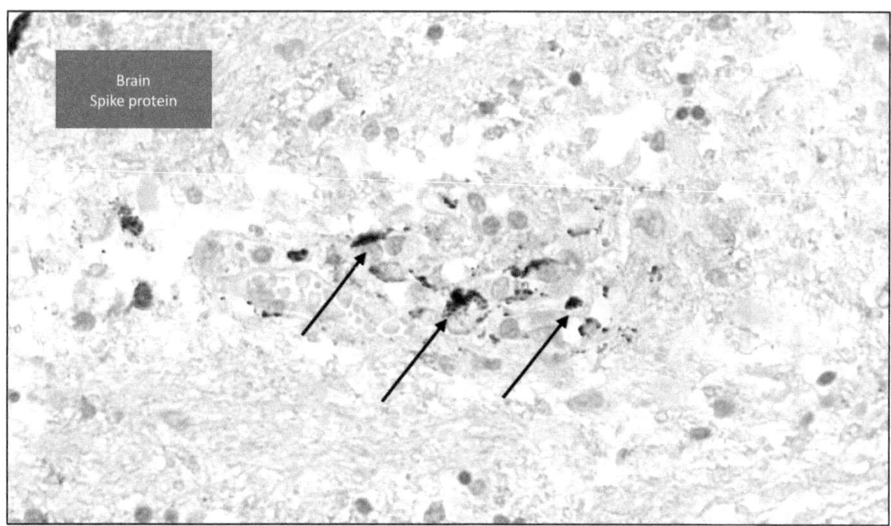

. . . you can see that in these vessel walls we can demonstrate the spike protein.

Large Vessels

Now, most alarming, is the finding in the large vessels.

Main pathological findings on blood vessels

Large vessels

- **Disrupted wall structure of aorta with lymphocytic vasculitis and perivasculitis**
- **Endothelial damage - Break-up of atheromatous plaques**
- **Media necrosis / Dissection**
- **Perforation (5 cases)**
- **Thrombotic casts without erosion of atherosclerotic lesions**

We see disrupted wall structure of the aorta with lymphocytic vasculation, vasculitis, and perivasculitis.

Again, the endothelial damage seems to be the leading adverse effect.

And, as I have also showed you, that this may lead to the breakup of atheromatous plaques.

Then in the deep media, we find necrosis and dissection.

Finally, perforation in five cases and thrombotic casts without erosion.

Aorta Aneurysm and Rupture (See SR 8.)

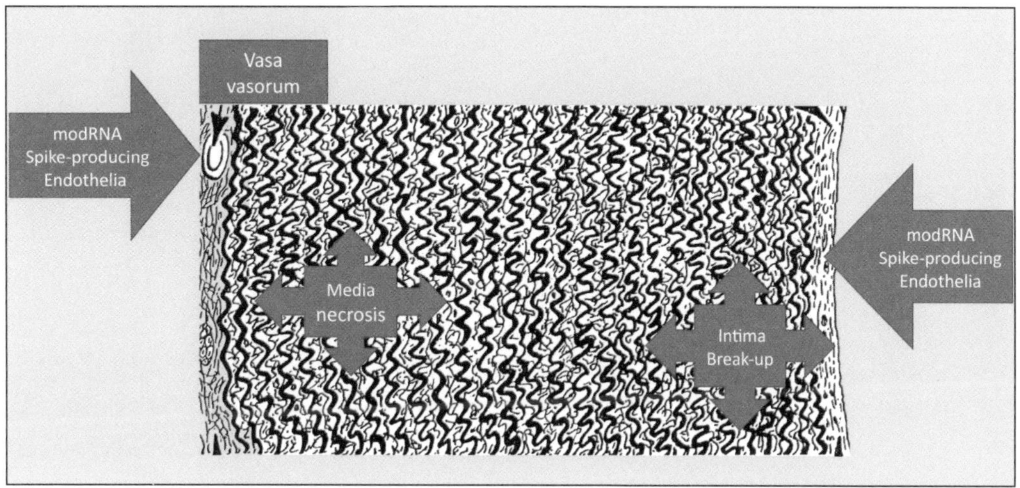

Now, again, this is a scheme of the order. And you can see there's **a two-front war**, so to say, by the spike protein. It's attacking the endothelium of the intima, and this may lead to break up. And they are attacking the endothelium of the vasa vasorum. And this apparently leads to **media necrosis**.

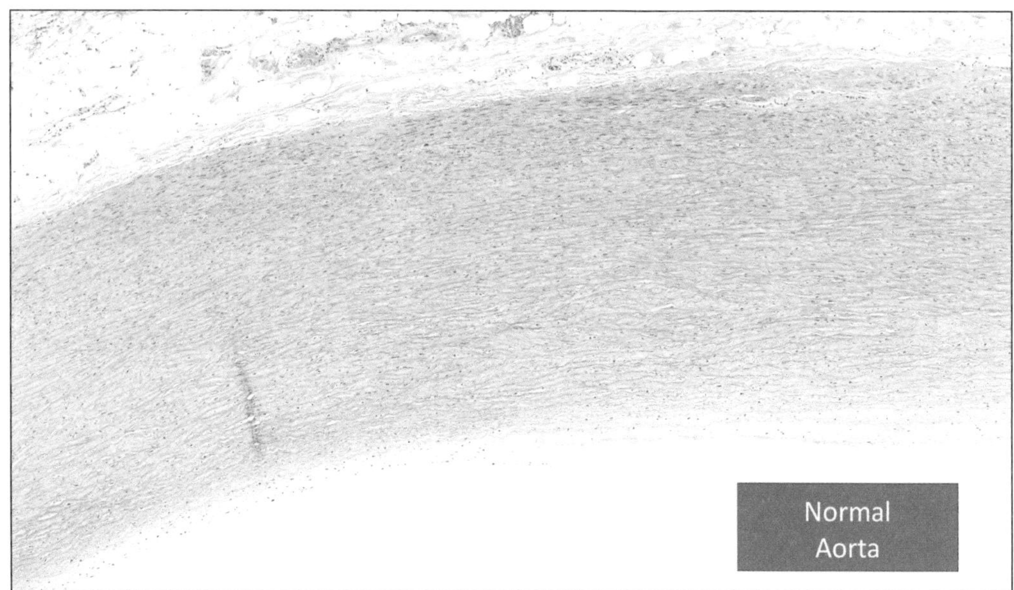

The normal aorta is just a very regular layers of elastic fibers and muscle cells. But here we observe media necrosis, and when do we find media necrosis or mesaortitis? It is with aortic rupture, like we have found in many, in so many of our cases. Now, it may be idiopathic, but this is usually, it's a genetic defect, but it does not go with inflammatory.

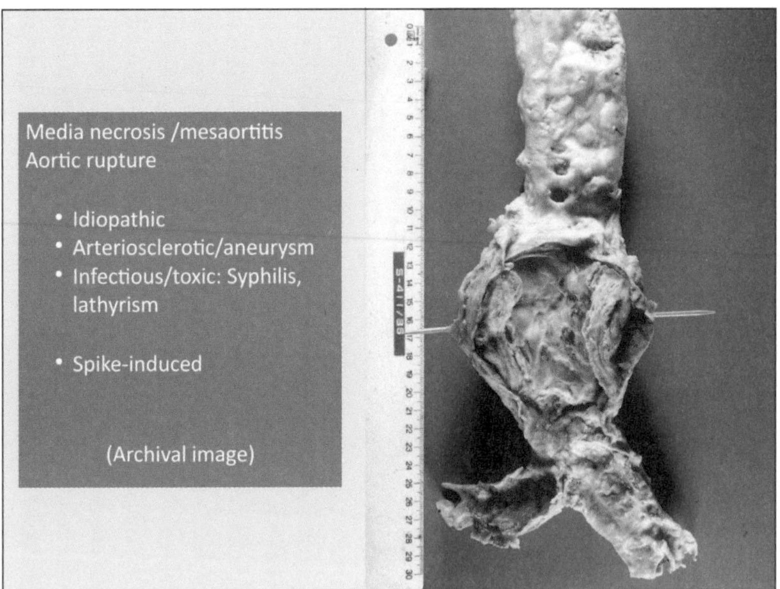

Then we have the atherosclerotic dissection, as we see here.

And finally, we have the toxic which we observe in Lues *(late stage syphilis)* and Lathyrism which is a rare poisoning from fava beans.

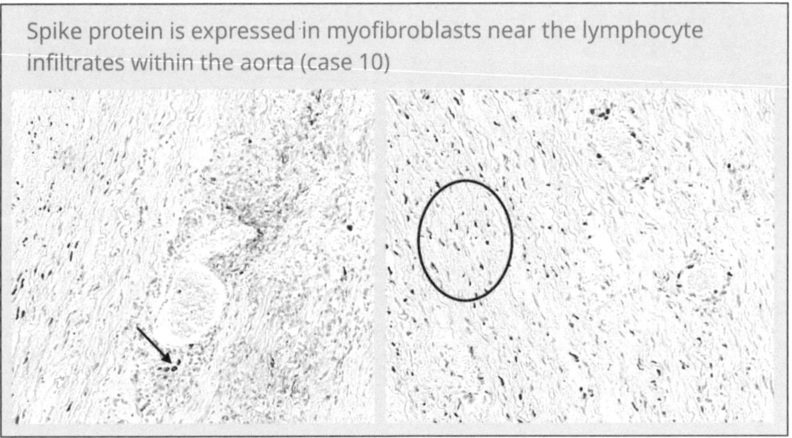

In our cases, apparently it is spike induced in these cases.

We could demonstrate the **spike protein in the myofibroblasts** of this damaged aorta *(circle)*, and you can see mostly around vasa vasorum *(black arrow)* we found, we found this **inflammatory infiltrate** positive and here the **myofibroblasts that are positive**.

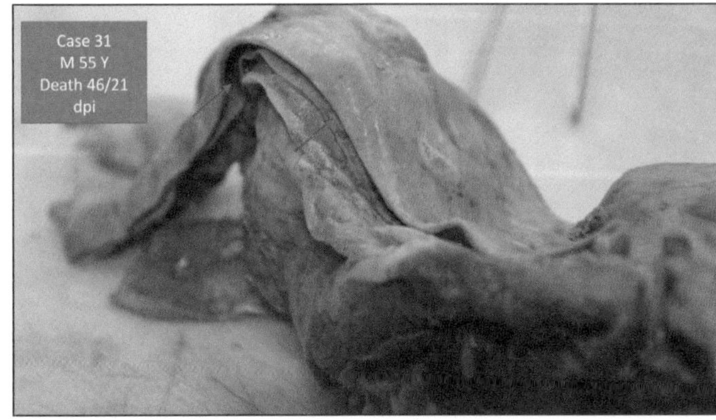

And this (previous image) is just to show you how this looks in the organ specimen, and then you can see this split *(red arrows)*, which is going all here; and you can see it's, here the blood would be flowing, and you can see that this aorta is split in half, and **this man died of the ruptured aorta**.

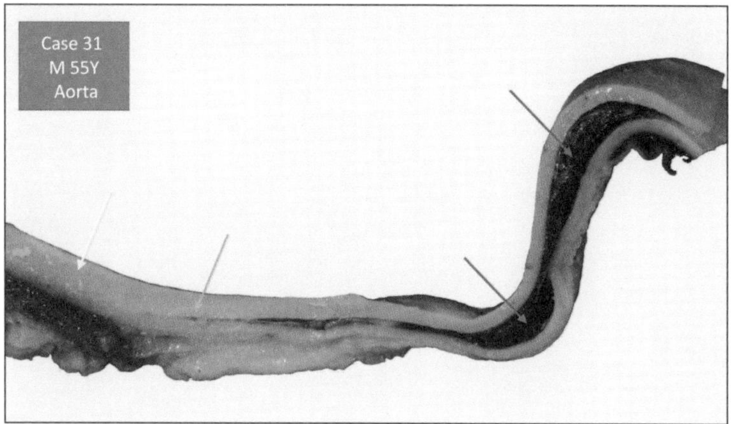

And you can see here *(yellow arrow)*, here it's still, the aorta is still intact in some areas, but here it starts to split *(orange arrow)*. And then you can see **the black**, this **is the blood** *(red arrows)* that has formed here.

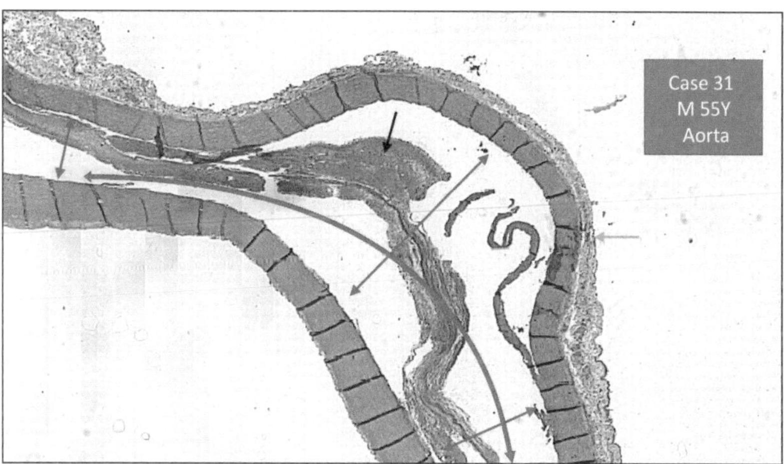

And in the tissue section you can see this also. The aortic wall is split *(blue arrows)* in two. And the blood flow would have, would have been in the path of the split wall (*red arrow)* of the aorta rather than where it should flow. This is the surrounding *(orange arrow vasa vasorum)*. Here you can see the bleeding in this lesion *(black arrow)*.

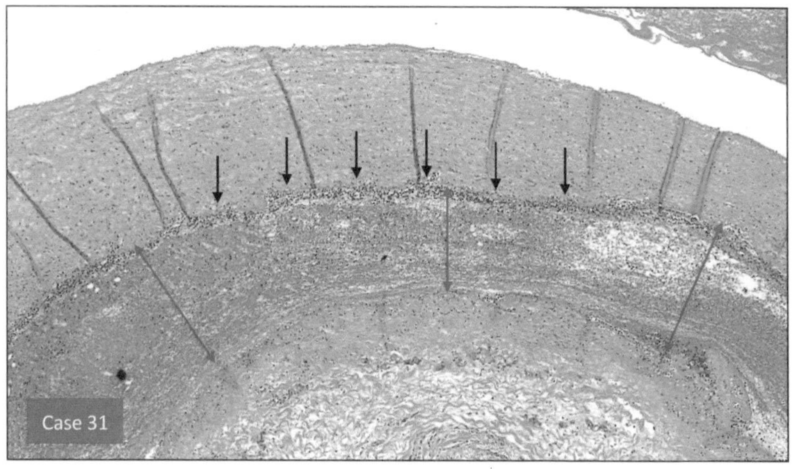

In this case, they thought it might be an idiopathic *(possible genetic versus unknown causes)* media necrosis; but, in contra distinction to the genetic media necrosis, we found a dense lymphocytic and macrophagic inflammatory infiltrate *(black arrows)*, which apparently is induced by the spike protein toxicity (previous image).

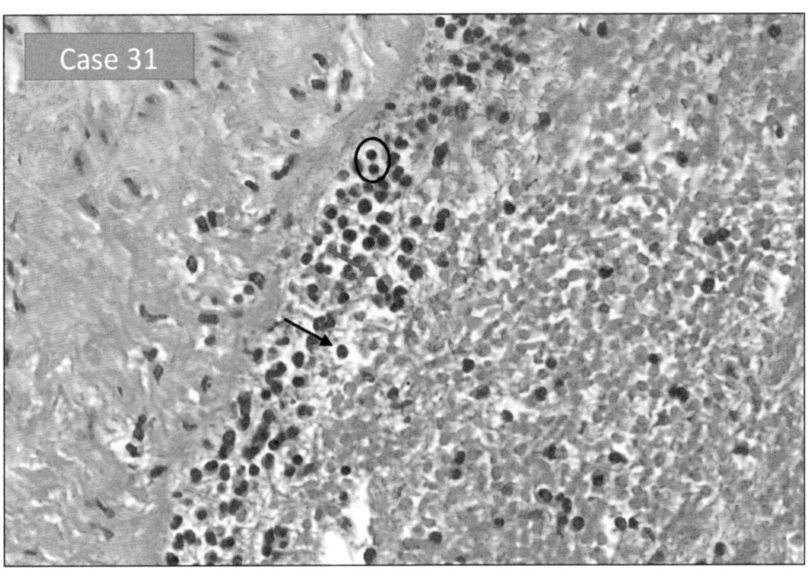

Here you can see a large magnification, there are lymphocytes *(circle),* some mast cells *(red arrow)* and some macrophages *(black arrow).* And here the bleeding has taken place.

We are not the only ones who have seen this. This is in a report of a single case from Japan, just to mark this.

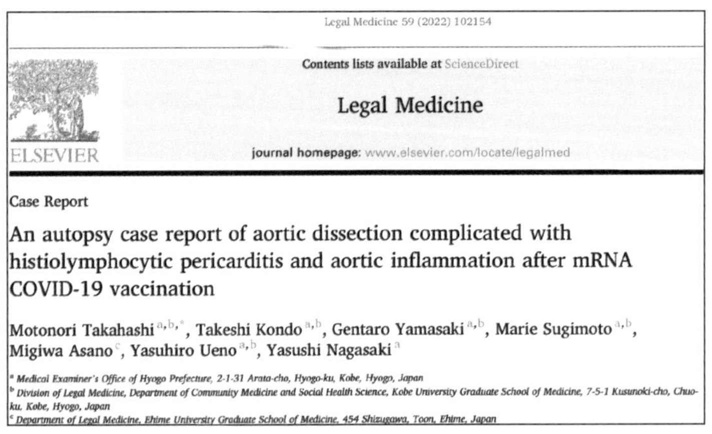

Legal Medicine 59 (2022) 102154

Contents lists available at ScienceDirect

Legal Medicine

ELSEVIER

journal homepage: www.elsevier.com/locate/legalmed

Case Report

An autopsy case report of aortic dissection complicated with histiolymphocytic pericarditis and aortic inflammation after mRNA COVID-19 vaccination

Motonori Takahashi[a,b,*], Takeshi Kondo[a,b], Gentaro Yamasaki[a,b], Marie Sugimoto[a,b], Migiwa Asano[c], Yasuhiro Ueno[a,b], Yasushi Nagasaki[a]

[a] Medical Examiner's Office of Hyogo Prefecture, 2-1-31 Arata-cho, Hyogo-ku, Kobe, Hyogo, Japan

[b] Division of Legal Medicine, Department of Community Medicine and Social Health Science, Kobe University Graduate School of Medicine, 7-5-1 Kusunoki-cho, Chuo-ku, Kobe, Hyogo, Japan

[c] Department of Legal Medicine, Ehime University Graduate School of Medicine, 454 Shizugawa, Toon, Ehime, Japan

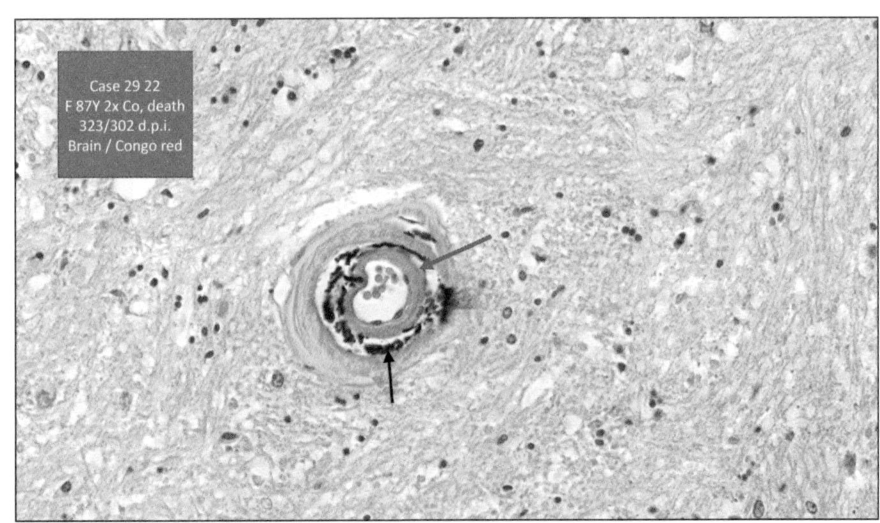

We can see (previous image) the destruction of some part of the vessel walls, especially in many cases in the brain.

First of all, here you can see that there's some deposition of amyloid-like *(red arrow)* substances at the innermost of the vessel.

And then you can see here where the elastic lamellae are situated *(black arrow)*. They are destroyed, and they are strangely black.

We didn't know at first what this was, but actually this is elastic lamellae incrustation, destroyed hemosiderin deposition caused by previous bleeding into the vessel wall.

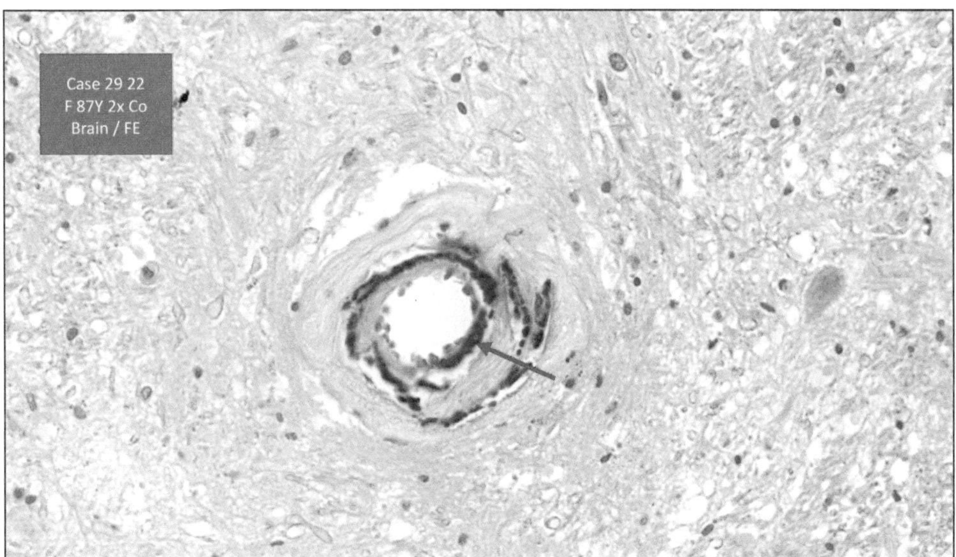

And here you can see it's close to the artery that I showed you before, and you can see that these elastic lamellas are incrusted by hemosiderin *(red arrow)* as a **residue of hemorrhage**.

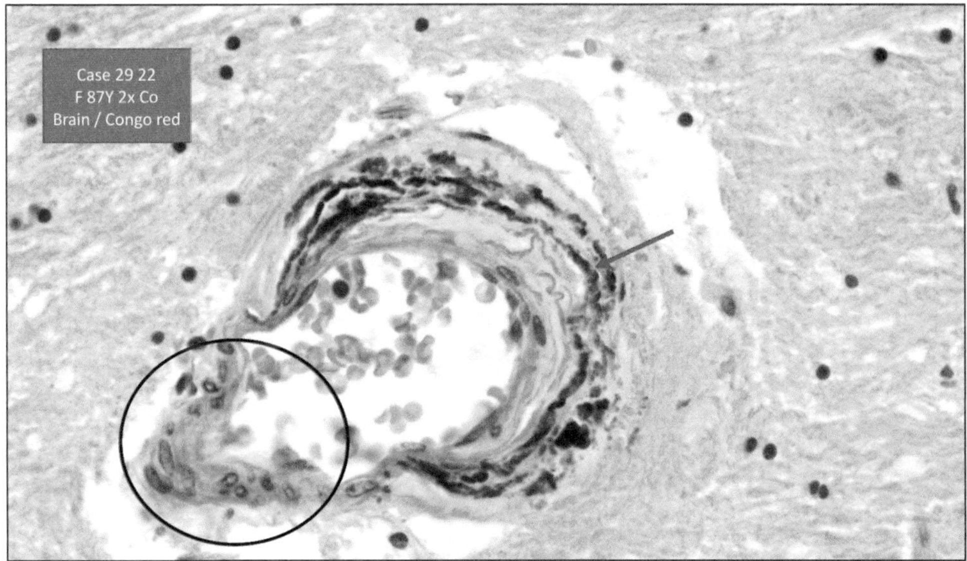

And here you can see again, these are the iron-positive, partly destroyed **elastic lamella** *(red arrow)*. Here they are completely destroyed, and there is a **micro-aneurysm** which at any time might lead to bleeding, to fatal bleeding. **(See SR 5.)**

And we find this phenomenon not only in the arteries of the brain, but mostly in the brain.

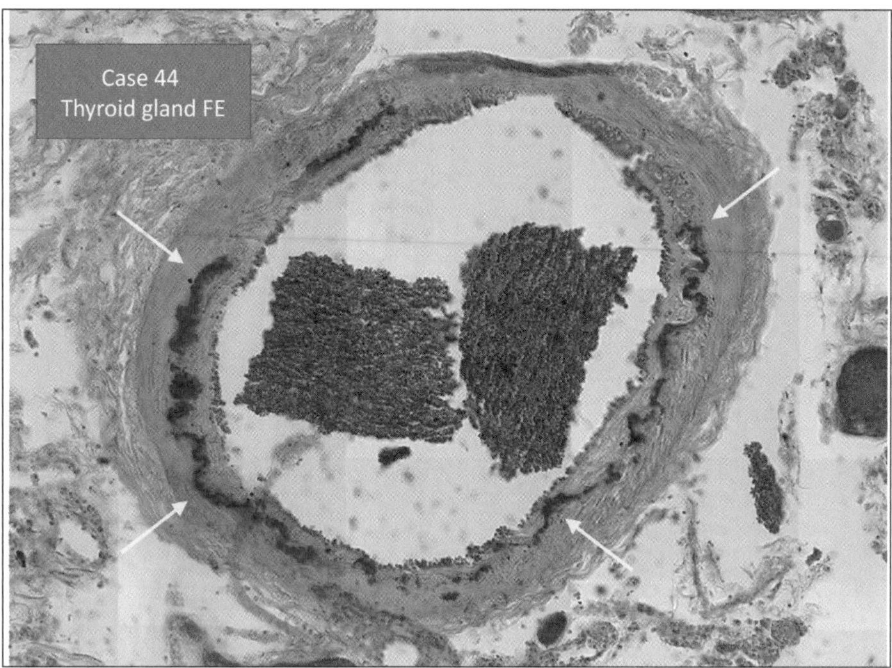

But here you can see a larger artery in the thyroid gland which has these deposits of iron associated with the elastic lamella. *(yellow arrows)*

Sudden Adult Death Syndrome

„SADS – Sudden Adult Death Syndrome" - death without conventionally detectable cause

- Focal media necrosis of coronary arteries with swelling luminal constriction, with or without thrombosis
- Spike expression / T-lymphocyte – macrophage - myofibroblast reaction
- Lymphocytic perivasculitis of the vasa vasorum
- Acute heart failure without microscopic manifestation of necrosis

„acute coronary syndrome", „rhythmogenic heart failure"

Now we consider that these lesions of the of the vessels may be one of the causes of the sudden death syndrome, **a death without conventional detectable causes, which has not been known before this vaccination campaign.** It usually is referred to as **arrhythmogenic heart failure**.

Now what we found might be the cause,

a. Focal media necrosis of the coronary artery and
   i. swelling with luminal constriction,
   ii. with thrombosis, or

iii. without thrombosis.

b. Spike expression, T lymphocyte macrophage, and myofibroblast reaction,
c. Lymphocytic perivasculitis. And this could be the cause of
d. Acute coronary syndromes without manifestation of an infection.

So, we come to the other main pathologic findings.

Main pathologic findings (other organs)

3. Myocarditis
- lymphocytic
- with/without destruction of muscle fibers
- scar formation

4. Alveolitis
- diffuse alveolar damage (DAD)
- lymphocytic interstitial pneumonia
- extrinsic-allergic ?

5. Lymphocytic follicles outside lymphatic organs
- associated with autoimmune diseases
- „lymphocyte amok"

Myocarditis (See SR 9.)

We had the small and the large vessels, and we come to the myocarditis, which definitely is one of the most common findings. It's lymphocytic. It is without destruction of most muscle fibers and, in some, and it leads to scar formation.

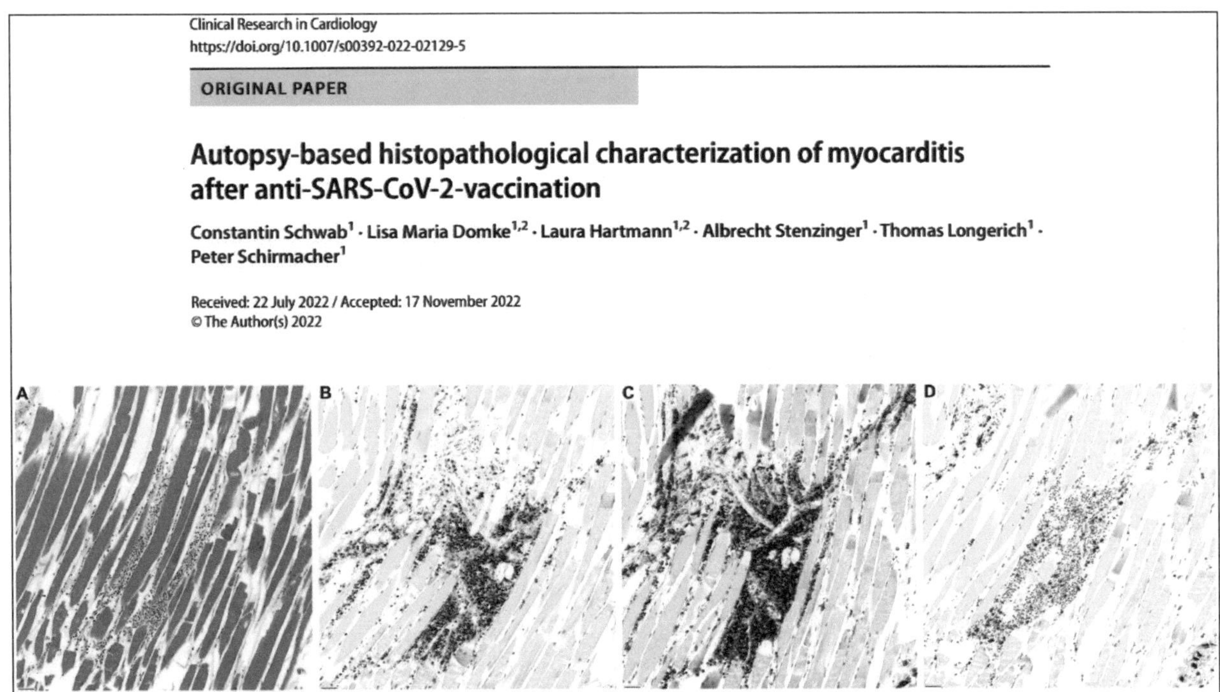

Clinical Research in Cardiology
https://doi.org/10.1007/s00392-022-02129-5

ORIGINAL PAPER

Autopsy-based histopathological characterization of myocarditis after anti-SARS-CoV-2-vaccination

Constantin Schwab[1] · Lisa Maria Domke[1,2] · Laura Hartmann[1,2] · Albrecht Stenzinger[1] · Thomas Longerich[1] · Peter Schirmacher[1]

Received: 22 July 2022 / Accepted: 17 November 2022
© The Author(s) 2022

And this (previous image), as you may know now, is recognized by the international literature to be connected to this spike from the Corona vaccination. *(Schwab, et al.)*

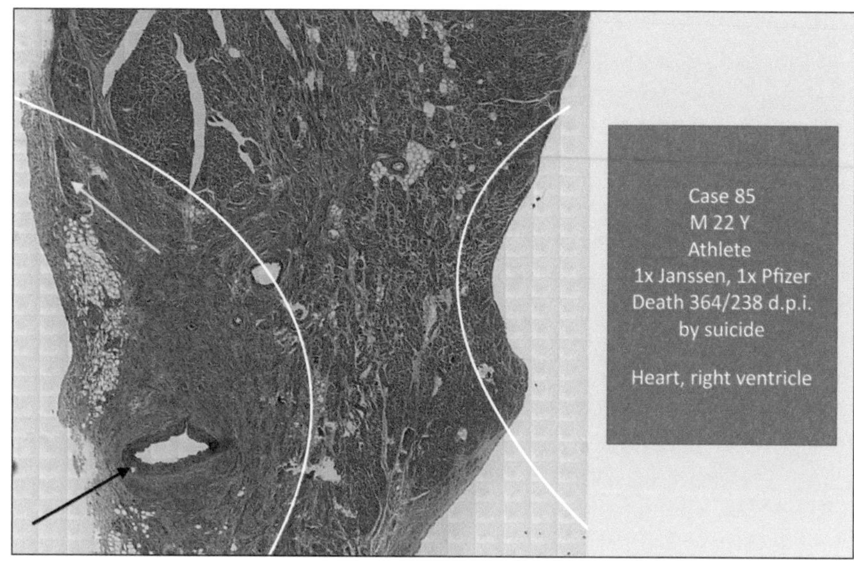

And here you can see a 22-year-old man, and he was an athlete; and he died by suicide because severe myocarditis was diagnosed.

And this is what we found at the autopsy.

1. This is the left (*right*) heart chamber, and you can see he doesn't have any pronounced **arteriosclerosis *(black arrow)*,**
2. **Inflammatory infiltrates** beneath the epicardium *(yellow arrow).*
3. He has this **large scar** (*white arrow and areas defined by white lines*).
4. Residual inflammatory infiltrates *(yellow arrow).*

This might be in the first injection, which was 364 days before he died. So, this could be formed after the first injection. The right side was in the healing process.

But in the left ventricle, it was still very active. And you can see these accumulations of lymphocytes *(black circle).*

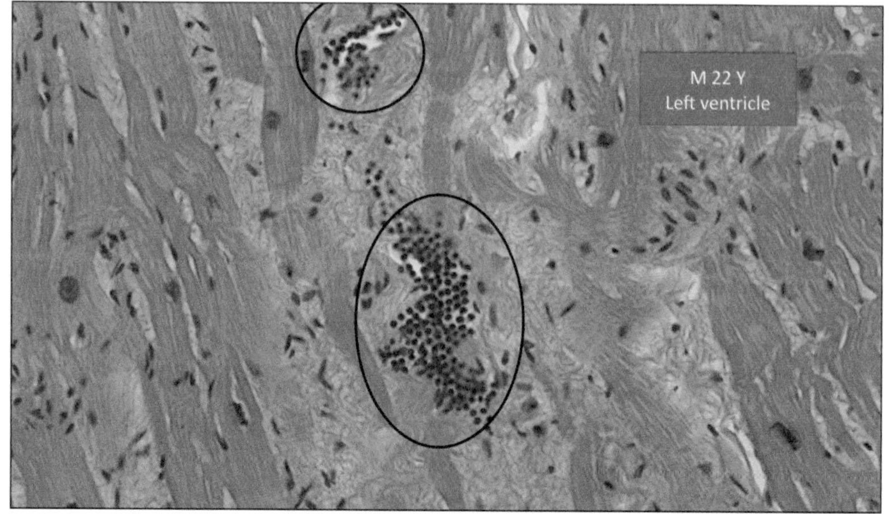

So probably in the **consecutive injections**, this **inflammation of the myocardium** was **triggered again, boosted**.

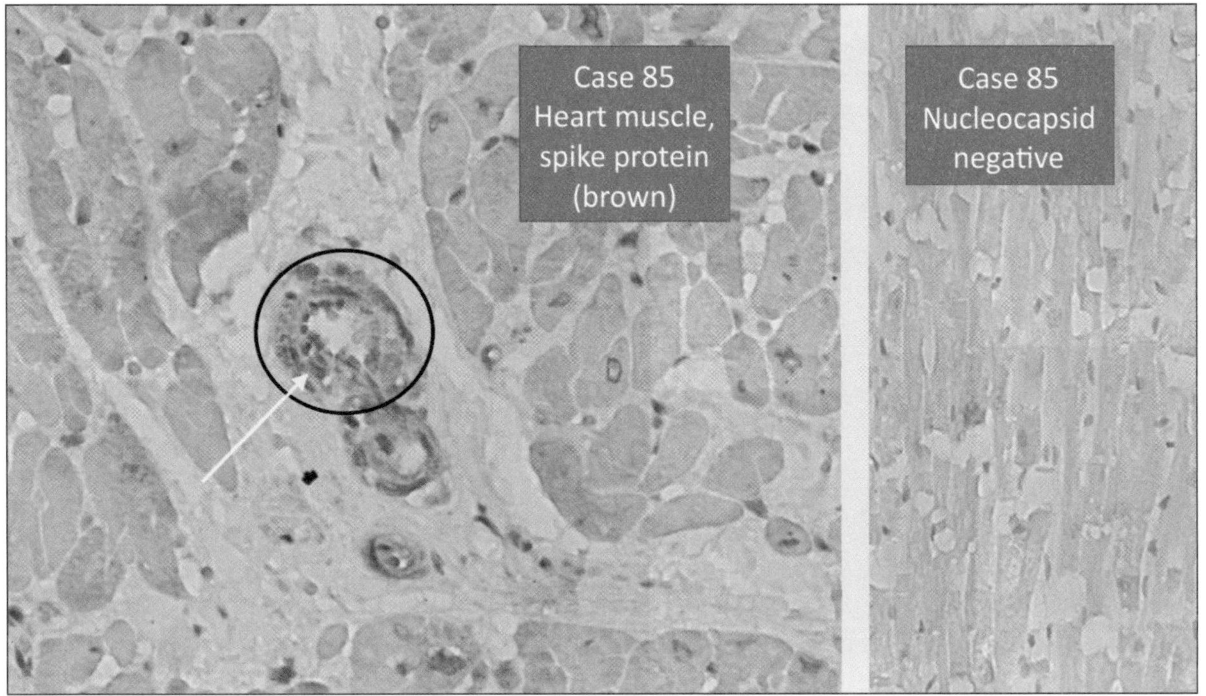

And in this case, you, we could show that the muscle cells expressed in variable amount the **spike protein**, but very **pronounced** (it was expressed) **in the vessel walls** of the small vessels. The vessel wall is swollen and has a disturbance of the structure *(yellow arrow)*. We did the nucleocapsid *stain* in this case, and it was, as you can see, negative. *(right image)*

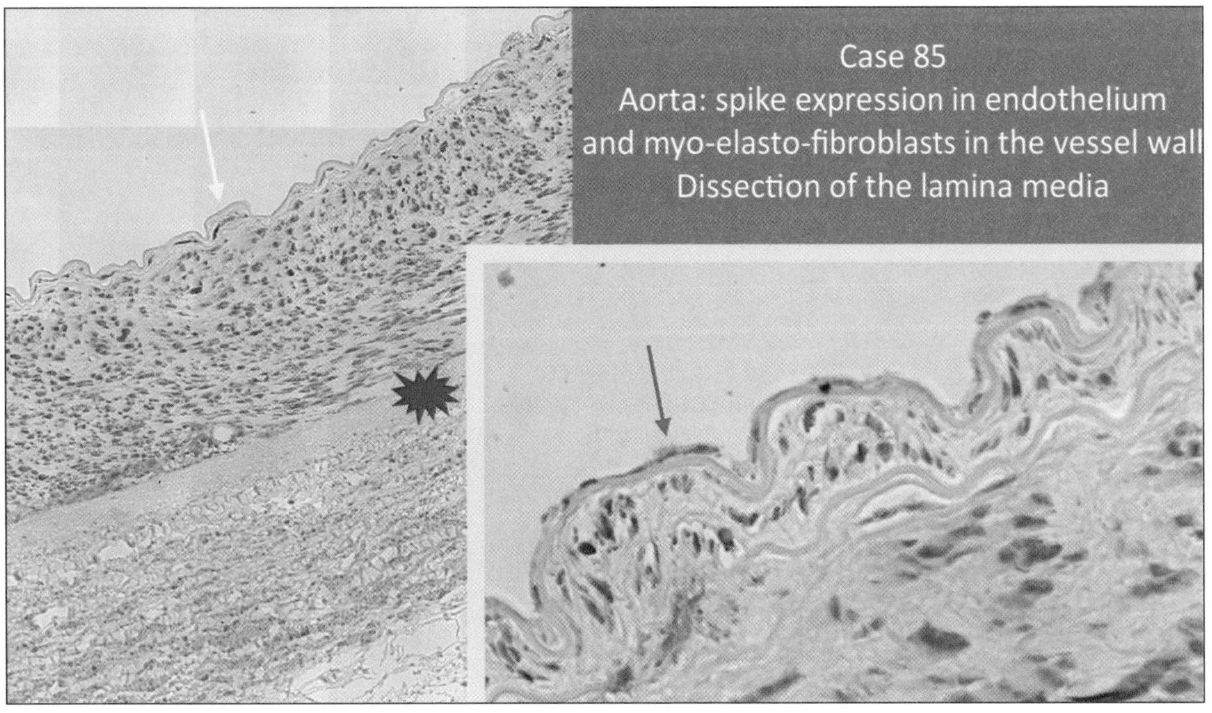

You can see here (previous image), this is the aorta, and you can see the endothelium here is intact *(yellow arrow)*. You can see this clear demonstration of **positive reaction for spike protein** *(red arrow)*. And you can see the **dissection of the aorta** *(spiked red oval)*.

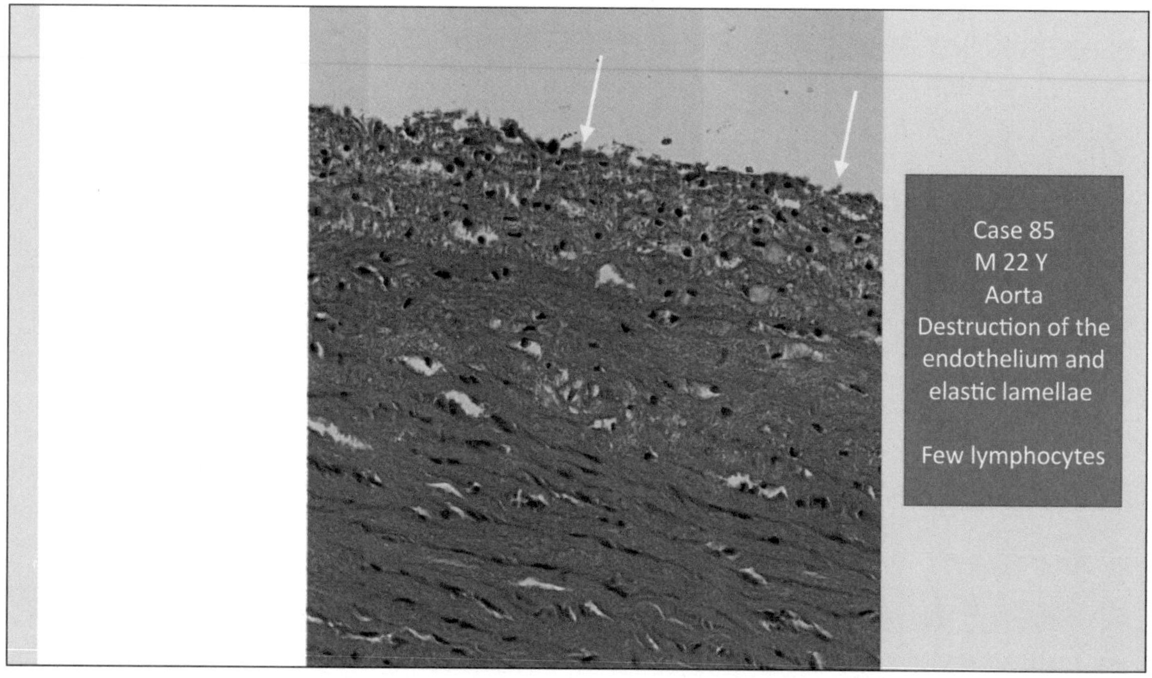

You can see another spot where the **endothelium is destroyed** *(yellow arrows)*.

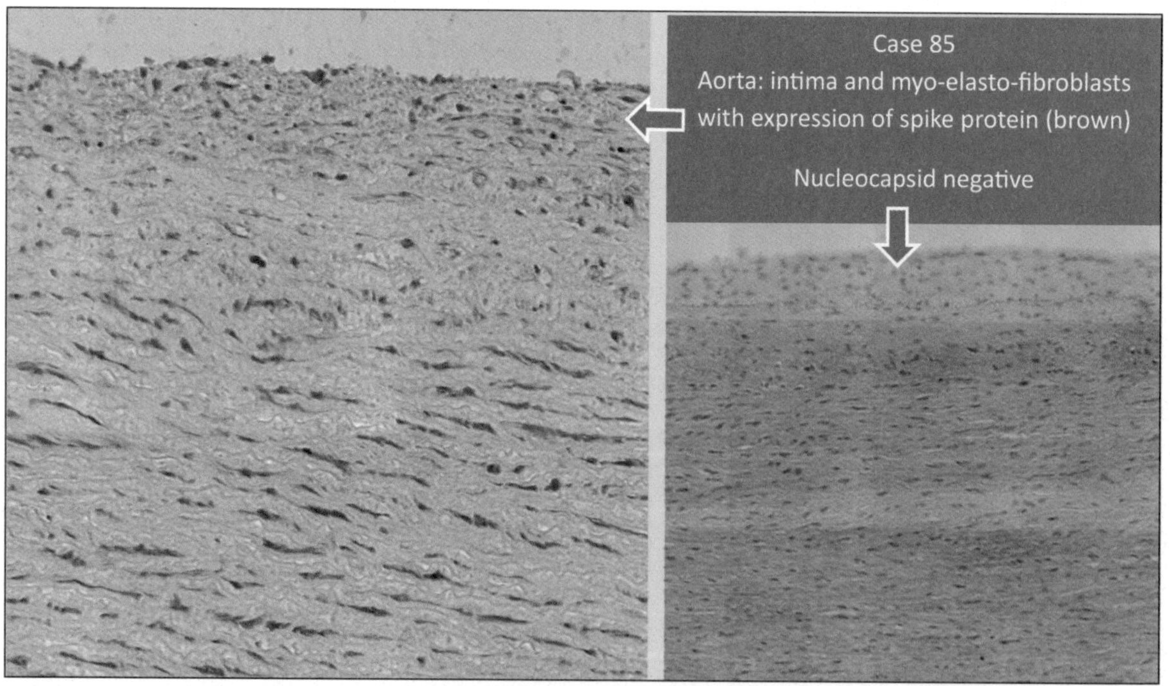

And here you can see the demonstration of spike protein and the negative control for nucleocapsid.

And, unfortunately, many doctors and medical persons state that the myocarditis usually is mild, and it is not of much concern.

HERZ MEDIZIN.

Late effects of myocarditis: many die within 10 years

‹ Für Ärzte und Fachpersonal

Spätfolgen einer Virus-Myokarditis: Viele sterben binnen 10 Jahre

Die 10-Jahres-Sterblichkeit nach einer durch Viren ausgelösten Myokarditis ist hoch. Deutsche Kardiologen haben nun untersucht, welche Patienten besonders gefährdet sind.

Von: **Veronika Schlimpert**

14.08.2020

But this study here showed that many of the myocarditis patients die within 10 years, so it's not something that should be taken easy.

Because, as an analysis by German cardiologists makes clear, the prognosis for viral myocarditis is generally quite unfavorable: almost 40% of the affected patients died within the next ten years, most of them from a cardiac cause, one in ten suffered sudden cardiac death. [Greulich, Simon, et al. "Predictors of Mortality in Patients With Biopsy-Proven Viral Myocarditis: 10-Year Outcome Data." *Journal of the American Heart Association*, ahajournals.org, 13 Aug. 2020, www.ahajournals.org/doi/10.1161/JAHA.119.015351.]

"Cardiac failure"

31 Cases: initially classified as natural death due to

- cardiac failure
- rhythmogenic heart failure
- myocardial infarct,
- cardial decompensation
- cardiomyopathy

Second opinion:

- 15 cases of peri-myocarditis
- 16 cases of microangiopathy with stenosis and dissection

In our cases that were diagnosed as a natural cause (previous graphic), we had **31 cases** where the primary pathologist or coroner said it was cardiac failure, they called it cardiac failure, arrhythmogenic heart failure, myocardial infarct, cardial decompensation, myocardiopathia.

But, of our second opinions, in **15 of these 31** cases there was a clear cut **peri-myocarditis**.

And in 60 cases we had this, what I showed you, microangiopathy with stenosis and dissection. So, this is one of the main causes of death after the vaccination.

Lung and Alveolitis

Now we come to the next organ, the lung.

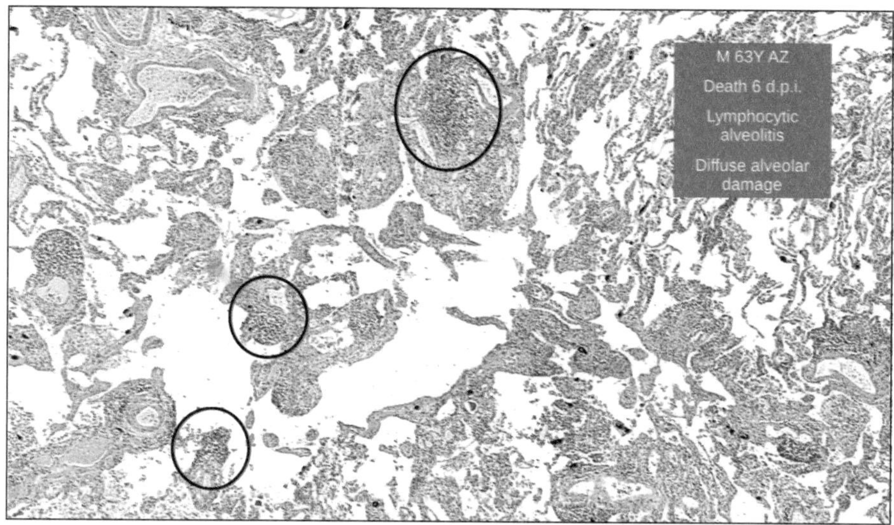

In the lung we have a lymphocytic alveolitis and focal infiltrations of lymphocytes, mainly perivascular and in some areas what we call diffuse alveolar damage.

Lymphatic Organs

Main pathological findings (lymphatic organs)

Lesions of the lymphatic organs – spleen and lymph nodes

- Depletion, activation, pseudolymphomas
- "Onion skin" alveolitis
- Defects of vessel walls with follicular prolaps
- Necrosis, infarction, perisplenitis, spleen rupture
- Central lymph node necrosis
- Unidentified objects within and outside of spleen vessels
- Malignant lymphomas

Now we come to the lymphatic organs, lesions of the lymphatic organs, spleen, and lymph nodes. We have,

1. Depletion, activation, and pseudolymphomas,
2. 'Onion skin' arteriolitis (typical of autoimmune diseases),
3. Defects of the vessel walls with follicular prolapse,

4. Necrosis, infarction, spleen rupture
5. Central lymph node necrosis,
6. Unidentified intravasal and extravasal objects.
7. Malignant lymphomas.

One of the main organs where the vaccination and the spike protein is active is the spleen.

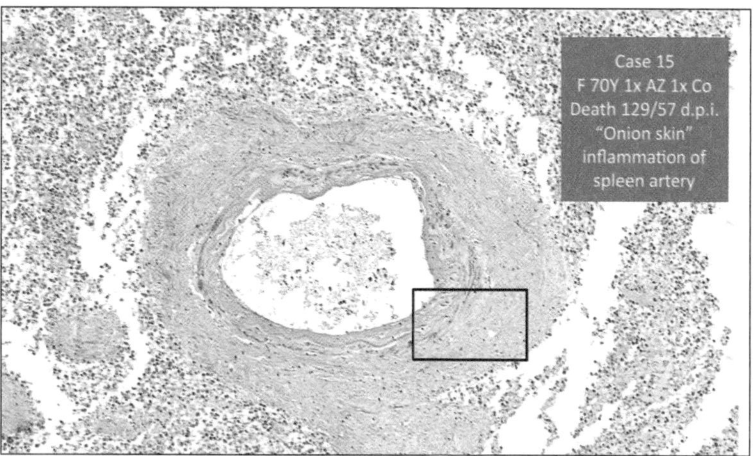

You can see this is what is called the **onion skin-like changes of the vessel walls** in the spleen, which, as I said, is, mainly seen in **autoimmune diseases** like lupus erythematosus.

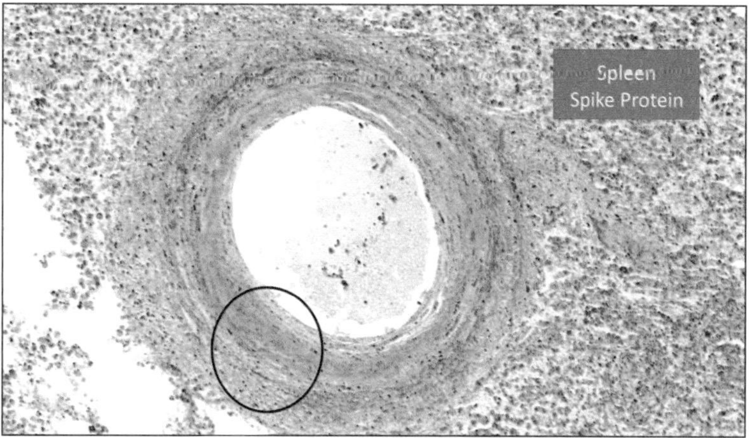

These disturbed small vessels also demonstrate the spike protein.

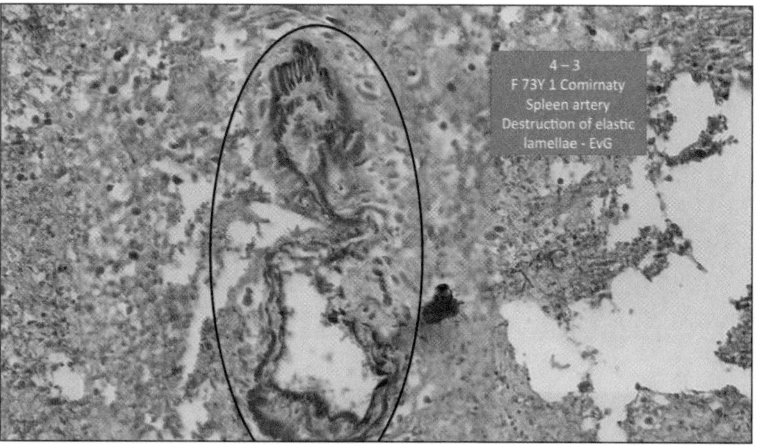

And here (previous image) you can see some completely destroyed vessels. You can see the elastic lamella are completely destroyed.

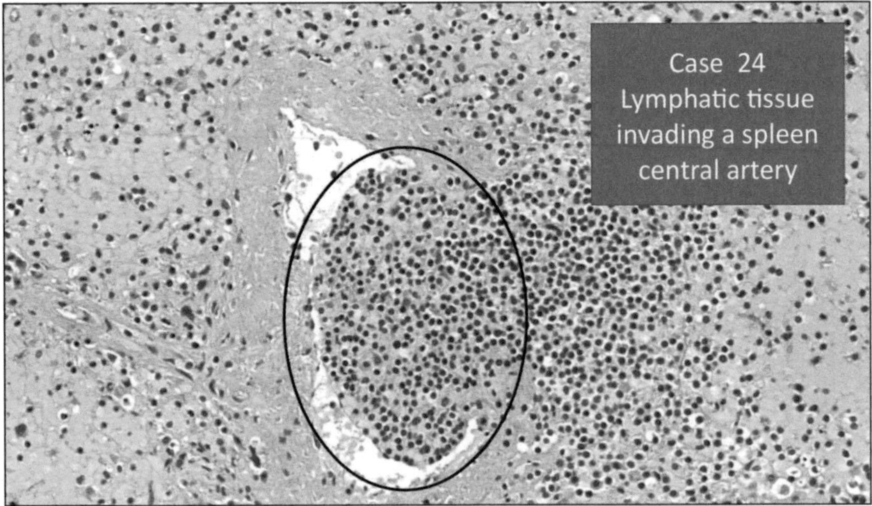

In some cases, the pressure of the proliferating lymphocytic follicles makes a disruption into these vessel walls.

Lymphocytic infiltration outside the heart muscle and lung

Thyroid Gland	2 Cases (5, 13)
Salivary Glands	2 Cases (7, 15)
Aorta, larger Vessels	7 Cases (3, 6, 7, 9, 13, 19, 20)
Skin	2 Cases (4, 20)
Liver /NASH	1 Case (8)
Kidney	3 Cases (12, 14, 19)
Lymphocytic Pyelonephritis/ Nephritis/Ureteritis	1 Case (19)
Testis	1 Case (10)

Autoimmunity and lymphocytic infiltration outside the heart muscle and lung

Autoimmune diseases observed after COVID-19 Vaccination

• Hashimoto thyroiditis	5 cases
• Sjoegren's syndrome	4 cases
• Atypical lichen planus with vasculitis	4 cases

And here (above) you can see where we find these lymphocytic infiltrates, thyroid, gland, salivary, aorta, skin, liver, kidney, testis, dura. And in these cases, we definitely observed autoimmune disease, Hashimoto's, thyroiditis in five cases, Sjogren's syndrome in four cases and atypical lichen planus with vasculitis in four cases.

Now, of course, this could be preexisting autoimmune diseases, but, apparently, they were not known before the deaths and before vaccination; and they might have been aggravated.

Now we come to the brain.

Brain

Main Pathological Findings: Brain

- Transfection-associated encephalitis
- Lymphocytic Infiltration and focal destruction of intracerebral and arachnoidal blood vessels
- Microthrombi
- Macro- and microaneurysms
- Subarachnoidal haemorrhage with and without aneurysms
- Focal lymphocytic infiltration of the dura mater
- Amyloid deposits

In the order of our main pathological findings, we have in a few cases,

1. Transfection associated encephalitis,
2. Lymphocytic infiltration and focal destruction of intracerebral and subarachnoid blood vessels,
3. Microthrombi,
4. Macro and microaneurysms,
5. Subarachnoid hemorrhage with and without aneurysms,
6. Focal lymphocytic infiltration in the dura matter,
7. Amyloid deposits.

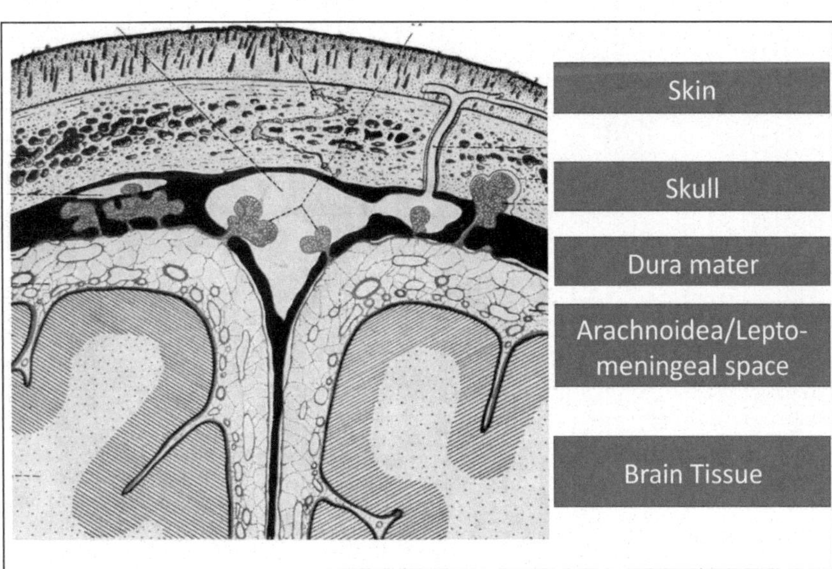

And this is the scheme where we have the lesions. We have the lesions, as I showed you, in the dura mater, where we have lymphocytic infiltrations. We have the arachnoidea. And we have the brain tissue. And, finally, I will show you this case report of this man to illustrate the findings in the brain. **(See SR 5.)**

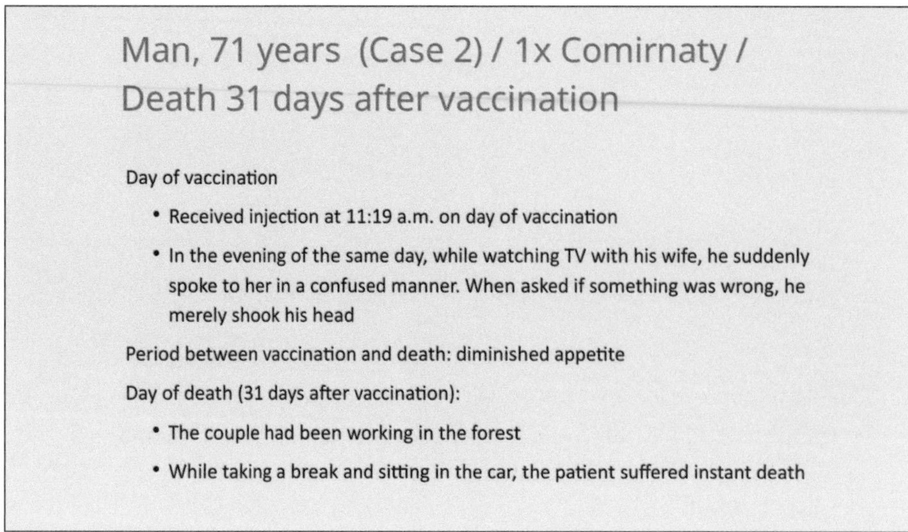

This is a man, 71-year-old. He was vaccinated one time with Comirnaty and died 31 days after this vaccination. But it's very interesting that on the day he was . . . Oh, excuse me. This is in German. I have to translate it.

On the day of the vaccination, he fainted shortly and was disturbed and could not speak, but he recovered. So probably had, he had some lesion in the brain; but then he had recovered, and he worked in the forest and to cut down trees.

And when he made a break and sat in the car, suddenly he died unexpectedly.

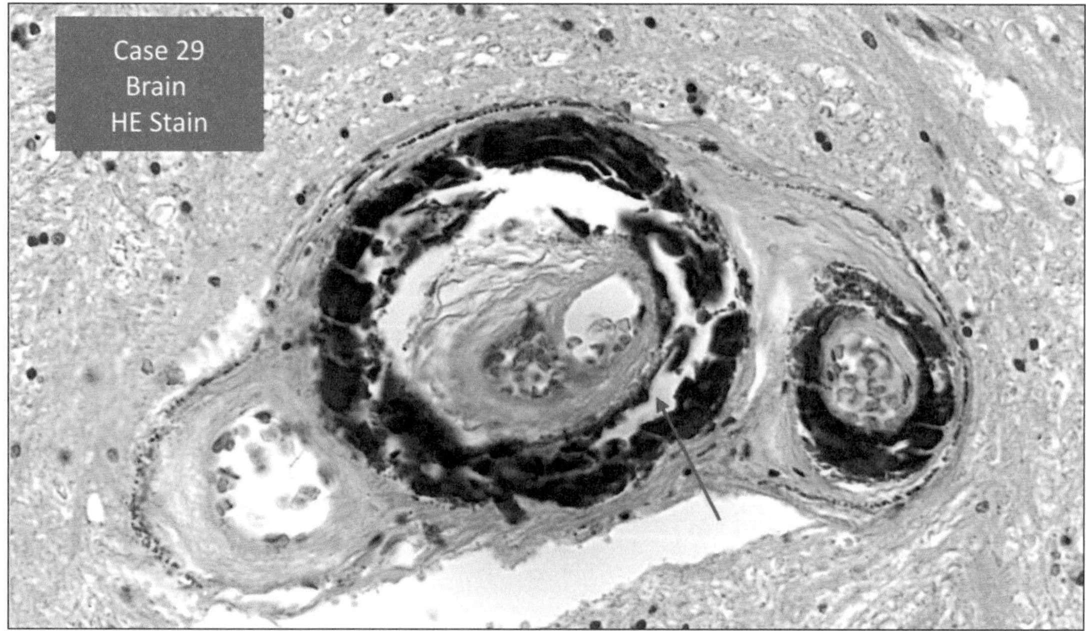

And this is what we found in the brain. This is a small artery, as you can see. There's a dissection in the media *(red arrow)*, and these elastic lamellae are destroyed, and they are strangely black stained by H&E.

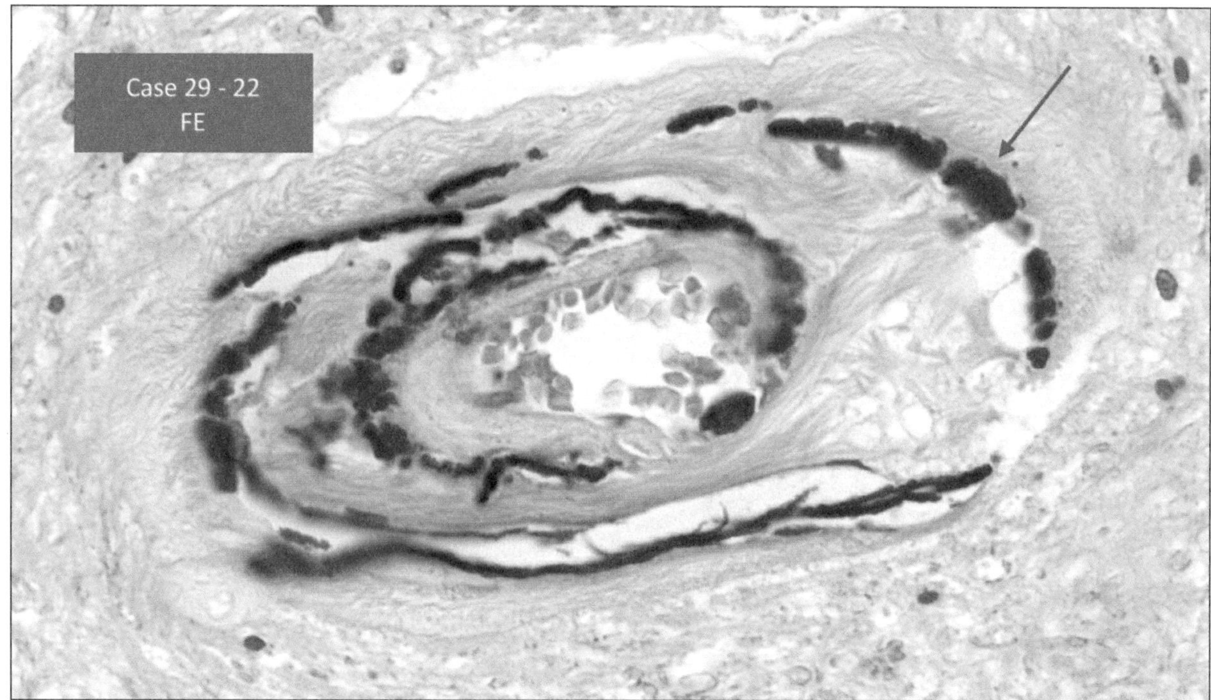

And as I have shown you before, this is all hemosiderin (*red arrow),* so he had probably on the day that he was vaccinated this bleeding inside the small vessels in the brain but recovered; and, days later, he had the fatal bleeding in the brain and died.

Pituitary Gland (See SR 10.)

So, finally, I just want to make one remark to the pituitary gland, which is situated near the brain, as you know, and has a very important function for the endocrine system, the same as the heart for the for the blood circulation. Now I found some very disturbing fact that of these 75 cases, only in eight cases the pituitary gland was examined by histology.

Contents lists available at ScienceDirect

Legal Medicine

journal homepage: www.elsevier.com/locate/legalmed

ELSEVIER

LEGAL MEDICINE

Check for updates

Post-mortem histopathology of pituitary and adrenals of COVID-19 patients

Antonia Fitzek[a,1], Moritz Gerling[a,1], Klaus Püschel[a], Wolfgang Saeger[b,*]

[a] Institute of Legal Medicine of the University Medical Center Hamburg-Eppendorf, Germany
[b] Institutes of Pathology and Neuropathology of the University Medical Center Hamburg-Eppendorf, Germany

And also, when you look into the literature, in practically all of the reports that I read, the pituitary glands had not been examined, neither in the cases of COVID-19 infection, nor in the cases after vaccination. Now, the only public publication on this the subject was by Fitzek et al. And they found that, in some cases, there was necrosis of the pituitary.

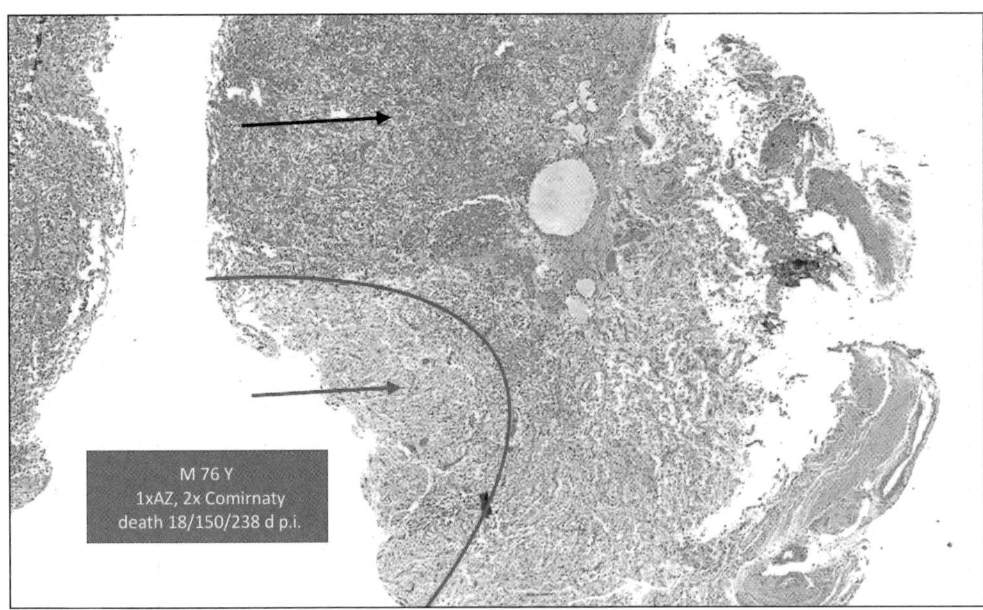

In one case, we could also find partial necrosis *(red demarcation and arrow)*. This is an anterior lobe of the pituitary gland. Here it's intact *(black arrow)*, and here you can see the necrosis. So, this is definitely one thing that we should go further into.

So, I tried to show you our results, our slides and our findings. Of course, this is an ongoing study, and we don't have the final answers to all of what we found or what I showed you. And I will be glad to discuss all of this with you.

Thank you !

Danke !

Dr. Burkhardt passed away the day before his 80th birthday. We are fortunate that Dr. Burkhardt was able to add immeasurably to our understanding of the many manifestations of harms from spike-producing drugs.

Thank you, dear Arne, and rest in peace

We have lost a brilliant colleague, a good and kind man, and a true warrior . . .

but you have left us an invaluable legacy with your work, and with your inspiring example of true scholarship.

You will be in our hearts forever.

Dear Doctors for Covid Ethics and friends:

With heavy hearts we announce that, on May 30, 2023, Prof. Dr. Arne Burkhardt passed away.

Arne was an accomplished pathologist who in 2021 came out of his well-earned retirement to investigate the injury and death caused by the gene-based COVID vaccines. Arne's tireless and expert work showed irrefutable evidence of vaccine-induced inflammation in blood vessels and in all major organs.

He presented his findings at our Interdisciplinary Symposiums II and III.

June 22, 2023
"It's Human Responsibility"—In Memory of Prof. Dr. Arne Burkhardt

Arne Burkhardt and Walter Lang at work in Arne's lab, Reutlingen

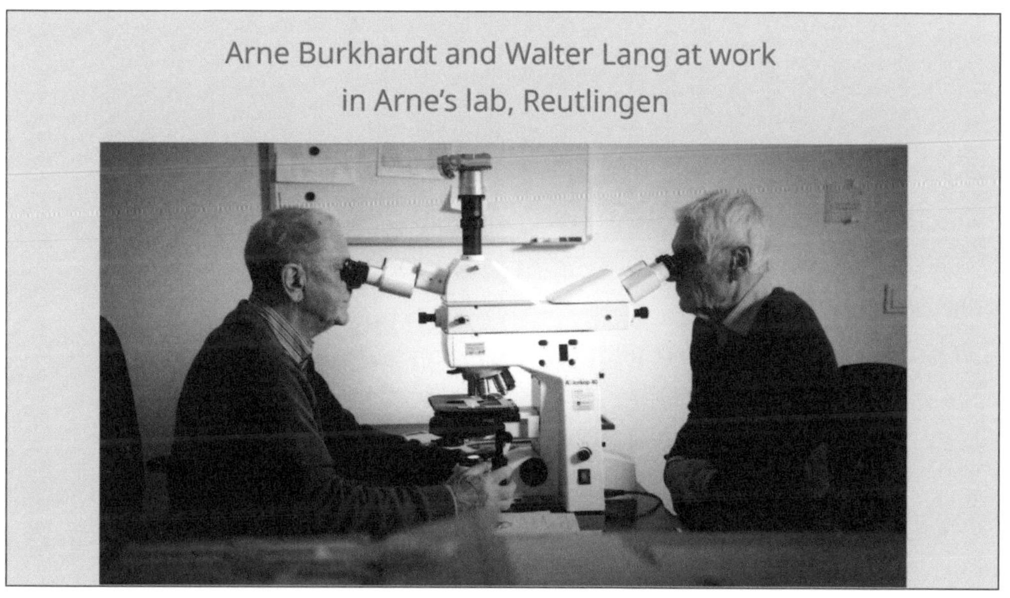

Dear Doctors for Covid Ethics subscribers and friends:

It is with heavy hearts and great difficulty that we announce the passing of Arne Burkhardt. On May 30, 2023, Prof. Dr. Arne Burkhardt passed away.
Arne was an accomplished pathologist who, in 2021, came out of his well-earned retirement to investigate the injury and death caused by the gene-based COVID vaccines. Arne's tireless and expert work showed irrefutable evidence of vaccine-induced inflammation in blood vessels and all major organs. He presented his findings at the Gold Standard Covid Science in Practice: An Interdisciplinary Symposium II and III, organized by Doctors For Covid Ethics.

This tribute to Dr. Burkhardt contains three videos lasting 29:09, 41:14, and 1:12 minutes.
(https://doctors4covidethics.org/its-human-responsibility-in-memory-of-prof-dr-arne-burkhardt/)

Supplemental Resources

Links to these instructional videos are provided for background purposes. None are specific to COVID-related drugs or CoVax (COVID Vaccine-related) Diseases.

Part I: Pathological Processes

SR 1 Central Dogma in the Life Sciences
https://www.youtube.com/watch?v=7Hk9jct2ozY

SR 2 Immune System Overview
https://www.youtube.com/watch?v=_jBpv9fYSU4

SR 3 Inflammation after Injury
https://www.youtube.com/watch?v=9bvMv5dQ7RU

SR 4 Inflammatory Process (-itis)
https://www.youtube.com/watch?v=yIMz9pkT9xQ

SR 5 Cytotoxicity
https://www.youtube.com/watch?v=rDEduT62Awc

SR 6 Amyloidosis
https://www.youtube.com/watch?v=kOv827pMBi8

Part II: Specific Organs

SR 7 Vascular Disease and Vasculitis (Endotheliitis)
https://www.youtube.com/watch?v=D9h6-LPySPI

SR 8 Aneurysm
https://www.youtube.com/watch?v=A5MEe0lb0YA

SR 9 Cardiac Scarring
https://www.youtube.com/watch?v=w3ZqeKcoxlw

SR 10 Pituitary and Hormones
https://youtu.be/3Lt9I5LrWZw?feature=shared

Twenty-Eight: "Pfizer's Clinical Trial 'Process 2' COVID Vaccine Recipients Suffered 2.4X the Adverse Events of Placebo Recipients; 'Process 2' Vials Were Contaminated with DNA Plasmids."

—Chris Flowers, MD; Erika Delph, Ed Clark; and Team 3 WarRoom/DailyClout Pfizer Documents Investigators

Process 2 was hidden all along in Pfizer's COVID "vaccine" clinical trial, and the WarRoom/DailyClout investigators' findings about it are mind-blowing. The Food and Drug Administration (FDA) knew that the Process 2 subjects had very high levels of adverse events, but there is no evidence that the agency acted on those alarming findings.

This Process 2 "trial within a trial" was discovered earlier this year in the tens of thousands of the Pfizer documents released by the FDA. The DailyClout teams were reviewing the expert testimony of Josh Guetzkow, Ph.D. of Hebrew University, Jerusalem, used in a lawsuit in the United Kingdom, and started looking for evidence of the approximate 250 subjects who may have taken part in an experiment on behalf of the European Medicines Agency (EMA). Further reporting by Dr. Guetzkow and Retsef Levi, in a *British Medical Journal* (*BMJ*) Rapid Response letter to the Editor, highlighted that the amended Pfizer clinical protocol (C4591001) in October 2020 included references to Processes 1 and 2 as well as to a trial subset (a trial within a trial) and that no public reporting of results was available: "The protocol amendment states that 'each lot of 'Process 2'-manufactured BNT162b2 would be administered to approximately 250 participants 16 to 55 years of age' with comparative immunogenicity and safety analyses conducted with 250 randomly selected 'Process 1' batch recipients. To the best of our knowledge, there is no publicly available report on this comparison of 'Process 1' versus 'Process 2' doses."

Eagle-eyed WarRoom/DailyClout volunteer, pharmacist Erika Delph, noted an anomaly in randomization numbers that matched the number and dates of this appended "trial within a trial." Our data team and medical experts analyzed the data. What they found is shocking: 502 subjects were in a Pfizer COVID-19 vaccine sub-trial and received a drug contaminated with unacceptably high levels of DNA plasmids. It may be tempting to write this off as an accident; however, the documentation notes show that Pfizer knew that it was giving 252 unfortunate trial subjects a completely different injection than that for which they had signed up. This fact alone violates the Nuremberg Code (1947), which states that it is unlawful to run human experiments without full informed consent.

What Is Process 2, and Why All the Fuss?

The terms "Process 1" and "Process 2" were mentioned by Pfizer in the different iterations of the clinical trial protocol for this novel drug platform that would be used worldwide. The "process" refers to the way the "vaccine" was manufactured.

The original manufacturing process of BNT162b2, Pfizer's COVID "vaccine," for the clinical trial used a messenger RNA duplication (amplification) technique known as PCR (polymerase chain reaction)—essentially like a photocopier, multiplying/cloning the original mRNA. This is known as "Process 1."

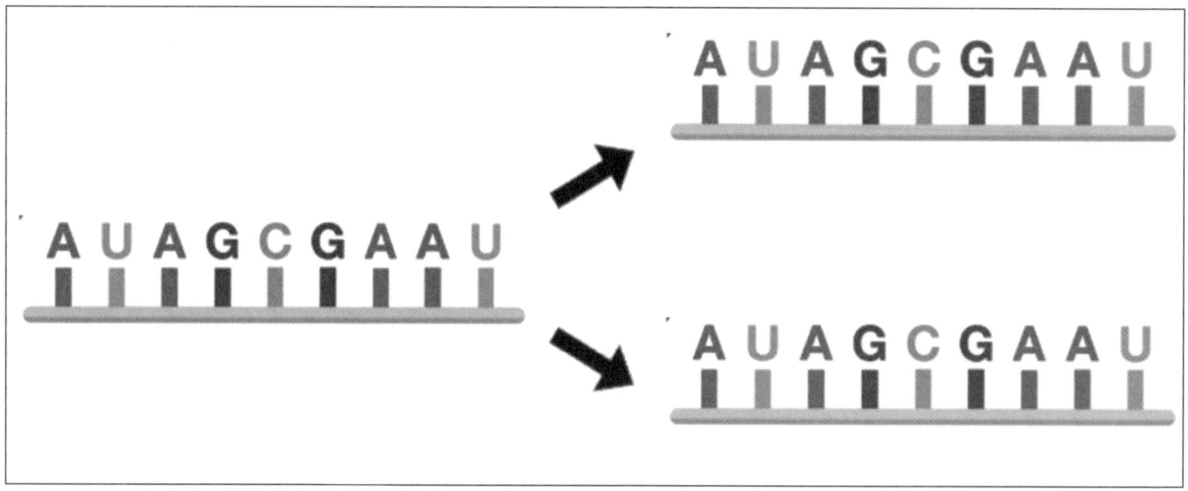

Commercially, this type of process is expensive and would have to be significantly ramped up to provide doses for the whole world. The commercial scaling of the product used a proven way of mass production using e. coli bacteria. This mass production technique is "Process 2." The thorny issue was that "Process 2" used a completely different manufacturing process than that used for the product in the clinical trial (Process 1), and the Emergency Use Authorization (EUA) for the "vaccine" was granted based on Process 1. Moreover, Process 2 was not compliant with Good Manufacturing Practice (GMP). Note the FOIAed national contracts with Pfizer from South Africa and Albania.

Revelations from gene sequencing of the residual product in the vials, produced using Process 2 by Kevin McKernan, confirm other groups' reporting of the determination that there is marked contamination of the modified mRNA with high levels of DNA plasmid fragments.

This contamination is attributed to the use of e-coli during manufacture. These bacteria are naturally found in human gut bacteria and are a regular means of mass-producing mRNA sequences. The required gene is inserted into a ring of DNA, and the bacteria continually replicates these plasmids.

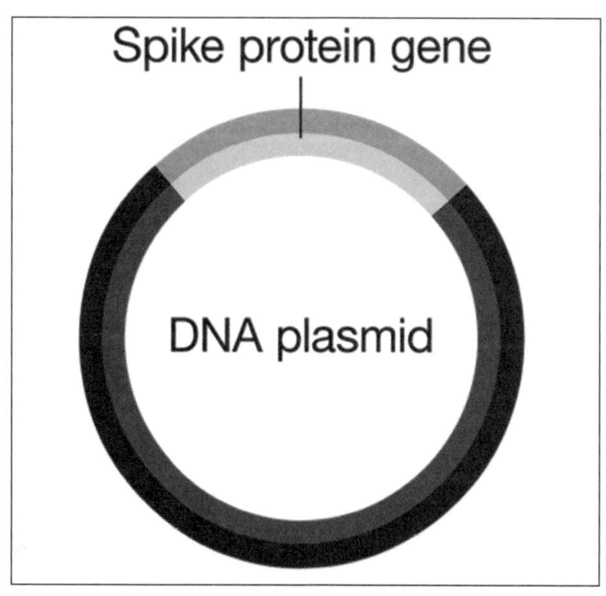

The plasmids produced by this process are purified using enzymes (DNAase) and have a regulated UPPER LIMIT in the end product due to the theoretical concerns about the incorporation of this DNA into the human host genome.

Despite the active ingredient being identical in Process 2 compared with Process 1, the European Medicines Agency (EMA) had noted the level of contamination by DNA plasmids and were concerned because it was well above previously published safety levels. EMA was sufficiently concerned to ask Pfizer and the FDA to incorporate the new process to the end of the clinical trial using around 250 subjects.

What Have the WarRoom/DailyClout Volunteers Found That Was Hiding in Plain Sight?

We have identified a distinct cohort, due to anomalous randomization numbers that otherwise made no sense, compared with the sequences used during the main part of the clinical trial. We have also identified the anomalous batch numbers that contain Process 2 developed products. These 502 subjects were tested at four sites in the United States, 250 of whom acted as placebo subjects and the other 252 who received the Process 2 product. (*Thank you to OpenVAET for providing data for the WarRoom/DailyClout volunteers to validate their findings about the 252 Process 2 subjects.*)

- **Site, Vaccine Category, Subject count:**
  - **1133 Hollywood, Florida: BNT162b2 (60); Placebo (60)**
  - **1135 Anaheim, California: BNT162b2 (64); Placebo (63)**
  - **1146 Raritan, New Jersey: BNT162b2 (64); Placebo (64)**
  - **1170 Dallas, Texas: BNT162b2 (64); Placebo (63)**
- **Randomization:**
  - **Unique sequence numbers: 400000-499999**
  - **Start: 19 Oct 2020; thru 02 Nov 2020**
    - **Site 1146: 19 Oct → 29 Oct 2020**
    - **Site 1170: 20 Oct → 30 Oct 2020**
    - **Site 1135: 21 Oct → 27 Oct 2020**
    - **Site 1133: 23 Oct → 2 Nov 20200**

The product had a unique Vendor Lot No. "**EE8493Z**," identified in Pfizer Batch/Lot inventory document (https://www.phmpt.org/wp-content/uploads/2022/06/125742_S1_M5_5351_c4591001-interim-mth6-patient-batches.pdf).

The cohort also was separated from the subjects receiving Process 1, as well as from the unblinded segment of the trial after the EUA was granted and where virtually all the placebo group from the main trial received the "vaccine."

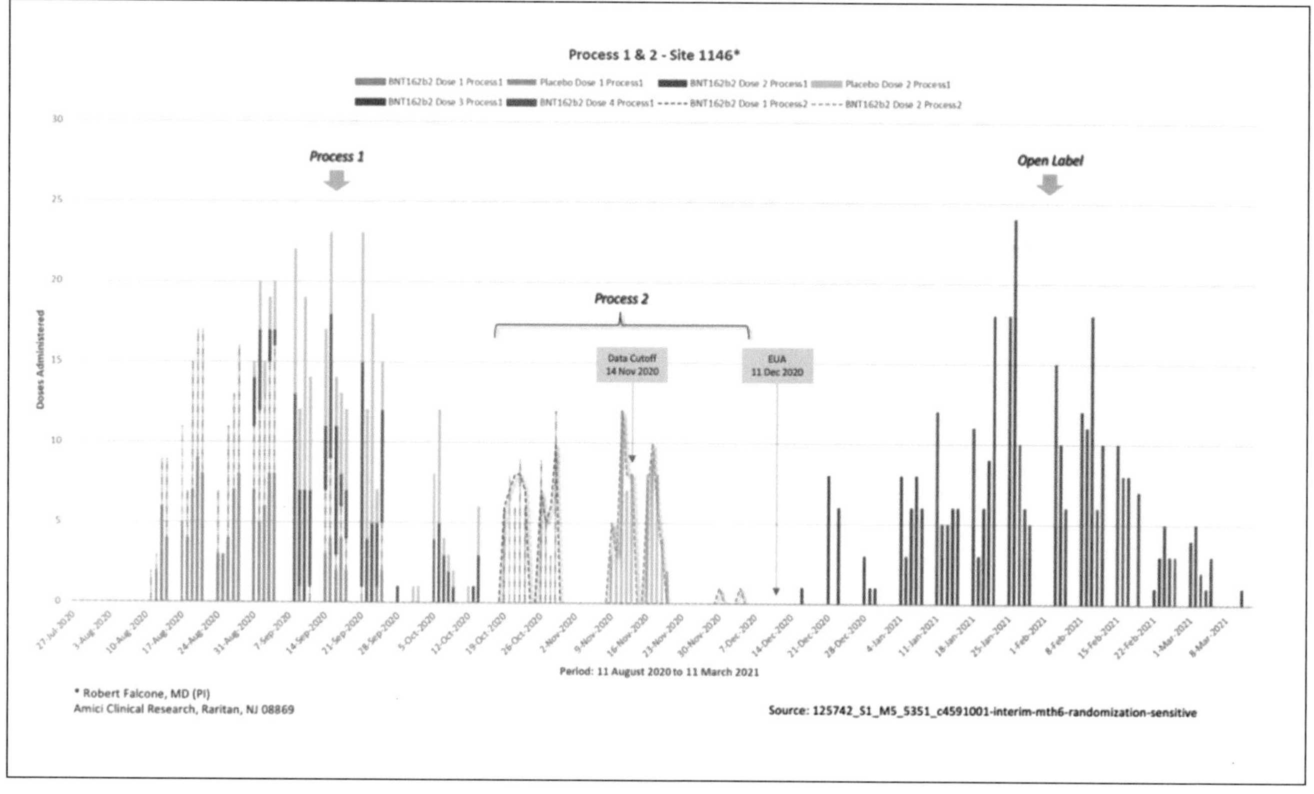

Zooming into the Process 2 data, the separation of subjects is easier to see.

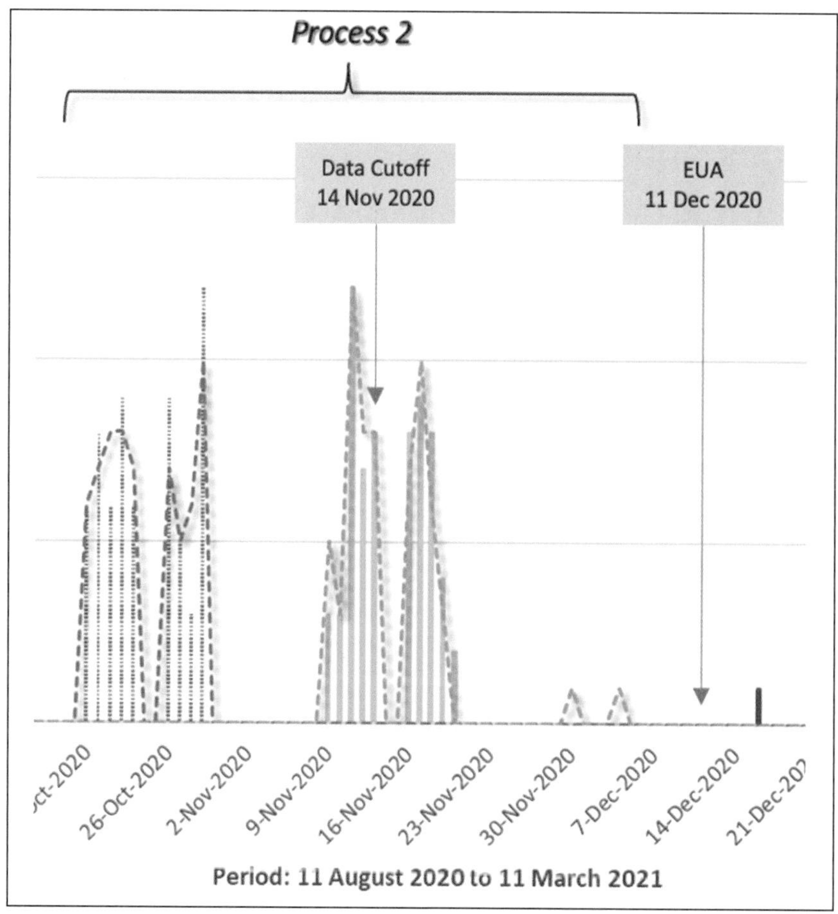

There was a significant difference in the number of adverse events in the Process 2 group of test subjects, which should have rung alarm bells in the regulator's heads as it was so much worse than the significant adverse events (AEs) found in the majority of the clinical trials.

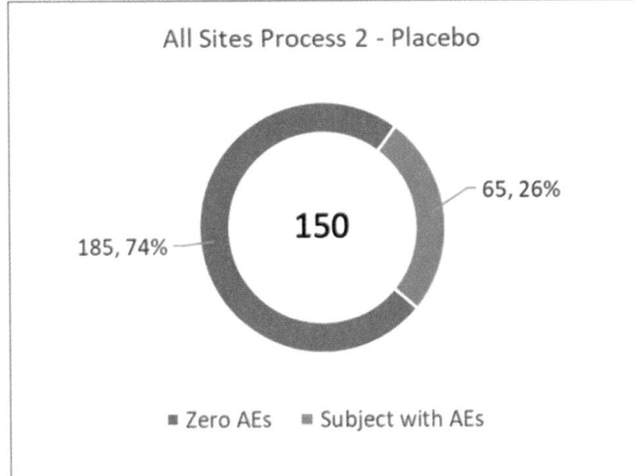

Of the 250 subjects in the placebo group, 185 had no adverse events and 65 had adverse events.
These were mainly minor with pain or swelling at the injection site.

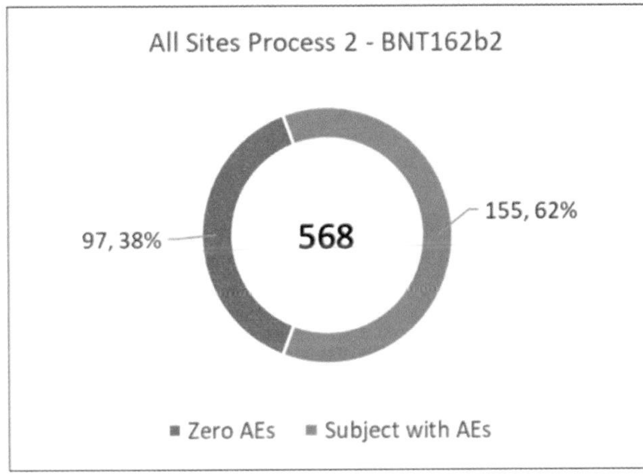

Of the 252 subjects in the Process 2 product arm, 97 had no adverse events, and 155 had adverse events.
These were mainly minor with pain or swelling at the injection site.

Although the adverse events were minor, there is such a big difference between the placebo and treatment arms, 65 versus 155 or 2.4 times more, that further scrutiny would be expected to determine the cause, since the NEW PROCESS was about to be used for the worldwide roll-out.

How Do these Findings Compare to Those Already Found and Reported on By Josh Guetzkow?

Reporting out of Europe was based on testimony about an 'emergency batch,' EJ0553, that was used in 11 subjects at four different sites than those in the mini clinical trial (sites 1001, 1002, 1003 and 1007). In the Pfizer lot number document, Process 2 product also has a "Z" designation that may have been used to identify product made with the new process. For the U.S., no product manufactured outside of the country was supposed to be used, but evidence from the Australian regulatory agency, the Therapeutic Goods Administration (TGA) FOI 3659 document 4, titled, "BNT162b2 (PF-07302048) Comparability Report for PPQ Drug Product Lots," the Lot EJ0553Z was manufactured in Puurs, Belgium, and released as an "emergency supply."

Our novel findings are based on empirical evidence found in the Pfizer documents, already released.

As a result, the *different adverse events profiles* demonstrated that product from Process 2 was different from product from Process 1. With that safety signal, the FDA should have taken note and determined that the clinical trial would need repeating, as it is a different product with a different safety profile.

Conclusions

Process 2 should have been the subject of a separate clinical trial due to the safety signals from the small number of subjects tested at the end of the clinical trial before the EUA was approved. The DNA plasmid fragment contamination found was multiple times more than the maximum allowed by the EMA.

Data found in the Pfizer documents show the 502 subjects who made up the additional trial within a trial all having a marked safety signal due to higher adverse events.

The process tested and approved was never publicly rolled out and given to the population of the world. Instead, the public only received the DNA plasmid tainted Process 2 product.

Twenty-Nine: "In Early 2021, Pfizer Documented Significant Harms and Deaths Following Vaccination with Its mRNA COVID Vaccine. The FDA Did Not Inform the Public."

—Lora Hammill BSMT, ASCP; David Shaw; Chris Flowers MD; Loree Britt; Joseph Gehrett MD; Barbara Gehrett MD; Michelle Cibelli RN, BSN; Margaret Turoski, RPh; Cassie B. Papillon; and Tony Damian

Introduction

"5.3.6 Cumulative Analysis of Post-Authorization Adverse Event Reports of PF-07302048 (BNT162b2) Received Through 28-Feb-2021," Pfizer's post-marketing report, a required U.S. Food and Drug Administration (FDA) compliance document, is one of the ways in which the FDA assessed patients' risks associated with Pfizer's COVID-19 vaccine, BNT162b2. The data within the Pfizer's 5.3.6 analysis are not indicative of a safe-for-humans biologic. (Worldwide Safety Pfizer. "5.3.6 Cumulative Analysis of POST-AUTHORIZATION ADVERSE Event Reports of PF-07302048 (BNT162B2) RECEIVED THROUGH 28-FEB-2021". *Public Health and Medical Professionals for Transparency*, 1 Apr. 2022, www.phmpt.org/wp-content/uploads/2022/04/reissue_5.3.6-postmarketing-experience.pdf.)

This Post-Marketing Experience (PME) Team's summary analysis describes adverse events (AEs) with percentages representing proportions of case reports received during Pfizer's BNT162b2 post-marketing period. Percentages should not be taken as incidence occurrence rates as these are observational data, not clinical trial data. The Pfizer post-marketing report is built from case submissions to Pfizer in the first 90 days, starting on December 1, 2020, of its vaccine being publicly available. ***Within this short time frame, there were 1,223 deaths reported.*** Categories within Pfizer's report have a combination of medically confirmed and non-medically confirmed adverse events (AEs).

An European Union (EU) Periodic Safety Update Report (PSUR #1), covering December 19, 2020, through June 18, 2021, confirms the PME Team summary showing alarming safety signals caused by the Pfizer COVID-19 vaccine. PSUR #3 was released August 18, 2022, covering December 19, 2021, to June 18, 2022, which as discussed below also confirms the findings of the PME Team summary. As of early August 2023, no publicly available follow-up reporting to FDA regulators regarding Pfizer's post-marketing report exists.

The PME Team will state arguments made by Pfizer within its 5.3.6 post-marketing report followed by the team's conclusions and key findings.

The 90-day, post-vaccine-rollout 5.3.6 post-marketing experience report, required by the FDA from Pfizer, is touted as showing that the BNT162b2 "vaccine" is safe. However, Pfizer's own data do not back up that claim.

There are multiple Serious Adverse Events (SAEs) reported across all age groups. Age groups were divided as follows: Adult = 18 to 64 years; Elderly = greater than or equal to 65 years; child = two (2) to 11 years;

Adolescent = 12 to less than 18 years (defined in 5.3.6) [https://www.phmpt.org/wp-content/uploads/2022/04/reissue_5.3.6-postmarketing-experience.pdf, p. 16]. Many of the adverse events occurred within days of receiving the injection and include ***permanent harms and fatalities***.

- It is important to note that the adverse events in the 5.3.6 document were reported to Pfizer for **only a 90-day period** starting on December 1, 2020. There were **1,223 deaths**, and Pfizer's concluded, "This cumulative case review **does not raise new safety issues**." (Emphasis added.) That conclusion was repeated 15 times within the document. [pp. 17, 18, 19, 20, 21, 22, 23, and 24]
- Pfizer states, "Reports are submitted voluntarily, and the **magnitude of underreporting is unknown**." (Emphasis added) [p. 5]
- Pfizer planned to increase personnel for data entry and case reporting to **2,400 additional full-time employees** to handle the large increase of adverse event reports. [p. 6] It is unknown how many persons were already employed for data entry purposes.
- Latency, the period of time between when the drug was given and an adverse event occurred, was often short, typically **zero (0) to four (4) days**. There were **four deaths on the same day of vaccination**.
- Of the 42,086 "cases" (i.e., patients), AEs for women (**29,914 or 71%**) were reported more than three times that of men (9,182 or 21.8%).
- Age demographics show that **54% of the cases were less than or equal to 50 years** old; the highest number of cases were in the age bracket of 31- to 50 years old. The range of years among the age brackets was variable. For example: less than 17 years old (a 17-year range), 18 to 30 years old (a 13-year range), 31 to 50 years old (a 20-year range), 51 to 64 years old (a 14-year range), 65 to 74 years old (a 10-year range), and greater than or equal to 75 years old.
- **One hundred and seventy-five (175) cases were under 17 years of age**. Unknown dosages were given to children under the age of 12 as there was **no authorization for children under 12** during this period. Harms to children included **Bell's palsy in a one-year-old, stroke in a seven-year-old,** and **renal failure in an infant less than 23 months of age**.
- The Pregnancy category reveals **56 fetuses and infants died. There were 54 pregnancy cases in which the baby was not carried to a live birth**. It should be noted that **BNT162b2 was not approved for use in pregnancy or during lactation at the time of the post-marketing data collection.**
- **Full informed consent** was **NOT** provided. Before the public rollout, the FDA and similar agencies in other countries knew about adverse events related to Pfizer's vaccine and yet did not provide that information to the general public. (https://www.phmpt.org/wp-content/uploads/2022/04/125742_S1_M5_5351_c4591001-interim-mth6-adverse-events-sensitive.pdf, https://www.phmpt.org/wp-content/uploads/2022/05/125742_S1_M5_5351_bnt162–01-interim3-adverse-events.pdf, https://pdata0916.s3.us-east-2.amazonaws.com/pdocs/060122/125742_S1_M5_5351_c4591001-fa-interim-adverse-events.pdf, and https://pdata0916.s3.us-east-2.amazonaws.com/pdocs/070122/125742_S1_M5_5351_c4591001-interim-mth6-adverse-events.zip)

- Outcomes were separated into the following categories: Unknown, **Recovered/Recovering** (inconsistent categories combined), Not Recovered at time of report, **Recovered with sequelae** (meaning, there is a pathological condition as a result of the initial adverse event), and **Fatal**.
- Pfizer reported that surveillance would continue, yet no **information on subsequent surveillance has been released to U.S. regulatory authorities. Periodic Safety Update Report (PSUR) #1 and #3 have been PARTIALLY released to the European Union.** A brief summary of the PSUR reports and reference links appear in this summary.

Safety: Fatalities and Lack of Informed Consent

Argument

This PME summary is a comprehensive review of Pfizer document "5.3.6 CUMULATIVE ANALYSIS OF POST-AUTHORIZATION ADVERSE EVENT REPORTS OF PF-07302048 (BNT162B2) RECEIVED THROUGH 28-FEB-2021." (https://www.phmpt.org/wp-content/uploads/2022/04/reissue_5.3.6-postmarketing-experience.pdf) The Pfizer report findings represent adverse events (AEs) submitted voluntarily to Pfizer's safety database from various sources worldwide, including medical providers and clinical studies. In just three months, ***42,086 case (or patient) reports*** were submitted to Pfizer representing ***158,895 adverse events*** [p. 6] among them. That averages to 3.78 adverse events per case/patient. Adverse events are broken into System Organ Classes (SOC) with each System Organ Class containing conditions found in the field.

Post-marketing represents the results of the first 90 days following the rollout of a drug to the public. Although not a scientific data set, this Pfizer analysis includes critical harms, reported to Pfizer by providers in the field, that were not relayed to the public until the post-marketing document was released under court order on November 17, 2021, and then again in an unredacted version on April 1, 2022. (https://www.phmpt.org/wp-content/uploads/2021/11/5.3.6-postmarketing-experience.pdf and https://www.phmpt.org/wp-content/uploads/2022/04/reissue_5.3.6-postmarketing-experience.pdf)

Review Team Conclusion

Key Finding: **Full informed consent was not provided**. Informed consent must list all potential harms, one of which is death.

"5.3.6 Cumulative Analysis of Post-Authorization Adverse Events Reports of PF-07302048 (BNT162b2) Received Through 28-FEB-2021" does not represent a safety report. Pfizer repeatedly concluded "no new safety issues" despite an analysis showing significant injuries and even fatalities.

With different risks and rates of risk noted, the singular issue at hand is informed consent. By the end of February 2021, the adverse events reported across several System Organ Classes show serious damages including deaths. Pfizer detected harms through its data collection that were not included in the original December 11, 2020, emergency use authorization (EUA) (https://www.fda.gov/media/144959) As such, the detection of these potential harmful side effects necessitates an updated list of potential risks as part of patient consent. The detection of adverse events and subsequent **omission of them in printed or online package inserts and/or fact sheets represents a violation of truth in medical ethics**. Most importantly, the concerning and, in many cases, potentially long-term adverse events listed within the Pfizer post-marketing report do not support assertions of product safety.

Informed consent requires the disclosure of known potential for adverse events on the package inserts. Patients should be given the opportunity to fully understand potential harmful side effects before receiving the drug. (https://www.verywellhealth.com/understanding-informed-consent-2615507)

Sufficient concerns are documented from review of Pfizer's post-marketing analysis to conclude its COVID-19 vaccine (BNT162b2) raises important safety concerns.

Key finding: Mandating the administration of the experimental COVID-19 vaccine without offering full informed consent to the public should be viewed as **unacceptable**. ***The FDA was negligent in its duty to protect the public when it allowed Pfizer to leave its package inserts blank, without written warnings or links to those known warnings, rather than listing the known adverse events found through post-marketing surveillance in its printed or online package inserts and/or fact sheets.*** (https://www.fda.gov/media/72139/download)

Pfizer claimed [p.6] that the reported cases could be used to look for signals, and yet ***Pfizer disregarded the signal of 1,223 deaths in a 90-day period***, during the early rollout of the "vaccine."

***Key Finding*:** Fatalities were seemingly obscured in the analysis by Pfizer spreading the deaths across many different System Organ Classes. The total **reported deaths were 1,223**, yet this important figure is not readily apparent in the post-marketing report.

The fatalities noted during this review are spread across all but three System Organ Classes. Though there were many routes to death, when it is viewed as a singular result, **the fatality rate of cases observed in the post-marketing patient population is 2.9% of the cases reported**. Pfizer analysis Table 2 [pp. 8–9] claimed to identify adverse events equal to or above a two percent threshold. However, Pfizer failed to include fatalities as a separate adverse event category even though it met Pfizer's criterion for equal to or greater than two percent. Deaths would have stood out as a significant issue in the Pfizer report Table 2 [pp. 8–9] if included as its own category.

In Pfizer's own words [p. 6], post-marketing data should be used for "signal detection." ***The Post-Marketing Experience Team authors of this summary suggest that 1,223 deaths and the level of other adverse events present large signals that should have triggered, at minimum, a pause in the vaccine rollout.*** Distribution and injection of Pfizer's mRNA vaccine should be suspended based on known, severe adverse events to date.

After a review with the Post-Marketing Experience Team, several points emerged as both relevant conclusions and possible larger questions moving forward. This summary acknowledges the fact that the data are not a scientific data set. It is unknown what number of injections were administered during the 90-day post-marketing timeframe and, thus, how that relates to the 42,086 cases reported.

Every System Organ Class had fatalities except for Dermatological adverse events, Facial Paralysis adverse events and Musculoskeletal (dealing with muscles, bones, and joints) adverse events.

How do Pfizer and the FDA consider 1,223 deaths—a 2.9% death rate—"safe?"
Who *(which individuals or groups) made the determination of safety?*
Why *does the FDA agree with Pfizer's assessment and continue to allow this drug to be given to patients?*

Female Preponderance of Adverse Events

Argument

Gender is a demographic tracked throughout the analysis. Adverse event female cases number 29,914 (71%), and male cases number 9,182 (21.8%); and the remaining 2,990 cases listed the gender as "unknown." Pfizer does not offer an explanation for this difference.

Review Team Conclusion

Key Finding: The number of adverse events reported for women is more than three times that reported for men. While it is possible that the large number of female adverse events may be due to sampling bias and not related to gender-specific effects, the disparity in the number of events women suffer is evident. ***This large discrepancy is a signal that females are likely at greater risk of an adverse event than males, and the signal should be investigated further.***

The ratio of women to men receiving injections is unknown. It is also unknown whether the persons reported in post-marketing, as receiving injections, were otherwise healthy individuals at the time of inoculation. The adverse events represent reports from the field alone without standardization.

What *does Pfizer know about gender-related risk of adverse events from its COVID vaccine?*
What *was the number of women versus men vaccinated during the timeframe of December 1, 2020, through February 28, 2021?*
Why *did Pfizer and the FDA seemingly ignore the signal indicating that women were much more likely to be harmed by this vaccine?*

Why Were Age Groupings Irregular?

Argument

Pfizer [p. 7] displays the events/age group categories as: less than or equal to 17, 18 to 30, 31 to 50, 51 to 64, 65 to 74, greater than or equal to 75, and unknown.

The age demographic showing the most adverse events is 31 to 50. This segment encompasses 20 years. Other age categories are 14 or fewer years in span except for the under 17-year-old group.

Review Team Conclusion

***Key Finding*:** Irregular age groupings obfuscate relevant risks. Age groups of standard intervals would show age-related side effects if they exist. If there are no age-related effects, the risk of adverse events would be constant at all ages.

Pfizer reports adverse events by age brackets that are not standardized in age range, leading to potential issues in understanding age-related side effects. There are groupings of 13 years, 20 years, 14 years, and 10 years. The bookend ranges are under-17 years and over-75 years. This unbalanced grouping approach obscures possible age-related relationships to adverse events. Most adverse events occurred in the 31- to 50-year age range, but this age range is also the broadest (encompassing 20 years). The large proportion of adverse events in the age group of 31 to 50 years, combined with the larger number of reports involving females, could suggest that women of childbearing age are greatly affected. This age-based analysis cannot be linked to adverse events without clear, high-quality data. As such, statistical conclusions are impossible to make based on the post-marketing analysis.

The number of reported adverse events for patients in their 30s or 40s may be similar or different, but it cannot be determined. If there are adverse events more commonly seen in patients in their 30s, the events would not be detectable due to the irregular interval of a combined age range of 31 to 50.

Ideally, Pfizer would have reported adverse events by standardized age intervals, for instance, five-year or 10-year intervals. It is, thus, more difficult to conclude the potential relationship between reported adverse events and age. There are an additional 6,876 adverse event reports that list age as "Unknown." This unknown set may also skew the data.

Mean age was calculated by Pfizer [p. 7] from an age range of 0.01 to 107 years to yield a mean age of 50.9 years. **If the range were one year to 78 years (life expectancy) the median age would be 39.5 years.** Of the number of cases where the age of the recipient is known, ***54% are less than or equal to 50-years-old.***

How many persons *aged 31 to 40 had adverse events?*
How many women *of childbearing age were negatively affected or permanently harmed?*

Why Did Pfizer Seemingly Hide 97 % of the Outcomes?

Argument

Table 1 is found on page seven of the Pfizer post-marketing report. Pfizer grouped the results into categories of "unknown," "fatal," "not recovered at the time of the report," "recovered with sequelae," and "recovered/recovering." ***Table 1 lists 1,223 fatalities or deaths (2.9% of cases).*** "Recovered/Recovering" is a combined category (19,582 cases). Additional outcome categories include "Recovered with Sequelae" (which includes 520 events), "Not Recovered at the time of report" (11,361 cases), and "Unknown" (9,400 cases).

Review Team Conclusion

***Key Finding*:** The proportion of cases with permanent harms (death) is 2.9 percent. The true proportion of harms that may become chronic conditions is unknown.

Table 1 data [p. 7] are important as they describe deaths in almost three percent of patients.

- "Not Recovered at the time of report" and "Recovered with Sequelae" were 11,361 cases and 520 cases respectively of the 42,086 total cases, or 27%.
- By February 28, 2021, 27% of case reports from a 90-day period were not resolved.
- "Recovered/Recovering" is 19,582 or 47% of cases.
- The "Unknown" population for outcomes is 9,400, which represents 22% of case reports.

The category of "Recovered/Recovering" involves two different outcomes since "Recovering," as a term, does not represent "Recovered." However, they are combined as a single outcome in the Pfizer report. **It is**

unknown if the "recovering" group is a small or large proportion of that 47%. Proportions of "unknown" cases (22%) and "recovering" cases (47%) may also represent unresolved side effects. The listing of outcomes in Table 1 represents harms that are permanent, temporary, or unknown.

"Recovered with sequelae" means that the patient may have recovered from the original ailment but now has a different, lingering health problem that was not present prior to being injected with Pfizer's drug. When totaled, the "unknown," "not recovered at the time of the report," "recovered with sequelae," and "recovered/recovering," relevant cases come to 97%.

We do NOT know the outcome of almost 97% or 40,863 of the total 42,086 relevant cases.
We do know 2.9% died.

Why did the FDA attempt to conceal this data, along with other Pfizer clinical trial data, for 75 years?

Why *were "Recovered/Recovering" combined? It seems Pfizer would want to highlight the number of recovered persons if this outcome category were favorable.*

How *is "Recovering" different from "Not recovered at time of report"?*

Confusing Representation of Pediatric Data

Argument

Pediatric individuals were considered less than 12 years old, and cases reported ranged in age from two months to nine years [p. 13]. There were 34 cases included in the report, ***24 (71%) of which were listed as "Serious"*** [p. 13]. There were two cases of facial paralysis in the young; one was a child, and one was an infant. The System Organ Class for Pediatric lists 34 cases, but Pfizer acknowledged that 28 additional cases were excluded due to height and/or weight. Forty-six cases were listed as under 16 years old, which leaves the remaining cases (out of 175) as 67 cases between 16 and 17 years old.

Review Team Conclusion

At the time of the Pfizer 5.3.6 report, there was no emergency use authorization for a COVID vaccine for individuals under the age of 16. However, there were post-marketing cases reported for individuals younger than 16. It is unclear if individuals who were 12 to 17 years old were inoculated during Phase 3 of the trial and if these cases were reported after a trial inoculation rather than being separated into "trial related data."

In the Pfizer report [p. 7], the age category lists 175 cases as less than 17 years with 46 cases less than 16 years, and 34 cases less than 12 years. The Under 12 category is misleading since page 13 shows 62 total cases under the age of 12.

After the Pfizer post-marketing report was released, Pfizer presented Phase 3 data to the FDA regarding 12 to 17-year-olds, https://www.pfizer.com/science/coronavirus/vaccine/about-our-landmark-trial. Pfizer and BioNTech announced positive topline results of the pivotal COVID-19 vaccine study in 2,260 adolescents ages 12 to 15. A claim is made by Pfizer and BioNTech that all participants in the trial will continue to be monitored for long-term protection and safety for an additional two years after their second dose. On April 9, 2021, the companies submitted these data to the FDA and requested an amendment to the BNT162b2 emergency use authorization to expand use to adolescents 12 to 15 years of age. The 16 to 17 age group is not addressed separately, and no public data has been released regarding this data from the FDA.

***Key Finding*:** Do fatalities occur in the young? The Pfizer post-marketing report does not address this clearly.

Although the Pfizer report section for the Pediatric System Organ Class listed 34 adverse event cases, there were 62 case reports with ages listed as under 12 years old. Pfizer excluded 28 cases of patients under 12 years old due to their height and/or weight. [p.15] Since this category was age related only and the drug was not administered based on body weight, these additional 28 cases seemingly should not have been excluded.

During the timeframe of these adverse event cases (December 1, 2020, through February 28, 2021), the emergency use authorization (EUA) dosage publicly available was only for persons older than 16. There was no EUA for a pediatric dose. It is possible that these children received adult doses of the Pfizer product. The dosing of these children warrants further investigation.

A one-year-old suffered Bell's palsy one day following vaccination, and it had not resolved at the conclusion of the post-marketing report.
A seven-year-old suffered a stroke, outcome unknown.
Children under 12 should not have received the Pfizer vaccine as the trial for the 5 to 11 age group did not begin until late March 2021.

Cases for those younger than 18 are scattered within the listed sections of System Organ Classes for Adverse Events of Special Interest (Anaphylaxis, Cardiovascular, COVID-19, Hematologic, Facial Paralysis, Immune-Mediated/Autoimmune, Musculoskeletal, Renal, Respiratory, Stroke, Other).

***What** dose or doses were children under age 12 given since there was no emergency use authorization for children that age during this the time?*

Selective Reporting of Deaths

Argument

Pfizer claimed [p. 8] that any adverse events that occurred in greater than or equal to (≥) two percent of adverse event reports would be listed in Table 2 [pp. 8–9]. Table 2 includes 93,473 events grouped by System Organ Class. Conditions listed are milder symptoms such as pain, malaise, fever, and/or nausea. Also, it is important to note that there were COVID-19 infections (4.6%); Paraesthesia (paresthesia), which is an abnormal sensation of tingling or prickling caused chiefly by pressure on or damage to peripheral nerves (3.6%); and Hypoaesthesia (hypoesthesia) or numbness (2.4%).

Review Team Conclusion:

Key Finding*: *Unresolved adverse event cases and fatalities are not included in Pfizer's Table 2 even though they exceed the two percent threshold required for adverse events' inclusion.

Why did Pfizer not list the deaths in the table and only listed them in the body of the text? In the strict sense of the wording, ***there were 1,223 "cases" resulting in death out of 42,086 total cases; this is 2.9% of the cases*** *and should have been included in the Table 2 per Pfizer's own criteria.*

"Serious" Versus "Non-Serious" Adverse Events

Argument

Pfizer provided adverse event groupings of "Serious" and "Non-Serious." When aggregated, there are 16,147 "Serious" adverse events and 11,617 "Non-Serious" adverse events.

Review Team Conclusion

Key Finding**:** On page 26 of the Pfizer report, the ***"non-serious"*** **event category included deaths related to medication errors.** Pfizer footnoted, *"All the medication errors reported in these cases were assessed as non-serious occurrences with an unknown outcome; based on the available information including the causes of death, the relationship between the medication error and the death is weak."* **Is it left only to Pfizer to determine if the link between a medication error and death is "weak"?** Pfizer still concluded product safety despite significant "Serious Adverse Events."

The FDA defines a Serious Adverse Event as: any untoward medical occurrence that: 1) ***results in death***; 2) is life threatening; 3) requires inpatient hospitalization or causes prolongation of existing hospitalization; 4) results in persistent or significant disability/incapacity; 5) may have caused a congenital anomaly/birth defect; 6) **requires intervention to prevent permanent impairment or damage**. (Emphasis added.) (https://www.fda.gov/safety/reporting-serious-problems-fda/what-serious-adverse-event and https://www.accessdata.fda.gov/scripts/cdrh/cfdocs/cfCFR/CFRSearch.cfm?fr=312.32)

According to Pfizer [p. 6], the company planned to hire an additional 2,400 full-time personnel for data entry and case processing. Pfizer indicated, "Pfizer has also taken multiple actions to help alleviate the large increase of Adverse Event reports." [p. 6] By its own admission, Pfizer expected to have large numbers of cases and adverse events reported. This expectation of increased adverse event reports supports a conclusion that Pfizer's vaccine is not as safe as was communicated to the general public. Pfizer also indicated, "Reports are submitted voluntarily, and the magnitude of underreporting is unknown." [p. 5]

A truly safe vaccine should not lead to a post-marketing report with large numbers of "Serious Adverse Events."

The scope of this PME Team summary will not document all individual findings within each System Organ Class. The Post-Marketing Team micro-reports are available for a more detailed investigation into some of the System Organ Class evidence, such as stroke, thromboembolic (clotting) and hematologic (blood), liver, cardiovascular, neurologic events and more [https://dailyclout.io/category/pfizer-reports/] (Also see references below).

Pfizer Concludes Its Covid-19 Vaccine Is "Safe"

Argument

At the end of each System Organ Class section, Pfizer drew a conclusion—"Conclusion: this cumulative case review does not raise new safety issues."

Review Team Conclusion

Key Finding: Incredibly, Pfizer's conclusion of "***no new safety issues" is repeated 15 times including for System Organ Class categories that included deaths.*** The conclusion for pediatrics is, "No new **significant** safety information was identified based on a review of these cases compared with the **non-pediatric population**." Based on the evidence in the report, the PME Team strongly disagrees with Pfizer's conclusion.

***How** can a 2.9% fatality rate among reported cases in just 90 days lead to a conclusion of drug safety?*

There are three ***Safety Concern Categories (SCC)*** in the Pfizer report [p. 9] which lead to the data included in Table 3. The first safety concern, Anaphylaxis, is considered an "Identified Risk." Anaphylaxis is a severe, potentially life-threatening allergic reaction. The second, Vaccine-Associated Enhanced Disease (VAED), is considered a "Potential Risk." The third category of "Missing Information" concerns: "Pregnancy and Lactation," "Use in Pediatric Individuals," and "Vaccine Effectiveness."

As detailed above, sufficient concerns are documented from the review of the Pfizer post-marketing data to conclude the Pfizer COVID-19 vaccine (BNT162b2) is **not safe.** Additional data and follow-up on the reported cases identified as "*unknown*" would be required to clarify significant unanswered questions about safety.

The only follow-up data for the remaining adverse event reporting to date are the Periodic Safety Update Reports (PSURs) #1 and #3 issued in the European Union (EU), as mentioned above. These two reports confirm the PME Team findings in this report, though it is unclear why the data appears to have not also been released to the FDA. That follow-up data is needed to begin to understand the scope of the harms to the human population. PSUR #3 states, "There were no marketing authorization (sic) withdrawal for safety reasons during the reporting interval." In their own words, "According to the European Risk Management Plan (EU-RMP) version 4.0 adopted 26 November 2021, in effect the beginning of the reporting period, safety concerns for BNT162b2 are: . . . "

Then, Pfizer proceeds to list serious safety concerns such as vaccine-associated enhanced respiratory disease (VAERD), anaphylaxis, myocarditis, and pericarditis, all of which affirm the findings in 5.3.6. Use in pregnancy and breastfeeding and use in immunocompromised patients is incomplete and noted as "missing information" in PSUR #3. A more comprehensive report comparing this PME Team summary and PSUR#3 data will be forthcoming. The EU PSURs beg the question of why no similar U.S. reports have been released to meet the stated reporting requirements.

Table 1: System Organ Class and "Serious" Versus "Non-Serious" Adverse Events According to Pfizer's Post-Marketing Report

System Organ Class	Page	Number, %	Serious	Non-Serious	Death/ Unknown	Female/Male/ Unknown	Notations
Anaphylaxis (Safety Concern Category)	10	2,958 7.0%	2,341	617	9/754	876/106/20	
Vaccine-Associated Enhanced Disease VAED (SCC)	11	138 0.33%	-	-	38/8	73/57/8	75 potential cases. Pfizer suggested: "None of the 75 cases could be definitively considered as VAED/ VAERD."
Pregnancy and Lactation (SCC)	12–13	413 0.98%	84	329	(Baby[2] =28)/242	Babies not listed	Spontaneous abortions and neonatal deaths reported; alterations to breastfeeding
Pediatric (SCC) (2 mo.—9 yr.)	13	34 0.08%	24	10	0/5	25/7/2	Two Facial Paralysis; one Stroke
Vaccine Effectiveness (SCC)	13–15	1,665 4.0%	1625	21	65/1230	Not listed	Contracting COVID is considered "Serious"; no immunity conferred; 16 cases of vaccine failure per Pfizer criteria
Cardiovascular	16	1,403 3.3%	946	495	136/380	1076/291/36	130 myocardial infarctions, 91 cardiac failures
COVID-19	17	3,067 7.4%	2,585	774	136/2110	1650/844/573	Unremarkable; deals with positive cases
Dermatological	17	20 0.05%	16	4	0/6	17/1/2	Unremarkable; Reactions
Hematological	18	932 2.2%	681	399	34/371	676/222/34	Numerous examples of spontaneous bleeding from mucous membranes
Hepatic	18–19	70 0.2%	53	41	5/47	43/26/1	Metabolic alterations within the liver
Facial Paralysis	19–20	449 1.07%	399	54	0/97	295/133/21	One infant; one child
Immune-Mediated and Autoimmune	20	1,050 2.5%	780	297	12/312	526/156/368	32 Pericarditis, 25 Myocarditis
Musculoskeletal	20–21	3,600 8.5%	1,614	2,026	0/853	2760/711/129	3,525 Arthralgia (joint pain)
Neurological	21	501 1.2%	515	27	16/161	328/150/23	204 Seizure, 83 Epilepsy
Other	21–22	8,152 19.4%	3,674	4,568	96/1685	5969/1860/323	7,666 Pyrexia, Herpetic conditions
Renal	22	69 0.17%	70	0	23/22	46/23/0	All serious: 40 acute kidney injury, 30 renal failure, includes one infant
Respiratory	22–23	130 0.3%	126	11	41/31	72/58/0	44 respiratory failure [1]
Thromboembolic Events	23	151 0.3%	165	3	18/42	89/55/7	60 Pulmonary Embolism, 39 thrombosis, 35 DVT

System Organ Class	Page	Number, %	Serious	Non-Serious	Death/ Unknown	Female/Male/ Unknown	Notations
Stroke	23–24	275 0.6%	300	0	61/83	182/91/2	All serious; Ischaemic and Haemorrhagic conditions reported
Vasculitis	24	32 0.08%	25	9	1/8	26/6/0	Specific condition leading to one fatality not noted
Medication Error	26	2056 4.9%	124	1932	7/1498	Not disclosed	7 fatalities not categorized as "Serious." Pfizer lack information leading to fatalities, considered noncontributory

- Note: 10 cases of ARDS (Acute Respiratory Distress Syndrome).
- Twenty-eight (28) deaths in babies/infants. The outcome for 242 of the pregnant mothers/babies was listed as "unknown."

As stated previously, the period of this set of adverse events reporting to Pfizer was 90 days from December 1, 2020, ending on February 28, 2021. Interestingly**, Pfizer changed the coding convention related to Vaccine Effectiveness (VE) on February 15, 2021**. (https://www.phmpt.org/wp-content/uploads/2022/04/reissue_5.3.6-postmarketing-experience.pdf, p. 13) This change, less than two weeks before the end of the report's data cycle as well as a month and a half *after* public rollout, allowed the company to shift at least three cases from the *Vaccine Failure* category to the *Drug Ineffective* category. That shift in category was enabled due to the addition of requiring a "confirmed laboratory test" to qualify a case as Vaccine Failure [p. 13].

How Many Fetuses/Infants Actually Died?

Argument

The numbers for the Pregnancy category are found in the Pfizer report on pages 12 and 13. The first bullet point lists 270 mother-related cases and "4 foetus/baby cases". **Pfizer lists that "no outcome was provided for 238 pregnancies . . . ", which is over 88% of the pregnancies** [p12].

Review Team Conclusion

Key Finding:

The Pregnancy category enumerates **56 cases of dead babies/infants** as follows: 48 spontaneous abortions; two premature births with neonatal death; two spontaneous abortions with intrauterine death; one spontaneous abortion with neonatal death, one abortion, one abortion missed, and one fetal death. 238 pregnancies were listed as "no outcome reported."

There was one "normal outcome" and, sadly, that was one baby from a set of twins in which the other twin died. Upon counting the numbers, woven through the report's text, there were 56 dead fetuses/infants among the classifications.

***Why** is the first bullet point listed as only four foetus/baby cases when many more than were affected?*
***How** do 56 dead fetuses/babies lead to a conclusion of a "safe and effective" drug for pregnant women?*
***What happened** to the 238 pregnancy cases where there was "**no outcome reported**"?*

Latency: Why Were So Many Adverse Events Occurring So Quickly?

Argument

Latency is the time span between vaccine dose and the emergence of an adverse event. Pfizer concluded, "***The data do not reveal any novel safety concerns or risks requiring label changes and support a favorable benefit risk profile of to [sic] the BNT162b2 vaccine."*** Pfizer continued, ***"Review of available data for this cumulative PM experience, confirms a favorable benefit: risk balance for BNT162b2."*** [p. 28]

Review Team Conclusion

Key Finding: Short latency suggests a dose-response relationship between Pfizer's COVID-19 "vaccine" (BNT162b2) and reported adverse events.

The occurrence of harms soon after doses, a temporal relationship, suggests the product causes harms. **There are four instances of patient deaths the same day** as the patient received Pfizer's COVID-19 vaccine (BNT162b2). *In the first emergency use authorization (EUA), the FDA considered that any adverse events within six weeks of product use could be plausibly related to the product itself.* [https://www.fda.gov/media/144416/download] Within the 5.3.6 Pfizer report, some **adverse events occurred within zero to four days** after product administration, and most occurred within six weeks. That pattern is consistent across all System Organ Classes. The 5.3.6 report data do not speak to any long-term adverse events which may be present beyond 90 days. After cataloguing events from clinicians in the field who made the connection between product use and adverse events, Pfizer repeatedly concluded, "**This cumulative case review does not raise safety issues.**" The use of this vaccine product led to a large number of adverse event reports that included significant harms and even fatalities. **Pfizer dismissed its own data to make that safety conclusion.**

Below, the PME Team reports median time, which is the midpoint value, for adverse event occurrences in each System Organ Class. When this value is within the immediate days after the vaccination, most adverse events develop quickly. These categories show there is a wide range of onset, with some adverse events not developing for over a month. Latencies for System Organ Classes are listed in PME Team summary Table 2 below.

Table 2: Latency: Time from dose to emergence of an Adverse Event (AE)

System Organ Class	AE Development Range	AE Development Median
Cardiovascular	<24 hours—21 days	<24 hours
COVID-19	<24 hours—50 days	5 days
Dermatological	<24 hours—17 days	3 days
Hematological	<24 hours—33 days	1 day
Hepatic	<24 hours—20 days	3 days
Facial Paralysis	<24 hours—46 days	2 days

System Organ Class	AE Development Range	AE Development Median
Immune-Mediated and Autoimmune	<24 hours—30 days	<24 hours
Musculoskeletal	<24 hours—32 days	1 day
Neurological	<24 hours—48 days	1 day
Other	<24 hours—61 days	1 day
Renal	<24 hours—15 days	4 days
Respiratory	<24 hours—18 days	1 day
Thromboembolic	<24 hours—28 days	4 days
Stroke	<24 hours—41 days	2 days
Vasculitic	<24 hours—19 days	3 days

References

1. Pfizer report, "5.3.6 Cumulative Analysis of Post-Authorization Adverse Events Reports of PF-07302048 (BNT162b2) Received Through 28-FEB-2021."
2. Report 47: Blood System-Related Adverse Events Following Pfizer COVID-10 mRNA Vaccination—DailyClout
3. Report 49: Clotting System-Related Adverse Events Following Pfizer COVID-19 mRNA Vaccination—DailyClout
4. Report 50: 20% of Post-Jab Strokes Fatal in the 90 Days 03Following Pfizer COVID mRNA Vaccine Rollout—DailyClout
5. Report 51: Liver Adverse Events—Five Deaths Within 20 Days of Pfizer's mRNA COVID Injection. 50% of Adverse Events Occurred Within Three Days.—DailyClout
6. Report 53: 77% of Cardiovascular Adverse Events from Pfizer's mRNA COVID Shot Occurred in Women, as Well as in People Under Age 65. Two Minors Suffered Cardiac Events.—DailyClout
7. Report 54: Infants and Children Under 12 Given the Pfizer mRNA COVID "Vaccine" Seven Months Before the Product Was Approved for Children—DailyClout
8. Report 57: 542 Neurological Adverse Events, 95% Serious, in First 90 Days of Pfizer mRNA Vaccine Rollout 16 Deaths Females Suffered AEs More Than Twice as Often as Males—DailyClout
9. Report 60: 449 Patients Suffer Bell's Palsy Following Pfizer—DailyClout
10. Report 62: Acute Kidney Injury and Acute Renal Failure—DailyClout
11. Report 65: In the First Three Months of Pfizer's mRNA "Vaccine" Rollout, Nine Patients Died of Anaphylaxis—DailyClout
12. Report 66: 1,077 Immune-Mediated/Autoimmune Adverse Events in First 90 Days of Pfizer mRNA "Vaccine" Rollout, Including 12 Fatalities. Pfizer Undercounted This Category of Adverse Events by 270 Occurrences.—DailyClout
13. Report 68: 34 Blood Vessel Inflammation, Vasculitis, Adverse Events Occurred in First 90 Days After Pfizer mRNA "Vaccine" Rollout, Including One Fatality. Half Had Onset Within Three Days of Injection. 81% of Sufferers Were Women.—DailyClout

14. Report 71: Musculoskeletal Adverse Events of Special Interest Afflicted 8.5% of Patients in Pfizer's Post-Marketing Data Set, Including Four Children and One Infant. Women Affected at a Ratio of Almost 4:1 Over Men.—DailyClout
15. Report 72: "Other AESIs" Included MERS, Multiple Organ Dysfunction Syndrome (MODS), Herpes Infections, and 96 DEATHS. 15 Patients Were Under Age 12, Including Six Infants.—DailyClout
16. Report 77: Women Suffered 94% of Dermatological Adverse Events Reported in First 90 Days of Pfizer COVID "Vaccine" Rollout. 80% of These Adverse Events Were Categorized As "Serious."
17. Report 78: Thirty-Two Percent of Pfizer's Post-Marketing Respiratory Adverse Event Patients Died, Yet Pfizer Found No New Safety Signals.
18. Report 83: 23% of Vaccinated Mothers' Fetuses or Neonates Died. Suppressed Lactation and Breast Milk Discoloration Reported.

Appendix 1: Report Authors

The Post-Marketing Experience Team

- LD LaLonde, M.S.
- Lora Hammill BSMT, ASCP
- David Shaw
- Dr. Chris Flowers MD
- Loree Britt
- Dr. Joseph Gehrett MD
- Dr. Barbara Gehrett MD
- Michelle Cibelli RN, BSN
- Margaret Turoski, RPh
- Cassie B. Papillon
- Tony Damian

Thirty: "2.5 Months After COVID Vaccine Rollout, Pfizer Changed Criteria for 'Vaccination Failure,' Causing 99% of Reported Cases to Not Meet That Definition. 3.9% of Reported 'Lack of Efficacy' Cases Ended in Death in First 90 Days of Public Vaccine Availability."

—Barbara Gehrett, MD; Joseph Gehrett, MD; Chris Flowers, MD; and Loree Britt

The WarRoom/DailyClout Pfizer Documents Analysis Project Post-Marketing Group (Team 1)—Barbara Gehrett, MD; Joseph Gehrett, MD; Chris Flowers, MD; and Loree Britt—penned a telling analysis of the "Vaccine Effectiveness" Safety Concern section found in Pfizer document *5.3.6 Cumulative Analysis of Post-Authorization Adverse Event Reports of PF-07302048 (BNT162B2) Received Through 28-FEB-2021* (a.k.a., "*5.3.6*").

It is important to note that the AESIs in the *5.3.6* document were reported to Pfizer for **only a 90-day period** starting on December 1, 2020, the date of the United Kingdom's public rollout of Pfizer's COVID-19 experimental mRNA "vaccine" product.

Key highlights from this important report include:

- During the first three months of vaccine rollout, 1,665 cases were submitted to Pfizer with a definition of **"lack of efficacy" (LOE)**.
  - There were **65 deaths (3.9%) among the lack of efficacy cases**.
- Lack of efficacy cases fell into two categories:
  - Drug ineffective
  - Vaccination failure
- Without explanation, **Pfizer revised the coding conventions (or criteria) for the "vaccination failure" category on February 15, 2021, two and a half months into the vaccine's public rollout and a mere two weeks before data collection for the report ended**.
  - The new definition of "**vaccination failure**" required **all three** of the following criteria to be met:
    - Both doses received per local regime.
    - At least seven days since the second dose.
    - Infection with confirmed lab test positive for SARS-CoV-2.
- Revising the criteria for vaccination failure likely allowed Pfizer to shift cases out of the "vaccination failure" category and into "drug ineffective" category. With the new definition, 1,649 cases, or 99%, of the 1,665 lack of efficacy cases met the "drug ineffective" criteria of:
  - Infection *not* confirmed by a lab test
  - Unknowns present:
    - Vaccine doses followed proper local regimen.
    - Number of days since first dose.

  - Whether seven days passed since second dose.
  - COVID onset between 14 days after first dose and through six days after second dose.
- However, based on data in *5.3.6*, **only 788 (47.8%) of the 1,649 drug ineffective cases can be stated categorically *not* to have been drug failure.**
- **1,625 (98.5%) of the 1,649 "drug ineffective" cases were labeled as "serious."**
- Using the revised coding conventions, only **16 lack of efficacy cases, under 1%, were categorized as "vaccination failure."**
- Up to another 861 cases may have been "vaccination failure" had missing data been collected.

Please read the full report below.

DAILYCLOUT

War Room/DailyClout Pfizer Document Analysis

Post-Marketing Team Micro-Report 18:

Vaccine Effectiveness Review of 5.3.6

SOURCE

https://www.phmpt.org/wp-content/uploads/2022/04/reissue_5.3.6-postmarketing-experience.pdf

5.3.6 AE REPORTING PERIOD:

"Since the first temporary authorization for emergency supply under Regulation 174 in the UK (01 December 2020) and through 28 February 2021."

Adverse Events were reported to Pfizer during a 90-day period, following the December 1, 2020, public rollout of its COVID-19 experimental "vaccine" product.

In the Pfizer 5.3.6 document, these AEs were categorized by System Organ Classes (SOC) – in other words, by systems in the body.

ABBREVIATIONS:

5.3.6 : Pfizer source document

SOC : System Organ Class

AE : Adverse Event

AESI : Adverse Event of Special Interest

EUA : Emergency Use Authorization by FDA

PM : Post-Marketing

BNT162b2 : Pfizer's mRNA COVID-19 vaccine

AUTHORS:

Dr Barbara Gehrett MD
Dr Joseph Gehrett MD
Dr Chris Flowers MD
Loree Britt

Post Marketing Team

09Oct23

Pfizer identified three categories of **Safety Concerns** in Table 3. These were:

1) Important identified risks: Anaphylaxis (Table 4)
2) Important potential risks: Vaccine-Associated Enhanced Disease (VAED), including Vaccine-Associated Enhanced Respiratory Disease (VAERD) (Table 5)
3) Missing Information: Use in Pregnancy and Lactation, Use in Paediatric Individuals < 12 years of Age, and **Vaccine Effectiveness** (Table 6)

Post Authorization Cases Evaluation (cumulative to 28 Feb 2021)
Total Number of Cases in the Reporting Period (N=42086)

1st dose (day 1-13)	From day 14 post 1st dose to day 6 post 2nd dose	Day 7 post 2nd dose
Code only the events describing the SARS-CoV-2 infection	Code "Drug ineffective"	Code "Vaccination failure"
Scenario Not considered LOE	Scenario considered LOE as "Drug ineffective"	Scenario considered LOE as "Vaccination failure"

https://www.phmpt.org/wp-content/uploads/2022/04/reissue_5.3.6-postmarketing-experience.pdf

In this section of 5.3.6, Pfizer reported 1,665 cases submitted with the diagnosis of "lack of efficacy" (LOE). They are broken into two categories:

1. Drug ineffective
2. Vaccination failure.

It takes time for the body to develop immunity to any vaccination. Pfizer asserted that the immune system needed a full 13 days to "respond" to the vaccine. That is why Pfizer did not consider it to be lack of efficacy if COVID-19 infection occurred within that 13-day period following the first injection. In Table 6, these cases are described simply as "COVID-19," and it appears that there were 383 of them (155 suspected and 228 confirmed).

Beginning 14 days after the first shot through Day Six after the second shot, Pfizer termed LOE as "drug ineffective." For BNT162b2 to be "ineffective," the Pfizer definition required that an individual must have received both doses of the vaccine with a documented COVID-19 infection developing seven days or longer after the second shot.

The cases represented in 5.3.6 were collected from multiple countries during the 90-day period (December 1, 2020, through February 28, 2021). **Interestingly, the coding conventions (or criteria) for the drug ineffective and vaccination failure categories were revised on February 15, 2021.** No information is supplied in 5.3.6 regarding the specific content of the *initial* coding conventions.

Why were the coding conventions changed?

What were the outcomes in the 1,665 individuals with lack of efficacy?

65 deaths (3.9%)

165 *resolved/resolving* **205 *not resolved*** **1230 *outcome unknown***

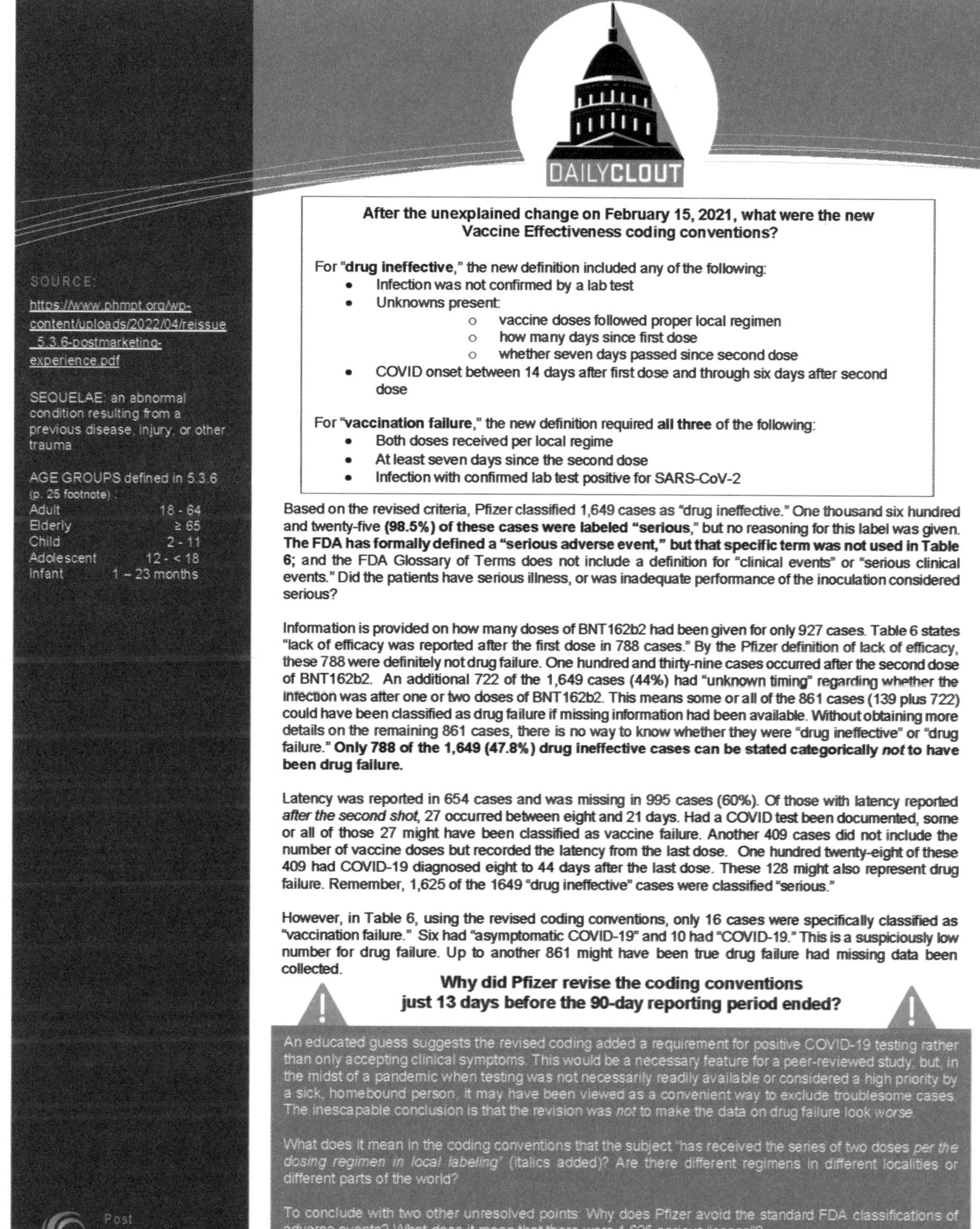

SOURCE:
https://www.phmpt.org/wp-content/uploads/2022/04/reissue_5.3.6-postmarketing-experience.pdf

SEQUELAE: an abnormal condition resulting from a previous disease, injury, or other trauma

AGE GROUPS defined in 5.3.6 (p. 25 footnote) :

Adult	18 - 64
Elderly	≥ 65
Child	2 - 11
Adolescent	12 - < 18
Infant	1 – 23 months

After the unexplained change on February 15, 2021, what were the new Vaccine Effectiveness coding conventions?

For "**drug ineffective**," the new definition included any of the following:

- Infection was not confirmed by a lab test
- Unknowns present:
    - vaccine doses followed proper local regimen
    - how many days since first dose
    - whether seven days passed since second dose
- COVID onset between 14 days after first dose and through six days after second dose

For "**vaccination failure**," the new definition required **all three** of the following:

- Both doses received per local regime
- At least seven days since the second dose
- Infection with confirmed lab test positive for SARS-CoV-2

Based on the revised criteria, Pfizer classified 1,649 cases as "drug ineffective." One thousand six hundred and twenty-five **(98.5%) of these cases were labeled "serious,"** but no reasoning for this label was given. **The FDA has formally defined a "serious adverse event," but that specific term was not used in Table 6;** and the FDA Glossary of Terms does not include a definition for "clinical events" or "serious clinical events." Did the patients have serious illness, or was inadequate performance of the inoculation considered serious?

Information is provided on how many doses of BNT162b2 had been given for only 927 cases. Table 6 states "lack of efficacy was reported after the first dose in 788 cases." By the Pfizer definition of lack of efficacy, these 788 were definitely not drug failure. One hundred and thirty-nine cases occurred after the second dose of BNT162b2. An additional 722 of the 1,649 cases (44%) had "unknown timing" regarding whether the infection was after one or two doses of BNT162b2. This means some or all of the 861 cases (139 plus 722) could have been classified as drug failure if missing information had been available. Without obtaining more details on the remaining 861 cases, there is no way to know whether they were "drug ineffective" or "drug failure." **Only 788 of the 1,649 (47.8%) drug ineffective cases can be stated categorically *not* to have been drug failure.**

Latency was reported in 654 cases and was missing in 995 cases (60%). Of those with latency reported *after the second shot*, 27 occurred between eight and 21 days. Had a COVID test been documented, some or all of those 27 might have been classified as vaccine failure. Another 409 cases did not include the number of vaccine doses but recorded the latency from the last dose. One hundred twenty-eight of these 409 had COVID-19 diagnosed eight to 44 days after the last dose. These 128 might also represent drug failure. Remember, 1,625 of the 1649 "drug ineffective" cases were classified "serious."

However, in Table 6, using the revised coding conventions, only 16 cases were specifically classified as "vaccination failure." Six had "asymptomatic COVID-19" and 10 had "COVID-19." This is a suspiciously low number for drug failure. Up to another 861 might have been true drug failure had missing data been collected.

Why did Pfizer revise the coding conventions just 13 days before the 90-day reporting period ended?

An educated guess suggests the revised coding added a requirement for positive COVID-19 testing rather than only accepting clinical symptoms. This would be a necessary feature for a peer-reviewed study; but, in the midst of a pandemic when testing was not necessarily readily available or considered a high priority by a sick, homebound person, it may have been viewed as a convenient way to exclude troublesome cases. The inescapable conclusion is that the revision was *not* to make the data on drug failure look *worse*.

What does it mean in the coding conventions that the subject "has received the series of two doses *per the dosing regimen in local labeling*" (italics added)? Are there different regimens in different localities or different parts of the world?

To conclude with two other unresolved points: Why does Pfizer avoid the standard FDA classifications of adverse events? What does it mean that there were 1,625 serious "cases"?

The evasive phrasing throughout renders this table even more obscure than those previously analyzed.

Post Marketing Team

Thirty-One: "WarRoom/DailyClout Researchers Find Pfizer Delayed Recording Vaccinated Deaths at Critical Juncture of EUA Process. Improper Delays in Reporting Deaths in the Vaccinated Led FDA to Misstate Vaccine's Effectiveness, Influenced EUA Grant Decision."

—Analysis by Jeyanthi Kunadhasan, MD, and Dan Perrier; Writing and Editing by Amy Kelly

On September 5, 2023, DailyClout reported that a WarRoom/DailyClout Research Team—Corinne Michels, PhD; Daniel Perrier; Jeyanthi Kunadhasan, MD; Ed Clark, MSE; Joseph Gehrett, MD; Barbara Gehrett, MD; Kim Kwiatek, MD; Sarah Adams; Robert Chandler, MD; Leah Stagno; Tony Damian; Erika Delph; and Chris Flowers, MD—broke a huge story about there being more cardiovascular deaths in the vaccinated than in the unvaccinated in Pfizer's clinical trial, as well as that Pfizer did not report the 3.7-fold cardiovascular adverse events signal and also delayed reporting deaths so that it favored the vaccinated arm of the trial. (https://dailyclout.io/report-84-warroom-dailyclout-research-team-breaks-huge-story/)

Now, as a follow up to that report, DailyClout reveals that Jeyanthi Kunadhasan, MD, part of that same Research Team, found even more damning evidence showing that **Pfizer delayed recording deaths in Case Report Forms (CRFs), which allowed the company to not report those deaths as part of its emergency use authorization (EUA) data filing with the Food and Drug Administration (FDA)**. Daniel Perrier from the Research Team and Dr. Kunadhasan found, analyzed, and verified the data supporting this discovery. *It is very likely that had the deaths been recorded when Pfizer became aware of them and then accurately filed as part of the EUA dataset, the public would not have accepted an EUA had been granted by the FDA.*

On December 10, 2020, one day before the FDA granted Pfizer's EUA, Susan Wollersheim, MD, of the FDA/Center for Biologics Evaluation and Research (CBER) Office of Vaccines Research and Review, Division of Vaccines and Related Products Applications, presented, "FDA Review of Efficacy and Safety of Pfizer-BioNTech COVID-19 Vaccine Emergency Use Authorization Request" at the CBER Vaccines and Related Biological Products Advisory Committee 162nd Meeting. On page 42 of the slide deck, Dr. Wollersheim presented:

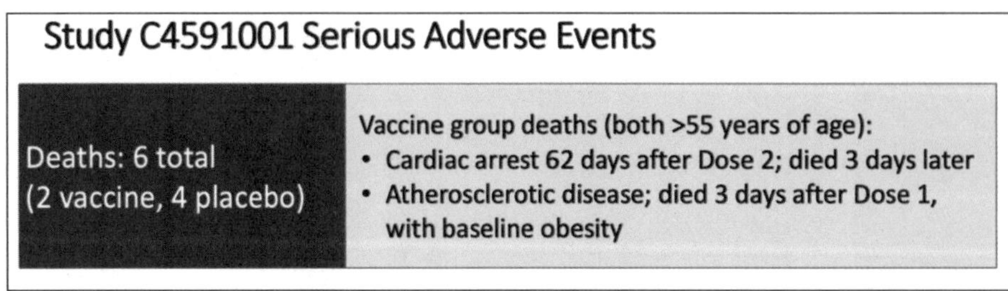

The data showing two vaccine deaths and four placebo deaths as of December 10, 2020, was incorrect at the time of the EUA request presentation, and the correct data was available in Pfizer's own documentation and, thus, also available to the FDA.

As of **November 14, 2020, the data cutoff for the EUA dataset**, Pfizer possessed data showing that **the vaccine arm of the COVID-19 mRNA vaccine clinical trial had the same number of deaths as the placebo arm**. In other words, **there was no evidence of Pfizer's COVID vaccine having a positive impact on death outcomes**.

The Pfizer narrative descriptions of the deaths show that, though that the clinical trial sites had been informed by the deceased patients' loved ones on the days of their deaths [October 19, 2020, and November 7, 2020 (pp. 71 and 75), well before the EUA data cutoff date in both cases], those two deaths from the vaccinated arm that were *not* included in the FDA presentation are listed below:

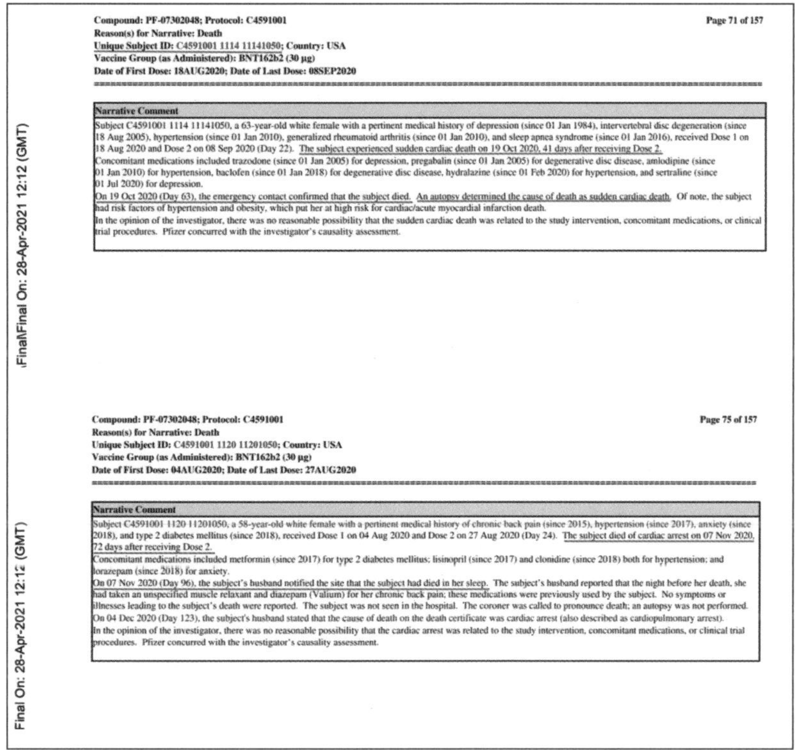

\Final\Final On: 28-Apr-2021 12:12 (GMT)

Compound: PF-07302048; Protocol: C4591001 **Page 71 of 157**
Reason(s) for Narrative: Death
Unique Subject ID: C4591001 1114 11141050; **Country: USA**
Vaccine Group (as Administered): BNT162b2 (30 µg)
Date of First Dose: 18AUG2020; Date of Last Dose: 08SEP2020

Narrative Comment

Subject C4591001 1114 11141050, a 63-year-old white female with a pertinent medical history of depression (since 01 Jan 1984), intervertebral disc degeneration (since 18 Aug 2005), hypertension (since 01 Jan 2010), generalized rheumatoid arthritis (since 01 Jan 2010), and sleep apnea syndrome (since 01 Jan 2016), received Dose 1 on 18 Aug 2020 and Dose 2 on 08 Sep 2020 (Day 22). The subject experienced sudden cardiac death on 19 Oct 2020, 41 days after receiving Dose 2.

Concomitant medications included trazodone (since 01 Jan 2005) for depression, pregabalin (since 01 Jan 2005) for degenerative disc disease, amlodipine (since 01 Jan 2010) for hypertension, baclofen (since 01 Jan 2018) for degenerative disc disease, hydralazine (since 01 Feb 2020) for hypertension, and sertraline (since 01 Jul 2020) for depression.

On 19 Oct 2020 (Day 63), the emergency contact confirmed that the subject died. An autopsy determined the cause of death as sudden cardiac death. Of note, the subject had risk factors of hypertension and obesity, which put her at high risk for cardiac/acute myocardial infarction death.

In the opinion of the investigator, there was no reasonable possibility that the sudden cardiac death was related to the study intervention, concomitant medications, or clinical trial procedures. Pfizer concurred with the investigator's causality assessment.

Final On: 28-Apr-2021 12:12 (GMT)

Compound: PF-07302048; Protocol: C4591001 **Page 75 of 157**
Reason(s) for Narrative: Death
Unique Subject ID: C4591001 1120 11201050; **Country: USA**
Vaccine Group (as Administered): BNT162b2 (30 µg)
Date of First Dose: 04AUG2020; Date of Last Dose: 27AUG2020

Narrative Comment

Subject C4591001 1120 11201050, a 58-year-old white female with a pertinent medical history of chronic back pain (since 2015), hypertension (since 2017), anxiety (since 2018), and type 2 diabetes mellitus (since 2018), received Dose 1 on 04 Aug 2020 and Dose 2 on 27 Aug 2020 (Day 24). The subject died of cardiac arrest on 07 Nov 2020, 72 days after receiving Dose 2.

Concomitant medications included metformin (since 2017) for type 2 diabetes mellitus; lisinopril (since 2017) and clonidine (since 2018) both for hypertension; and lorazepam (since 2018) for anxiety.

On 07 Nov 2020 (Day 96), the subject's husband notified the site that the subject had died in her sleep. The subject's husband reported that the night before her death, she had taken an unspecified muscle relaxant and diazepam (Valium) for her chronic back pain; these medications were previously used by the subject. No symptoms or illnesses leading to the subject's death were reported. The subject was not seen in the hospital. The coroner was called to pronounce death; an autopsy was not performed.

On 04 Dec 2020 (Day 123), the subject's husband stated that the cause of death on the death certificate was cardiac arrest (also described as cardiopulmonary arrest).

In the opinion of the investigator, there was no reasonable possibility that the cardiac arrest was related to the study intervention, concomitant medications, or clinical trial procedures. Pfizer concurred with the investigator's causality assessment.

Case Report Forms (CRFs) capture patient data during a clinical trial. The CRFs for these two patients, Subjects 11141050 and 11201050, do not mention that Pfizer knew of their deaths on their dates of death. Rather, Subject 1141050's death was first entered into the CRF on November 25, 2020, at 18:51:46 Central Time, *37 days after the patient's death.*

Header Text: c4591001
Visit: Logs - Unscheduled | **Form:** ADVERSE EVENT REPORT - eCRF Audit Trail History
Form Version: 22-Apr-2020 21:02 | **Form Status:** Data Complete, Frozen, Verified
Site No: 1114 | **Site Name:** (1114) Alliance for Multispecialty Research Inc
Subject No: 11141050 | **Subject Initials:** ---
Generated By: (b) (4) | **Generated Time (GMT):** 29-Mar-2021 10:58

(US & Canada)				death as the AE term with an outcome of FATAL or clarify.
Nov-25-2020 18:51:46 (UTC-06:00) Central Time (US & Canada)	ACV0PFEINFP6000	(b) (4), (b) (6)	**Data Entry :** death-cause unknown	Initial Entry

And, Subject 11201050's death was first entered into the CRF on December 3, 2020, at 12:48:29 Eastern Time, *26 days after the patient died*:

Header Text: c4591001
Visit: Follow-Up - Unscheduled
Form Version: 15-Sep-2020 21:53
Site No: 1120
Subject No: 11201050
Generated By: (b) (4)

Form: DISPOSITION - FOLLOW-UP - eCRF Audit Trail History
Form Status: Data Complete, Frozen, Verified
Site Name: (1120) Meridian Clinical Research
Subject Initials: ---
Generated Time (GMT): 29-Mar-2021 11:09

Time (US & Canada)

3. Status:

Date	Location	User	Value	Reason
Dec-03-2020 12:48:29 (UTC-05:00) Eastern Time (US & Canada)	ACV0PFEINFP6000	(b) (4), (b) (6)	Data Entry: DEATH	Initial Entry

It appears that Pfizer not entering the death dates of the above two Subjects into their CRFs when Pfizer first learned of them—i.e., on the dates of death—allowed the company to hide the deaths until the EUA data cutoff date had passed. To summarize what was known and important dates:

Site No	Subject Id	Age	Sex	Date Died	Date Site Notified	Clinical Trial Data Cutoff	Date Death Reported in CRF	Reporting Delay (Days)	VRBPAC Meeting	EUA Granted
1114	11141050	63	Female	19-Oct-20	19-Oct-20	14-Nov-20	25-Nov-20	37	10-Dec-20	11-Dec-20
1120	11201050	58	Female	07-Nov-20	07-Nov-20	14-Nov-20	03-Dec-20	26	10-Dec-20	11-Dec-20

According to Pfizer's own protocol, reporting of deaths should have occurred within 24 hours:

- Section 8.3.1.1. "Reporting SAEs to Pfizer Safety" states, "All SAEs occurring in a participant during the active collection period as described in Section 8.3.1 are reported to Pfizer Safety on the Vaccine SAE Report Form immediately upon awareness and **under no circumstance should this exceed 24 hours**, as indicated in Appendix 3 [pp. 1271–1277]. The investigator will submit any updated SAE data to the sponsor within 24 hours of it being available." [Emphasis added.] [p. 74]
- Additionally, Section 8.3.1.2. "All nonserious AEs and SAEs occurring in a participant during the active collection period, which begins after obtaining informed consent as described in Section 8.3.1, will be recorded on the AE section of the CRF. **The investigator is to record on the CRF all directly observed and all spontaneously reported AEs and SAEs reported by the participant.**" [Emphasis added.] [p. 74]

This graphic below shows the accurate and available serious adverse events data which should have been presented at the December 10, 2020, EUA Authorization Request meeting:

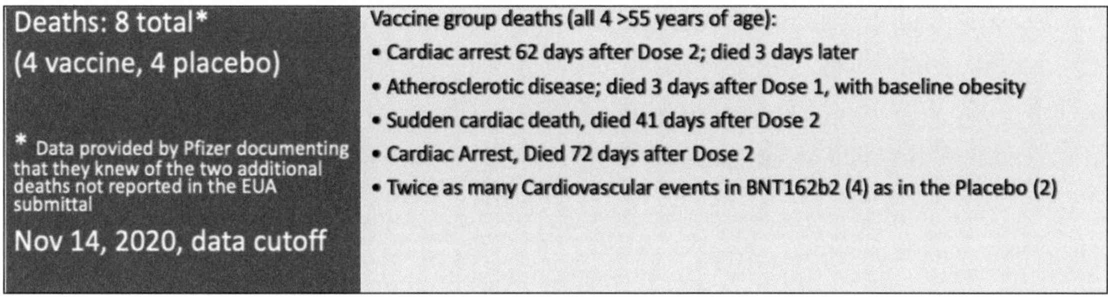

It shows an equal chance of death in vaccine and placebo arms of the trial and, thus, no benefit from Pfizer's mRNA COVID vaccine. Moreover, it also clearly shows Pfizer knew and, therefore, the FDA should have known that there were **twice as many cardiovascular adverse events in the vaccinated arm of the clinical trial** versus the placebo arm.

The charts below show the delays in recording deaths in the vaccinated ("BNT") arm trial subjects. The chart on the left is based on CRF data, which is the data that was used to request EUA approval. The chart on the right shows marked improvement in the speed of reporting deaths when the newly discovered date in the report narrative (https://www.phmpt.org/wp-content/uploads/2023/09/125742_S1_M5_5351_c4591001-interim-mth6-narrative-sensitive.pdf), outlined above, is included.

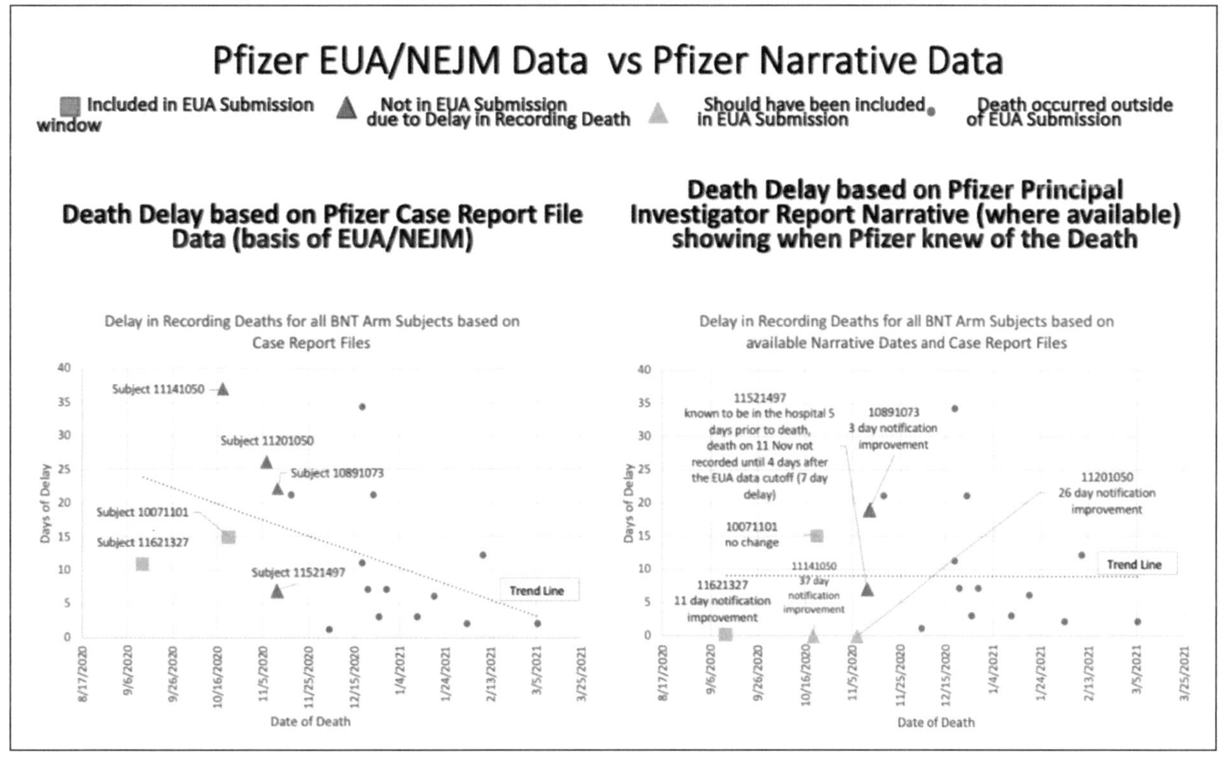

The NEJM article mentioned in the chart is available at https://www.nejm.org/doi/10.1056/NEJMc2036242.

Had the death data that was available to Pfizer before November 14, 2020, and which showed no positive impact on death outcomes, been presented at the December 10, 2020, Vaccines and Related Biologics Products Advisory Committee (VRBPAC) Meeting for Pfizer's COVID vaccine emergency use authorization request, it would have been very difficult, if not impossible, for the VRBPAC members to vote to authorize Pfizer's EUA.

Instead, Pfizer seemingly buried two "inconvenient" deaths from the vaccinated arm until after the data cutoff date, thus inaccurately showing twice as many deaths in the placebo arm of its trial on December 10, 2020. **On December 11, 2020, the FDA granted the EUA for Pfizer's vaccine based on inaccurate data, and that data-related negligence has negatively impacted the health of countless people worldwide.**

Dr. Kunadhasan will be presented about this finding, as well as on the Forensic Analysis of the 38 Subject Deaths in the 6-Month Interim Report of the Pfizer/BioNTech BNT162b2 mRNA Vaccine Clinical Trial preprint, both in the Australian Parliament in Canberra, Australia, and at the Australian Medical Professionals' Society's "Inquiry into Australia's Excess Mortality" in Kambah, Australia, on October 18, 2023. Please read more about the inquiry at https://www.trialsitenews.com/a/australian-medical-society-investigating-excess-deathssuspects-the-population-faces-an-iatrogenic-crisis-ff734f27. *The results of the inquiry will be published as a book, and book-related information is available on* https://amps.redunion.com.au/too-many-dead.

Thirty-Two: "Pfizer's 'Post-Marketing Surveillance' Shows mRNA-Vaccinated Suffered 1000s of COVID Cases in 1st 90 Days of Vaccine Rollout. Most Infections in the Vaccinated Categorized as 'Serious Adverse Events.'"

—Barbara Gehrett, MD; Joseph Gehrett, MD; Chris Flowers, MD; and Loree Britt

Though spokespeople assured us that the COVID-19 injection stops—well—COVID, Pfizer and the Food and Drug Administration (FDA) both knew that during the first three months of Pfizer's mRNA COVID-19 vaccine rollout, thousands of COVID cases were reported to Pfizer—among *vaccinated* people. The 'COVID-19 Adverse Events of Special Interest' (AESI) category in Pfizer's internal report includes **3,067 cases of vaccinated patients infected with COVID**. Two of these were infants, and one which was a child, though at that time no Pfizer COVID vaccine was yet authorized for children or infants.

Among those 3,067 cases of vaccinated people infected with COVID, there were 3,359 COVID-19-related adverse events (AEs); that is to say, there was more than one adverse event related to COVID per vaccinated patient in this category. **Of the 42,086 total cases of which Pfizer was notified during that period, seven percent (or 3,067 out of 42,086) were COVID-related adverse events.**

The WarRoom/DailyClout researchers' analysis of the shockingly high levels of COVID-19-related AESIs is based on Pfizer document '*5.3.6 Cumulative Analysis of Post-Authorization Adverse Event Reports of PF-07302048 (BNT162B2) Received Through 28-FEB-2021*' (a.k.a., "*5.3.6*").

The AESIs in the *5.3.6* document were reported to Pfizer for a 90-day period—starting on December 1, 2020, the date of the United Kingdom's public rollout of Pfizer's COVID-19 experimental mRNA "vaccine" product—alone. So many more thousands, or hundreds of thousands if not millions, of vaccinated people can be expected to have suffered serious adverse events (SAEs) related to COVID in the months that followed.

Key points from this report:

- The COVID-19-related AESIs in the post-marketing report show that **at least 2,391 (71%) of the adverse events in this category were COVID-19 *infection*.**
- **One hundred and thirty-six patients (4.4%) experiencing COVID-related adverse events died.**
- The non-infection COVID-related adverse events (AEs) were either COVID-19 exposures or COVID-19 test results, neither of which can be considered SAEs.
- All **2,585 AEs categorized as "serious,"** 77% **of total AEs for this category, were related to COVID-19 *infections*.**
  - The FDA considers an adverse event as "serious" when a patient dies or had a life-threatening injury, is hospitalized, or has a pre-existing hospitalization prolonged, disability or permanent damage, experiences a birth defect, or requires medical or surgical intervention to prevent permanent impairment or damage.

- **Fifty percent of the COVID-related adverse events began within five days** of the injection, with a range of onset between 24 hours and 374 days.
  - Pfizer's post-marketing report covered only the first 90 days of the vaccine availability, and the report was received by the FDA in late April; therefore, the maximum range should have been 150 days or less (December 1, 2020, through April 30, 2021). Given that, how is an onset of 374 days captured in the post-marketing surveillance report? It does not make sense.
- There were **2,110 (63%) "unknown" adverse event outcomes**, and **at least 1,300 of those were *serious* adverse events.**
  - This is an unusually large number of unknown outcomes compared to other post-marketing categories in Table 7. What happened to these patients experiencing COVID-related AEs?
- Ages across this post-marketing AESI category were: 1,315 adults, 560 elderly individuals, two adolescents, **two infants, and one child.**
  - ***Pfizer's BNT162b2 COVID vaccine was not approved for infants and children at the time of the data collection for Pfizer's post-marketing surveillance report.***
- The gender breakdown for this category includes 1,650 females, 844 males, and 573 unknown. Among patients with their gender noted, **women suffered almost twice as many adverse events as men.**
- There were **505 adverse events that were positive COVID-19 tests**, which included 31 patients who were reported to have "asymptomatic COVID-19."
- COVID-19 cases are referenced in three places in *5.3.6*: Table 2 reports **1,927** cases (4.6% of the 42,086 cases); Table 6 reports **2,211** cases (1,665 loss of efficacy cases and 546 COVID-19 cases excluded because they occurred so early after the first vaccine dose); and Table 7 with at least **2,391** cases. **Which of those figures is correct?** Or should the three be combined? **The numbers don't add up.**

Please read the full report below.

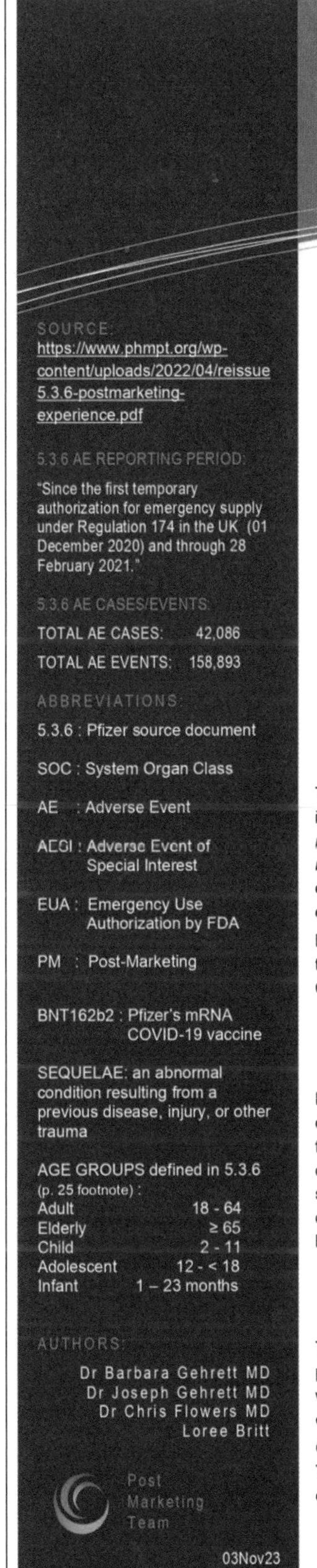

SOURCE:
https://www.phmpt.org/wp-content/uploads/2022/04/reissue 5.3.6-postmarketing-experience.pdf

5.3.6 AE REPORTING PERIOD:
"Since the first temporary authorization for emergency supply under Regulation 174 in the UK (01 December 2020) and through 28 February 2021."

5.3.6 AE CASES/EVENTS:
TOTAL AE CASES: 42,086
TOTAL AE EVENTS: 158,893

ABBREVIATIONS:
5.3.6 : Pfizer source document
SOC : System Organ Class
AE : Adverse Event
AESI : Adverse Event of Special Interest
EUA : Emergency Use Authorization by FDA
PM : Post-Marketing
BNT162b2 : Pfizer's mRNA COVID-19 vaccine
SEQUELAE: an abnormal condition resulting from a previous disease, injury, or other trauma

AGE GROUPS defined in 5.3.6 (p. 25 footnote) :

Adult	18 - 64
Elderly	≥ 65
Child	2 - 11
Adolescent	12 - < 18
Infant	1 – 23 months

AUTHORS:
Dr Barbara Gehrett MD
Dr Joseph Gehrett MD
Dr Chris Flowers MD
Loree Britt

Post Marketing Team

03Nov23

War Room/DailyClout Pfizer Document Analysis

Post-Marketing Team Micro-Report 19 :

COVID-19 AESI Review of 5.3.6

COVID-19 AESIs *Search criteria: Covid-19 SMQ (Narrow and Broad) OR PTs Ageusia; Anosmia*	• Number of cases: 3067 (7.3% of the total PM dataset), of which 1013 are medically confirmed and 2054 are non-medically confirmed. • Country of incidence: US (1272), UK (609), Germany (360), France (161), Italy (94), Spain (69), Romania (62), Portugal (51), Poland (50), Mexico (43), Belgium (42), Israel (41), Sweden (30), Austria (27), Greece (24), Denmark (18), Czech Republic and Hungary (17 each), Canada (12), Ireland (11), Slovakia (9), Latvia and United Arab Emirates (6 each); the remaining 36 cases were distributed among 16 other different countries; • Subjects' gender: female (1650), male (844) and unknown (573); • Subjects' age group (n= 1880): Adult (1315), Elderly (560), Infant[a] and Adolescent (2 each), Child (1); • Number of relevant events: 3359, of which 2585 serious, 774 non-serious; • Most frequently reported relevant PTs (>1 occurrence): COVID-19 (1927), SARS-CoV-2 test positive (415), Suspected COVID-19 (270), Ageusia (228), Anosmia (194), SARS-CoV-2 antibody test negative (83), Exposure to SARS-CoV-2 (62), SARS-CoV-2 antibody test positive (53), COVID-19 pneumonia (51), Asymptomatic COVID-19 (31), Coronavirus infection (13), Occupational exposure to SARS-CoV-2 (11), SARS-CoV-2 test false positive (7), Coronavirus test positive (6), SARS-CoV-2 test negative (3) SARS-CoV-2 antibody test (2); • Relevant event onset latency (n = 2070): Range from <24 hours to 374 days, median 5 days; • Relevant event outcome: fatal (136), not resolved (547), resolved/resolving (558), resolved with sequelae (9) and unknown (2110). Conclusion: This cumulative case review does not raise new safety issues. Surveillance will continue

https://www.phmpt.org/wpcontent/uploads/2022/04/reissue 5.3.6-postmarketexperience.pdf

- **Adverse Events were reported to Pfizer during a 90-day period, following the December 1, 2020, public rollout of its COVID-19 experimental "vaccine" product.**
- **In the Pfizer 5.3.6 document, these AEs were categorized by System Organ Classes (SOC) – in other words, by systems in the body.**
- **Gender: 1,650 females and 844 males, with 573 unknown. Age: adult 1,315, elderly 560, two infant, two adolescent, and one child.** ***(BNT162b2 was not approved for infants and children at the time of this report).***

The COVID-related AESI category consists of 3,067 vaccinated cases (patients), including one child and two infants despite no authorized vaccine for those ages, who experienced a total of 3,359 adverse events (AEs). ***This means that 7% (3,067/42,086) of all vaccinated patients of which Pfizer was notified during the post-marketing time frame suffered COVID-related AEs.*** **Those include 2,585** ***serious*** **adverse events (SAEs), all of which (77% of total AEs) appear to be COVID** ***infections.*** The other adverse events (AEs) were COVID-19 exposures or COVID-19 test results.

Five hundred and five adverse events were positive COVID-19 tests; including 31 patients who were reported to have "asymptomatic COVID-19." Pfizer did not clarify which test was used, other than 138 who had a SARS-CoV-2 antibody test.

> The FDA definition of "serious" means patients died or had a life-threatening injury, were hospitalized, or had a pre-existing hospitalization prolonged, disability or permanent damage, experienced a birth defect, or required medical or surgical intervention to prevent permanent impairment or damage.

Fifty percent of the AEs began within five days of injection (after first or second shot is unknown), with an onset range of under 24 hours to 374 days. This means that half of the AEs occurred before the vaccination was fully protective by Pfizer's definition (i.e., starting day seven after the second shot). The other half, however, occurred more than five days post-injection; so at least some of them likely occurred seven or more days post-second injection. **An onset range of 374 days** in this report does not make sense since it covers the first 90 days of public vaccine availability and was received by the FDA on April 30,2021. The maximum onset range should have been 150 days or less.

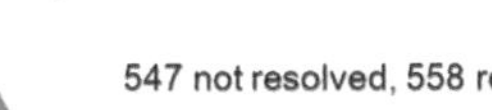

136 deaths

547 not resolved, 558 resolved/resolving, nine resolved with sequelae.

2,110 unknown (63%)

The number of unknown outcomes is disproportionately large compared to the other categories in Table 7. If one postulates that all the fatalities and all the "not resolved," "resolved/resolving," and "resolved with sequelae" AEs were serious, that leaves **over 1,300** ***serious*** **adverse events with** ***unknown*** **outcomes. Why were so many outcomes unknown? What happened to those patients?**

COVID-19 cases are referenced in three places in 5.3.6: Table 2 reports **1,927** cases (4.6% of the 42,086 cases); Table 6 reports **2,211** cases (1,665 loss of efficacy cases and 546 COVID-19 cases excluded because they occurred so early after the first vaccine dose); and Table 7 with at least **2,391** cases.

Which figure is correct? Or should the three figures be combined? **The numbers don't add up.**

RECALL this unsafe "vaccine."

Thirty-Three: "FDA Based Moderna's mRNA COVID Vaccine Approval on Test of a Completely Different Non-COVID Vaccine. Only Males Included in Test."

—Linnea Wahl, MS, Team 5

Moderna researchers did not test their COVID-19 mRNA drug, called SPIKEVAX, to find out where it would go in our bodies (biodistribution). Instead, their biodistribution study was for a completely different vaccine. Despite this substitution of one drug study for another, the U.S. Food and Drug Administration (FDA) approved SPIKEVAX for both emergency and routine use by Americans.

Introduction

The FDA declared in its own words that Moderna researchers did not test their COVID-19 mRNA drug SPIKEVAX to determine its distribution throughout the body. In the FDA reviewers' own words: "A biodistribution study was not performed with mRNA-1273 vaccine [SPIKEVAX]. Results from the biodistribution study *of a different vaccine* . . . were submitted in support of SPIKEVAX" (FDA 2022a, p. 14). (Italics added.) Incredibly, instead of testing their COVID vaccine for biodistribution, Moderna researchers substituted tests *from a different mRNA drug*. The researchers tested this completely different drug in the blood of rats and examined its distribution in 13 different body tissues (Moderna Therapeutics n.d.). In spite of this substitution, on December 18, 2020, the FDA authorized emergency use of Moderna's COVID-19 mRNA drug (FDA 2020), and on January 31, 2022, the FDA fully approved Moderna to manufacture and distribute SPIKEVAX (FDA 2022b). It does not appear that the FDA challenged the substitution of a completely different drug biodistribution study for the Moderna COVID-19 injection, SPIKEVAX.

What is a biodistribution study?

Why is it so important that one drug was substituted for another in Moderna's biodistribution study? A biodistribution is an important hurdle in the drug approval process. Before the FDA approves a new drug for use in humans, drug manufacturers need to show where the drug goes in the body, how long it stays there, and how it is removed. Test animals, such as mice, rats, monkeys, are used for this. Typically, a group of test animals receives the same drug at the same dosage that is proposed for use in humans. Then at regular intervals, researchers sacrifice a subset of the animals and examine their tissues to see how much of the drug has reached each tissue. A drug that biodistributes to inappropriate organs—such as to the heart, in the case of a drug designed to move through the lymph system—may not be safe for human use.

The FDA certainly recommends biodistribution studies for proposed drugs that involve gene therapy; drugs which include genetic materials such as RNA and messenger RNA (mRNA) (FDA 2023). As a result, Moderna officials would have fully expected that the FDA would regulate their mRNA drugs as gene therapy drugs (SEC 2019). This is why Moderna researchers had performed a biodistribution study in 2017 on their experimental mRNA drug against cytomegalovirus (CMV) (Moderna Therapeutics n.d.).

So, bizarrely, the FDA recommends biodistribution studies for mRNA gene therapies in general, but the agency excludes "vaccines" from this guidance even if they include gene therapy mRNA (Vervaeke et al. 2022).

No biodistribution study on SPIKEVAX

So, even though Moderna's mRNA-1273 drug for COVID-19 is a gene therapy, this wordplay loophole allowed Moderna to avoid a biodistribution study. Consequently, the researchers submitted the results of a study on a different mRNA drug (mRNA-1647) and asked FDA reviewers to accept it instead (ModernaTX n.d.).

Moderna researchers told FDA reviewers that SPIKEVAX would distribute throughout our bodies in the same way as Moderna's mRNA drug for CMV (a common virus that usually is not a problem for adults but can result in hearing loss in infants). Because Moderna claims that the two drugs use the same lipid nanoparticle (LNP) formulation to encapsulate mRNA, Moderna researchers argued that they would spread to tissues and organs in the body in the same way.

Substitute biodistribution study

But the substituted biodistribution study itself did not show anything like safety. The materials injected into the rats went everywhere.

The substitute biodistribution study was conducted from August 23, 2017, through September 7, 2017 (Moderna Therapeutics n.d.). Moderna researchers injected 35 male rats with an experimental mRNA drug that they hoped would cause the rats to produce antibodies to CMV. At seven timed intervals (0, 2, 8, 24, 48, 72, and 120 hours), researchers sacrificed five rats and analyzed their blood and tissues for the injected mRNA.

Researchers analyzed tissues from 13 organs after each group of rats was euthanized: lung, liver, heart, kidney, lymph nodes, spleen, brain, stomach, testes, eye, bone marrow, intestine, and injection site muscle. Blood samples were also collected.

Moderna's rat study showed that the CMV mRNA drug quickly distributed to 12 of 13 organs and blood. Researchers measured maximum concentrations of the injected mRNA within 2 to 8 hours after injection, and they found mRNA in blood and all tissues except the rats' kidneys. In the spleen and the eye tissues, researchers found mRNA at greater levels than in the blood, suggesting that the drug concentrated in those two organs. Despite the wide distribution of this drug throughout rats' bodies, the FDA reviewers accepted this study as proof that Moderna's SPIKEVAX was safe for human use (FDA 2020; FDA 2022b).

Faulty comparison

Are the two drugs similar enough to support Moderna's claim that they distribute throughout the body in the same way? No. There are important differences between the two mRNA drugs. Table 1 summarizes the components of Moderna's two mRNA drugs, as well as those of Pfizer's COVID-19 mRNA drug, BNT162b2.

Table 1. Comparison of CMV and COVID-19 mRNA Drug Components

	Moderna CMV Drug	Moderna COVID-19 Drug	Pfizer COVID-19 Drug
Name	mRNA-1647	mRNA-1273 (SPIKEVAX)	BNT162b2 (Comirnaty)
		LNP Capsule	
Ionizable Lipid	SM-102[a]	SM-102[b]	ALC-0315[b]
Helper Lipid	?	DSPC[b]	DSPC[b]
PEG Lipid	?	DMG-PEG2000[b]	ALC-0159[b]
		Genetic Payload	
Protein	glycoprotein B[c]	SARS-CoV-2 spike[b]	SARS-CoV-2 spike[b]
	glycoprotein H[c]		
	glycoprotein L[c]		
	UL128[c]		
	UL130[c]		
	UL131A[c]		

a ModernaTX n.d.
b Wilson and Geetha 2022
c Moderna Therapeutics n.d.

In SPIKEVAX, the LNP capsule has four lipid components (a lipid is a fatty compound): 1) an ionizable lipid called SM-102, 2) a helper lipid called DSPC, 3) a polyethylene glycol (PEG) lipid called PEG2000-DMG, and 4) a cholesterol molecule, which is essentially the same in all mRNA drugs that use LNP capsules (Barbier et al. 2022). Much less is known about the LNP capsule of Moderna's CMV mRNA drug. Moderna admits only that the CMV mRNA drug is encased in an LNP shell that contains their "standard proprietary" SM-102-ionizable lipid (ModernaTX n.d.).

No information is available to prove that the other three lipids in Moderna's CMV drug are the same as lipids in SPIKEVAX. Yet LNP identicalness is the basis for Moderna's claim that the two drugs spread throughout the body in the same way, and that Moderna researchers did not need a separate study of the biodistribution of SPIKEVAX.

More information is known about the mRNA that is inside Moderna's LNP capsule—the genetic payload. In SPIKEVAX, the contents of the LNP shell are the genetic sequence that codes for the SARS-CoV-2 spike protein. By contrast, Moderna's CMV mRNA drug consists of six different genetic sequences, all of which code for CMV proteins. While Moderna researchers argued that the contents of the LNP shells would not affect where the drug goes in the human body, no data are available to show this.

Are the LNP shells of the two drugs the same size, given that their contents are radically different mRNA molecules? Researchers have shown that the size of drug particles affects tissue distribution and clearance; for example, liver uptake and accumulation and kidney excretion depend significantly on LNP size (Danaei et al. 2018). Do the LNP particles of Moderna's CMV and COVID-19 mRNA drugs biodistribute similarly? The FDA reviewers did not ask.

Where are the females?

Because the study included only male rats, no female rats were injected, and no female organs were analyzed. Nor did the FDA reviewers ask whether SPIKEVAX, for its sake, would concentrate in ovaries. The CMV mRNA drug biodistribution study, which excluded females, was not designed either to ask or answer questions about the drug's impact on female organs. How can the FDA thus conclude that SPIKEVAX will be safe for females?

Conclusion

The FDA approved Moderna's COVID-19 mRNA drug SPIKEVAX for both emergency use and routine use by Americans without adequate testing for biodistribution: Moderna replaced apples with oranges. As researchers have noted: "The wide and persistent biodistribution of mRNAs and their protein products, incompletely studied due to their classification as vaccines, raises safety issues." (Banoun 2023, p. 1)

Biodistribution studies must be done in animals of both sexes—using the correct formula—*before* FDA approval and not in humans after FDA approval.

References

Banoun 2023. Banoun H. mRNA: Vaccine or gene therapy? The safety regulatory issues. International Journal of Molecular Sciences, 24, 10514. doi: 10.3390/ijms241310514

Barbier et al. 2022. Barbier AJ, Jiang AY, Zhang P, Wooster R, Anderson DG. The clinical progress of mRNA vaccines and immunotherapies. Nature Biotechnology 40: 840—854. doi: 10.1038/s41587–022-01294–2

Danaei et al. 2018. Danaei M, Dahghankhold M, Ataei S, Hasanzadeh Davarani F, Javanmard R, Dokhani A, Khorasani S, Mozafari MR. Impact of particle size and polydispersity index on the clinical applications of lipid nanocarrier systems. Pharmaceutics10(2): 57. doi: 10.3390/pharmaceutics10020057

FDA 2020. U.S. Food and Drug Administration. FDA takes additional action in fight against COVID-19 by issuing emergency use authorization for second COVID-19 vaccine. (https://www.fda.gov/news-events/press-announcements/fda-takes-additional-action-fight-against-covid-19-issuing-emergency-use-authorization-second-covid)

FDA 2022a. U.S. Food and Drug Administration. Summary basis for regulatory action: SPIKEVAX. (https://www.fda.gov/media/155931/download)

FDA 2022b. U.S. Food and Drug Administration. BLA approval: SPIKEVAX. (https://www.fda.gov/media/155815/download?attachment)

FDA 2023. U.S. Food and Drug Administration. S12 nonclinical biodistribution considerations for gene therapy products: guidance for industry. (https://www.fda.gov/regulatory-information/search-fda-guidance-documents/s12-nonclinical-biodistribution-considerations-gene-therapy-products)

ModernaTX no date. 2.4 Nonclinical overview FDA-CBER-2021–4379-0001131. Obtained by FOIA by Judicial Watch, Inc. (https://www.judicialwatch.org/wp-content/uploads/2022/12/JW-v-HHS-Biodistribution-Prod-4–02418-pgs-671–701.pdf)

Moderna Therapeutics, Inc. no date. A single dose intramuscular injection tissue distribution study of mRNA-1647 in male Sprague-Dawley rats. (https://phmpt.org/wp-content/uploads/2023/08/125752_S1_M4_report-body-5002121-amend-1.pdf)

NIH 2021. U.S. National Institutes of Health, Division of Microbiology and Infection Disease. Phase I, open-label, dose-ranging study of the safety and immunogenicity of 2019-nCoV vaccine (mRNA-1273) in healthy adults. DMID Protocol 20–0003, version 7.0.

SEC 2019. U.S. Securities Exchange Commission. Moderna, Inc: Annual report pursuant to section 13 or 15(d) of the Securities Exchange Act of 1934. (https://www.sec.gov/Archives/edgar/data/1682852/000168285220000006/moderna10-k12312019.htm)

Vervaeke et al. 2022. Vervaeke P., Borgos SE, Sanders NN, Combes F. Regulatory guidelines and preclinical tools to study the biodistribution of RNA therapeutics. Advanced Drug Delivery Reviews, 184, 114236. doi: 10.1016/j.addr.2022.114236

Wilson and Geetha. 2022. Wilson B, Geetha KM. Lipid nanoparticles in the development of mRNA vaccines for COVID-19. Journal of Drug Delivery Science and Technology 74: 103553. doi: 10.1016/j.jddst.2022.103553

Thirty-Four: "100s of Possible Vaccine-Associated Enhanced Disease (VAED) Cases in First 3 Months of Pfizer's mRNA COVID Vaccine Rollout, Yet Public Health Spokespeople Minimized Their Severity by Calling Them 'Breakthrough Cases.'"

—Barbara Gehrett, MD; Joseph Gehrett, MD; Chris Flowers, MD; and Loree Britt

By late February 2021, a group of experts called Brighton Collaboration released a paper, which was published in *Vaccine*, clearly defining **Vaccine-Associated Enhanced Disease (VAED), which is a more severe clinical presentation of a disease *in a person vaccinated against that disease than would normally be seen in an unvaccinated person*.** Yet, Pfizer, public health, and media spokespeople only referred to post-COVID-vaccination COVID-19 infections as "breakthrough cases"—i.e., normal, not more severe, COVID infections—without explaining the possibility of VAED in such cases. Not explaining VAED kept the public in the dark about how receiving initial and additional COVID vaccine doses may cause worse COVID illness than remaining unvaccinated.

Public health entities, the medical community, and media constantly assured the public that the COVID-19 mRNA shots were "safe and effective;" yet, during the first three months of the Pfizer's public vaccine rollout, Pfizer was informed about *thousands of cases of post-vaccination COVID-19* as covered by WarRoom/DailyClout researchers in Report 90. Those COVID cases are very important, because contracting COVID post-vaccination opens the door to suffering **VAED** or **Vaccine-Associated Enhanced Respiratory Disease (VAERD)**, which is "disease with predominant involvement of the lower respiratory tract." [https://www.ncbi.nlm.nih.gov/pmc/articles/PMC7901381/] **In other words, getting a COVID vaccine opened the door to vaccinated individuals getting more serious COVID illness than they may have experienced if they remained unvaccinated.** Despite Table 5 of Pfizer's document '*5.3.6 Cumulative Analysis of Post-Authorization Adverse Event Reports of PF-07302048 (BNT162B2) Received Through 28-FEB-2021*' (a.k.a., "*5.3.6*") stating, "No post-authorized AE [adverse event] reports have been identified as cases of VAED/VAERD, therefore, there is no observed data at this time," the actual data in Pfizer's post-marketing report paint a very different picture showing that thousands of the COVID cases and COVID-related serious adverse events (SAEs) appear to qualify as VAED or VAERD.

As Pfizer received post-marketing surveillance reports in December 2020 and early 2021, it knew its COVID vaccine was not stopping people from contracting COVID. In the VAED/VAERD table on page 11 of *5.3.6*, 138 patients had 317 "relevant events," i.e., conditions or diseases, which included 101 confirmed and 37 suspected COVID-19 cases. **Pfizer categorized all 138 cases, including *38 deaths*, as "serious."** The company stated that, **of the subjects with confirmed COVID-19, "75 of the 101 cases were severe, resulting in hospitalization, disability, life-threatening consequences or death."** Seemingly inexplicably, Pfizer went on to conclude, "None of the 75 cases could be definitively considered as VAED/VAERD. In this review of subjects with COVID-19 following vaccination, based on the current evidence, VAED/VAERD remains a theoretical risk for the vaccine. Surveillance will continue." In light of that, the definition of VAED and the different diagnostic levels of the condition, as defined by the Brighton Collaboration experts, becomes critically important.

In March 2020, the Brighton Collaboration gathered to define VAED. The group concluded that it is "difficult to separate vaccine failure (also called breakthrough disease) from VAED in vaccinated individuals. ***All cases of vaccine failure should be investigated for VAED.***" [Bold and italics added.] Despite this expert recommendation, the FDA, Centers for Disease Control and Prevention (CDC), and other public health spokespeople told the public that COVID infections after vaccination were "breakthrough" cases and did not mention VAED or VAERD as possibilities. The Brighton authors also asserted, "**VAED always involves a memory response primed by vaccination** and, in the experiences best characterized until now, targets the same organs as wild-type infections." [Bold added.] They concluded, "The broad spectrum of natural disease manifestations in different populations and age groups makes it **very difficult, if not impossible, to determine how severe COVID-19 infection *would have been* in the absence of vaccination in the individual case."** (Italics in the original, bold added.) However, the Brighton Collaboration provide a means to make such a determination easier.

The Brighton experts wrote, "Identifying cases of VAED/VAERD might be impossible when assessing individual patients, however, in clinical studies, **a control group is helpful to compare the frequency of cases and the severity of illness in vaccinees vs. controls**, including the occurrence of specific events of concern such as hospitalization and mortality." (Bold added.) Unfortunately, and perhaps conveniently, Pfizer quickly eliminated its clinical trial control group. Pfizer's Phase 2/3 randomized controlled clinical trial started in July 2020 with a vaccinated group and a placebo (unvaccinated) group, both of which were to be followed for two years. However, in December 2020, when the Food and Drug Administration (FDA) granted emergency use authorization (EUA) for Pfizer's COVID-19 vaccine, Pfizer asked for and received permission from the FDA to "unblind" the study—meaning, to offer the vaccine to the placebo participants. Most of placebo group accepted the offer and were vaccinated by March 2021, at which time the control group ceased to exist.

When Pfizer concluded on page 11 of *5.3.6*, "None of the 75 cases could be definitively considered as VAED/VAERD," the company appeared to be referencing Brighton Collaboration's "Level 1 of Diagnostic Certainty (Definitive Case)"—i.e., "The working group considers that a Definitive Case (LOC 1) of VAED *cannot be ascertained with current knowledge of the mechanisms of pathogenesis of VAED.*" [Italics added.] However, Brighton Collaboration defined multiple levels of diagnostic certainty, including **LEVEL 3B**:

> **A possible case of VAED** is defined by the occurrence of disease in vaccinated individual with *no prior history of infection and unknown serostatus*, with:
> **Laboratory confirmed infection** with the pathogen targeted by the vaccine
> AND
> Clinical findings of disease involving one or more organ systems **(a case of VAERD if the lung is the primarily affected organ)**
> AND
> Severe disease as evaluated by a **clinical severity index/score (systemic in VAED or specific to the lungs in VAERD)**

AND
Increased frequency of severe outcomes (including severe disease, hospitalization and mortality) when compared to a non-vaccinated population (control group or background rates)
AND
No identified alternative etiology." (https://www.ncbi.nlm.nih.gov/pmc/articles/PMC7901381/table/t0025/?report=objectonly)

The aforementioned 75 cases of severe COVID to which Pfizer admits fit the Level 3B criteria for possible VAED and should not have been dismissed because they did not fit the impossible to ascertain "Level 1" VAED definition. Pfizer failed to do due diligence when it did not investigate the 75 cases for VAED, yet that was just the tip of the iceberg of cases needing investigation.

As WarRoom DailyClout Report 90 reveals, there are ***at least*** **several hundred more COVID-19 cases, and as many as 2,391 cases, in Table 7 of Pfizer's *5.3.6* report which are suspicious of VAED**. Pfizer and the FDA owe the public—the vast majority of whom are COVID vaccinated—a thorough and fair assessment of possible COVID-vaccine-related VAED cases—an assessment that does not attempt to diminish the cases' severity by referring to them only as "breakthrough infections." We are left to wonder how many vaccinees suffered more severe cases of COVID-19 than they would have without taking Pfizer mRNA COVID vaccine and were not given the correct diagnosis of VAED.

Please read the full report below.

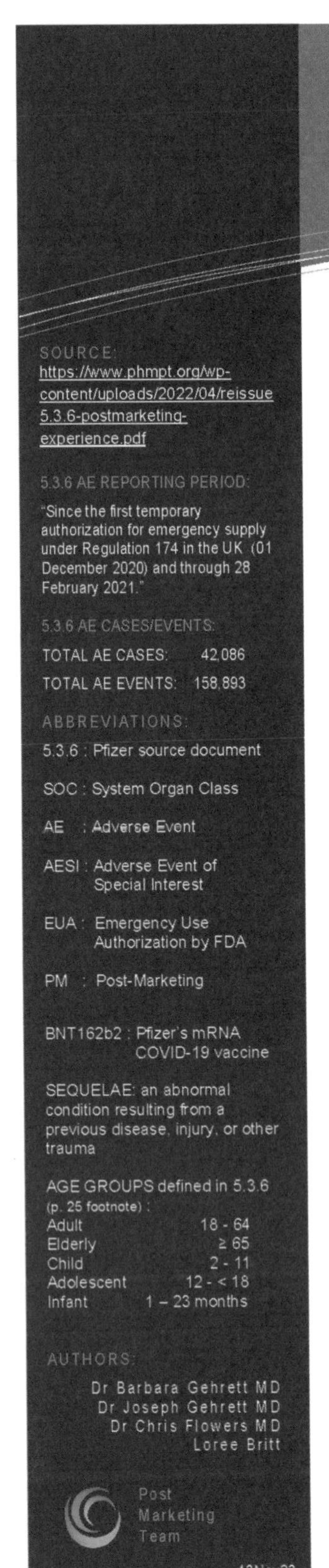

War Room/DailyClout Pfizer Document Analysis

Post-Marketing Team Micro-Report 20:

Vaccine-Associated Enhanced Disease (VAED)* Review of 5.3.6

*including Vaccine-Associated Enhanced Respiratory Disease (VAERD)

Vaccine-Associated Enhanced Disease (VAED), including Vaccine-Associated Enhanced Respiratory Disease (VAERD)	No post-authorized AE reports have been identified as cases of VAED/VAERD, therefore, there is no observed data at this time. An expected rate of VAED is difficult to establish so a meaningful observed/expected analysis cannot be conducted at this point based on available data. The feasibility of conducting such an analysis will be re-evaluated on an ongoing basis as data on the virus grows and the vaccine safety data continues to accrue. The search criteria utilised to identify potential cases of VAED for this report includes PTs indicating a lack of effect of the vaccine and PTs potentially indicative of severe or atypical COVID-19ª. Since the first temporary authorization for emergency supply under Regulation 174 in the UK (01 December 2020) and through 28 February 2021, 138 cases [0.33% of the total PM dataset], reporting 317 potentially relevant events were retrieved: Country of incidence: UK (71), US (25), Germany (14), France, Italy, Mexico, Spain, (4 each), Denmark (3); the remaining 9 cases originated from 9 different countries; Cases Seriousness: 138; Seriousness criteria for the total 138 cases: Medically significant (71, of which 8 also serious for disability), Hospitalization required (non-fatal/non-life threatening) (16, of which 1 also serious for disability), Life threatening (13, of which 7 were also serious for hospitalization), Death (38). Gender: Females (73), Males (57), Unknown (8); Age (n=132) ranged from 21 to 100 years (mean = 57.2 years, median = 59.5); Case outcome: fatal (38), resolved/resolving (26), not resolved (65), resolved with sequelae (1), unknown (8); Of the 317 relevant events, the most frequently reported PTs (≥2%) were: Drug ineffective (135), Dyspnoea (53), Diarrhoea (30), COVID-19 pneumonia (23), Vomiting (20), Respiratory failure (8), and Seizure (7). Conclusion: VAED may present as severe or unusual clinical manifestations of COVID-19. Overall, there were 37 subjects with suspected COVID-19 and 101 subjects with confirmed COVID-19 following one or both doses of the vaccine; 75 of the 101 cases were severe, resulting in hospitalisation, disability, life-threatening consequences or death. None of the 75 cases could be definitively considered as VAED/VAERD. In this review of subjects with COVID-19 following vaccination, based on the current evidence, VAED/VAERD remains a theoretical risk for the vaccine. Surveillance will continue.

https://www.phmpt.org/wpcontent/uploads/2022/04/reissue 5.3.6-postmarketiexperience.pdf

- **Adverse Events were reported to Pfizer during a 90-day period, following the December 1, 2020, public rollout of its COVID-19 experimental "vaccine" product.**
- **In the Pfizer 5.3.6 document, these AEs were categorized by System Organ Classes (SOC) – in other words, by systems in the body.**

VAED refers to a case of post-vaccination COVID-19 that is unusual or more severe than normal. Table 5 of *5.3.6* starts by stating, "No post-authorized AE reports have been identified as cases of VAED/VAERD, therefore, there is no observed data at this time." However, Pfizer reported a long list of illnesses experienced by patients who had had one or both doses of its vaccine and subsequently developed suspected or proven COVID. Thousands of those COVID cases and COVID-related serious adverse events appear to qualify as VAED or VAERD.

In the VAED/VAERD table on page 11, 138 patients had 317 "relevant events,", i.e., conditions or diseases. Of the 138 patients from 17 countries, 101 patients had confirmed COVID-19. Thirty-seven had suspected COVID-19. Times from vaccine administration to onset of COVID-19 illness and associated conditions were not reported. There were 130 patients with known gender: 73 females, and 57 males. Ages ranged from 21 years to 100 years with half of the patients over 59.5 years of age. Outcomes included **38 deaths, 65 cases were not resolved**, 26 resolved or resolving, **one resolved with sequelae, and eight unknown.**

All 138 cases, including 38 deaths, were deemed serious. Of the 100 non-fatal cases, 13 were considered life threatening, 16 were hospitalized with non-life-threatening conditions (including one disability), and the remaining **71 patients were "medically significant,"** with eight of those "serious for disability". **These were sick patients.** Specific conditions named more than six times were: vaccine ineffective (135), shortness of breath (53), diarrhea (30), COVID-19 pneumonia (23), vomiting (20), respiratory failure (8), and seizure (7). Another 41 conditions were not stated.

Pfizer stated that, **of the subjects with confirmed COVID-19, "75 of the 101 cases were severe, resulting in hospitalization, disability, life-threatening consequences or death."**

Pfizer concluded:

"None of the 75 cases could be definitively considered as VAED/VAERD. In this review of subjects with COVID-19 following vaccination, based on the current evidence, VAED/VAERD remains a theoretical risk for the vaccine. Surveillance will continue."

Until now, there have been various terms to describe **a vaccine worsening a viral illness when compared to the natural infection in an unvaccinated individual**. Some of these terms are vaccine-mediated enhanced disease, enhanced respiratory disease, and antibody-dependent enhancement. In the Pfizer document *5.3.6*, the term "Vaccine Associated Enhanced Disease" (VAED) is used. **This syndrome is not theoretical** and has been identified since at least the 1960s when it was found with early Respiratory Syncytial Virus (RSV) and measles vaccines, and also later with a trial vaccine for HIV.

In March 2020, the Coalition for Epidemic Preparedness Innovations (CEPI) and the Brighton Collaboration Safety Platform for Emergency Vaccines met to discuss safety and design issues and risk assessment in rapidly developed vaccines. The particular issue arose because "some Middle East respiratory syndrome (MERS) and SARS-CoV-1 vaccines have shown evidence of disease enhancement in some animal models," which raised concerns about SARS-CoV-2 vaccines. Consensus summary report for CEPI/BC March 12–13, 2020 meeting: Assessment of risk of disease enhancement with COVID-19 vaccines - PMC (nih.gov)

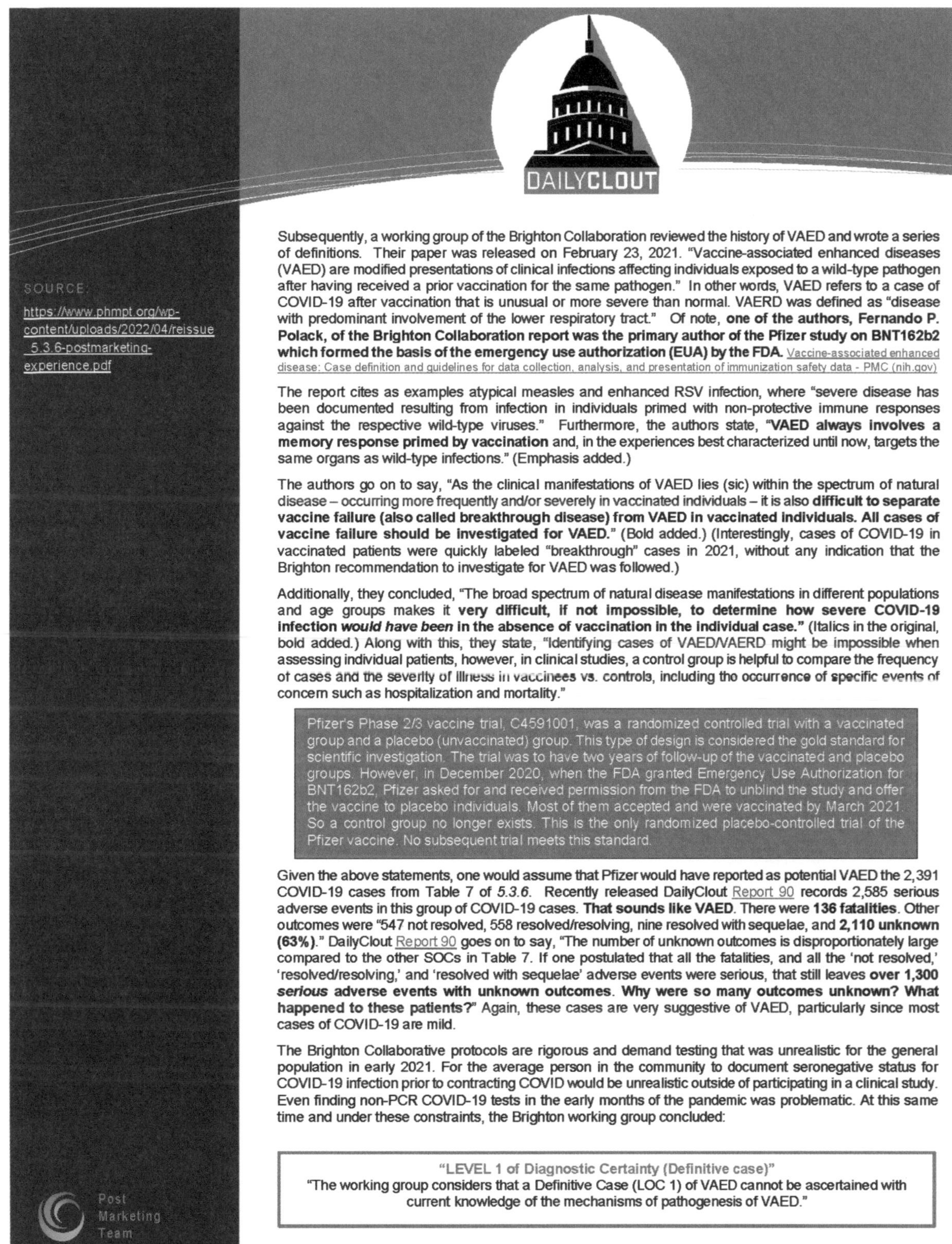

DAILYCLOUT

SOURCE:
https://www.phmpt.org/wp-content/uploads/2022/04/reissue_5.3.6-postmarketing-experience.pdf

Subsequently, a working group of the Brighton Collaboration reviewed the history of VAED and wrote a series of definitions. Their paper was released on February 23, 2021. "Vaccine-associated enhanced diseases (VAED) are modified presentations of clinical infections affecting individuals exposed to a wild-type pathogen after having received a prior vaccination for the same pathogen." In other words, VAED refers to a case of COVID-19 after vaccination that is unusual or more severe than normal. VAERD was defined as "disease with predominant involvement of the lower respiratory tract." Of note, **one of the authors, Fernando P. Polack, of the Brighton Collaboration report was the primary author of the Pfizer study on BNT162b2 which formed the basis of the emergency use authorization (EUA) by the FDA.** Vaccine-associated enhanced disease: Case definition and guidelines for data collection, analysis, and presentation of immunization safety data - PMC (nih.gov)

The report cites as examples atypical measles and enhanced RSV infection, where "severe disease has been documented resulting from infection in individuals primed with non-protective immune responses against the respective wild-type viruses." Furthermore, the authors state, "**VAED always involves a memory response primed by vaccination** and, in the experiences best characterized until now, targets the same organs as wild-type infections." (Emphasis added.)

The authors go on to say, "As the clinical manifestations of VAED lies (sic) within the spectrum of natural disease – occurring more frequently and/or severely in vaccinated individuals – it is also **difficult to separate vaccine failure (also called breakthrough disease) from VAED in vaccinated individuals. All cases of vaccine failure should be investigated for VAED.**" (Bold added.) (Interestingly, cases of COVID-19 in vaccinated patients were quickly labeled "breakthrough" cases in 2021, without any indication that the Brighton recommendation to investigate for VAED was followed.)

Additionally, they concluded, "The broad spectrum of natural disease manifestations in different populations and age groups makes it **very difficult, if not impossible, to determine how severe COVID-19 infection *would have been* in the absence of vaccination in the individual case.**" (Italics in the original, bold added.) Along with this, they state, "Identifying cases of VAED/VAERD might be impossible when assessing individual patients, however, in clinical studies, a control group is helpful to compare the frequency of cases and the severity of illness in vaccinees vs. controls, including the occurrence of specific events of concern such as hospitalization and mortality."

Pfizer's Phase 2/3 vaccine trial, C4591001, was a randomized controlled trial with a vaccinated group and a placebo (unvaccinated) group. This type of design is considered the gold standard for scientific investigation. The trial was to have two years of follow-up of the vaccinated and placebo groups. However, in December 2020, when the FDA granted Emergency Use Authorization for BNT162b2, Pfizer asked for and received permission from the FDA to unblind the study and offer the vaccine to placebo individuals. Most of them accepted and were vaccinated by March 2021. So a control group no longer exists. This is the only randomized placebo-controlled trial of the Pfizer vaccine. No subsequent trial meets this standard.

Given the above statements, one would assume that Pfizer would have reported as potential VAED the 2,391 COVID-19 cases from Table 7 of *5.3.6*. Recently released DailyClout Report 90 records 2,585 serious adverse events in this group of COVID-19 cases. **That sounds like VAED.** There were **136 fatalities**. Other outcomes were "547 not resolved, 558 resolved/resolving, nine resolved with sequelae, and **2,110 unknown (63%)**." DailyClout Report 90 goes on to say, "The number of unknown outcomes is disproportionately large compared to the other SOCs in Table 7. If one postulated that all the fatalities, and all the 'not resolved,' 'resolved/resolving,' and 'resolved with sequelae' adverse events were serious, that still leaves **over 1,300 *serious* adverse events with unknown outcomes. Why were so many outcomes unknown? What happened to these patients?**" Again, these cases are very suggestive of VAED, particularly since most cases of COVID-19 are mild.

The Brighton Collaborative protocols are rigorous and demand testing that was unrealistic for the general population in early 2021. For the average person in the community to document seronegative status for COVID-19 infection prior to contracting COVID would be unrealistic outside of participating in a clinical study. Even finding non-PCR COVID-19 tests in the early months of the pandemic was problematic. At this same time and under these constraints, the Brighton working group concluded:

"LEVEL 1 of Diagnostic Certainty (Definitive case)"
"The working group considers that a Definitive Case (LOC 1) of VAED cannot be ascertained with current knowledge of the mechanisms of pathogenesis of VAED."

Post Marketing Team

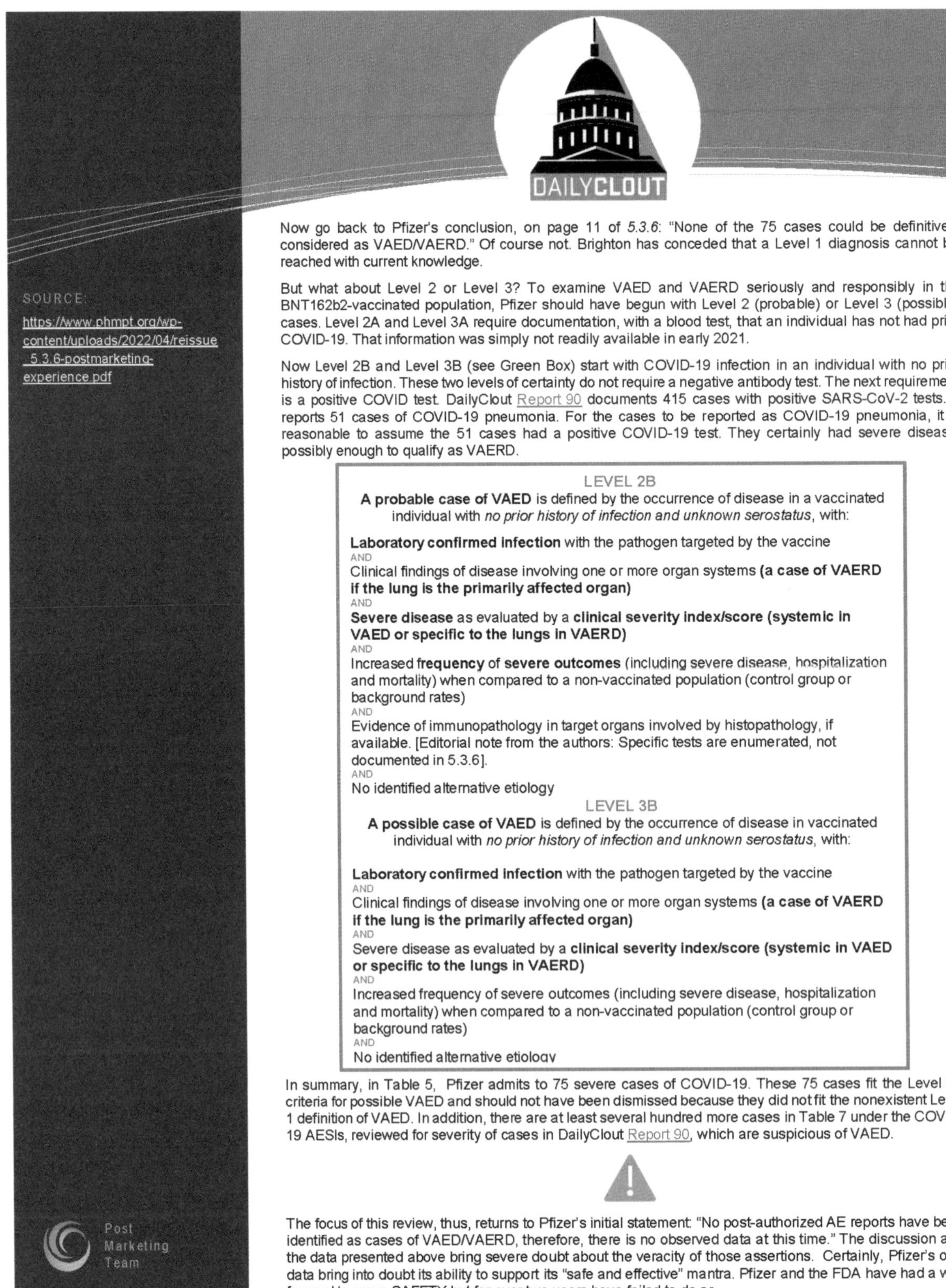

SOURCE:
https://www.phmpt.org/wp-content/uploads/2022/04/reissue_5.3.6-postmarketing-experience.pdf

Now go back to Pfizer's conclusion, on page 11 of *5.3.6*: "None of the 75 cases could be definitively considered as VAED/VAERD." Of course not. Brighton has conceded that a Level 1 diagnosis cannot be reached with current knowledge.

But what about Level 2 or Level 3? To examine VAED and VAERD seriously and responsibly in the BNT162b2-vaccinated population, Pfizer should have begun with Level 2 (probable) or Level 3 (possible) cases. Level 2A and Level 3A require documentation, with a blood test, that an individual has not had prior COVID-19. That information was simply not readily available in early 2021.

Now Level 2B and Level 3B (see Green Box) start with COVID-19 infection in an individual with no prior history of infection. These two levels of certainty do not require a negative antibody test. The next requirement is a positive COVID test. DailyClout Report 90 documents 415 cases with positive SARS-CoV-2 tests. It reports 51 cases of COVID-19 pneumonia. For the cases to be reported as COVID-19 pneumonia, it is reasonable to assume the 51 cases had a positive COVID-19 test. They certainly had severe disease, possibly enough to qualify as VAERD.

LEVEL 2B

A probable case of VAED is defined by the occurrence of disease in a vaccinated individual with *no prior history of infection and unknown serostatus*, with:

Laboratory confirmed infection with the pathogen targeted by the vaccine
AND
Clinical findings of disease involving one or more organ systems **(a case of VAERD if the lung is the primarily affected organ)**
AND
Severe disease as evaluated by a **clinical severity index/score (systemic in VAED or specific to the lungs in VAERD)**
AND
Increased **frequency** of **severe outcomes** (including severe disease, hospitalization and mortality) when compared to a non-vaccinated population (control group or background rates)
AND
Evidence of immunopathology in target organs involved by histopathology, if available. [Editorial note from the authors: Specific tests are enumerated, not documented in 5.3.6].
AND
No identified alternative etiology

LEVEL 3B

A possible case of VAED is defined by the occurrence of disease in vaccinated individual with *no prior history of infection and unknown serostatus*, with:

Laboratory confirmed infection with the pathogen targeted by the vaccine
AND
Clinical findings of disease involving one or more organ systems **(a case of VAERD if the lung is the primarily affected organ)**
AND
Severe disease as evaluated by a **clinical severity index/score (systemic in VAED or specific to the lungs in VAERD)**
AND
Increased frequency of severe outcomes (including severe disease, hospitalization and mortality) when compared to a non-vaccinated population (control group or background rates)
AND
No identified alternative etiology

In summary, in Table 5, Pfizer admits to 75 severe cases of COVID-19. These 75 cases fit the Level 3B criteria for possible VAED and should not have been dismissed because they did not fit the nonexistent Level 1 definition of VAED. In addition, there are at least several hundred more cases in Table 7 under the COVID-19 AESIs, reviewed for severity of cases in DailyClout Report 90, which are suspicious of VAED.

The focus of this review, thus, returns to Pfizer's initial statement: "No post-authorized AE reports have been identified as cases of VAED/VAERD, therefore, there is no observed data at this time." The discussion and the data presented above bring severe doubt about the veracity of those assertions. Certainly, Pfizer's own data bring into doubt its ability to support its "safe and effective" mantra. Pfizer and the FDA have had a way forward to prove SAFETY but for over two years have failed to do so.

Post Marketing Team